Orbital and Celestial Mechanics

Orbital and Celestial Mechanics

John P. Vinti

Edited by
Gim J. Der
TRW
Los Angeles, California

Nino L. Bonavito
NASA Goddard Space Flight Center
Greenbelt, Maryland

Volume 177
PROGRESS IN ASTRONAUTICS AND AERONAUTICS

Paul Zarchan, Editor-in-Chief
Charles Stark Draper Laboratory, Inc.
Cambridge, Massachusetts

Published by the
American Institute of Aeronautics and Astronautics, Inc.
1801 Alexander Bell Drive, Reston, Virginia 20191-4344

ISBN 1-56347-256-2

Foreword

John Vinti was one of the few surviving figures from the American Golden Age of Science that began during the 1930s. After entering the Massachusetts Institute of Technology (MIT) on a scholarship, he received an S.B. Degree in mathematics. Awarded a James Savage Fellowship at MIT, he pursued graduate studies in physics. It was during this time that he became interested in Hamiltonian mechanics. Then, as now, the Hamilton–Jacobi equation was regarded by most physicists as only a point of departure for quantum mechanics. Years later, he was to be the first to apply it effectively to an important practical problem in orbital mechanics. He began his doctoral dissertation on atomic wave functions under the physicist Rudolf Langer and finished his thesis under Philip Morse, who is famous for the Morse potential for diatomic molecules. It was the approach of finding a "solvable problem" suggested by Morse that became a dominant factor in Vinti's later scientific career.

After receiving a Doctor of Science degree in physics from MIT, Vinti spent two years in postdoctoral research at the University of Pennsylvania and produced a number of research papers. The most important of these papers for space science was the calculation of the continuous absorption spectrum of helium; this extraordinary contribution is referenced in the *Encyclopedia of Physics*. Several of his publications in electromagnetic wave propagation and gamma-ray scattering, which appeared in the *Physical Review* during this period, are still widely quoted. Although the devastating effect of the Great Depression on America's academic institutions halted a well-deserved rapid rise of his professional career, his scientific work is nevertheless noted for its creative versatility. First as a theoretical physicist, he made fundamental contributions to atomic and molecular physics as well as related fields, resulting in more than 70 important papers in physics, mathematics, and engineering. These unique accomplishments earned him the following honors: Fellow of the American Physics Society in 1936; Fellow of the British Interplanetary Society in 1960; Fellow of the Royal Astronomical Society (London) in 1961; Member of the Cosmos Club, Washington, D.C., in 1961; Fellow of the Washington Academy of Sciences in 1963; and Fellow of the American Association for the Advancement of Science in 1967.

With the advent of World War II and the effects of the Great Depression beginning to recede in the early 1940s, Vinti moved to the Aberdeen Proving Ground in Maryland. The genesis of Vinti's interest in celestial mechanics began at Aberdeen. It was while working on interior ballistics of rockets that he met Boris Garfinkel, an astronomer, and Joel Brenner, a mathematician, both of whom had a major influence on his subsequent career. Garfinkel helped direct his efforts in celestial mechanics, while Brenner reinvigorated his focus in finding a solvable solution of the Hamilton–Jacobi equation in orbital mechanics. It was also during his stay at Aberdeen that he developed a close association with giants such as John von Neumann, Martin Schwarzchild, Subramanyan Chandrasekhar, and Josef and Maria Goeppert-Maier.

In 1957, Vinti was invited by Robert Dressler to join his Mathematical-Physics Division at the National Bureau of Standards (NBS), Washington, D.C., where Vinti was free to choose his own research areas. This gave Vinti the opportunity to work on his orbital ideas. In 1959, he produced his first series of papers on the motion of a close-Earth, drag-free satellite by means of separable Hamiltonian. By introducing a gravitational potential in oblate spheroidal coordinates, Vinti was able simultaneously to satisfy Laplace's equation and to separate the Hamilton–Jacobi equation. Since the assumed potential is very close to that of the Earth, the resulting equations of motion, which are solved in closed form, yield very accurate and rapid results. Until that time, standard general perturbation methods used in orbit determination were both computationally intensive and relatively low in accuracy for use in orbit prediction. In a single brilliant effort, this changed overnight. Scientists and engineers especially in the Soviet, French, Japanese, and Chinese space communities were quick to recognize this work and adapt it to their needs in both research and applications.

In 1968, Vinti returned to MIT where he had started his career, combining the teaching of celestial mechanics and research at the Measurement Systems Laboratory. Several papers emerged during this period. Work on the problem of the stability of free rotation of a rigid body led to new quantitative results. Another paper showed the feasibility of representing the higher harmonics of the Earth's gravitational field by means of a monopole layer on a spherical surface just containing the Earth. These higher harmonics amount to perturbations of only a few parts in a million, but there are hundreds of them that have to be accounted for in calculating an accurate satellite orbit as a baseline for satellite geodesy. At the urging of scientist-astronaut Dr. Philip Chapman, Vinti and colleague Leonard Wilk completed an analysis of an experimental method for determining the gravitational constant G in a large manned orbiting laboratory. The motivation was to search for possible variations in G with gravitational potential to test Robert Dicke's modification of general relativity.

As a teacher, Vinti was acclaimed by both his students and fellow researchers. While at Aberdeen he resumed his academic career in 1940, serving at various times as lecturer in physics and mathematics for the Universities of Delaware and Maryland. From a course in theoretical mechanics he delivered at Aberdeen for the University of Maryland, two-thirds of the students went on to receive doctoral degrees in physics, and each of them pointed out that more than half of the material on their written comprehensives had been covered in Vinti's course. In the academic environment, Vinti always put his students' concerns above all else. His teaching method was unique: He made his students lecture to him from the blackboard. Invariably, that they said that his courses were the most valuable they had ever experienced.

G. J. Der
TRW, Los Angeles, California

Table of Contents

Preface

This book presents one of the many extraordinary contributions given to the aerospace sciences by the late Professor John Pascal Vinti. It contains the text of lecture notes that Vinti used in a course first given at the Catholic University of America in 1966, and which was later refined for a similar course he taught at MIT. The step-by-step derivations could have been shortened by drastically reducing the number of equations, but Vinti endeavored to achieve, above all, clarity and rigor, as well as elegance and practicality.

As both a researcher and a professor of physics, Vinti is able toaddress and relate the various topics in orbital and celestial mechanics starting from the first principle. The text is organized to bring together work from different areas of satellite astronomy so as to examine critically the discipline from the viewpoint of classical mechanics. Advanced courses in classical mechanics have long been a time-honored part of the graduate physics curriculum. As such, it remains an indispensable component of a student's education. In one or another of its advanced formulations, it serves as a springboard to various branches of physics including the applications to celestial and orbital mechanics. Thus, the technique of action-angle variables, which was needed for the older quantum mechanics, is invaluable for the discussion of conditionally periodic Staeckel systems. The Hamilton–Jacobi equation, which in modern physics provided the transition to wave mechanics, is now seen as the starting point for the Vinti spheroidal method for satellite orbits and ballistic trajectories. Lagrange and Poisson brackets, and canonical transformations, which also were of signal importance in modern physics, are indispensable in the theory of general perturbations. Moreover, the approach to celestial and orbital mechanics affords both the student and researcher the opportunity to master many of the mathematical techniques necessary for this discipline while still working in terms of the familiar universal concepts of classical physics.

With these objectives in mind, the traditional treatment of the subject, which was in large measure fixed in the latter part of the 19th century, is no longer adequate. The present book is an exposition of celestial and orbital mechanics that fulfills the new requirements. Those formulations that are of importance to this field have received emphasis, and mathematical techniques have been introduced whenever they result in increased elegance, compactness, and understanding. For both students and workers in celestial and orbital mechanics, a great deal of effort was made to keep the book self-contained. Much of Chapters 1–4 is devoted, therefore, to material usually covered in preliminary courses. Until now, no connected account was available on the classical foundations arising from forces that are not derivable from a potential. This powerful concept is included in Chapters 12 and 13 on the Gaussian variational equations for both the Jacobi and Keplerian elements. A natural followup to this is the effect of drag on the orbits of Earth satellites, which is covered in Chapter 18.

The Vinti spheroidal method, which is many years ahead of its time, predicts position and velocity vectors for satellites and ballistic missiles almost as accurately as numerical integration. Those nonspecialists who may not be familiar with the underlying mathematics or who may not have access to sophisticated numerical integration routines can simply use one of the available Vinti computer routines to obtain accurate solutions for a satellite orbit or ballistic trajectory. To save memory and improve numerical integration efficiency, the Vinti spheroidal method was implemented onboard one of our ballistic missile targeting programs with great success more than 20 years ago. The targeting portion of the computer code has deliberately been deleted for clarity. It is simple to apply a Vinti trajectory computer routine to solve a targeting problem. The important routines are commented, to help interested readers who wish to understand the Vinti spheroidal method in detail. Helpful hints and clarifying details are presented in various appendices.

A debt of gratitude is expressed to Professor Robert F. Dressler, Paul Janiczek of the U.S. Naval Observatory, Felix Hoots of General Research Corporation, Thomas Lang of The Aerospace Corporation, Steve Madden of C. S. Draper Laboratory, Albert Monuki of TRW, Professor William Vander Velde of MIT, Harvey Walden and Stan Watson of Goddard Space Flight Center for reviewing and supporting the manuscript, Professor Boris Garfinkel for his official encouragement, the late mathematician Bassford Getchell for his generosity and guidance, and Vinti's many students whose favorable reaction and active interest provided the continuing impetus for this book.

G. J. Der
TRW, Los Angeles, California

N. L. Bonavito
NASA Goddard Space Flight Center
Greenbelt, Maryland

Introduction

PROFESSOR John Pascal Vinti is an example of a brilliant American scientist whose outstanding works have gone essentially unrecognized. This book is a belated tribute to the extraordinary contributions of Vinti in the fields of orbital and celestial mechanics. Until Vinti, standard general perturbations methods and semi-analytic satellite theories applied to orbit determination were computationally intensive and low in accuracy. With a single brilliant stroke, this changed overnight. Vinti was the first physicist to apply effectively the Hamiltonian–Jacobi equation to solve analytically the orbit prediction problem in mechanics. His revolutionary method for the orbits of satellites about an oblate Earth is, to this day, yet to be fully acknowledged. This was due, in part, to the advanced nature of his techniques, as well as his lack of self-promotion for his work.

The first eight chapters of this book provide the fundamentals of orbital and celestial mechanics: Newton's Laws, The Two-Body Problem, Langrangian Dynamics, The Hamiltonian Equations, Canonical Transformations, Hamilton–Jacobi Theory, Hamilton–Jacobi Perturbation Theory, and Vinti Spheroidal Method for Satellite Orbits and Ballistic Trajectories. By introducing a gravitational potential in oblate spheroidal coordinates, Vinti was able simultaneously to satisfy Laplace's equation and to separate the Hamilton–Jacobi equation. Since the assumed potential is very close to that of the Earth, the resulting equations of motion, which are solved in closed form, rapidly yield very accurate results. Today's extremely fast computers motivate numerical integration of trajectories in almost every application. Very often, numerical techniques are not well understood, making the numerical solutions erroneous and/or computationally inefficient. Analytic methods for long-term satellite orbit prediction and short-term ballistic missile impact-point prediction are indispensable. A Vinti trajectory propagator has the same input and output formats as a Kepler routine but gives solutions that approach the accuracy of numerical integration in most cases, especially for a drag-free satellite and a long-range ballistic missile. A Vinti trajectory propagator is difficult to implement, and once developed, it is usually guarded as proprietary software. Through the generosity of his friends and students, this book includes six Vinti trajectory propagators that have been independently developed by Wadsworth, Izsak–Borchers, Bonavito, Lang, Getchell, and Der–Monuki. Appendix A describes the coordinate systems and coordinate transformations used in the Vinti spheroidal method. Appendix B provides the computational procedures of two Vinti trajectory algorithms. Appendix C presents a set of examples to address the accuracy and robustness of the Vinti spheroidal method.

The remaining chapters of this book consist of additional topics of several important elements of orbital and celestial mechanics: Delaunay Variables, The Lagrange Planetary Equations, The Planetary Disturbing Function, Gaussian Variational Equations for the Jacobi Elements, Gaussian Variational Equations for the Keplerian Elements, Potential Theory, The Gravitational Potential of a Planet, Elementary Theory of Satellite Orbits with Use of the Mean Anomaly, Elementary Theory of Satellite Orbits with Use of the True Anamoly, The Effects of Drag on Satellite Orbits, The Brouwer–von Zeipel Method I, The Brouwer–von Zeipel Method II, Lagrange and Poisson Brackets, Lie Series, Perturbations by Lie Series, The General Three-Body Problem, The Restricted Three-Body Problem, and

Staeckel Systems. This latter part is based on Vinti's lecture notes used at the Catholic University of America and MIT.

The equations of motion of orbital and celestial mechanics can be traced to the works of Newton, D'Alembert, Lagrange, Hamilton, Jacobi, and many others. They formulated the kinematical problem by providing the equations of motion expressed in either a set of N second-order, ordinary differential equations or $2N$ first-order, ordinary differential equations. Few of these great mathematicians and physicists were able to provide even a single analytic solution to the equations of motion of orbital and celestial objects.

The primary purpose of this book is to describe Vinti's potential theory in orbital mechanics and his interpretation of the elements of celestial mechanics. Vinti's potential theory leads to the best analytic solution to the equations of motion for the satellite orbits and ballistic trajectories about an oblate Earth. By analytic, we mean that the algorithm does not involve any numerical integration. Vinti's interpretation of the elements of orbital and celestial mechanics provides refreshing, yet simple and logical, reading. A secondary purpose is to provide several practical Vinti trajectory algorithms that are included on the floppy disk. A Vinti trajectory algorithm, which gives an accurate analytic solution to Kepler's problem, computes the position and velocity vectors $\boldsymbol{r}(t)$ and $\boldsymbol{v}(t)$ at a given final time t, from the given initial position vector $\boldsymbol{r}(t_0)$, the initial velocity vector $\boldsymbol{v}(t_0)$, and the initial time (t_0).

Figure 1 shows that the equations of motion can be solved by special perturbations or general perturbations. Special perturbations methods, which employ numerical integration, theoretically provide the most accurate solution at the expense of computational time. General perturbations methods, whose solutions are analytic, can be represented by three basic methods: Kepler, Brouwer, and Vinti. Other general perturbations methods that employ a reference orbit, power series, averaging process, and special rectangular coordinates are usually application-specific and, thus, omitted from this discussion. A conceptual comparison of typical numerical and analytic solutions for Kepler's problem is depicted in Fig. 2. The Vinti solution is usually very close to the numerically integrated solution for the satellite state prediction or the ballistic missile impact-point prediction.

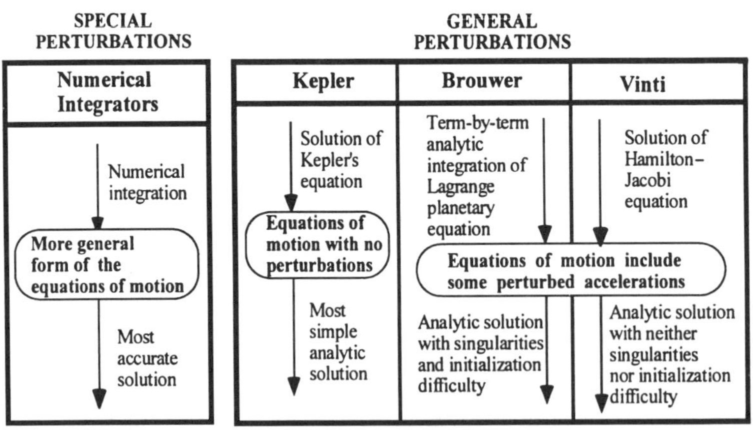

Fig. 1 Methods of solution for the equations of motion in orbital mechanics and celestial mechanics.

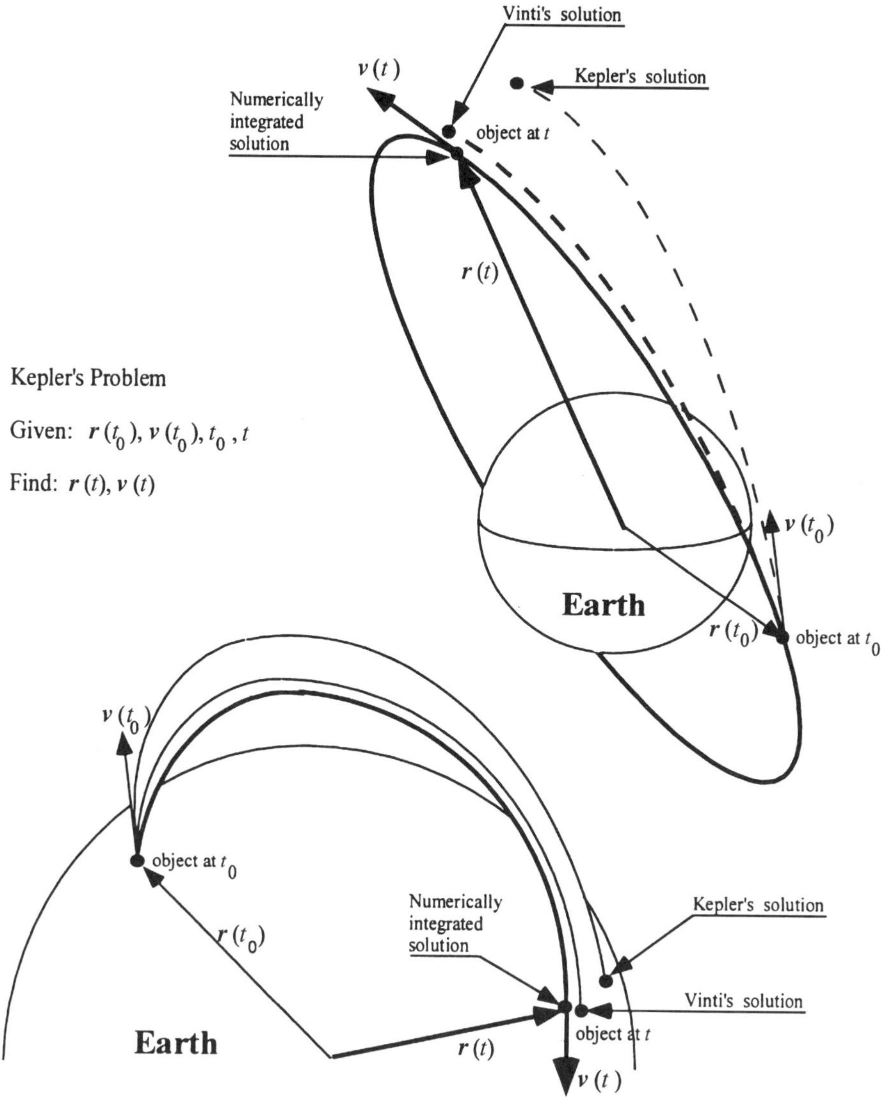

Fig. 2 A conceptual comparison of numerical and analytic methods for satellite-state prediction and ballistic missile impact-point prediction.

Kepler and Newton provided the most simple analytic solution for the unperturbed problem, in which the equations of motion are reduced to three homogeneous second-order, ordinary differential equations. Brouwer performed successive canonical transformations and analytic term-by-term integration using the von Zeipel averaging technique. A Brouwer (or Kozai) method often encounters numerical difficulties in the neighborhood of the singularities of zero eccentricity, zero inclination, or critical inclination. Vinti formulated the equations of motion with the oblate spheroidal coordinate system (the Earth is an

NEWTONIAN MECHANICS

Classical Formulation

Equations of motion

$$\frac{d^2\boldsymbol{r}}{dt^2} = -\frac{\mu}{r^3}\boldsymbol{r} + \boldsymbol{a}_d$$

Given: $\boldsymbol{r}(t_0), \boldsymbol{v}(t_0), t_0, t$
Find: $\boldsymbol{r}(t), \boldsymbol{v}(t)$

Kepler's method:

(1) Assume no perturbations or zero disturbed acceleration: $\boldsymbol{a}_d = \boldsymbol{0}$

(2) Solve the unperturbed Kepler's equation:
$F(x) = 0$
where x is the universal variable.

(3) Express solution in the form:

$$\begin{bmatrix} \boldsymbol{r}(t) \\ \boldsymbol{v}(t) \end{bmatrix} = \begin{bmatrix} fI & gI \\ \dot{f}I & \dot{g}I \end{bmatrix} \begin{bmatrix} \boldsymbol{r}(t_0) \\ \boldsymbol{v}(t_0) \end{bmatrix}$$

where $f, g, \dot{f}, \dot{g}$ are functions of x.

Hamilton–Jacobian Formulation

Equations of motion

$$\dot{p}_k = -\frac{\partial H(q,p,t)}{\partial q_k} \qquad \dot{q}_k = \frac{\partial H(q,p,t)}{\partial p_k}$$

where q's and p's are respectively coordinates and momenta, and $k = 1, 2, 3$.

Given: $\boldsymbol{r}(t_0), \boldsymbol{v}(t_0), t_0, t$
Find: $\boldsymbol{r}(t), \boldsymbol{v}(t)$

Vinti's method:

(1) Define Hamiltonian and generating function:
$H = T + V$ and $S = S(q, P)$
where T is the kinetic energy and and V is the potential energy that includes perturbations.

(2) Define the spheroidal gravitation potential:

$$V = -\frac{\mu(\rho + \delta\eta)}{\rho^2 + c^2\eta^2}$$

which simultanuously satisfies the Laplace's equation and separates the Hamiltonian–Jacobi equation

$$H + \frac{\partial S}{\partial t} = 0$$

resulting in three kinematical equations

$$\begin{aligned} t + \beta_1 &= R_1 + c^2 N_1 \\ \beta_2 &= -\alpha_2 R_2 + \alpha_2 N_2 \\ \beta_3 &= \phi + c^2 \alpha_3 R_3 - \alpha_3 N_3 \end{aligned}$$

where α's, R's, N's and β's can be computed at t_0.

(3) Substitute the β's back into the kinematical equations and solve for ρ, η, ϕ and then $\dot{\rho}, \dot{\eta}, \dot{\phi}$ at t, which then transform into $\boldsymbol{r}(t)$ and $\boldsymbol{v}(t)$.

Fig. 3 Computational procedures of Kepler and Vinti methods of solution for the equations of motion from the Newtonian mechanics point of view.

oblate spheroid) and then took advantage of separation of variables to solve analytically the Hamilton–Jacobi partial differential equations while simultaneously satisfying the Laplace equation. Even though Vinti's method includes only the second-, third-, and about 70% of the fourth-order zonal gravitational harmonics in the perturbed accelerations, his method is not only the most computationally efficient (fastest and most accurate), but also demonstrates no singularity behavior whatsoever.

Figure 3 depicts the computational procedures of the Kepler and Vinti methods from the classical mechanics point of view. Kepler's method of solution is a classical formulation of Newtonian mechanics by directly solving the second-order, ordinary differential equation. The Brouwer's method, which is not included in Fig. 3, uses the Delaunay form of the canonical equations of motion and eliminates the lower case variables from the Hamiltonian by means of successive canonical transformations. The canonical equations are essentially the Lagrange equations of motion. Kozai used the classical element form of the canonical equations of motion and developed almost the same solutions as Brouwer's. In a programmable (first-order) Brouwer's algorithm, only the first-order short periodic terms, second-order secular terms, and long periodic terms can be kept. Using the von Zeipel averaging technique and analytic term-by-term integration by brute force, Brouwer's solution must also begin with a set of mean (averaged) orbital elements. Vinti's method, which is a Hamilton–Jacobian formulation of Newtonian mechanics, is straightforward and elegant. The equations of motion for the classical and Hamilton–Jacobian formulations are expressed in terms of force and energy, respectively.

The trajectory propagation algorithms of Brouwer and Kozai, which are represented by the simplified general perturbations (SGP) and its derivatives (SGP4, SDP4, SGP8, SDP8), have been developed by the North American Aerospace Defense Command (NORAD) and used for over 30 years. For comparison purposes, an unofficial version of these SGP algorithms and the necessary conversion algorithms are also included on the floppy disk. These SGP algorithms, which were downloaded from a computer at the U.S. Air Force Institute of Technology via the Internet, are slightly modified for true double precision computing.

The singularity problems that we have described are insignificant when compared with the difficulty of initialization or starting procedure. The input state vector for a term-by-term analytic integration method such as Brouwer's requires a six-dimensional mean vector (the six mean elements in the NORAD two-card element set are $\bar{n}$, $\bar{e}$, $\bar{i}$, $\bar{\Omega}$, $\bar{\omega}$, $\bar{M}$). The mean anomaly $\bar{n}$ is used instead of the mean semi-major axis $\bar{a}$. Thus, all SGP propagators start with a given mean vector, and their output is the predicted (osculating) position and velocity vectors $\boldsymbol{r}(t)$ and $\boldsymbol{v}(t)$, which can be transformed to the osculating elements (a, e, I, Ω, ω, M), if desired. Osculating elements are the ones that are usually available, and the reconstruction of mean elements must begin with osculating elements. Therefore, the SGP propagators that accept only mean elements as input are difficult to use because they require an additional step of converting osculating elements to mean elements. Conversion is unnecessary if the input and output are initial and final position and velocity vectors. Although Vinti's method starts with the given osculating position and velocity vectors, it actually computes a set of mean elements and then outputs the predicted position and velocity vectors $\boldsymbol{r}(t)$ and $\boldsymbol{v}(t)$. That is, the input and output formats of Vinti's method are identical to those of Kepler's method, and this transparency of mean elements alone presents a formidable advantage of Vinti's method over any term-by-term analytic integration method.

The Hamilton–Jacobi equation was regarded by most physicists only as the point of departure for quantum mechanics. Vinti mathematically solved the Kepler problem by separating the Hamilton–Jacobi equation and simultaneously satisfying the Laplace equation and exploited the spheroidal Earth to provide the physical meaning. The Vinti spheroidal method relies not just on a solid mathematical foundation, but also on the laws of physics. Formulating this potential in terms of oblate spheroidal coordinates is in itself a combination of masterful insight and hard work. The editors' objective is to make this elegant theory understandable and to make its great practical utility for satellite orbit and ballistic missile launch and impact-point prediction accessible to a new generation of astronomers, physicists, applied mathematicians, and engineers. Goddard Space Flight Center was an active center for the development of Vinti's work. Vinti and scientists at Goddard published numerous reports that extended Vinti's analytic method to include drag and perform differential correction in orbit determination.

Chapter 1

Newton's Laws

I. Newton's Laws of Motion

ORBITAL and celestial mechanics are based almost entirely on the mechanics of Newton. According to this, we can attach a number, called the *inertial mass*, to any given particle, and this number governs its response to its environment. Let the position vector of the particle be $\boldsymbol{r}$, its vector displacement from the origin O of some reference system that we call *inertial*. Such an inertial system is said to be at rest relative to the "fixed stars," or more accurately, with respect to the universe as a whole.

If t is time, we denote a time derivative by a superscript dot. The velocity $\boldsymbol{v}$ of any particle is then given by $\boldsymbol{v} = \dot{\boldsymbol{r}}$, and this is the quantity that the ancients supposed to be directly responsive to the environment for all objects below the moon. Galileo and Newton gave up this idea and assumed that it is the second derivative $\ddot{\boldsymbol{r}}$, the acceleration, that plays this role. Thus, $\ddot{\boldsymbol{r}}$ is some function of position (and sometimes velocity) that governs the motion.

Newton's first two laws of motion can be expressed as

$$m\,\ddot{\boldsymbol{r}} = \boldsymbol{F}$$

where the environmental function $\boldsymbol{F}$ is called the force acting on the particle and where m is called the inertial mass.

Newton's third law of motion, of action and reaction, is concerned with the interaction of two particles A and B. It states that they exert equal and opposite forces on each other, not necessarily along the line joining them. The caveat, important only when the forces are electromagnetic and the relative velocity is high, does not affect orbital and celestial mechanics.

II. Newton's Law of Gravitation

If two particles A and B are separated by a distance r, Newton's law of gravitation states that they attract each other, along the line joining them, with a force proportional to $(M_A M_B)/r^2$. Here M_A and M_B are numbers called the gravitational masses of the particles. As an equation

$$\boldsymbol{F} = -G M_A M_B \boldsymbol{r}/r^3$$

where $\boldsymbol{r}$ is their separation vector and G is a gravitational constant very nearly equal to $(2/3)10^{-20}\ \mathrm{km}^3/(\mathrm{kg\,s}^2)$.

At a given point in space, the gravitational field strength is defined as the gravitational force per unit gravitational mass on a test particle placed at the point. If

M is the gravitational mass of the test particle and the field strength is f, the force on the test particle is

$$\boldsymbol{F} = M\boldsymbol{f}$$

It is well known that all bodies fall to the Earth with the same acceleration $\boldsymbol{g}$ if atmospheric resistance is eliminated. Thus, for any two particles with inertial masses m_k and gravitational masses M_k $(k = 1, 2)$, we have $m_1\boldsymbol{g} = M_1\boldsymbol{f}$ and $m_2\boldsymbol{g} = M_2\boldsymbol{f}$, where $\boldsymbol{f}$ is the gravitation field strength at the place of fall. Thus, $m_1/M_1 = m_2/M_2$ so that m is proportional to M. By a suitable choice of units they may be treated as equal. With such a choice of units the law of gravitation becomes

$$\boldsymbol{F} = -Gm_1m_2\boldsymbol{r}/r^3$$

and the gravitational field strength produced by a particle of mass m at a vector distance $\boldsymbol{r}$ is

$$\boldsymbol{f} = -Gm\boldsymbol{r}/r^3$$

III. The Gravitational Potential

Consider a source point of mass m_k, with position vector $\boldsymbol{r}_k$ relative to some origin O, and a field point at P, with position vector $\boldsymbol{r}$ (Fig. 1.1). If $\boldsymbol{\rho}_k = \boldsymbol{r} - \boldsymbol{r}_k$, the source point produces at P the field strength

$$\boldsymbol{f}_k = -Gm_k\boldsymbol{\rho}_k/\rho_k^3$$

Suppose we keep the source mass fixed at $\boldsymbol{r}_k$ and vary the field point P. Then $\mathrm{d}\boldsymbol{r} = \mathrm{d}\boldsymbol{\rho}_k$ and

$$\boldsymbol{f}_k \cdot \mathrm{d}\boldsymbol{r} = -\frac{Gm_k}{\rho_k^3}\boldsymbol{\rho}_k \cdot \mathrm{d}\boldsymbol{\rho}_k = -\frac{Gm_k}{\rho_k^2}\mathrm{d}\rho_k = \mathrm{d}\left(\frac{Gm_k}{\rho_k}\right)$$

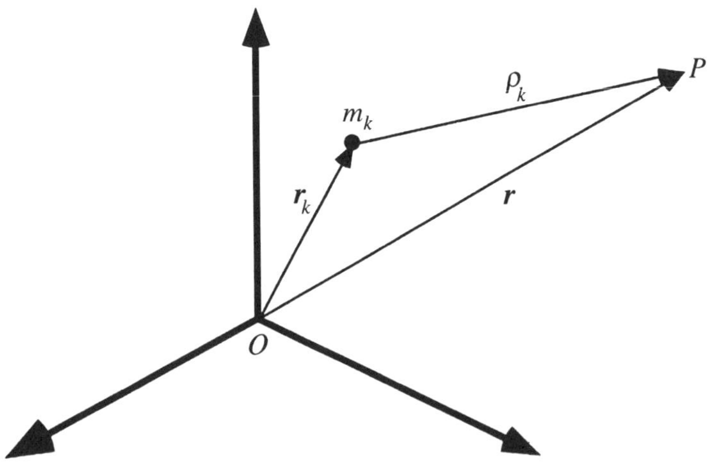

Fig. 1.1 Gravitational potential.

Let us now consider a gravitational field to be produced by n source masses $m_1, m_2, \ldots, m_n$. The total field at point P will then be

$$\boldsymbol{f} = \sum_1^n \boldsymbol{f}_k = -\sum_1^n \frac{Gm_k}{\rho_k^3}\boldsymbol{\rho}_k$$

If we keep the sources fixed but move the field point by $\mathrm{d}\boldsymbol{r}$, then $\mathrm{d}\boldsymbol{\rho}_k = \mathrm{d}\boldsymbol{r}$ and

$$\boldsymbol{f} \cdot \mathrm{d}\boldsymbol{r} = -\sum_1^n \frac{Gm_k}{\rho_k^3}\boldsymbol{\rho}_k \cdot \mathrm{d}\boldsymbol{\rho}_k = \mathrm{d}\sum_1^n \left(\frac{Gm_k}{\rho_k}\right) = -\mathrm{d}V$$

where

$$V = -\sum_1^n \left(\frac{Gm_k}{\rho_k}\right)$$

is called the gravitational potential at P.

In rectangular coordinates $\boldsymbol{f} \cdot \mathrm{d}\boldsymbol{r} = -dV$ becomes

$$\sum_{xyz} f_x \,\mathrm{d}x = -\sum_{xyz} \frac{\partial V}{\partial x}\,\mathrm{d}x$$

Since $\mathrm{d}x$, $\mathrm{d}y$, and $\mathrm{d}z$ are independent, we find

$$f_x = -\frac{\partial V}{\partial x} \qquad f_y = -\frac{\partial V}{\partial y} \qquad f_z = -\frac{\partial V}{\partial z}$$

so that

$$\boldsymbol{f} = -\nabla V$$

We thus represent the vector field $\boldsymbol{f}$ by a scalar potential field V. The potential produced by a point mass m at a distance r is then

$$V = -Gm/r$$

The potentials produced at a field point by a number of point sources are scalar additive.

The equation for the potential produced by a number of point sources is readily generalized to the case of a continuum of sources. If $\mathrm{d}\tau'$ is a volume element, ε the mass density, and $\boldsymbol{r}'$ the position vector of a volume element, the potential at a field point at $\boldsymbol{r}$ outside a distribution D is

$$V = -G\int \frac{\varepsilon\,\mathrm{d}\tau'}{|\boldsymbol{r} - \boldsymbol{r}'|}$$

It is a simple matter to show that in free space V satisfies Laplace's equation

$$\nabla^2 V = 0$$

IV. Gravitational Flux and Gauss' Theorem

The integral $\int \boldsymbol{f} \cdot \mathrm{d}\boldsymbol{S}$ over a closed surface S is called the flux from S. Here $\mathrm{d}\boldsymbol{S}$ is a vector surface element pointing along the outward normal. If m is the total mass enclosed by S, Gauss' theorem states that

$$\int \boldsymbol{f} \cdot \mathrm{d}\boldsymbol{S} = -4\pi G m$$

The proof for the case of discrete particles inside S is as follows: Surround each particle by a small sphere of radius a_k, with m_k at its center. Consider the free space R bounded by S and the totality of spherical surfaces Σ. The outward normal for R is outward from S and inward into each small sphere. Then

$$\int_S \boldsymbol{f} \cdot \mathrm{d}\boldsymbol{S} + \int_\Sigma \boldsymbol{f} \cdot \mathrm{d}\boldsymbol{S} = \int_R \boldsymbol{f} \cdot \mathrm{d}\boldsymbol{S}$$

Since V has no singularities in R, the divergence theorem holds:

$$\int_R \boldsymbol{f} \cdot \mathrm{d}\boldsymbol{S} = \int_R \nabla \cdot \boldsymbol{f}\, \mathrm{d}t = -\int_R \nabla^2 V\, \mathrm{d}t = 0$$

since $\boldsymbol{f} = -\nabla V$ and $\nabla^2 V = 0$ in free space. Thus,

$$\int_S \boldsymbol{f} \cdot \mathrm{d}\boldsymbol{S} + \int_\Sigma \boldsymbol{f} \cdot \mathrm{d}\boldsymbol{S} = 0$$

Since Σ consists of a number of spheres $\Sigma_1, \Sigma_2, \ldots, \Sigma_m$, this becomes

$$\int_S \boldsymbol{f} \cdot \mathrm{d}\boldsymbol{S} = -\Sigma_k \int_{\Sigma_k} \boldsymbol{f} \cdot \mathrm{d}\boldsymbol{S}$$

If we let each $a_k \to 0$, the value of $\boldsymbol{f}$ over the sphere Σ_k is

$$\boldsymbol{f}_k = \frac{G m_k}{a_k^2} \boldsymbol{n}_k + O\left(a_k^0\right)$$

the quantity $O(a_k^0)$ being produced by the sources other than m_k and $\boldsymbol{n}_k$ is the unit vector along $\boldsymbol{f}_k$. Then

$$\int_{\Sigma_k} \boldsymbol{f} \cdot \mathrm{d}\boldsymbol{S} = \frac{G m_k}{a_k^2} 4\pi a_k^2 + O\left(a_k^2\right)$$

As $a_k \to 0$

$$\int_{\Sigma_k} \boldsymbol{f} \cdot \mathrm{d}\boldsymbol{S} \to 4\pi G m_k$$

Thus

$$\int_S \boldsymbol{f} \cdot \mathrm{d}\boldsymbol{S} = -4\pi G \sum_1^n m_k = -4\pi G m$$

This is Gauss' theorem.

V. Gravitational Properties of a True Sphere

Define a true sphere as a body with a spherical surface and with density $\varepsilon(r)$, a function only of the distance r from the center of the sphere. By symmetry the field outside the sphere is then

$$\boldsymbol{f} = \psi(r)\boldsymbol{l}_r$$

where $\boldsymbol{l}_r$ is the unit vector $\boldsymbol{r}/r$. Thus

$$\int_S \boldsymbol{f} \cdot \mathrm{d}\boldsymbol{S} = 4\pi r^2 \psi(r) = -4\pi G m$$

m being the total mass of the sphere. Then $\psi(r) = -Gm/r^2$ and

$$\boldsymbol{f} = -Gm\,\boldsymbol{r}/r^3$$

just as though all the mass were concentrated at the center of the sphere. The *active* gravitational behavior of a true sphere is the same as that of a particle.

The *passive* behavior of a true sphere in a gravitational field is the same as that of a particle. The relevant theorem is

$$\boldsymbol{F} = m\boldsymbol{f}_c$$

where m is the sphere's mass, $\boldsymbol{f}_c$ the gravitational field at its center, and $\boldsymbol{F}$ the resulting force. To prove this, consider the external field as arising from n point masses m_k ($k = 1, \ldots, n$). The sphere attracts each point mass m_k with the force $Gmm_k\boldsymbol{r}_k/r_k^3$, where $\boldsymbol{r}_k$ is the vector from the mass m_k to the center C of m. By Newton's third law, each m_k exerts a force $-Gmm_k\boldsymbol{r}_k/r_k^3$ on the sphere. The total force on the sphere is thus $-\Sigma_k Gmm_k\boldsymbol{r}_k/r_k^3$, which equals $m\boldsymbol{f}_c$. Here

$$\boldsymbol{f}_c = -\Sigma_k Gm_k\boldsymbol{r}_k/r_k^3$$

the total field strength produced at C by the external particles. Thus, $\boldsymbol{F} = m\boldsymbol{f}_c$, as stated.

It is now a matter of simple integration to prove that a single external particle exerts zero gravitational torque on a true sphere. By addition, any external distribution of mass produces zero gravitational torque on it. For orbital motion, we may treat a true sphere as a mass point, both actively and passively. Moreover, its spin motion can never be coupled with its orbital motion.

Chapter 2

The Two-Body Problem

I. Reduction to the One-Center Problem

LET m_1 and m_2 be the masses of two true spheres. They may be the sun and a planet, a planet and a satellite (natural or artificial), or a double star (see Fig. 2.1). Let the reference system $Oxyz$ be inertial; let $\boldsymbol{r}_1$ and $\boldsymbol{r}_2$ be the position vectors of m_1 and m_2, $\boldsymbol{R}$ that of their center of mass C; and let $\boldsymbol{s}_1$ and $\boldsymbol{s}_2$ be the position vectors of m_1 and m_2 relative to C. With m_1 as the primary, let $\boldsymbol{r}$ be the position vector of m_2 relative to m_1.

Then

$$\boldsymbol{r} = \boldsymbol{s}_2 - \boldsymbol{s}_1 = \boldsymbol{r}_2 - \boldsymbol{r}_1 \qquad \boldsymbol{R} = (m_1 + m_2)^{-1}(m_1\boldsymbol{r}_1 + m_2\boldsymbol{r}_2)$$

The equations of motion are

$$m_1\ddot{\boldsymbol{r}}_1 = Gm_1m_2\boldsymbol{r}/r^3 \qquad m_2\ddot{\boldsymbol{r}}_2 = -Gm_1m_2\boldsymbol{r}/r^3$$

so that

$$m_1\ddot{\boldsymbol{r}}_1 + m_2\ddot{\boldsymbol{r}}_2 = \boldsymbol{0}$$

from which $\ddot{\boldsymbol{R}} = \boldsymbol{0}$, $\boldsymbol{R} = \boldsymbol{C}_0 + \boldsymbol{C}_1 t$.

Now consider motion relative to C. We have

$$\boldsymbol{r}_1 = \boldsymbol{R} + \boldsymbol{s}_1 \qquad \boldsymbol{r}_2 = \boldsymbol{R} + \boldsymbol{s}_2$$

With use of the definition of $\boldsymbol{R}$, these give

$$\begin{aligned}\boldsymbol{s}_1 &= \boldsymbol{r}_1 - (m_1 + m_2)^{-1}(m_1\boldsymbol{r}_1 + m_2\boldsymbol{r}_2) = (m_1 + m_2)^{-1}m_2(\boldsymbol{r}_1 - \boldsymbol{r}_2)\\ &= -(m_1 + m_2)^{-1}m_2\boldsymbol{r}\\ \boldsymbol{s}_2 &= \boldsymbol{r}_2 - (m_1 + m_2)^{-1}(m_1\boldsymbol{r}_1 + m_2\boldsymbol{r}_2) = (m_1 + m_2)^{-1}m_1(\boldsymbol{r}_2 - \boldsymbol{r}_1)\\ &= (m_1 + m_2)^{-1}m_1\boldsymbol{r}\end{aligned}$$

These equations show that the orbits of m_1 and m_2 relative to the center of mass have the same behavior, both in regard to shape and time, as the orbit of m_2 relative to m_1. The only difference is a distance scale factor in each case. Any characteristic length for the relative orbit will be multiplied by $m_2(m_1 + m_2)^{-1}$ for the orbit of m_1 relative to C or by $m_1(m_1 + m_2)^{-1}$ for the orbit of m_2.

The orbit of m_2 relative to m_1 is characterized by

$$\boldsymbol{r}(t) = \boldsymbol{r}_2 - \boldsymbol{r}_1 \qquad \ddot{\boldsymbol{r}}(t) = \ddot{\boldsymbol{r}}_2 - \ddot{\boldsymbol{r}}_1$$

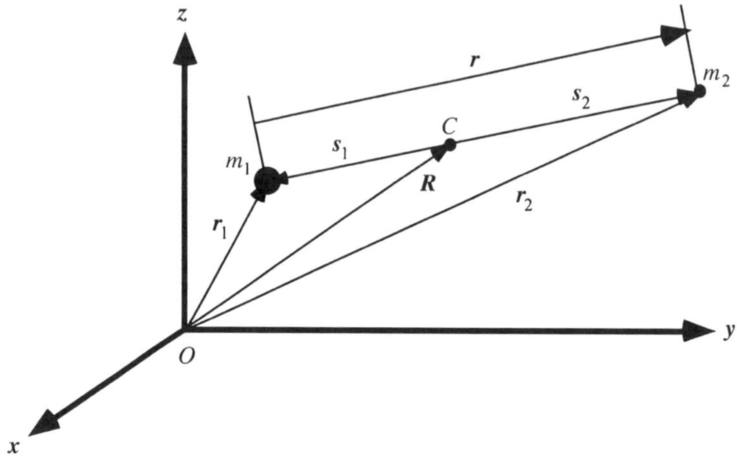

Fig. 2.1 The two-body problem.

but

$$\ddot{r}_1 = Gm_2 \boldsymbol{r}/r^3 \qquad \ddot{r}_2 = -Gm_1 \boldsymbol{r}/r^3$$

so that

$$\ddot{\boldsymbol{r}}(t) = -G(m_1 + m_2)\boldsymbol{r}/r^3 = -\mu \boldsymbol{r}/r^3$$

where $\mu \equiv G(m_1+m_2)$. This is the same as for a particle of unit mass moving under the attraction of a center with gravitational mass m_1+m_2 and infinite inertial mass.

II. The One-Center Problem

Before integrating $\ddot{\boldsymbol{r}}(t) = -\mu \boldsymbol{r}/r^3$, let us consider the more general problem of a particle moving in a field derivable from a potential $V(q, t)$. Such a potential depends not only on the coordinates, but also explicitly on the time t. Then $\ddot{\boldsymbol{r}}(t) = -\nabla V(q, t)$.

Such a system is called monogenic; if t does not appear explicitly, it is called conservative. An example for $V(q, t)$ would be the drag-free motion of a satellite around a spinning planet with equatorial ellipticity. An example for $V(q$ only) would be the drag-free motion of a satellite around an axially symmetric planet.

If V depends only on the distance r from the planet, then

$$\ddot{\boldsymbol{r}}(t) = -\nabla V(r) = -V'(r)\boldsymbol{l}_r$$

where $\boldsymbol{l}_r$ is the unit vector $\boldsymbol{r}/r$.

Then

$$\boldsymbol{0} = \boldsymbol{r} \times \ddot{\boldsymbol{r}}(t) = \frac{d}{dt}(\boldsymbol{r} \times \dot{\boldsymbol{r}})$$

so that if $\boldsymbol{L}$ is the angular momentum per unit mass

$$\boldsymbol{L} = \boldsymbol{r} \times \dot{\boldsymbol{r}} = \text{constant vector}$$

The total angular momentum is conserved, and the orbit lies in a fixed plane. To see this, note that $\boldsymbol{L}$ is perpendicular to both $\boldsymbol{r}$ and $\dot{\boldsymbol{r}}$, which determine the instantaneous plane of the orbit. Since $\boldsymbol{L}$ is constant, the normal to the orbital plane remains fixed in direction, and the orbital plane remains fixed for such a central field.

If the field is not central but is symmetric with respect to the axis Oz, then $V = V(r, z)$ and the z-component L_z of angular momentum is constant. To show this, note that per unit mass

$$L_z = x\dot{y} - y\dot{x} \qquad \dot{L}_z = x\ddot{y} - y\ddot{x}$$

with

$$\ddot{x} = -\frac{\partial V}{\partial x} = -\frac{\partial V}{\partial r}\frac{x}{r} \qquad \ddot{y} = -\frac{\partial V}{\partial y} = -\frac{\partial V}{\partial r}\frac{y}{r}$$

so that

$$\dot{L}_z = -\frac{xy}{r}\frac{\partial V}{\partial r} + \frac{xy}{r}\frac{\partial V}{\partial r} = 0$$

and L_z is constant. If V depends only on q, then

$$\dot{V} = V_x\dot{x} + V_y\dot{y} + V_z\dot{z} = \nabla V \cdot \dot{\boldsymbol{r}}$$

On scalar multiplication of $\ddot{\boldsymbol{r}}(t) = -\nabla V$ by $\dot{\boldsymbol{r}}$, we obtain

$$\dot{\boldsymbol{r}} \cdot \ddot{\boldsymbol{r}} = -\nabla V \cdot \dot{\boldsymbol{r}} = -\dot{V}$$

so that

$$\frac{1}{2}\frac{d}{dt}(\dot{\boldsymbol{r}}^2) = -\dot{V}$$

and

$$\tfrac{1}{2}(\dot{\boldsymbol{r}}^2) + V(q) = \text{const} = W$$

Here W is the energy integral, so that this theorem is the conservation of energy. Thus, $\ddot{\boldsymbol{r}}(t) = -\nabla V(q)$ is called a conservative system.

For the two-body problem $\ddot{\boldsymbol{r}}(t) = -G(m_1 + m_2)\boldsymbol{l}_r/r^2$, so that the energy integral becomes

$$\frac{1}{2}\boldsymbol{v}^2 - \frac{G(m_1 + m_2)}{r} = W$$

where $\boldsymbol{v}$ is the relative velocity.

By using the relations that reduced the two-body problem to a one-center problem, it is easy to show that

$$\frac{1}{2}m_1 v_1^2 + \frac{1}{2}m_2 v_2^2 - \frac{Gm_1m_2}{r} = \frac{m_1m_2}{(m_1 + m_2)}W$$

where W is the constant just met and $\boldsymbol{v}_1$ and $\boldsymbol{v}_2$ are the velocities of m_1 and m_2 relative to the center of mass.

III. The Laplace Vector

If $\boldsymbol{L}$ is the angular momentum per unit mass, the vector

$$\boldsymbol{R} = \dot{\boldsymbol{r}} \times \boldsymbol{L} - \mu\boldsymbol{l}_r$$

is constant. It is known by various names: Laplace, Runge-Lenz, perifocus vector, or e-vector. To prove its constancy, we begin with motion in a general central field $V(r)$ and show that $V(r)$ must be $-\mu/r$ for the theorem to hold.

Write

$$\ddot{\boldsymbol{r}}(t) = -\nabla V(r) = -V'(r)\boldsymbol{l}_r = -V'(r)\boldsymbol{r}/r$$

Then since $\boldsymbol{L} \equiv \boldsymbol{r} \times \dot{\boldsymbol{r}}$ is constant, it follows that

$$\begin{aligned}
\frac{d}{dt}(\dot{\boldsymbol{r}} \times \boldsymbol{L}) = \ddot{\boldsymbol{r}} \times \boldsymbol{L} &= -V'(r)r^{-1}\boldsymbol{r} \times \boldsymbol{L} \\
&= -V'(r)r^{-1}\boldsymbol{r} \times (\boldsymbol{r} \times \dot{\boldsymbol{r}}) \\
&= -V'(r)r^{-1}\,[\boldsymbol{r}(\boldsymbol{r}\cdot\dot{\boldsymbol{r}}) - r^2\dot{\boldsymbol{r}}] \\
&= -V'(r)r^{-1}[\boldsymbol{r}(r\dot{r}) - r^2\dot{\boldsymbol{r}}] \\
&= -V'(r)\dot{r}\boldsymbol{r} + V'(r)r\dot{\boldsymbol{r}} \\
&= -\dot{V}(r)\boldsymbol{r} + V'(r)r\dot{\boldsymbol{r}}
\end{aligned}$$

However,

$$\dot{V}\boldsymbol{r} = \frac{d}{dt}(V\boldsymbol{r}) - V\dot{\boldsymbol{r}}$$

thus

$$\frac{d}{dt}(\dot{\boldsymbol{r}} \times \boldsymbol{L}) = -\frac{d}{dt}(V\boldsymbol{r}) + V\dot{\boldsymbol{r}} + rV'(r)\dot{\boldsymbol{r}}$$

and

$$\frac{d}{dt}(\dot{\boldsymbol{r}} \times \boldsymbol{L} + V\boldsymbol{r}) = \dot{\boldsymbol{r}}\frac{d}{dr}(rV)$$

This equation yields an integral of the motion if and only if

$$\frac{d}{dr}(rV) = k$$

is a constant. In such a case

$$rV = kr - \mu$$

or

$$V = k - \frac{\mu}{r}$$

Since k vanishes for a planet (potential vanishing at infinity), we obtain such an integral of the motion if $V = -\mu/r$. This corresponds to the two-body problem if $\mu \equiv G(m_1 + m_2)$. Then

$$\frac{d}{dt}\left(\dot{\boldsymbol{r}} \times \boldsymbol{L} - \frac{\mu}{r}\boldsymbol{r}\right) = \boldsymbol{0}$$

or

$$\dot{\boldsymbol{r}} \times \boldsymbol{L} - \mu \boldsymbol{l}_r = \boldsymbol{R}$$

where $\boldsymbol{R}$ is the Laplace vector, now proved constant.

Any function of the coordinates and momenta, and possibly of the time t, is called an integral of the motion if it remains constant. In rectangular coordinates, the momenta are simply $\dot{x}$, $\dot{y}$, $\dot{z}$ per unit mass of the orbiter.

We have found seven integrals for the two-body problem: the energy W, the three components of the angular momentum $\boldsymbol{L}$, and the three components of the Laplace vector $\boldsymbol{R}$. They are not all independent, however, because there are two relations connecting them. One of these is $\boldsymbol{R} \cdot \boldsymbol{L} = 0$; we shall write down the other one later. This leaves five independent integrals. Later, we shall discover a sixth independent integral.

IV. The Conic Section Solutions

Since the angular momentum $\boldsymbol{L}$ is perpendicular to the orbital plane and since the Laplace vector $\boldsymbol{R}$ is perpendicular to $\boldsymbol{L}$, it follows that $\boldsymbol{R}$ lies in the orbital plane. If f is the angle from $\boldsymbol{R}$ to the position vector $\boldsymbol{r}$, then

$$\boldsymbol{r} \cdot \boldsymbol{R} = rR \cos f = \boldsymbol{r} \cdot (\dot{\boldsymbol{r}} \times \boldsymbol{L} - \mu \boldsymbol{l}_r) = L^2 - \mu r$$

Solution for r gives

$$r = \frac{L^2/\mu}{1 + (R/\mu)\cos f}$$

This is the equation of a conic section

$$r = \frac{p}{1 + e\cos f}$$

with the semi-latus rectum $p = L^2/\mu$, the eccentricity $e = R/\mu \geq 0$, and the true anomaly f. Note the relations $L^2 = \mu p$ and $R = \mu e$.

A conic section may be defined as the locus of a point A, the ratio of whose distances to a focus F and a directrix dd remains constant (see Fig. 2.2). Let FC be a perpendicular from the focus F to the directrix dd and FB a perpendicular to FC intersecting the conic at B. From the definition

$$r/D = \text{const} = e = (D + r\cos f)^{-1} p$$

Then $D = r/e$ and

$$r = \frac{p}{1 + e\cos f}$$

For the two-body problem, $L^2 = \mu p$ and $R = \mu e$. This second relation explains the occasional use of the term e-vector for $\boldsymbol{R}$. The point P, for which r is a minimum, is called the pericenter, and we denote by $\boldsymbol{i}$ a unit vector pointing from F toward P.

We next prove that

$$\boldsymbol{R} = \mu e \boldsymbol{i} \qquad L^2 = \mu(1 + e) r_p$$

where $r_p = FP$. To do so, we may evaluate $\boldsymbol{R}$ and $\boldsymbol{L}$ at P, since they are constants.

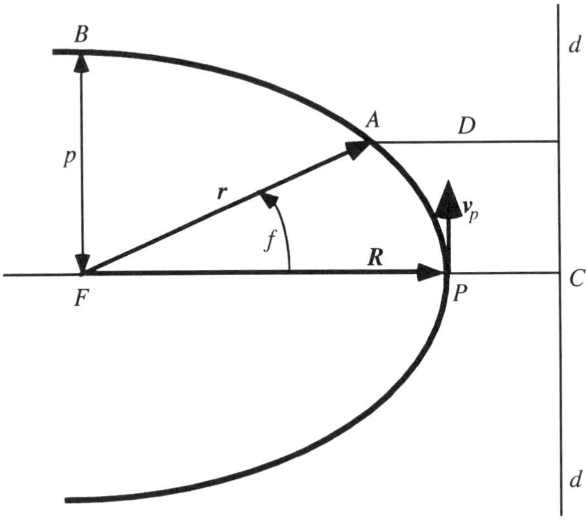

Fig. 2.2 Conic section.

At P, since $\dot{\boldsymbol{r}} \perp \boldsymbol{r}$, with the orbiter moving counterclockwise, it follows that $\boldsymbol{L} = \boldsymbol{r} \times \dot{\boldsymbol{r}}$ points out from the figure at P. Then $\dot{\boldsymbol{r}} \times \boldsymbol{L}$ points from F toward P. However, at P, with $\dot{\boldsymbol{r}} = \boldsymbol{v}$

$$|\boldsymbol{L}| = L = r_p v_p \qquad |\dot{\boldsymbol{r}} \times \boldsymbol{L}| = r_p\, v_p^2 = \frac{L^2}{r_p}$$

Then

$$(\dot{\boldsymbol{r}} \times \boldsymbol{L})_p = \left(L^2/r_p\right)\boldsymbol{i}$$

From $\boldsymbol{R} = \dot{\boldsymbol{r}} \times \boldsymbol{L} - \mu \boldsymbol{l}_r$ we find

$$\boldsymbol{R}_p = (\dot{\boldsymbol{r}} \times \boldsymbol{L})_p - \mu\boldsymbol{i} = \left(\frac{L^2}{r_p} - \mu\right)\boldsymbol{i}$$

Then

$$\boldsymbol{R} = \boldsymbol{R}_p = \left(\frac{L^2}{r_p} - \mu\right)\boldsymbol{i}$$

Since $L^2 = \mu p$ and $r_p = p(1+e)^{-1}$, we find that

$$\boldsymbol{R} = [\mu(1+e) - p]\boldsymbol{i} = \mu e\boldsymbol{i}$$

as was to be shown. On eliminating p between the two equations for L^2 and r_p, we obtain

$$L^2 = \mu(1+e)\, r_p$$

which also was to be shown.

If $\boldsymbol{v} = \dot{\boldsymbol{r}}$, the energy W per unit mass is $W = \frac{1}{2}v^2 - (\mu/r)$, a constant that may also be evaluated at P. From $L = r_p v_p$, $r_p = p(1+e)^{-1}$, and $L^2 = \mu(1+e)r_p$, it follows simply that $W = (\mu/2p)(e^2 - 1)$. If $e > 1$, then $W > O$ and the curve is a hyperbola. Comparison with $r = p(1 + e\cos f)^{-1}$ shows that $\cos f \geq -1/e$, so that f cannot exceed $\cos^{-1}(-1/e)$, and this reveals the asymptotes. If $e = 1$, the speed v vanishes as $r \to \infty$, and the curve is a parabola. If $e < 1$, then $W < O$, and only those values of r occur for which $\mu/r > -W$, i.e., for which $r < -\mu/W$.

With $0 \leq e < 1$, the orbit is an ellipse, and we can define a quantity a by

$$W = -\frac{\mu}{2a} = \frac{\mu}{2p}(e^2 - 1)$$

where $a > 0$ and $p = a(1 - e^2)$. Here a will be the semi-major axis.

At this point, it is easy to find the remaining relation connecting the seven integrals already found. From

$$W = \frac{\mu}{2p}(e^2 - 1) \qquad R = \mu e \qquad L^2 = \mu p$$

elimination of e and p yields

$$R^2 = \mu^2 + 2WL^2$$

Before going into elliptic orbits in detail, it should be mentioned here that the inverse square law of gravitation has led us to Kepler's first law: The planets move around the sun in elliptic orbits with the sun at one focus. This conclusion follows from the finiteness of only those orbits with $e < 1$. It has also led to Kepler's second law, since we have shown the constancy of angular momentum. Specifically, consider $\boldsymbol{L} = \boldsymbol{r} \times \dot{\boldsymbol{r}} = \text{const}$. With $\boldsymbol{r} = r\boldsymbol{l}_r$ we have

$$\dot{\boldsymbol{r}} = \dot{r}\boldsymbol{l}_r + r\frac{d}{dt}\boldsymbol{l}_r$$

But $(d/dt)\boldsymbol{l}_r = \dot{f}\boldsymbol{l}_f$, where $\boldsymbol{l}_f$ is a unit vector along the transverse. Then $\boldsymbol{r} \times \dot{\boldsymbol{r}} = r^2\boldsymbol{l}_r \times \dot{f}\boldsymbol{l}_f = r^2\dot{f}\boldsymbol{k}$, where $\boldsymbol{k}$ is a unit vector normal to the orbital plane. However, $r^2\dot{f}$ is twice the rate at which area is swept out by the planetary vector. This is constant, and we have Kepler's second law.

V. Elliptic Orbits

Here, the orbit is a closed curve, periodic in the time t by the law of equal areas and symmetric about $f = 0$. The quantity f is the true anomaly in the equation

$$r = p(1 + e\cos f)^{-1} \qquad e < 1$$

The energy equation is

$$\frac{1}{2}v^2 - \frac{\mu}{r} = \frac{\mu}{2p}(e^2 - 1) = -\frac{\mu}{2a}$$

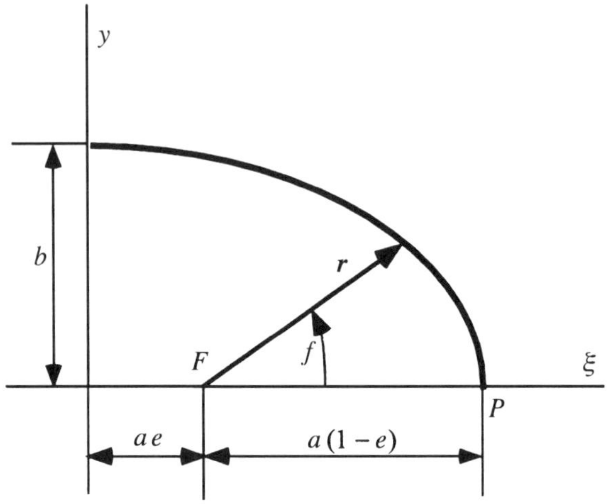

Fig. 2.3 Elliptic orbit.

Since $p = a(1 - e^2)$, we find

$$r_{\min} = p(1+e)^{-1} = a(1-e)$$

$$r_{\max} = p(1-e)^{-1} = a(1+e)$$

therefore

$$r_{\min} + r_{\max} = 2a$$

a being called the semi-major axis or "mean distance." It is only the arithmetic mean of the extreme distances and not the time mean. To put the equation in rectangular coordinates ξ and y, with the center of the ellipse as origin, write $r = p(1 + e\cos f)^{-1}$, $p = a(1 - e^2)$, and note that $FP = a(1 - e)$ as shown in Fig. 2.3. Then

$$\xi = a\,e + \frac{a(1-e^2)\cos f}{1+e\cos f} = \frac{a(e+\cos f)}{1+e\cos f}$$

$$y = r\sin f$$

It is a simple exercise to show that

$$\left(\frac{\xi}{a}\right)^2 + \left(\frac{y}{a\sqrt{1-e}}\right)^2 = 1$$

so that the semi-minor axis

$$b = a\sqrt{1-e}$$

The Eccentric Anomaly E

We next introduce an important variable, the eccentric anomaly E. To do so, circumscribe an auxiliary circle around the ellipse, and draw a perpendicular from

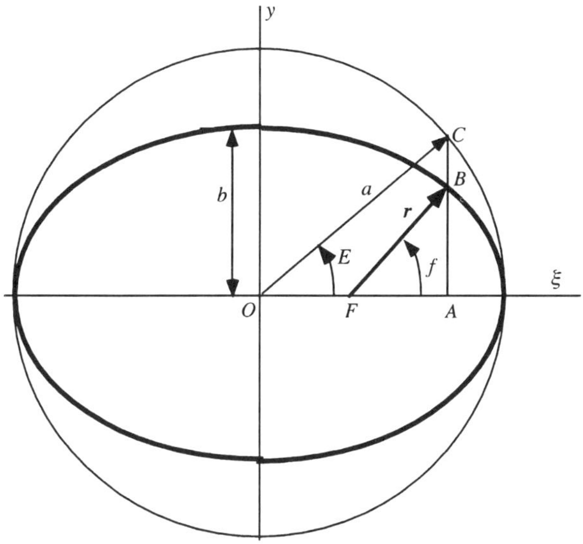

Fig. 2.4 Eccentric anomaly *E*.

the orbiter at B, intersecting the circle at C as shown in Fig. 2.4. Draw OC from the center of the ellipse to C, and define the eccentric anomaly E as the counterclockwise angle from the major axis to OC. (We shall always view an orbiter so that pericenter is at the right and so that the motion is counterclockwise.)

To relate E to f, we first derive an important lemma,

$$b \sin E = r \sin f$$

To do so, regard CA and BA as signed quantities, plus when C and B are above the major axis and minus when below. Then

$$(CA)^2 = a^2 - \xi^2$$

$$(BA)^2 = y^2 = (b^2/a^2)(a^2 - \xi^2)$$

from the equation of the ellipse. Then

$$\frac{CA}{BA} = \frac{a}{b}$$

because CA and BA always have the same sign. However,

$$CA = a \sin E \qquad BA = r \sin f$$

so that

$$\frac{a \sin E}{r \sin f} = \frac{a}{b}$$

The lemma follows immediately. It should be remarked that the anomalies f and E are to be thought of as always increasing, so that $\dot{f} > 0$ and $\dot{E} > 0$ for all time t.

Cosine Relation

$$\xi = a\cos E = ae + \frac{a(1-e^2)\cos f}{1+e\cos f} = \frac{a(e+\cos f)}{1+e\cos f}$$

Thus

$$\cos E = \frac{e+\cos f}{1+e\cos f}$$

Sine Relation

Rewrite the lemma $b\sin E = r\sin f$ as

$$a\sqrt{1-e^2}\sin E = \frac{a(1-e^2)\sin f}{1+e\cos f}$$

Then

$$\sin E = \frac{\sqrt{1-e^2}\sin f}{1+e\cos f}$$

Before inverting these relations, note that $r = a\,(1-e\cos E)$. This follows from

$$1-e\cos E = 1-\frac{e(e+\cos f)}{1+e\cos f} = \frac{1-e^2}{1+e\cos f}$$

since

$$r = \frac{a(1-e^2)}{1+e\cos f}$$

The inverted relations are

$$\cos f = \frac{\cos E - e}{1-e\cos E} = \frac{a}{r}(\cos E - e)$$

$$\sin f = \frac{\sqrt{1-e^2}\sin E}{1-e\cos E} = \frac{a}{r}\sqrt{1-e^2}\sin E$$

Note that, as the orbiter goes round and round, f and E agree at all multiples of π, so that $\dot{f} = \dot{E}$.

There is an important relation connecting the half-angles $f/2$ and $E/2$. To derive it, note that

$$\begin{aligned}\sin f &= 2\sin(f/2)\cos(f/2) = \sqrt{1-e^2}\sin E(1-e\cos E)^{-1}\\ &= 2\sqrt{1-e^2}\sin(E/2)\cos(E/2)(1-e\cos E)^{-1}\end{aligned}$$

$$\begin{aligned}2\cos^2(f/2) &= 1+\cos f = 1+(\cos E - e)(1-e\cos E)^{-1}\\ &= (1-e)(1-\cos E)(1-e\cos E)^{-1}\\ &= 2(1-e)\cos^2(E/2)(1-e\cos E)^{-1}\end{aligned}$$

By division

$$\tan\frac{f}{2} = \sqrt{\frac{1+e}{1-e}}\tan\frac{E}{2}$$

Kepler's Third Law

We next show that $\mu = n^2a^3$, where n is the "mean motion," defined by $n = 2\pi/T$, T being the period. Because the area of the ellipse is πab, we have

$$|\boldsymbol{L}| = r^2\dot{f} = 2\dot{A} = 2\pi ab/T = n\,a\,b = n\,a^2\sqrt{1-e^2}$$

Here A is the area swept out in time t. However,

$$|\boldsymbol{L}| = \sqrt{\mu p} = \sqrt{\mu a(1-e^2)}$$

Thus, $na^2 = \sqrt{\mu a}$, so that $\mu = n^2a^3$. This is essentially Kepler's third law, which states that among the planets the square of the period is proportional to the cube of the semi-major axis. If m_s is the sun's mass and m_1 and m_2 are the masses of two planets, we have $\mu_1 = G(m_s + m_1)$ and $\mu_2 = G(m_s + m_2)$. Then

$$\frac{\mu_1}{\mu_2} = \frac{m_s + m_1}{m_s + m_2} = \left(\frac{T_2}{T_1}\right)^2\left(\frac{a_1}{a_2}\right)^3$$

Kepler's third law is thus an approximation. It would be rigorously true if the planets all had equal masses and if there were no planetary interactions.

Kepler's Equation

If τ is the time of passage through pericenter, this states that

$$E - e\sin E = n(t-\tau)$$

where $n(t-\tau) = \ell$ is called the mean anomaly. To prove it, begin with

$$\frac{r}{a} = 1 - e\cos E = \frac{1-e^2}{1+e\cos f}$$

Differentiate with respect to t to find

$$e\dot{E}\sin E = \frac{(1-e^2)e\dot{f}\sin f}{(1+e\cos f)^2} = \frac{e\sin f\,r^2\dot{f}}{a^2(1-e^2)}$$

$$= \frac{e\,n\,a\,b\sin f}{a^2(1-e^2)} = \frac{e\,n\sin f}{\sqrt{1-e^2}} = \frac{e\,n\sin E}{1-e\cos E}$$

Thus

$$\dot{E} = \frac{n}{1-e\cos E} \qquad (1-e\cos E)\,\dot{E} = n$$

Integration with respect to time gives

$$E - e\sin E = n(t-\tau)$$

where $-n\tau$ is the constant of integration. Here τ is the sixth independent integral of the motion. Unlike the other integrals, it is not algebraic:

$$\tau = t - \frac{1}{n}\cos^{-1}\left[\frac{1}{e}\left(1 - \frac{r}{a}\right)\right] + \frac{e}{n}\sin\left\{\cos^{-1}\left[\frac{1}{e}\left(1 - \frac{r}{a}\right)\right]\right\}$$

where

$$a = -\mu/2W \qquad e = \sqrt{1 + 2WL^2/\mu} \qquad n = \sqrt{\mu a^{-3}}$$

VI. Spherical Trigonometry

Before putting the orbit in three-space, it is desirable to state here the two laws of spherical trigonometry that will be of use. Let A, B, C be the three angles of a spherical triangle and a, b, c be the respective opposite sides.

Law of cosines:

$$\cos c = \cos a \cos b + \sin a \sin b \cos C$$

Law of sines:

$$\frac{\sin A}{\sin a} = \frac{\sin B}{\sin b} = \frac{\sin C}{\sin c}$$

There are simple vector derivations of these two laws.

VII. Orbit in Space

We draw an octant of the celestial sphere; its radius is arbitrary. For the case of a planet moving around the sun, we take its center at the center of mass of the sun. For motion of a satellite around the Earth, we take its center at the center of mass of the Earth. Ox points toward the vernal equinox of some fixed date, say 1950.0. For a planet around the sun, Oz points toward the pole of the ecliptic and, for a satellite around the Earth, toward the north pole of the equator. A line from O to the orbiter intersects the celestial sphere at the suborbital point; the orbit is represented on the celestial sphere by the locus of its suborbital points, of which NPS is an arc. In Fig. 2.5, S is the orbiter, P the pericenter, N the ascending node, ON the line of nodes, ω the argument of pericenter, f the true anomaly, and I the inclination of the orbit to the xy plane. The latter is the plane of the ecliptic for a planet or the equatorial plane of the Earth for a satellite of the Earth.

If we draw a meridian through the suborbital point, the position of the orbiter is fixed by the angles θ and ϕ and the radial distance r. For a planet, θ is the ecliptic latitude λ and ϕ the ecliptic longitude β; for a satellite, θ is the declination δ (same as geocentric latitude) and ϕ the right ascension α.

Let Ω be the longitude or right ascension of the node. To put the orbit in space, we need to find the rectangular coordinates as functions of r, Ω, ω, I, and f. Call $\omega + f = \psi$, the argument of latitude, and apply spherical trigonometry to the spherical triangle SQN. We have

$$\sin\theta = \sin I \sin\psi \tag{2.1}$$

$$\cos\theta = \cos\chi\cos\psi + \sin\chi\sin\psi\cos I \tag{2.2}$$

$$\cos\psi = \cos\chi\cos\theta \tag{2.3}$$

where $\chi = \phi - \Omega$. Multiply Eq. (2.2) by $\sin\chi$ to find

$$\cos\theta\sin\chi = \sin\chi\cos\chi\cos\psi + \sin^2\chi\sin\psi\cos I \tag{2.4}$$

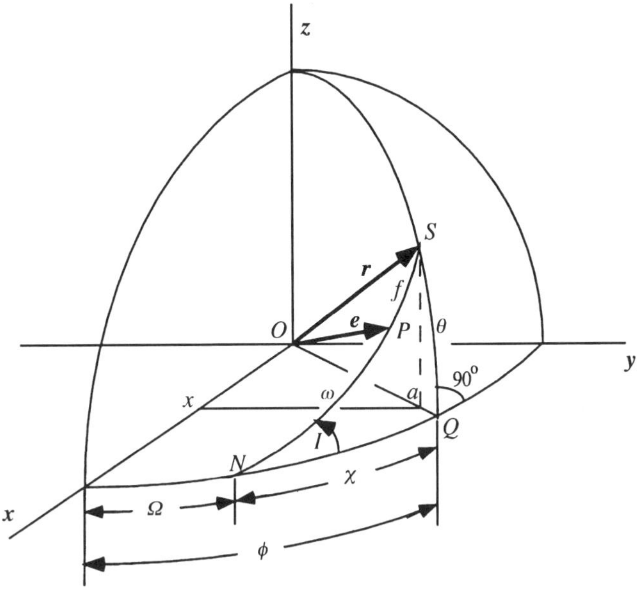

Fig. 2.5 Octant of the celestial sphere with $\psi = \omega + f$ (argument of latitude), $\phi = \Omega + \chi$ (right ascension), $O_a = r\cos\theta$, and $O_x = O_a \cos\phi = r\cos\theta\cos\phi$.

Now apply Eq. (2.3) in $\sin\chi\cos\chi\cos\psi$ to find

$$\cos\theta\sin\chi = \sin\chi\cos^2\chi\cos\theta + \sin^2\chi\sin\psi\cos I$$

A transposition gives

$$\cos\theta\sin\chi(1 - \cos^2\chi) = \sin^2\chi\sin\psi\cos I$$

and cancellation of $\sin^2\chi$ throughout yields

$$\cos\theta\sin\chi = \sin\psi\cos I \tag{2.5}$$

The rectangular coordinates satisfy

$$x = r\cos\theta\cos\phi = r\cos\theta\cos(\Omega + \chi) = r\cos\theta\,[\cos\Omega\cos\chi - \sin\Omega\sin\chi]$$

$$y = r\cos\theta\sin\phi = r\cos\theta\sin(\Omega + \chi) = r\cos\theta\,[\sin\Omega\cos\chi + \cos\Omega\sin\chi]$$

$$z = r\sin\theta = r\sin I\sin\psi \tag{2.6}$$

In Eqs. (2.6), insert $\cos\chi\cos\theta = \cos\psi$ and $\cos\theta\sin\chi = \sin\psi\cos I$ and replace ψ by $\omega + f$. The result is

$$x = r[\cos\Omega\cos(\omega + f) - \sin\Omega\cos I\sin(\omega + f)]$$

$$y = r[\sin\Omega\cos(\omega + f) + \cos\Omega\cos I\sin(\omega + f)]$$

$$z = r\sin I\sin(\omega + f)$$

Then

$$\boldsymbol{r} = \boldsymbol{i}x + \boldsymbol{j}y + \boldsymbol{k}z$$

where $\boldsymbol{i}, \boldsymbol{j}, \boldsymbol{k}$ are unit vectors along the Cartesian axes.

Eccentric Anomaly

To find $\boldsymbol{r}$ in terms of the eccentric anomaly, we use

$$r\cos f = a(\cos E - e)$$

$$r\sin f = b\sin E$$

derived previously, and write

$$\boldsymbol{r} = \boldsymbol{l}_A r\cos f + \boldsymbol{l}_B r\sin f$$

where $\boldsymbol{l}_A$ is a unit vector pointing from the force center O to pericenter and $\boldsymbol{l}_B$ is a unit vector pointing from O parallel to the semi-minor axis as shown in Fig. 2.6.

Then

$$\boldsymbol{r} = \boldsymbol{A}(\cos E - e) + \boldsymbol{B}\sin E$$

where $\boldsymbol{A} = \boldsymbol{l}_A\, a$ and $\boldsymbol{B} = \boldsymbol{l}_B b = \boldsymbol{l}_B a\sqrt{1-e^2}$.

Comparison of the expression for $\boldsymbol{r}$ in terms of E with that for $\boldsymbol{r}$ in terms of f yields

$$A_x = a[\cos\Omega\cos\omega - \sin\Omega\cos I\sin\omega]$$

$$A_y = a[\sin\Omega\cos\omega + \cos\Omega\cos I\sin\omega]$$

$$A_z = a\sin I\sin\omega$$

$$B_x = -b[\cos\Omega\sin\omega + \sin\Omega\cos I\cos\omega]$$

$$B_y = b[-\sin\Omega\sin\omega + \cos\Omega\cos I\cos\omega]$$

$$B_z = b\sin I\cos\omega$$

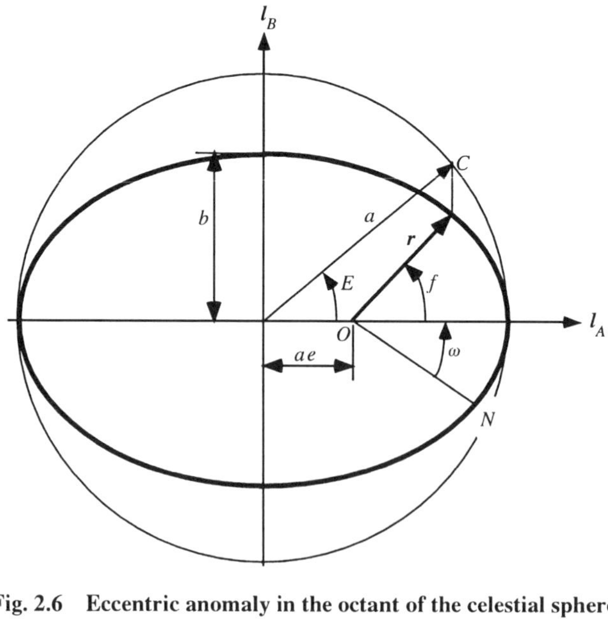

Fig. 2.6 Eccentric anomaly in the octant of the celestial sphere.

A useful form for $\boldsymbol{r}$ is

$$\boldsymbol{r} = \mathrm{Re}\left[(\boldsymbol{l}_A + i\boldsymbol{l}_B) r \varepsilon^{-if}\right]$$

where ε is the base of natural logarithms.

The velocity $\dot{\boldsymbol{r}}$ is obtained most easily in terms of E.

$$\dot{\boldsymbol{r}} = (-\boldsymbol{A}\sin E + \boldsymbol{B}\cos E)\dot{E}$$

Here $\dot{E}$ is to be found by using Kepler's equation

$$E - e\sin E = n(t - \tau)$$

We have

$$(1 - e\cos E)\dot{E} = n$$

so that

$$(r/a)\dot{E} = n \qquad \text{and} \qquad \dot{E} = (an/r)$$

Thus

$$\dot{\boldsymbol{r}} = (an/r)(-\boldsymbol{A}\sin E + \boldsymbol{B}\cos E)$$

where $r = a(1 - e\cos E)$.

Derivation of *A* and *B* by Use of Rotations

Examination of Fig. 2.6 shows that if we take the orbital plane as an $\boldsymbol{xy}$ plane, the position vector $\boldsymbol{r}$ is expressible as the column matrix CM, where

$$CM = \begin{pmatrix} r\cos f \\ r\sin f \\ 0 \end{pmatrix} = \begin{pmatrix} a(\cos E - e) \\ b\sin E \\ 0 \end{pmatrix}$$

If we perform a rotation about the normal through O to the orbital plane through the angle $(-\omega)$, we obtain ON as a new $\boldsymbol{x}$ axis. The square matrix $[-\omega]$ for this rotation is

$$[-\omega] = \begin{bmatrix} \cos\omega & -\sin\omega & 0 \\ \sin\omega & \cos\omega & 0 \\ 0 & 0 & 1 \end{bmatrix}$$

If we now form the matrix product $[-\omega]CM$, we obtain a second column matrix for $\boldsymbol{r}$, with a new $\boldsymbol{x}$ axis along ON and a z axis still perpendicular to the orbital plane.

Next, examine Fig. 2.5. If we perform a rotation about ON as $\boldsymbol{x}$ axis through the angle $(-I)$, we obtain a new representation for $\boldsymbol{r}$ as a column matrix with ON as $\boldsymbol{x}$ axis and a new z axis in the inertial direction Oz. The square matrix $[-I]$ for this rotation is

$$[-I] = \begin{bmatrix} 1 & 0 & 0 \\ 0 & \cos I & -\sin I \\ 0 & \sin I & \cos I \end{bmatrix}$$

The result $[-I][-\omega]CM$ is again a column matrix.

Finally, if we rotate the axes through the angle $(-\omega)$ about the inertial axis Oz, we obtain a column matrix for $\boldsymbol{r}$ in the actual inertial system. The square matrix $[-\Omega]$ for this rotation is

$$[-\Omega] = \begin{bmatrix} \cos\Omega & -\sin\Omega & 0 \\ \sin\Omega & \cos\Omega & 0 \\ 0 & 0 & 1 \end{bmatrix}$$

The final result for $\boldsymbol{r}$ is the column matrix

$$[-\Omega][-I][-\omega] \begin{pmatrix} a(\cos E - e) \\ b\sin E \\ 0 \end{pmatrix}$$

The reader should carry out the preceding matrix multiplication, always multiplying a column matrix by the adjacent square matrix so as to diminish the labor of calculation. One obtains another derivation of $\boldsymbol{A}/a$ and $\boldsymbol{B}/b$ as functions of Ω, ω, and I.

There are some other orbital elements that are often used in celestial mechanics, especially in planetary theory. The first of these is $\tilde{\omega} = \omega + \Omega$, called the longitude of pericenter. It has the peculiarity of being the sum of two angles in different planes, i.e., a "broken angle." Variables based on it are $\tilde{\omega} + f$, called the "true longitude," and $\tilde{\omega} + \ell$, called the "mean longitude"; these are also broken angles. To see how they might appear, consider a term in a perturbing function, the product of $\cos\Omega$ and $\cos(\omega + f)$. On writing this out one obtains cosines of $\tilde{\omega} + f$ and $\omega + f - \Omega$. The mean rates of change of the true and mean longitudes are both equal to the mean rate of change of the longitude ϕ. To see this, divide Eq. (2.5) by Eq. (2.3). The result is

$$\tan\chi \equiv \tan(\phi - \Omega) = \cos I \tan\psi = \cos I \tan(\omega + f)$$

Whenever $\omega + f$ increases by π, so does $\phi - \Omega$, so that

$$\dot{\phi} - \dot{\Omega} = \dot{\omega} + \dot{f}$$

$$\dot{\phi} = \dot{\omega} + \dot{f} + \dot{\Omega} = \dot{\tilde{\omega}} + \dot{f} = \dot{\tilde{\omega}} + \dot{\ell}$$

Here we are anticipating the later use of ω and Ω, like the other Keplerian elements, as time variable quantities when pertubations are considered.

Algorithm for the Orbit Generator

Given μ, a, e, I, ω, Ω, and τ, calculate $\boldsymbol{r}$ and $\dot{\boldsymbol{r}}$ at time t. Calculate $n = \sqrt{\mu a^{-3}}$, $\ell = n(t - \tau)$, and E from $E - e\sin E = \ell$. Then calculate $\boldsymbol{l}_A$ and $\boldsymbol{l}_B$ from their preceding formulations as functions of ω, I, and Ω. With $\boldsymbol{A} = \boldsymbol{l}_A a$ and $\boldsymbol{B} = \boldsymbol{l}_B a\sqrt{1 - e^2}$, then

$$\boldsymbol{r} = \boldsymbol{A}(\cos E - e) + \boldsymbol{B}\sin E$$

$$\dot{\boldsymbol{r}} = (an/r)(-\boldsymbol{A}\sin E + \boldsymbol{B}\cos E)$$

where $r = a(1 - e\cos E)$.

VIII. Orbit Determination from Initial Values

Given initial coordinates x_i, y_i, z_i and velocities $\dot{x}_i$, $\dot{y}_i$, $\dot{z}_i$, calculate a, e, I, ω, Ω, and τ. It will simplify matters to drop the subscript i, understanding that all the x and $\dot{x}$ are for the same initial time.

For a, calculate $v^2 = \dot{x}^2 + \dot{y}^2 + \dot{z}^2$, $r = \sqrt{x^2 + y^2 + z^2}$, and $W = \frac{1}{2}v^2 - (\mu/r)$. Then $a = -(\mu/2W)$.

For p, calculate $L^2 = (y\dot{z} - z\dot{y})^2 + (z\dot{x} - x\dot{z})^2 + (x\dot{y} - y\dot{x})^2$. Then $p = (L^2/\mu)$.

For e, since $p = a(1 - e^2)$, $e = \sqrt{(1 - p/a)}$.

For I, since $L_3 = L\cos I$, $\cos I = L_3/L = (x\dot{y} - y\dot{x})/L$, where L^2 is given in a preceding equation. It is useful to find $\sin I$ as a check. If $\boldsymbol{l}_N$ is a unit vector pointing from 0 toward the node,

$$\boldsymbol{k} \times \boldsymbol{L} = \boldsymbol{l}_N L \sin I$$

On writing $\boldsymbol{L} = \boldsymbol{i}L_1 + \boldsymbol{j}L_2 + \boldsymbol{k}L_3$, we find

$$\boldsymbol{l}_N L \sin I = \boldsymbol{j}L_1 - \boldsymbol{i}L_2$$

so that

$$\sin I = \sqrt{L_1^2 - L_2^2}/L$$

where $L_1 = y\dot{z} - z\dot{y}$ and $L_2 = z\dot{x} - x\dot{z}$.

Of course, $\cos I$ alone determines I, which ranges from 0–180°, $\cos I$ being plus for direct orbits and minus for retrograde orbits. (A direct orbit goes from west to east.) However, $\sin I$ is a useful check.

For τ, from Kepler's equation,

$$E - e\sin E = n(t - \tau)$$

we have $(1 - e\cos E)\dot{E} = n$. Since $r = a(1 - e\cos E)$, we find $\dot{E} = n\,a/r$.

Thus

$$\dot{r} = (a\,e\sin E)\dot{E} = n\,a^2(e/r)\sin E$$

Since

$$n = \sqrt{\mu a^{-3}}$$

$$r\dot{r} = \sqrt{\mu a}\,e\sin E$$

then

$$\sin E = \frac{r\dot{r}}{e\sqrt{\mu a}} = \frac{x\dot{x} + y\dot{y} + z\dot{z}}{e\sqrt{\mu a}}$$

Also

$$\cos E = \frac{1}{e}\left(1 - \frac{r}{a}\right)$$

From $\sin E$ and $\cos E$, determine E. Then τ is found by putting $t = 0$ in Kepler's equation:

$$\tau = -(E - e\sin E)/n$$

For ω, use the Laplace vector

$$\boldsymbol{R} = \boldsymbol{v} \times \boldsymbol{L} - \mu \boldsymbol{l}_r = \mu e \boldsymbol{l}_A$$

Here $A_z = a \sin I \sin \omega$, so that

$$R_z = \dot{x} L_2 - \dot{y} L_1 - \mu(z/r) = \mu e \sin I \sin \omega$$

Thus

$$e \sin I \sin \omega = \frac{\dot{x} L_2 - \dot{y} L_1}{\mu} - \frac{z}{r}$$

To find $e \sin I \sin \omega$, use

$$\boldsymbol{L} \times \boldsymbol{R} = \boldsymbol{L} \times (\boldsymbol{v} \times \boldsymbol{L}) - \mu \boldsymbol{L} \times \boldsymbol{l}_r = \mu e \, \boldsymbol{L} \times \boldsymbol{l}_A$$

This gives

$$\boldsymbol{v} L^2 - \mu \frac{\boldsymbol{L} \times \boldsymbol{r}}{r} = \mu e \, L \boldsymbol{l}_B$$

since $\boldsymbol{l}_A$, $\boldsymbol{l}_B$, and $\boldsymbol{l}_L$ form a cyclic orthonormal triad of vectors. Now, $B_z = b \sin I \cos \omega$, so that the z component of the preceding equation gives

$$L^2 \dot{z} - \frac{\mu}{r}(L_1 y - L_2 x) = \frac{\mu e L}{b} b \sin I \cos \omega$$

Thus

$$e \sin I \cos \omega = \frac{L \dot{z}}{\mu} + \frac{(L_1 y - L_2 x)}{L r}$$

This equation, along with the one for $e \sin I \sin \omega$, permits the evaluation of $\sin \omega$ and $\cos \omega$, and thus ω.

For Ω, use $\boldsymbol{k} \times \boldsymbol{l} = \boldsymbol{l}_N L \sin I$. Scalar multiply by $\boldsymbol{i}$ to find

$$\boldsymbol{i} \cdot \boldsymbol{k} \times \boldsymbol{L} = \boldsymbol{i} \cdot \boldsymbol{l}_N L \sin I$$

However, $\boldsymbol{i} \cdot \boldsymbol{k} \times \boldsymbol{L} = \boldsymbol{i} \times \boldsymbol{k} \cdot \boldsymbol{L} = -\boldsymbol{j} \cdot \boldsymbol{L} = -L_2$. Also $\boldsymbol{i} \cdot \boldsymbol{I}_N = \cos \Omega$. Thus

$$\cos \Omega = -\frac{L_2}{L \sin I}$$

To find $\sin \Omega$, form

$$\boldsymbol{i} \times (\boldsymbol{k} \times \boldsymbol{L}) = \boldsymbol{i} \times \boldsymbol{l}_N L \sin I = \boldsymbol{k} \sin \Omega L \sin I$$

$$\boldsymbol{k}(\boldsymbol{i} \cdot \boldsymbol{L}) - \boldsymbol{L}(\boldsymbol{i} \cdot \boldsymbol{k}) = \boldsymbol{k} \sin \Omega L \sin I$$

Here $\boldsymbol{i} \cdot \boldsymbol{k} = 0$ and $\boldsymbol{i} \cdot \boldsymbol{L} = L_1$, so that

$$\sin \Omega = \frac{L_1}{L \sin I}$$

Having $\cos \Omega$ and $\sin \Omega$, one then finds Ω.

Chapter 3

Lagrangian Dynamics

I. Variations

THE purpose of this chapter is to develop some general formulations of dynamics that will be useful in treating nondissipative systems. Let a dynamical system be characterized by N generalized coordinates q_i, $i = 1, \ldots, N$, and let $f(q_i, \dot{q}_i, t)$ be any function of the q's and the generalized velocities $\dot{q}_i$. Call it $f(q_i, \dot{q}_i, t)$ for short. There may or may not be constraints among the q's; if there are k constraints, the number of degrees of freedom is $N - k$.

We call the space of the q's the configuration space (Fig. 3.1); this would be ordinary space if $N = 3$. During the motion, the system proceeds in configuration space from point A with coordinates q_{iA}, $i = 1, \ldots, N$ at time $t = 0$ to some point B at time t with coordinates q_{iB}, $i = 1, \ldots, N$. The system goes through a succession of points in the configuration space that we call the dynamical path. Let us next imagine a varied path, permitted by the constraints, that would take the system from A to B in the same time. Let P be a point reached at time t on the dynamical path and P' be the point supposedly reached at the same time on the varied path; here P and P' are corresponding points. Also let $f(q, \dot{q}, t)$ be any function of the q's, $\dot{q}$'s, and t at P and $F(q, \dot{q}, t)$ its value at P' at the same time. Define the variation δf by

$$\delta f = F - f$$

Then

$$\dot{F} - \dot{f} = \frac{d}{dt}(\delta f)$$

However,

$$\dot{f} - \dot{f} = \delta \dot{f}$$

so that

$$\delta \dot{f} = \delta \left(\frac{df}{dt} \right) = \frac{d}{dt}(\delta f)$$

That is,

$$\delta d = d\delta$$

so that d and δ are commuting operators. The function f may be either a scalar or a vector.

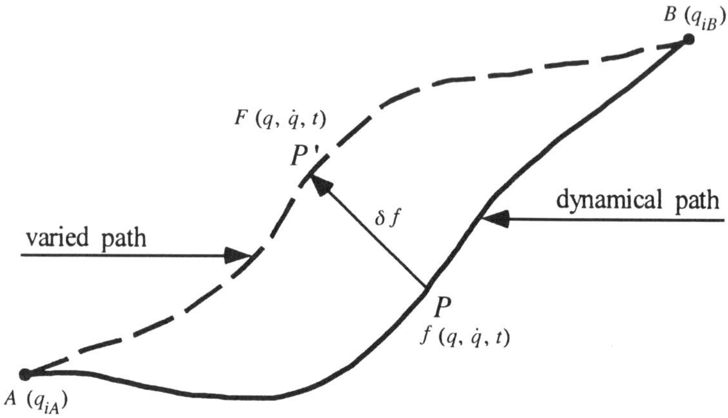

Fig. 3.1 Variations in the configuration space.

II. D'Alembert's Principle

Let us consider the system to be made up of a number of mass points, the k'th having a mass m_k. A given mass point k will be acted on by some applied force $\boldsymbol{F}_k$ and a constraint force $\boldsymbol{C}_k$. Constraint forces are forces that do no work. An example would be the normal force produced on a particle constrained to move on a surface; the frictional force, being tangential and doing work, would be called an applied force, but we shall soon rule out such dissipative forces.

If $\boldsymbol{r}_k$ is the position vector of particle k in some inertial system, then

$$m_k \ddot{\boldsymbol{r}}_k = \boldsymbol{F}_k + \boldsymbol{C}_k \tag{3.1}$$

If we now imagine the particle to be displaced by a vector amount $\delta \boldsymbol{r}_k$, in a way compatible with the constraints, we call $\delta \boldsymbol{r}_k$ a virtual displacement of k; this is the displacement to a varied path. On forming the scalar product of $\delta \boldsymbol{r}_k$ with Eq. (3.1) and summing over all the particles, it follows that

$$\Sigma_k m_k \ddot{\boldsymbol{r}} \cdot \delta \boldsymbol{r}_k = \Sigma_k \boldsymbol{F}_k \cdot \delta \boldsymbol{r}_k \tag{3.2}$$

since the constraint force $\boldsymbol{C}_k$ is normal to $\delta \boldsymbol{r}_k$. Now Eq. (3.2) can be written as

$$\Sigma_k (\boldsymbol{F}_k - m_k \ddot{\boldsymbol{r}}_k) \cdot \delta \boldsymbol{r}_k = 0 \tag{3.3}$$

an equation that is known as D'Alembert's principle.

If the applied forces are monogenic, then

$$\Sigma_k \boldsymbol{F}_k \cdot \delta \boldsymbol{r}_k = -\delta V(q, t) \tag{3.4}$$

Here the q's may be generalized coordinates. In applications to artificial satellites, V will be the gravitational potential energy of a satellite; it will depend explicitly on t when the departure of the Earth from axial symmetry is taken into account.

III. Hamilton's Principle

Theorem: $\delta \dot{\boldsymbol{r}}_k$ and $\dot{\boldsymbol{r}}_k$ are parallel, therefore $\dot{\boldsymbol{r}}_k \cdot \delta \dot{\boldsymbol{r}}_k = \delta(\frac{1}{2}\dot{\boldsymbol{r}}_k^2)$. (See Fig. 3.2.) Hamilton's principle selects the *correct* dynamical path from all possible varied paths and gives $\int_0^t \delta(T - V)\,dt = 0$ for a conservative system.

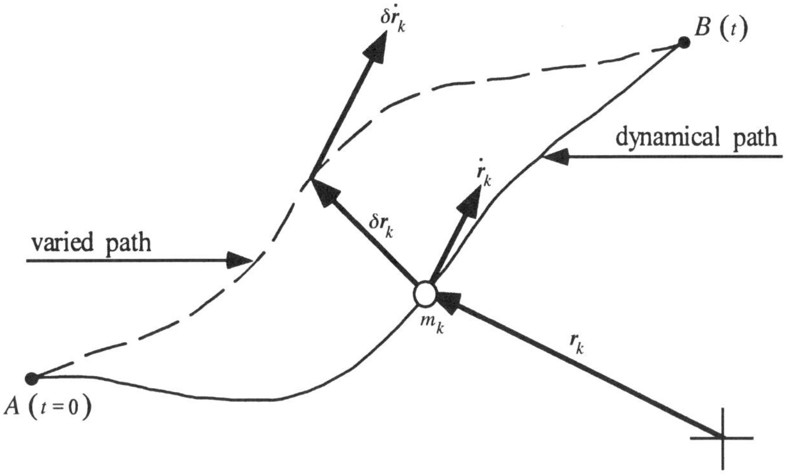

Fig. 3.2 Hamilton's principle selects the correct dynamical path.

Proof: Let us integrate Eq. (3.2) from 0 to t:

$$\Sigma_k \int_0^t \boldsymbol{F}_k \cdot \delta \boldsymbol{r}_k \,\mathrm{d}t = \Sigma_k \int_0^t m_k \ddot{\boldsymbol{r}}_k \cdot \delta \boldsymbol{r}_k \,\mathrm{d}t \tag{3.5}$$

Here

$$\int_0^t \ddot{\boldsymbol{r}}_k \cdot \delta \boldsymbol{r}_k \,\mathrm{d}t = \int_0^t \delta \boldsymbol{r}_k \cdot \mathrm{d}\dot{\boldsymbol{r}}_k = \dot{\boldsymbol{r}}_k \cdot \delta \boldsymbol{r}_k \Big|_0^t - \int_0^t \dot{\boldsymbol{r}}_k \cdot \mathrm{d}\delta \boldsymbol{r}_k$$

Since $\delta \boldsymbol{r}_k = 0$ at the endpoints, the first term on the right vanishes. Also $d(\delta \boldsymbol{r}_k) = \delta(\mathrm{d}\,\boldsymbol{r}_k) = \delta(\dot{\boldsymbol{r}}_k)\,\mathrm{d}t$, so that

$$\int_0^t \ddot{\boldsymbol{r}}_k \cdot \delta \boldsymbol{r}_k \,\mathrm{d}t = -\int_0^t \dot{\boldsymbol{r}}_k \cdot \delta \dot{\boldsymbol{r}}_k \,\mathrm{d}t = -\frac{1}{2}\int_0^t \delta \dot{\boldsymbol{r}}_k^2 \,\mathrm{d}t$$

and

$$\Sigma_k \int_0^t m_k \ddot{\boldsymbol{r}}_k \cdot \delta \boldsymbol{r}_k \,\mathrm{d}t = -\int_0^t \delta T \,\mathrm{d}t \tag{3.6}$$

where

$$T = \tfrac{1}{2}\Sigma_k m_k \dot{\boldsymbol{r}}_k^2$$

If the system is monogenic

$$\Sigma_k \boldsymbol{F}_k \cdot \delta \boldsymbol{r}_k = -\delta V \tag{3.7}$$

On inserting Eqs. (3.6) and (3.7) into Eq. (3.5), we find

$$\int_0^t \delta(T - V)\,\mathrm{d}t = 0 \tag{3.8}$$

This is then the property of the dynamical path that distinguishes it from all possible varied paths. It is one form of Hamilton's principle.

At this stage it is customary to take the δ outside the integral sign. This is possible if the system is holonomic, but not otherwise.[1] A holonomic system is one with integrable constraints. For a system without constraints—and we shall consider only such systems—the δ always commutes with the integral sign. The question does not really concern us very much because if we took the δ outside the integral sign, we should later find ourselves always putting it back inside. Thus, Eq. (3.8) expresses Hamilton's principle as we shall use it for unconstrained systems.

IV. Lagrange's Equations

Define the Lagrangian function L by

$$L \equiv T(q, \dot{q}, t) - V(q, t) \tag{3.9}$$

Here t is inserted as an argument of T in case we decide to use a rotating frame of reference. To apply Hamilton's principle

$$\int_0^t \delta L \, \mathrm{d}t = 0 \tag{3.10}$$

we must form

$$\delta L = \Sigma_k \left(\frac{\partial L}{\partial q_k} \delta q_k + \frac{\partial L}{\partial \dot{q}_k} \delta \dot{q}_k \right) \tag{3.11}$$

there being no term in $\partial L / \partial t$ because varied points are reached at the same times as the corresponding dynamical points. Since

$$\int_0^t \frac{\partial L}{\partial \dot{q}_k} \delta \dot{q}_k \, \mathrm{d}t = \int_0^t \frac{\partial L}{\partial \dot{q}_k} \frac{d}{\mathrm{d}t} (\delta q_k) \, \mathrm{d}t$$
$$= \frac{\partial L}{\partial q_k} \delta q_k \bigg|_0^t - \int_0^t \frac{d}{\mathrm{d}t} \left(\frac{\partial L}{\partial \dot{q}_k} \right) \delta q_k \, \mathrm{d}t$$

with the δq_k vanishing at the endpoints, we find

$$\int_0^t \delta L \, \mathrm{d}t = \int_0^t \Sigma_k \left[\frac{\partial L}{\partial q_k} - \frac{d}{\mathrm{d}t} \left(\frac{\partial L}{\partial \dot{q}_k} \right) \right] \delta q_k \, \mathrm{d}t \tag{3.12}$$

Consider only the case of no constraints. We may then choose

$$\delta q_k = Q_k \varepsilon_k(t) \qquad k = 1, \ldots, N$$

where

$$Q_k \equiv \frac{\partial L}{\partial q_k} - \frac{d}{\mathrm{d}t} \left(\frac{\partial L}{\partial \dot{q}_k} \right)$$

and where $\varepsilon_k(t) \geq 0$, always small and vanishing at $A(t = 0)$ and $B(t)$. Then Eq. (3.12) becomes

$$\int_0^t \Sigma_k Q_k^2 \, \varepsilon_k(t) \, \mathrm{d}t = 0$$

If we choose each $\varepsilon_k(t)$ to be continuous and assume Q_k to be continuous, then each Q_k must vanish over the whole range from 0 to t. It follows that

$$\frac{d}{\mathrm{d}t}\left(\frac{\partial L}{\partial \dot{q}_k}\right) = \frac{\partial L}{\partial q_k} \qquad k = 1, \ldots, N$$

These are Lagrange's equations of motion, sometimes called the Euler–Lagrange equations. There are cases where such a Lagrangian function L can be found, even though L may not be $T - V$. An example would be the mechanics of special relativity with electromagnetic forces. In general, any function $L(q, \dot{q}, t)$ that satisfies these equations is called a Lagrangian and can be used to set up the so-called Hamiltonian formulation of dynamics. We next proceed to this Hamiltonian form.

Reference

[1]Pars, L. A., *A Treatise on Analytical Dynamics,* Wiley, New York, 1963, p. 528.

Chapter 4

The Hamiltonian Equations

THE Lagrangian equations contain generalized coordinates q_k and generalized velocities $\dot{q}_k$. The Hamiltonian equations contain generalized coordinates q_k and generalized momenta p_k.

Here

$$p_k = \frac{\partial L(q, \dot{q}, t)}{\partial \dot{q}_k} \tag{4.1}$$

If the q's are rectangular coordinates,

$$L = \Sigma_k \tfrac{1}{2} m_k \left(\dot{x}_k^2 + \dot{y}_k^2 + \dot{z}_k^2\right) - V(x, t) \tag{4.2}$$

in which case

$$p_{x_k} = m_k \dot{x}_k \qquad p_{y_k} = m_k \dot{y}_k \qquad p_{z_k} = m_k \dot{z}_k \tag{4.3}$$

The reason for the name is thus apparent. In this special case the p's are dimensionally ordinary physical momenta, but this will not be true in general.

Next, introduce the Hamiltonian function $H(q, p, t)$ by means of the Legendre transformation.

$$H(q, p, t) = \Sigma_k p_k \dot{q}_k - L(q, \dot{q}, t) \tag{4.4}$$

Since H is to depend on the q's and p's and not on the q's and $\dot{q}$'s, we must regard the $\dot{q}$'s in Eq. (4.4) as functions of the q's and p's. Then from Eq. (4.4)

$$\begin{aligned} \frac{\partial H(q, p, t)}{\partial p_j} &= \dot{q}_j + \Sigma_k p_k \frac{\partial \dot{q}_k}{\partial p_j} - \Sigma_k \frac{\partial L}{\partial \dot{q}_k} \frac{\partial \dot{q}_k}{\partial p_j} \\ &= \dot{q}_j + \Sigma_k \left(p_k - \frac{\partial L}{\partial \dot{q}_k} \right) \frac{\partial \dot{q}_k}{\partial p_j} \\ &= \dot{q}_j \end{aligned} \tag{4.5}$$

by virtue of the definition (4.1) of p_k. It is thus a purely algebraic result, with no use of dynamics, that

$$\dot{q}_k = \frac{\partial H(q, p, t)}{\partial p_j} \tag{4.6}$$

To obtain the equation for $\dot{p}_k$ as a derivative of the Hamiltonian, we have to apply some dynamics, in the form of the Lagrangian equations, along the dynamical path. Begin with the Legendre transformation (4.4), applying $\partial/\partial q_j$ to it.

We find

$$\frac{\partial H(q, p, t)}{\partial q_j} = \Sigma_k p_k \frac{\partial \dot{q}_k}{\partial q_j} - \frac{\partial L}{\partial q_j} - \Sigma_k \frac{\partial L}{\partial \dot{q}_k} \frac{\partial \dot{q}_k}{\partial q_j}$$

$$= \Sigma_k \left(p_k - \frac{\partial L}{\partial \dot{q}_k} \right) \frac{\partial \dot{q}_k}{\partial q_j} - \frac{\partial L}{\partial q_j}$$

$$= -\frac{\partial L(q, \dot{q}, t)}{\partial q_j} \tag{4.7}$$

with use of Eqs. (4.1) and (4.3) for p_k. Now return to the Lagrangian equations, they state that

$$\frac{\partial L(q, \dot{q}, t)}{\partial q_j} = \frac{d}{\mathrm{d}t}\left(\frac{\partial L}{\partial \dot{q}_j} \right) = \frac{\mathrm{d}p_j}{\mathrm{d}t} \tag{4.8}$$

again with use of the definition (4.1) of p_k. Thus, by Eqs. (4.7) and (4.8)

$$\frac{\mathrm{d}p_k}{\mathrm{d}t} = -\frac{\partial H(q, p, t)}{\partial q_k} \tag{4.9a}$$

We also had

$$\frac{\mathrm{d}q_k}{\mathrm{d}t} = \frac{\partial H(q, p, t)}{\partial p_k} \tag{4.9b}$$

Equations (4.9) are the Hamiltonian or canonical equations of motion.

To get some idea of the physical meaning of the Hamiltonian H, we need to consider the kinetic energy

$$T = \frac{1}{2} \Sigma_k m_k \dot{\boldsymbol{r}}_k^2(q, t) \tag{4.10}$$

The velocity vector $\dot{\boldsymbol{r}}$ is expressed here not only as a function of the generalized coordinates q_k, but also as an explicit function of the time t. This is to take care of the possibility that we may be using a rotating coordinate system. Thus

$$\dot{\boldsymbol{r}}_k = \Sigma_k \left(\frac{\partial \boldsymbol{r}_k}{\partial q_j} \right) \dot{q}_j + \frac{\partial \boldsymbol{r}_k}{\partial t} \tag{4.11}$$

On squaring Eq. (4.11) and inserting the result into Eq. (4.10), we find that

$$T = T_0(\dot{q}, t) + T_1(\dot{q}, t) + T_2(\dot{q}, t) \tag{4.12}$$

where $T_n(\dot{q}, t)$ is a homogeneous function of the $\dot{q}$'s of degree N. Such a function has the property

$$T_n(\lambda \dot{q}_1, \lambda \dot{q}_2, \ldots, \lambda \dot{q}_N) = \lambda^n T_n(\dot{q}_1, \dot{q}_2, \ldots, \dot{q}_N) \tag{4.13}$$

and thus satisfies Euler's equation

$$\sum_{j=1}^{N} \dot{q}_j \frac{\partial T_n}{\partial \dot{q}_j} = n T_n \tag{4.14}$$

Now consider

$$H(q, p, t) = \Sigma_k p_k \dot{q}_k - L = \Sigma_k p_k \dot{q}_k - T + V \tag{4.15}$$

Here

$$\Sigma_k p_k \dot{q}_k = \Sigma_k \dot{q}_k \frac{\partial L(q, \dot{q}, t)}{\partial \dot{q}_k} = \Sigma_k \dot{q}_k \frac{\partial T}{\partial \dot{q}_k} \tag{4.16}$$

since $L = T - V$ and V does not depend on the $\dot{q}$'s. However,

$$\Sigma_k \dot{q}_k \frac{\partial T}{\partial \dot{q}_k} = \Sigma_k \dot{q}_k \left(\frac{\partial T_0}{\partial \dot{q}_k} + \frac{\partial T_1}{\partial \dot{q}_k} + \frac{\partial T_2}{\partial \dot{q}_k} \right) = T_1 + 2T_2 \tag{4.17}$$

by Eq. (4.14). Thus

$$\Sigma_k p_k \dot{q}_k = T_1 + 2T_2 \tag{4.18}$$

From Eqs. (4.15) and (4.18)

$$\begin{aligned} H &= T_1 + 2T_2 - (T_0 + T_1 + T_2) + V \\ H &= T_2 - T_0 + V \end{aligned} \tag{4.19}$$

In the usual case where the position vectors do not depend explicitly on the time t, i.e., in a nonrotating reference system, T_0 and T_1 vanish, so that $T = +T_2$. In this usual case

$$H = T + V \tag{4.20}$$

the total energy.

Even in this case, however, V may depend explicitly on t, and if so, H also does.

I. An Important Theorem

$$\frac{\mathrm{d}H}{\mathrm{d}t} = \frac{\partial H}{\partial t} \tag{4.21}$$

To prove this theorem, write

$$\frac{\mathrm{d}H}{\mathrm{d}t} = \Sigma_k \left(\frac{\partial H}{\partial q_k} \dot{q}_k + \frac{\partial H}{\partial q_k} \dot{p}_k \right) + \frac{\partial H}{\partial t} \tag{4.22}$$

Insertion of the canonical equations (4.9) then gives

$$\frac{\mathrm{d}H}{\mathrm{d}t} = \Sigma_k(-\dot{p}_k \dot{q}_k + \dot{q}_k \dot{p}_k) + \frac{\partial H}{\partial t} = \frac{\partial H}{\partial t}$$

and the theorem is proved. It follows that H is constant if it does not depend explicitly on t.

II. Ignorable Variables

If H does not contain q_j explicitly, q_j is called an ignorable or cyclic coordinate, and

$$\dot{p}_j = -\frac{\partial H(p, t)}{\partial q_j} = 0$$

so that $p_j = \alpha_j$, a constant. If all the q's were ignorable, we should have

$$H = H(p_1, p_2, \ldots, p_N)$$

and each

$$p_j = \alpha_j$$

$$\dot{q}_j = \frac{\partial H(p, t)}{\partial p_j} = \nu_j(\alpha_1, \alpha_2, \ldots, \alpha_N) \tag{4.23}$$

where each ν_j is a constant depending only on the constants $p_j = \alpha_j$, $j = 1, \ldots, N$. Then,

$$q_j = \nu_j t + \beta_j \tag{4.24}$$

We should have a complete solution of the canonical equations, with $2N$ constants of integration, $\alpha_1, \alpha_2, \ldots, \alpha_N$ and $\beta_1, \beta_2, \ldots, \beta_N$.

We shall use this idea to try to solve the canonical equations, introducing new canonical variables Q_k, P_k, $k = 1, \ldots, N$, which will make all the Q's ignorable. Otherwise we must do it piecemeal, one at a time. To do so, we have to consider the theory of transformations from canonical variables q_k, p_k to new ones Q_k, P_k, i.e., the theory of canonical transformations.

Chapter 5

Canonical Transformations

I. The Condition of Exact Differentials

GIVEN the set of canonical equations

$$\dot{p}_k = -\frac{\partial H(q, p, t)}{\partial q_k} \qquad \dot{q}_k = \frac{\partial H(q, p, t)}{\partial p_k} \qquad k = 1, \ldots, N \tag{5.1}$$

we wish to find which time-dependent mappings to new variables $P_1, P_2, \ldots, P_N$, $Q_1, Q_2, \ldots, Q_N$ will preserve the canonical form of these equations. That is, we map by means of

$$p_k = p_k(Q, P, t) \qquad q_k = q_k(Q, P, t) \tag{5.2}$$

to find conditions on this mapping and on a new function $K(Q, P, t)$, so that

$$\dot{P}_k = -\frac{\partial K(Q, P, t)}{\partial Q_k} \qquad \dot{Q}_k = \frac{\partial K(Q, P, t)}{\partial P_k} \tag{5.3}$$

To do so, we begin afresh, with the q's and p's defined only by Eqs. (5.1) and the Hamiltonian $H(q, p, t)$. Regarding the q's and p's as independent variables, with given initial values, we look for a variational principle that will take them from their initial values at $t = 0$ to the same final values at time t that would be produced by Eqs. (5.1). The form of the variational principle will resemble Hamilton's principle but will not really be the same.

We call the space of the q's and p's the phase space, as is usual in mechanics. Theorems of existence and uniqueness of solution then show that for given initial values $q_k(0)$, $p_k(0)$, $k = 1, \ldots, N$, the system (5.1) follows a unique path in the phase space from the initial point $P_0[q_k(0), p_k(0)]$ to the final point $P[q_k(t), p_k(t)]$; this is the dynamical path D. Other paths might be geometrically possible but would violate Eqs. (5.1). Any other adjacent path with the same endpoints, traversed in our imagination in the same time, is called a varied path V. For such a varied path, denote the variations at the same time from the dynamical path by

$$\delta q_k = q_{kV}(t) - q_{kD}(t) \qquad \delta p_k = p_{kV}(t) - p_{kD}(t)$$

Theorem 1: For arbitrary variations q_k, p_k, $k = 1, \ldots, N$, the condition

$$\int_0^t \delta[\Sigma_k p_k \dot{q}_k - H(q, p, t)] \mathrm{d}t = 0 \tag{5.4}$$

is necessary and sufficient that the q's and p's satisfy Eqs. (5.1), i.e., that the q's and p's be canonical with respect to H as Hamiltonian.

To show this, note that

$$\delta[\Sigma_k p_k\dot{q}_k - H(q,p,t)] = \Sigma_k\left[\dot{q}_k\delta p_k + p_k\delta\dot{q}_k - \frac{\partial H(q,p,t)}{\partial q_k}\delta q_k - \frac{\partial H(q,p,t)}{\partial p_k}\delta p_k\right] \tag{5.5}$$

However, $\delta\dot{q}_k = (d/\mathrm{d}t)(\delta q_k)$, so that

$$p_k\delta\dot{q}_k = \frac{d}{\mathrm{d}t}(p_k\delta q_k) - \dot{p}_k\delta q_k \tag{5.6}$$

Insertion of Eqs. (5.5) and (5.6) into Eq. (5.4) then yields

$$\int_0^t \delta[\Sigma_k p_k\dot{q}_k - H(q,p,t)]\,\mathrm{d}t = \int_0^t \Sigma_k\left[-\left(\dot{p}_k + \frac{\partial H(q,p,t)}{\partial q_k}\right)\delta q_k + \left(\dot{q}_k - \frac{\partial H(q,p,t)}{\partial p_k}\right)\delta p_k\right]\mathrm{d}t + \Sigma_k p_k\delta q_k\Big|_0^t \tag{5.7}$$

Since the endpoints are fixed, however,

$$\Sigma_k p_k\delta q_k\Big|_0^t = 0$$

Thus

$$\int_0^t \delta[\Sigma_k p_k\dot{q}_k - H]\,\mathrm{d}t = \int_0^t \Sigma_k\left[-\left(\dot{p}_k + \frac{\partial H}{\partial q_k}\right)\delta q_k + \left(\dot{q}_k - \frac{\partial H}{\partial p_k}\right)\delta p_k\right]\mathrm{d}t \tag{5.8}$$

If Eqs. (5.1) hold, the integral vanishes, so that condition (5.4) is necessary.

To prove sufficiency, assume that Eq. (5.4) holds, so that the integral in Eq. (5.8) vanishes. If we should assume that some of the terms $\dot{p}_k + (\partial H/\partial q_k)$ and $\dot{q}_k - (\partial H/\partial p_k)$ fail to vanish, we may choose our variations so that

$$\delta q_k = -\left(\dot{p}_k + \frac{\partial H}{\partial q_k}\right)\varepsilon_k(t) \qquad \delta p_k = \left(\dot{q}_k - \frac{\partial H}{\partial p_k}\right)\eta_k(t) \tag{5.9}$$

where the $\varepsilon_k(t)$ and $\eta_k(t)$ are small arbitrary nonnegative functions of t, vanishing at the endpoints. Then

$$\int_0^t \Sigma_k\left(\dot{p}_k + \frac{\partial H}{\partial q_k}\right)^2 \varepsilon_k(t)\,\mathrm{d}t + \int_0^t \Sigma_j\left(\dot{q}_j - \frac{\partial H}{\partial p_j}\right)^2 \eta_j(t)\,\mathrm{d}t = 0 \tag{5.10}$$

the summations being taken over those values of k and j for which the corresponding terms have been assumed nonvanishing. However, Eq. (5.10) is false unless Eqs. (5.1) hold. This completes the proof of sufficiency and thus of Theorem 1.

If we map from the q's and p's to Q's and P's, Theorem 1 shows that the condition

$$\int_0^t \delta[\Sigma_k P_k\dot{Q}_k - K(Q,P,t)]\,\mathrm{d}t = 0 \tag{5.11}$$

is necessary and sufficient that the Q's and P's be canonical with respect to $K(Q, P, t)$ as Hamiltonian.

Suppose now that the mapping of Eqs. (5.2) between q, p and Q, P has the Jacobian determinant.

$$M = \begin{vmatrix} A & B \\ C & D \end{vmatrix} \tag{5.12}$$

where A, B, C, and D are square matrices such that

$$A_{ij} = \frac{\partial q_i}{\partial Q_j} \qquad B_{ij} = \frac{\partial q_i}{\partial P_j} \qquad C_{ij} = \frac{\partial p_i}{\partial Q_j} \qquad D_{ij} = \frac{\partial p_i}{\partial P_j} \tag{5.13}$$

If the Jacobian does not vanish at $q_1, q_2, \ldots, q_N, p_1, p_2, \ldots, p_N$, then Eqs. (5.2) determine $Q_1, Q_2, \ldots, Q_N, P_1, P_2, \ldots, P_N$ at any such point. (A very simple example would be the point transformation $x = r\cos\theta$, $y = r\sin\theta$, with the Jacobian r; in this case, θ is determined for all x and y, except $x = y = 0$, where $r = 0$.) With the nonvanishing of the Jacobian, any function $F(Q, P, t)$ can be expressed, at least in principle, by

$$F(Q, P, t) = G(q, p, t) \tag{5.14}$$

Theorem 2: If there exist functions $H(q, p, t)$, $K(Q, P, t)$, and $F(Q, P, t)$ such that

$$\Sigma_k p_k \dot{q}_k - H(q, p, t) - [\Sigma_k P_k \dot{Q}_k - K(Q, P, t)] = \frac{d}{dt} F(Q, P, t) \tag{5.15}$$

then, if the q_k, p_k are canonical with respect to $H(q, p, t)$ as Hamiltonian, the Q_k, P_k will be canonical with respect to $K(Q, P, t)$ as Hamiltonian.

To prove this theorem, we form the time integral of the variation of Eq. (5.15) from a fixed path in the phase space of Q_k, P_k. The varied path is to have the same endpoints and be traversed in the same time as the fixed path. Since

$$\delta \frac{dF(Q, P, t)}{dt} = \frac{d}{dt}[\delta F(Q, P, t)] \tag{5.16}$$

we find

$$\int_0^t \delta[\Sigma_k p_k \dot{q}_k - H(q, p, t)]\,dt = \int_0^t \delta[\Sigma_k P_k \dot{Q}_k - K(Q, P, t)]\,dt \tag{5.17}$$

since $\delta F(Q, P, t)|_0^t = 0$.

The nonvanishing of the Jacobian guarantees no singularities in the mapping, so that δq_k and δp_k exist for any δQ_k and δP_k at any point in the phase space of the Q's and P's. If we now impose the condition that the q_k, p_k are to be canonical with respect to $H(q, p, t)$, the fixed path in the phase space of the q's and p's is the dynamical path. The integral on the left side of Eq. (5.17) vanishes by the necessity feature of Theorem 1. Since the integral on the right side of Eq. (5.17) also vanishes, the Q_k, P_k are canonical with respect to $K(Q, P, t)$ as Hamiltonian, by the sufficiency feature of Theorem 1. This completes the proof of Theorem 2.

If the Jacobian does not vanish, we may replace $F(Q, P, t)$ in Eq. (5.15) by $G(q, p, t)$ by virtue of Eq. (5.14). On reversing the roles of the q_k, p_k and Q_k, P_k in the preceding argument, we find that if Eq. (5.15) is satisfied, and if Q_k, P_k are known to be canonical with respect to $K(Q, P, t)$ as Hamiltonian, then q_k, p_k

will be canonical with respect to $H(q, p, t)$ as Hamiltonian. This is a corollary of Theorem 2. The latter and its corollary can be combined into one statement, as follows.

If the Jacobian from q_k, p_k to Q_k, P_k does not vanish in the region of phase space with which we are concerned, the condition

$$\Sigma_k(p_k \, dq_k - P_k \, dQ_k) + [K(Q, P, t) - H(q, p, t)] \, dt = dF(Q, P, t) \tag{5.18}$$

is sufficient for a canonical property of either set (q_k, p_k) or (Q_k, P_k) to ensure the canonical property of the other.

This is *not* a necessary condition, as a simple example will show. Let $Q_k = p_k$, $P_k = q_k$, and $K = -H$. It is verifiable at once that this is a canonical transformation, but it does not satisfy the perfect differential condition (5.18).

A condition that is both necessary and sufficient is

$$\lambda \Sigma_k(p_k \, dq_k - H \, dt) - \Sigma_k(P_k \, dQ_k - K \, dt) = \text{perfect differential} \tag{5.18a}$$

where λ is a constant and not necessarily equal to 1.[1,2]

II. Canonical Generating Functions

a) Suppose that

$$q = q(Q, P, t) \qquad p = p(Q, P, t) \tag{5.19}$$

is such a mapping that

$$p_k = \frac{\partial S}{\partial q_k} \qquad P_k = -\frac{\partial S}{\partial Q_k} \tag{5.20}$$

where S is a so-called generating function of the form

$$S = S(q, Q, t) \tag{5.21}$$

With use of the summation convention on $k = 1, \ldots, N$, it follows that

$$p_k \dot{q}_k - P_k \dot{Q}_k = \frac{\partial S}{\partial q_k} \dot{q}_k + \frac{\partial S}{\partial Q_k} \dot{Q}_k = \frac{dS}{dt} - \frac{\partial S}{\partial t} \tag{5.22}$$

We may write this as

$$(p_k \dot{q}_k - H) - (P_k \dot{Q}_k - K) = \frac{dS}{dt} + K - H - \frac{\partial S}{\partial t} \tag{5.23}$$

$$= \frac{dS}{dt} \qquad \text{if } K = H + \frac{\partial S}{\partial t} \tag{5.24}$$

By the sufficiency criterion of Sec. I, if the q, p are canonical with respect to H as Hamiltonian, the Q, P will be canonical with K as Hamiltonian if

$$K(Q, P, t) = H(q, p, t) + \frac{\partial S}{\partial t} \tag{5.25}$$

b) With

$$S = S(p, P, t) \tag{5.26}$$

if the mapping is such that

$$q_k = -\frac{\partial S}{\partial p_k} \qquad Q_k = \frac{\partial S}{\partial P_k} \tag{5.26a}$$

then

$$p_k\dot{q}_k - P_k\dot{Q}_k = \frac{d}{\mathrm{d}t}(p_kq_k - P_kQ_k) - q_k\dot{p}_k + Q_k\dot{P}_k \tag{5.27}$$

$$= \frac{d}{\mathrm{d}t}(p_kq_k - P_kQ_k) + \frac{\partial S}{\partial p_k}\dot{p}_k + \frac{\partial S}{\partial P_k}\dot{P}_k \tag{5.27a}$$

$$= \frac{d}{\mathrm{d}t}(p_kq_k - P_kQ_k + S) - \frac{\partial S}{\partial t} \tag{5.28}$$

and

$$(p_k\dot{q}_k - H) - (P_k\dot{Q}_k - K) = \frac{d}{\mathrm{d}t}(p_kq_k - P_kQ_k + S) + K - H - \frac{\partial S}{\partial t} \tag{5.29}$$

The sufficiency criterion then shows that if q, p are canonical with H as Hamiltonian, the Q, P will be canonical with

$$K = H + \frac{\partial S}{\partial t} \tag{5.29a}$$

as Hamiltonian.

c) With

$$S = S(q, P, t) \tag{5.30}$$

$$p_k = \frac{\partial S}{\partial q_k} \qquad Q_k = \frac{\partial S}{\partial P_k} \tag{5.31}$$

we have

$$p_k\dot{q}_k - P_k\dot{Q}_k = p_k\dot{q}_k + Q_k\dot{P}_k - \frac{d}{\mathrm{d}t}(P_kQ_k)$$

$$= \frac{\partial S}{\partial q_k}\dot{q}_k + \frac{\partial S}{\partial P_k}\dot{P}_k - \frac{d}{\mathrm{d}t}(P_kQ_k)$$

$$= \frac{\mathrm{d}S}{\mathrm{d}t} - \frac{d}{\mathrm{d}t}(P_kQ_k) - \frac{\partial S}{\partial t} \tag{5.32}$$

Then

$$(p_k\dot{q}_k - H) - (P_k\dot{Q}_k - K) = \frac{d}{\mathrm{d}t}(S - P_kQ_k) + K - H - \frac{\partial S}{\partial t} \tag{5.33}$$

The sufficiency criterion shows that, if q, p are canonical relative to H, then Q, P will be canonical relative to K as Hamiltonian if

$$K = H + \frac{\partial S}{\partial t} \tag{5.33a}$$

d) With

$$S = S(p, Q, t) \tag{5.34}$$

$$q_k = -\frac{\partial S}{\partial p_k} \qquad P_k = -\frac{\partial S}{\partial Q_k} \tag{5.35}$$

we find

$$\begin{aligned} p_k\dot{q}_k - P_k\dot{Q}_k &= -\dot{p}_k q_k + \frac{d}{dt}(p_k q_k) - P_k\dot{Q}_k \\ &= \frac{\partial S}{\partial p_k}\dot{p}_k + \frac{\partial S}{\partial Q_k}\dot{Q}_k + \frac{d}{dt}(p_k q_k) \\ &= \frac{dS}{dt} + \frac{dS}{dt}(p_k q_k) - \frac{\partial S}{\partial t} \end{aligned}$$

$$(p_k\,\dot{q}_k - H) - (P_k\dot{Q}_k - K) = \frac{d}{dt}(S + p_k q_k) + K - H - \frac{\partial S}{\partial t} \tag{5.36}$$

Thus, if q, p are canonical relative to H, then Q, P will be canonical relative to K as Hamiltonian if

$$K = H + \frac{\partial S}{\partial t} \tag{5.36a}$$

In case $S = S(p, Q)$, the reader can readily verify that the minus signs can be dropped in Eqs. (5.35) and the Q, P will still be canonical relative to the same Hamiltonian $K = H$. The only reason for using the minus signs in Eqs. (5.35) is to obtain $K = H + (\partial S/\partial t)$ when S depends explicitly on t. Case d falls into line with cases a, b, and c, which yield $K = H + (\partial S/\partial t)$. Not all canonical transformations can be derived from the preceding four generating functions. An example is $p_1 = Q_1, q_1 = -P_1, p_2 = P_2, q_2 = Q_2$, which satisfies

$$p_k\dot{q}_k - P_k\dot{Q}_k = -Q_1\dot{P}_1 - P_1\dot{Q}_1 = -\frac{d}{dt}(Q_1 P_1) \tag{5.37}$$

Such a mapping is canonical, without change of Hamiltonian, but it cannot be produced by means of any of the above generating functions.

As seen in Table 5.1, case c will be useful in the Hamilton–Jacobi theory and case d, without the explicit dependence of S on t, in the von Zeipel perturbation

Table 5.1 Summary of canonical generating functions

Case a: q, Q	Case b: p, P	Case c: q, P	Case d: p, Q
$p_k = \frac{\partial S(q, Q, t)}{\partial q_k}$	$q_k = -\frac{\partial S(p, P, t)}{\partial p_k}$	$p_k = -\frac{\partial S(q, P, t)}{\partial q_k}$	$q_k = \mp\frac{\partial S(p, Q, t)}{\partial p_k}$
$p_k = \frac{\partial S(q, Q, t)}{\partial Q_k}$	$Q_k = -\frac{\partial S(p, P, t)}{\partial P_k}$	$Q_k = -\frac{\partial S(q, P, t)}{\partial P_k}$	$P_k = \pm\frac{\partial S(p, Q, t)}{\partial Q_k}$
$K = H + \frac{\partial S}{\partial t}$	$K = H + \frac{\partial S}{\partial t}$	$K = H + \frac{\partial S}{\partial t}$	$K = H \pm \frac{\partial S}{\partial t}$

method. A simple example of case c is $S = \Sigma_k p_k q_k$. This gives the identity transformation $p_k = P_k$ and $q_k = Q_k$. In case d, without the explicit dependence on t and with use of the plus signs, $S = \Sigma_k p_k Q_k$ also gives the identity transformation.

III. Extended Point Transformation

Suppose we have q's and p's canonical relative to $H(q, p, t)$ as Hamiltonian. A point transformation is one in which the new Q's are functions only of the q's (and perhaps of t), but not of the p's. The new P's can be found by expressing the kinetic energy T in terms of the Q_k, $\dot{Q}_k$, t and then using

$$P_k = \partial T(Q, \dot{Q}, t)/\partial \dot{Q}_k$$

There is another method of doing this, however. Suppose

$$Q_k = f_k(q_1, q_2, \ldots, q_N, t) \tag{5.38}$$

Choose a generating function of case c, viz.,

$$S = \Sigma_j P_j f_j(q, t) \tag{5.39}$$

with

$$Q_k = \frac{\partial S}{\partial P_k} = f_k(q, t) \tag{5.40}$$

The new P's are to be found from

$$p_k = \frac{\partial S}{\partial q_k} = \Sigma_j P_j \frac{\partial f_j(q, t)}{\partial q_k} \tag{5.41}$$

Such a transformation is called an extended point transformation. It results in a new Hamiltonian

$$K(Q, P, t) = H(q, p, t) + \frac{\partial S}{\partial t} \tag{5.42}$$

Just to show how the method works, we shall devote the rest of this section to a simple example, which is not really very fruitful. Then, in the next section, we shall consider an example where an extended point transformation yields an important result.

IV. Transformation from Plane Rectangular to Plane Polar Coordinates

For a particle of mass m with rectangular coordinates x, y momenta $p_1 = m\dot{x}$, $p_2 = m\dot{y}$, potential energy $V(x, y)$, we have

$$H = (1/2m)\left(p_1^2 + p_2^2\right) + V(x, y) \tag{5.43}$$

The equations of point transformation are

$$x = r\cos\theta \qquad y = r\sin\theta \tag{5.44}$$

We could transform directly to plane polar coordinates by writing

$$T = (m/2)(\dot{r}^2 + r^2\dot{\theta}^2)$$

Then

$$\frac{\partial T}{\partial \dot{r}} = P_1 = m\dot{r} \qquad \frac{\partial T}{\partial \dot{\theta}} = P_2 = mr^2\dot{\theta}$$

so that

$$T = \frac{1}{2m}\left(P_1^2 + \frac{P_2^2}{r^2}\right)$$

$$H = \frac{1}{2m}\left(P_1^2 + \frac{P_2^2}{r^2}\right) + V(r,\theta)$$

The method of the extended point transformation goes as follows.

$$Q_1 = r = f_1(x,y) = (x^2 + y^2)^{\frac{1}{2}}$$

$$Q_2 = \theta = f_2(x,y) = \tan^{-1}(y/x)$$

$$\frac{\partial f_1}{\partial x} = \frac{x}{r} = \cos\theta \qquad \frac{\partial f_2}{\partial x} = -\frac{y}{x^2+y^2} = -\frac{\sin\theta}{r}$$

$$\frac{\partial f_1}{\partial y} = \frac{y}{r} = \sin\theta \qquad \frac{\partial f_2}{\partial y} = \frac{x}{x^2+y^2} = -\frac{\cos\theta}{r}$$

The equations

$$\Sigma_j P_j \frac{\partial f_j}{\partial q_k} = p_k$$

become

$$P_1\cos\theta - \frac{P_2\sin\theta}{r} = p_1$$

$$P_1\sin\theta + \frac{P_2\cos\theta}{r} = p_2$$

with the solution

$$P_1 = p_1\cos\theta + p_2\sin\theta$$

$$P_2 = r(-p_1\sin\theta + p_2\cos\theta)$$

To verify their correctness, use $p_1 = m\dot{x}$, $p_2 = m\dot{y}$ and form

$$P_1 + (iP_2/r) = m(\dot{x} + i\dot{y})\varepsilon^{-i\theta}$$

However,

$$x + iy = r\varepsilon^{i\theta}$$

$$\dot{x} + i\dot{y} = (\dot{r} + ir\dot{\theta})\varepsilon^{i\theta}$$

so that

$$P_1 + (iP_2/r) = m(\dot{r} + ir\dot{\theta})$$

and

$$P_1 = m\dot{r} \qquad P_2 = mr^2\dot{\theta}$$

which are correct.

V. The Jacobi Integral

Consider an artificial satellite in orbit about the Earth as shown in Fig. 5.1. If r = geocentric distance, θ = geocentric latitude, and ϕ = right ascension, its kinetic energy per unit mass in the usual inertial system is

$$T = \tfrac{1}{2}(\dot{r}^2 + r^2\dot{\theta}^2 + r^2\cos^2\theta\dot{\phi}^2)$$

Then $p_1 = \dot{r}$, $p_2 = r^2\dot{\theta}$, and $p_3 = r^2\cos^2\theta\dot{\phi}$, so that

$$T = \frac{1}{2}\left(p_1^2 + \frac{p_2^2}{r^2} + \frac{p_3^2}{r^2\cos^2\theta}\right)$$

With neglect of drag, the system is monogenic and the potential

$$V = V(r, \theta, \lambda)$$

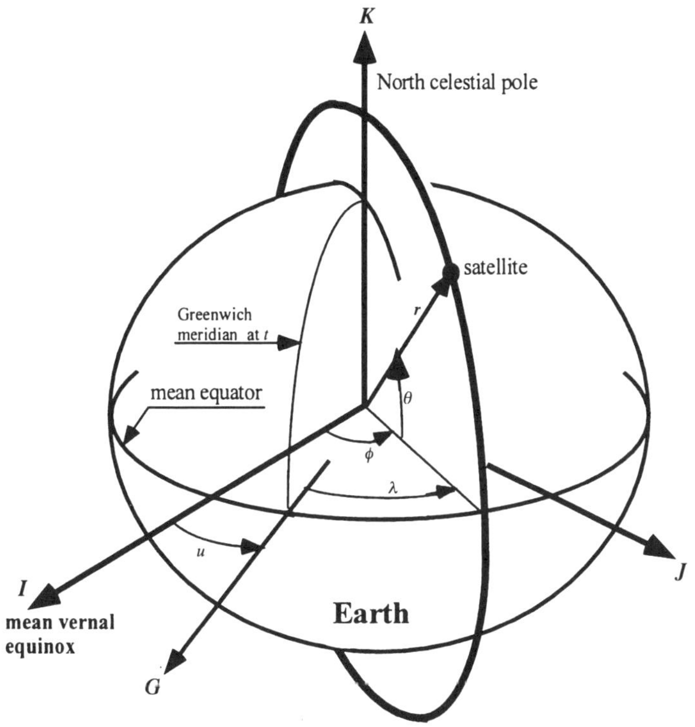

Fig. 5.1 Artificial satellite in orbit about the Earth.

where λ = geographic longitude (or geocentric longitude). However,

$$\lambda = \phi - u(t)$$

where u is the angle from the meridian through the vernal equinox to the meridian through Greenwich. It is called the Greenwich sidereal time and satisfies

$$\dot{u} = \omega_e \qquad u = \omega_e t + u_0$$

where ω_e is the sidereal rate of rotation of the Earth and u_0 is the Greenwich sidereal time at $t = 0$. Thus

$$V = V[r, \theta, \phi - u(t)]$$

The Hamiltonian is then

$$H = \frac{1}{2}\left(p_1^2 + \frac{p_2^2}{r^2} + \frac{p_3^2}{r^2 \cos^2\theta}\right) + V(r, \theta, \phi - u)$$

depending explicitly on the time. Thus

$$\frac{\mathrm{d}H}{\mathrm{d}t} = \frac{\partial H}{\partial t} = -\dot{u}\frac{\partial V}{\partial \lambda} = -\omega_e \frac{\partial V}{\partial \phi}$$

since t is kept fixed in evaluating $\partial V/\partial\lambda$. From Eqs. (5.1), $\dot{p}_3 = -(\partial H/\partial\phi) = -(\partial V/\partial\phi)$, thus

$$\frac{\mathrm{d}H}{\mathrm{d}t} = \omega_e \dot{p}_3$$

This suggests finding a transformation that will take us to a constant Hamiltonian

$$K = H - \omega_e p_3$$

To do so, introduce an extended point transformation

$$Q_1 = r \qquad Q_2 = \theta \qquad Q_3 = \phi - u = \lambda$$

$$S = \Sigma_j P_j f_j(q, t)$$

In the notation of the previous section,

$$f_1 = q_1 = r \qquad f_2 = q_2 = \theta \qquad f_3 = q_3 - u$$

where $u = \omega_e t + u_0$. The equations

$$p_k = \frac{\partial S}{\partial q_k}$$

give

$$\Sigma_j P_j \frac{\partial f_j}{\partial q_k} = p_k$$

which become

$$P_1 = p_1 \qquad P_2 = p_2 \qquad P_3 = p_3$$

Also

$$\frac{\partial S}{\partial t} = \Sigma_j P_j \frac{\partial f_j}{\partial t} = P_3 \frac{\partial f_3}{\partial t} = -P_3 \frac{\partial u}{\partial t} = -\omega_e P_3$$

The new Hamiltonian

$$K = H + \frac{\partial S}{\partial t} = H - \omega_e P_3 = \text{const}$$

where

$$H = \frac{1}{2}\left(p_1^2 + \frac{p_2^2}{r^2} + \frac{p_3^2}{r^2 \cos^2\theta}\right) + V(r, \theta, \phi - u)$$

The net result of this extended point transformation is that simply by changing from right ascension to geographic longitude as a new Q, we find that the corresponding Hamiltonian K is a constant. This new Hamiltonian is called the Jacobi integral. In the special case that the Earth is considered to be axially symmetric, $H - \omega_e P_3$ would be constant, but so would H and P_3 separately.

References

[1]Breves Filho, J. A., *Celestial Mechanics 6,* 1972, pp. 108–110.

[2]Goldstein, H., *Classical Mechanics,* 2nd ed., Addison–Wesley, Reading, MA, 1980, p. 380.

Chapter 6

Hamilton–Jacobi Theory

I. The Hamilton–Jacobi Equation

SUPPOSE we have a system with Hamiltonian $H(q, p, t)$ where the q's and p's are to be solutions of

$$\dot{p}_k = -\frac{\partial H(q, p, t)}{\partial q_k} \qquad \dot{q}_k = \frac{\partial H(q, p, t)}{\partial p_k} \qquad k = 1, \ldots, N$$

Transform to new variables Q, P by means of a generating function $S(q, P, t)$. The appropriate equations are

$$p_k = \frac{\partial S(q, P, t)}{\partial q_k} \qquad Q_k = \frac{\partial S(q, P, t)}{\partial P_k}$$

from case c in Table 5.1. If the q, p are canonical with H as Hamiltonian, the Q, P will then be canonical with

$$K = H + \frac{\partial S}{\partial t}$$

as Hamiltonian. Also, if Q, P are canonical with K as Hamiltonian, the q, p will be canonical with H as Hamiltonian.

The bold step to the Hamilton–Jacobi equation is to require that the transformation be such that

$$K(Q, P, t) = 0$$

If we can find such a transformation, then

$$\dot{Q}_k = \frac{\partial K}{\partial P_k} = 0 \qquad \dot{P}_k = \frac{\partial K}{\partial Q_k} = 0$$

so that

$$Q_k = \beta_k \qquad P_k = \alpha_k \qquad k = 1, \ldots, N$$

where the α's and β's are all constant. The original problem will then be solved, since we can then find q_k, p_k from

$$p_k = \frac{\partial S(q, \alpha, t)}{\partial q_k} \qquad \beta_k = \frac{\partial S(q, \alpha, t)}{\partial \alpha_k} \qquad k = 1, \ldots, N$$

The key step is putting

$$K = H(q, p, t) + \frac{\partial S(q, P, t)}{\partial t} = 0$$

If we replace p_k by $\partial S/\partial q_k$, we obtain

$$H\left(q, \frac{\partial S}{\partial q}, t\right) + \frac{\partial S(q, P, t)}{\partial t} = 0$$

a partial differential equation for *S*, called the Hamilton–Jacobi equation. If we can solve this equation for *S*, we can find the required canonical transformation. The integration constants arising in the solution will serve as the new *P*'s, which will be the same as the constants α's.

II. An Important Special Case

If *H* is explicitly independent of *t*, then

$$H = \alpha_1$$

a constant. Then

$$\frac{\partial S}{\partial t} = -\alpha_1$$

and

$$S = -\alpha_1 t + W(q)$$

Thus

$$p_k = \frac{\partial S}{\partial q_k} = \frac{\partial W(q)}{\partial q_k}$$

The HJ equation reduces to

$$H\left(q, \frac{\partial W}{\partial q}, t\right) = \alpha_1$$

To write down this equation, construct the Hamiltonian $H(q, p, t)$, replace each p_k by $\partial W/\partial q_k$, and set *H* equal to the constant α_1.

In most cases one cannot solve this equation in closed form or by quadratures. In some cases, however, one can solve it by separation of variables, and these cases are important. If $N = 3$ and we can separate variables, we shall find two separation constants α_2 and α_3, which along with α_1, will be the new *P*'s. We can find

$$W = W(q_1, q_2, q_3, \alpha_1, \alpha_2, \alpha_3)$$

so that

$$S = -\alpha_1 t + W$$

and

$$p_k = \frac{\partial S}{\partial q_k} = \frac{\partial W}{\partial q_k}$$

$$\beta_k = \frac{\partial S}{\partial P_k} = \frac{\partial S(q, \alpha, t)}{\partial \alpha_k} = -t\delta_{1k} + \frac{\partial W(q, \alpha)}{\partial \alpha_k}$$

We thus obtain, as the kinematical equations of motion,

$$\frac{\partial W(q,\alpha)}{\partial \alpha_1} = t + \beta_1$$

$$\frac{\partial W(q,\alpha)}{\partial \alpha_2} = \beta_2$$

$$\frac{\partial W(q,\alpha)}{\partial \alpha_3} = \beta_3$$

To find the q's as functions of t, we have to invert these equations, obtaining

$$q_k = q_k(\alpha_k, \beta_k, t) \qquad k = 1, 2, 3$$

To find the p's, we use

$$p_k = \frac{\partial W}{\partial q_k} \qquad k = 1, 2, 3$$

The $\dot{q}$'s then follow from the p's by means of

$$\dot{q}_k = \frac{\partial H(q, p, t)}{\partial p_k} = f_k(q, \alpha, \beta, t) \qquad k = 1, 2, 3$$

III. The Hamilton–Jacobi Equation for the Kepler Problem

If r is the radial distance, θ = latitude, and ϕ = longitude or the right ascension, the Hamiltonian for a unit mass is

$$H = \frac{1}{2}\left(p_r^2 + \frac{p_\theta^2}{r^2} + \frac{p_\phi^2}{r^2\cos^2\theta}\right) + V(r)$$

where $V(r) = -\mu/r$, $\mu = G(m_1 + m_2)$. On replacing p_k by $\partial W/\partial q_k$, the *HJ* equation becomes

$$H\left(q, \frac{\partial W}{\partial q}, t\right) = \alpha_1$$

or

$$\frac{1}{2}\left[\left(\frac{\partial W}{\partial r}\right)^2 + \frac{1}{r^2}\left(\frac{\partial W}{\partial \theta}\right)^2 + \frac{1}{r^2\cos^2\theta}\left(\frac{\partial W}{\partial \phi}\right)^2\right] - \frac{\mu}{r} = \alpha_1$$

We try to separate variables by placing

$$W = W_1(r) + W_2(\theta) + W_3(\phi)$$

If a prime denotes the derivative with respect to the indicated argument, this becomes

$$(W_1')^2 + \frac{1}{r^2}(W_2')^2 + \frac{1}{r^2\cos^2\theta}(W_3')^2 - \frac{2\mu}{r} = 2\alpha_1$$

Then

$$W_3'^2 = 2\alpha_1 r^2\cos^2\theta + 2\mu r\cos^2\theta - r^2\cos^2\theta W_1'^2 - \cos^2\theta W_2'^2 = \alpha_3^2 \qquad (6.1)$$

a constant, because the left side depends only on ϕ and the right side only on r and θ. Thus

$$W_3' = \alpha_3$$

where α_3 may have either sign, because

$$p_\phi = \frac{\partial W}{\partial \phi} = W_3'$$

and

$$p_\phi = r^2 \cos^2 \theta \dot{\phi}$$

which is L_z, the z component of angular momentum. To show this, note that

$$L_z = x\dot{y} - y\dot{x} = \text{Im}(x - iy)(\dot{x} + i\dot{y})$$

However, if $\rho = r \cos \theta$

$$x - iy = \rho \varepsilon^{-i\phi} \qquad x + iy = \rho \varepsilon^{i\phi} \qquad \dot{x} + i\dot{y} = (\dot{\rho} + i\rho\dot{\phi})\,\varepsilon^{i\phi}$$

Thus

$$\begin{aligned} L_z &= \text{Im}[(\rho\varepsilon^{-i\phi})(\dot{\rho} + i\rho\dot{\phi})\varepsilon^{i\phi}] = \text{Im}[\rho\dot{\rho} + i\rho^2\dot{\phi}] = \rho^2\dot{\phi} \\ &= r^2 \cos^2 \theta \dot{\phi} = p_\phi \end{aligned}$$

On dividing Eq. (6.1) by $\cos^2 \theta$ and transposing, we find

$$2\alpha_1 r^2 + 2\mu r - r^2 W_1'^2 = W_2'^2 + \alpha_3^2 \sec^2 \theta = \alpha_2^2$$

a constant, because the left side depends only on r and the right side only on θ. We may assume $\alpha_2 > 0$ without loss of generality.

Then

$$W_1'^2 = r^{-2}\left(-\alpha_2^2 + 2\mu r + 2\alpha_1 r^2\right)$$

$$W_2'^2 = \alpha_2^2 - \alpha_3^2 \sec^2 \theta$$

and

$$W_1' = \pm r^{-1}\left(-\alpha_2^2 + 2\mu r + 2\alpha_1 r^2\right)^{\frac{1}{2}}$$

$$W_2' = \pm\left(\alpha_2^2 - \alpha_3^2 \sec^2 \theta\right)^{\frac{1}{2}}$$

Since $W_1' = p_r = \dot{r}$ and $W_2' = p_\theta = r^2\dot{\theta}$, the plus sign holds for W_1' when $\dot{r} > 0$ and the minus sign when $\dot{r} < 0$. Similarly the plus sign holds for W_2' when $\dot{\theta} > 0$ and the minus sign when $\dot{\theta} < 0$.

From $W_3' = \alpha_3$, we obtain

$$W_3 = \alpha_3 \phi$$

In integral form

$$W_1 = \int_{r_1}^{r} \pm r^{-1}\left(-\alpha_2^2 + 2\mu r + 2\alpha_1 r^2\right)^{\frac{1}{2}} \mathrm{d}r$$

$$W_2 = \int_0^\theta \pm\left(\alpha_2^2 - \alpha_3^2 \sec^2 \theta\right)^{\frac{1}{2}} \mathrm{d}\theta$$

where the integrands are always nonnegative. The lower limit r_0 allows for a constant of integration.

Note that for real motion, $\alpha_2^2 - \alpha_3^2 \sec^2\theta \geq 0$, so that

$$\sec^2\theta \leq \alpha_2^2/\alpha_3^2 \qquad \cos^2\theta \geq \alpha_3^2/\alpha_2^2$$

Since $\cos^2\theta \leq 1$, we find $\alpha_3^2 \leq \alpha_2^2$. The minimum value of $\cos\theta$ is $|\alpha_3|/\alpha_2$. As θ increases from 0 toward $\pi/2$, $\cos\theta$ diminishes until it equals $|\alpha_3|/\alpha_2$. As θ diminishes from 0 toward $-\pi/2$, $\cos\theta$ again diminishes until it equals $|\alpha_3|/\alpha_2$. Thus

$$\theta_{\max} = \cos^{-1}(|\alpha_3|/\alpha_2)$$

$$\theta_{\min} = -\theta_{\max} = -\cos^{-1}(|\alpha_3|/\alpha_2)$$

Total energy is

$$\alpha_1 = \frac{1}{2}v^2 - \frac{\mu}{r}$$

where v is magnitude of the velocity. For a bounded orbit

$$\alpha_1 < 0$$

else v would remain real as $r \to \infty$.

The integrals W_1 and W_2 are difficult to evaluate, but we need only $\partial W_1/\partial\alpha_1$, $\partial W_1/\partial\alpha_2$, $\partial W_2/\partial\alpha_2$, and $\partial W_2/\partial\alpha_3$ since they are the quantities that appear in the kinematic equations

$$t + \beta_1 = \frac{\partial W(q,\alpha)}{\partial\alpha_1} = \frac{\partial W_1}{\partial\alpha_1}$$

$$\beta_2 = \frac{\partial W(q,\alpha)}{\partial\alpha_2} = \frac{\partial W_1}{\partial\alpha_2} + \frac{\partial W_2}{\partial\alpha_2}$$

$$\beta_3 = \frac{\partial W(q,\alpha)}{\partial\alpha_3} = \phi + \frac{\partial W_2}{\partial\alpha_3}$$

We shall see that we can express the derivatives of W_1 and W_2 with respect to the α's as integrals. Before evaluating these integrals, it is well to say what the α's and β's will turn out to be in terms of the Keplerian elements a, e, I, ω, Ω, and τ. We shall see that

$$\begin{aligned} \alpha_1 &= -\frac{\mu}{2a} & \beta_1 &= -\tau \\ \alpha_2 &= [\mu a(1-e^2)]^{\frac{1}{2}} & \beta_2 &= \omega \\ \alpha_3 &= \alpha_2 \cos I & \beta_3 &= \Omega \end{aligned} \tag{6.2}$$

At this point the question may arise: Since we have already solved the Kepler problem, why solve it again with such a complicated piece of machinery as the *HJ* procedure? The answer is this: The *HJ* solution will yield a canonical transformation of the Cartesian q's and p's or the spherical coordinate q's and p's to the α's and β's, which are so closely related to the Keplerian elements. Most problems in orbital mechanics and celestial mechanics are solved by a method of perturbations, beginning with a solution of a problem already solved, such as the Kepler problem. If we begin with the Keplerian solution, we use the Keplerian elements as variables in the perturbed problem. Once we have solved the perturbed problem by finding the variable Keplerian elements as functions of time, we can write down the solutions for the position vector $\boldsymbol{r}$ and the velocity $\dot{\boldsymbol{r}}$, as we did before, viz.,

$$\begin{aligned}\boldsymbol{r} &= \boldsymbol{A}(\cos E - e) + \boldsymbol{B}\sin E \\ \dot{\boldsymbol{r}} &= \tfrac{an}{r}(-\boldsymbol{A}\sin E + \boldsymbol{B}\cos E)\end{aligned} \tag{6.3}$$

where

$$n = \sqrt{\mu a^{-3}}$$

$$r = a(1 - e\cos E)$$

$$E - e\sin E = n(t - \tau)$$

Note that $\boldsymbol{A}$ and $\boldsymbol{B}$ are functions of a, e, Ω, ω, and I as derived in Chapter 2, Sec. VII, for an elliptic orbit. Equations (6.2) and (6.3) together always hold; they express a canonical transformation from the old p's and q's to the new ones, which are simply the α's and β's. As such they hold for the perturbed problem as well as for the unperturbed (Kepler) problem. Moreover, the HJ procedure will get us started on the perturbation calculations to find the perturbed α's and β's.

IV. The Integrals for the Kepler Problem

Integrals Involving Only W_1

The α_1 Integral

Consider

$$W_1 = \int_{r_1}^{r} \pm r^{-1}\left(-\alpha_2^2 + 2\mu r + 2\alpha_1 r^2\right)^{\frac{1}{2}}\, \mathrm{d}r$$

where that d$r > 0$ is for the upper sign and $dr < 0$ is for the lower sign. Let

$$F(r) \equiv -\alpha_2^2 + 2\mu r + 2\alpha_1 r^2 = 2\alpha_1(r - r_1)(r - r_2) = -2\alpha_1(r - r_1)(r_2 - r)$$

having the real positive zeros r_1 and r_2 for $\alpha_1 < 0$, satisfying $r_1 \le r \le r_2$. Solution of the quadratic equation $F(r) = 0$ gives

$$r_1 = \frac{-\mu}{2\alpha_1}\left(1 - \sqrt{1 + \frac{2\alpha_1\alpha_2^2}{\mu^2}}\right)$$

$$r_2 = \frac{-\mu}{2\alpha_1}\left(1 + \sqrt{1 + \frac{2\alpha_1\alpha_2^2}{\mu^2}}\right)$$

where r_1 is the pericenter distance and r_2 the apocenter distance. For a satellite of the Earth, the names are perigee and apogee; for a planet going around the sun, they are perihelion and aphelion. Here r_1 and r_2 satisfy

$$r_1 \le r \le r_2 \qquad a = \frac{1}{2}(r_1 + r_2) = \frac{-\mu}{2\alpha_1}$$

giving the integral

$$\alpha_1 = -\frac{\mu}{2a}$$

The α_2 Integral

To find $\partial W_1/\partial\alpha_1$

$$W_1 = \int_{r_0}^{r} \pm r^{-1} F^{\frac{1}{2}}(r, \alpha_1, \alpha_2)\,\mathrm{d}r$$

where that $\mathrm{d}r > 0$ is for the upper sign and $\mathrm{d}r < 0$ is for the lower sign.

Then

$$\frac{\partial W_1}{\partial\alpha_1} = \int_{r_0}^{r} \pm\frac{1}{2r}F^{-\frac{1}{2}}\frac{\partial F}{\partial\alpha_1}\,\mathrm{d}r \mp \frac{1}{r_0}F^{\frac{1}{2}}(r_0, \alpha_1, \alpha_2)\frac{\partial r_0}{\partial\alpha_1}$$

This equation follows from the theorem

$$\frac{\partial}{\partial\alpha}\int_a^b f(x,\alpha)\,\mathrm{d}x = \int_a^b \frac{\partial f}{\partial\alpha}\,\mathrm{d}x + f(b,\alpha)\frac{\partial b}{\partial\alpha} - f(a,\alpha)\frac{\partial a}{\partial\alpha}$$

If we choose $r_0 = r_1$, the term $F^{1/2}(r_0, \alpha_1, \alpha_2)$ will vanish, because r_1 is a zero of $F(r, \alpha_1, \alpha_2)$. Another choice of r_0 would give different β's, but $r_0 = r_1$ is the most convenient choice, because it will lead to $\beta_1 = -\tau$. Thus

$$W_1 = \int_{r_1}^{r} \pm r^{-1}\left(-\alpha_2^2 + 2\mu r + 2\alpha_1 r^2\right)^{\frac{1}{2}}\,\mathrm{d}r$$

$$\frac{\partial W_1}{\partial\alpha_1} = \int_{r_1}^{r} \pm r\left(-\alpha_2^2 + 2\mu r + 2\alpha_1 r^2\right)^{-\frac{1}{2}}\,\mathrm{d}r$$

$$= \int_{r_1}^{r} \pm r[-2\alpha_1(r - r_1)(r_2 - r)]^{-\frac{1}{2}}\,\mathrm{d}r$$

$$= (-2\alpha_1)^{-\frac{1}{2}}\int_{r_1}^{r} \pm r[(r - r_1)(r_2 - r)]^{-\frac{1}{2}}\,\mathrm{d}r$$

Now define a and e by

$$a = \frac{1}{2}(r_1 + r_2) = \frac{-\mu}{2\alpha_1} > 0 \quad \text{since} \quad \alpha_1 < 0$$

$$e = \frac{r_2 - r_1}{r_2 + r_1} = \left(1 + \frac{2\alpha_1\alpha_2^2}{\mu^2}\right)^{\frac{1}{2}} < 1 \quad \text{so that} \quad 0 \le e < 1$$

giving the integral

$$\alpha_2 = [\mu a(1 - e^2)]^{\frac{1}{2}}$$

The β_1 Integral

Then

$$r_1 = a(1 - e) \qquad r_2 = a(1 + e) \tag{6.4}$$

To avoid the double-valued function in the integrand, introduce a uniformizing variable E defined by

$$r = a(1 - e\cos E) \qquad \dot{E} > 0 \qquad \text{for all } t \tag{6.5}$$

Then

$$\dot{r} = ae\dot{E}\sin E$$

so that the sign of $\sin E$ is always the same as that of $\dot{r}$. Now by Eqs. (6.4) and (6.5)

$$(r - r_1)(r_2 - r) = a^2e^2\sin^2 E$$

$$[(r - r_1)(r_2 - r)]^{\frac{1}{2}} = ae|\sin E|$$

In the integrand of $\partial W_1/\partial\alpha_1$,

$$\pm r[(r - r_1)(r_2 - r)]^{-\frac{1}{2}} = \pm r\frac{ae\sin E}{ae|\sin E|} = r$$

since the upper sign is for $\dot{r} > 0$ and $\sin E > 0$ and the lower sign is for $\dot{r} < 0$ and $\sin E < 0$. Note also that $r = r_1$ gives $\cos E_1 = 1$ or $E_1 = 2\pi q$ $(q = 0, 1, 2, \ldots)$. Since $\partial W/\partial\alpha_1 = \partial W_1/\partial\alpha_1$,

$$\begin{aligned} t + \beta_1 &= \frac{\partial W_1}{\partial\alpha_1} = (-2\alpha_1)^{-\frac{1}{2}}\int_{r_1}^{r} r\,\mathrm{d}r \\ &= (-2\alpha_1)^{-\frac{1}{2}}\int_{2\pi q}^{E} a(1 - e\cos E)\,\mathrm{d}E \\ &= (-2\alpha_1)^{-\frac{1}{2}}a\,[E - e\sin E]\Big|_{2\pi q}^{E} \end{aligned}$$

Since $-2\alpha_1 = \mu/a$, therefore $(-2\alpha_1)^{-1/2}a = 1/n$, where $n = \sqrt{\mu a^{-3}}$. Thus

$$t + \beta_1 = n^{-1}(E - 2\pi q - e\sin E) \tag{6.6}$$

or

$$E - e\sin E = n(t + \beta_1)$$

If we let $E = 0$, then $t = \tau$, giving the integral

$$\beta_1 = -\tau$$

Now by Eq. (6.5) r is periodic in E with period 2π. Also, by Eq. (6.6), when $\Delta E = 2\pi$, we have $\Delta t = 2\pi/n$. The motion is periodic in t with period

$$T = 2\pi/n \tag{6.7}$$

and $n = \sqrt{\mu a^{-3}}$ is the mean motion. Equation (6.6) is Kepler's equation, which was discussed earlier, and $n = \sqrt{\mu a^{-3}}$ can be written

$$\mu = n^2a^3 \tag{6.8}$$

essentially Kepler's third law.

The summary of results of $t + \beta_1 = \partial W(q, \alpha)/\partial\alpha_1 = \partial W_1/\partial\alpha_1$ is as follows.

$$r = a(1 - e\cos E)$$

$$a = \frac{-\mu}{2\alpha_1}$$

$$e = \left(1 + \frac{2\alpha_1\alpha_2^2}{\mu^2}\right)^{\frac{1}{2}} < 1$$

leading to

$$\alpha_2^2 = \frac{-\mu^2}{2\alpha_1}(1 - e^2) = \mu a(1 - e^2) = \mu p$$

where

$$p = a(1 - e^2)$$

$$E - e\sin E = n(t + \beta_1)$$

$$\mu = n^2 a^3$$

We thus recognize the orbit as a Keplerian ellipse, where a is the semi-major axis, e the eccentricity, p the semi-latus rectum, n the mean motion, $\beta_1 = -\tau$, τ is the time of perigee passage, and E is the eccentric anomaly.

Integrals Involving Both W_1 and W_2

The α_3 Integral

$$\beta_2 = \frac{\partial W(q, \alpha)}{\partial\alpha_2} = \frac{\partial W_1}{\partial\alpha_2} + \frac{\partial W_2}{\partial\alpha_2}$$

We had

$$W_1 = \int_{r_1}^{r} \pm r^{-1}\left(-\alpha_2^2 + 2\mu r + 2\alpha_1 r^2\right)^{\frac{1}{2}}\,\mathrm{d}r$$

where that $\mathrm{d}r > 0$ is for the upper sign and $\mathrm{d}r < 0$ is for the lower sign.

Then

$$\frac{\partial W_1}{\partial\alpha_2} = \alpha_2 \int_{r_1}^{r} \mp r^{-1}\left(-\alpha_2^2 + 2\mu r + 2\alpha_1 r^2\right)^{-\frac{1}{2}}\,\mathrm{d}r$$

$$= \alpha_2 \int_{r_1}^{r} \mp r^{-1}[-2\alpha_1(r - r_1)(r_2 - r)]^{-\frac{1}{2}}\,\mathrm{d}r$$

$$= \alpha_2(-2\alpha_1)^{-\frac{1}{2}} \int_{r_1}^{r} \mp\frac{1}{r}[(r - r_1)(r_2 - r)]^{-\frac{1}{2}}\,\mathrm{d}r$$

To eliminate the double sign in the integrand, introduce a new uniformizing variable

f, defined by

$$\dot{f} > 0 \qquad \text{for all } t$$
$$r = \frac{a(1-e^2)}{1+e\cos f} \tag{6.9}$$

With $r_1 = a(1-e)$ and $r_2 = a(1+e)$, Eq. (6.9) covers the physical range $r_1 \le r \le r_2$. Then

$$\dot{r} = \frac{a(1-e^2)\, e\dot{f}\sin f}{(1+e\cos f)^2} \tag{6.10}$$

Note that Eq. (6.9) fixes $\cos f$ and that the sign of $\dot{r}$ is the same as the sign of $\sin f$, so that $\sin f$ and $\cos f$ are thus both determined. From Eq. (6.5) and Eq. (6.9), we then deduce that

$$\cos f = \frac{\cos E - e}{1 - e\cos E} \qquad \sin f = \frac{\sqrt{1-e^2}\sin f}{1-e\cos E}$$

$$\cos E = \frac{e + \cos f}{1+e\cos f} \qquad \sin E = \frac{\sqrt{1-e^2}\sin f}{1+e\cos f}$$

It is evident that f is the true anomaly.

Return to the preceding integrand. From Eq. (6.9) and $r_1 = a\,(1-e), r_2 = a\,(1+e)$, we find

$$r - r_1 = \frac{ae(1-e)(1-e\cos f)}{1+e\cos f} \qquad r_2 - r = \frac{ae(1+e)(1+e\cos f)}{1+e\cos f}$$

Thus

$$(r-r_1)(r_2-r) = \frac{a^2e^2(1-e^2)\sin^2 f}{(1+e\cos f)^2}$$
$$[(r-r_1)(r_2-r)]^{\frac{1}{2}} = \frac{ae(1-e^2)^{\frac{1}{2}}|\sin f|}{1+e\cos f} \tag{6.11a}$$

Also

$$\mathrm{d}r = \frac{a(1-e^2)\, e\sin f}{(1+e\cos f)^2}\,\mathrm{d}f$$

so that

$$r^{-1}\,\mathrm{d}r = \frac{e\sin f}{1+e\cos f}\,\mathrm{d}f \tag{6.11b}$$

and

$$\mp\frac{1}{r}[(r-r_1)(r_2-r)]^{-\frac{1}{2}}\,\mathrm{d}r = \mp\frac{1}{a(1-e^2)^{\frac{1}{2}}}\frac{\sin f}{|\sin f|}\,\mathrm{d}f \tag{6.12}$$

Here the upper sign goes with $\sin f > 0$ and the lower with $\sin f < 0$, so that

Eq. (6.12) becomes $-\mathrm{d}f/[a(1-e^2)^{\frac{1}{2}}]$. Thus

$$\frac{\partial W_1}{\partial \alpha_2} = \frac{-\alpha_2(-2\alpha_1)^{-1/2}}{a(1-e^2)^{\frac{1}{2}}} \int_{f_1}^{f} \mathrm{d}f \tag{6.13}$$

Here f_1 is the value of f corresponding to $r = r_1$. By Eq. (6.9) we then have

$$f_1 = 2\pi q \qquad q = 0, 1, 2, \ldots$$

We can take f_1 to be zero, or else it could be absorbed into the β_2, since $\beta_2 = \partial W/\partial \alpha_2$. Thus

$$\frac{\partial W_1}{\partial \alpha_2} = \frac{-\alpha_2(-2\alpha_1)^{-\frac{1}{2}}}{a(1-e^2)^{\frac{1}{2}}} f \tag{6.13a}$$

By

$$\alpha_2 = \sqrt{\mu a(1-e^2)} \qquad \alpha_1 = -\mu/2a$$

we find

$$\frac{\partial W_1}{\partial \alpha_2} = -f \tag{6.14}$$

Next we need $\partial W_2/\partial \alpha_2$. From

$$W_2 = \int_0^\theta \pm\left(\alpha_2^2 - \alpha_3^2 \sec^2\theta\right)^{\frac{1}{2}} \mathrm{d}\theta$$

where that $\mathrm{d}\theta > 0$ is for the upper sign and $\mathrm{d}\theta < 0$ is for the lower sign, we find

$$\frac{\partial W_2}{\partial \alpha_2} = \alpha_2 \int_0^\theta \pm\left(\alpha_2^2 - \alpha_3^2 \sec^2\theta\right)^{-\frac{1}{2}} \mathrm{d}\theta$$

To evaluate this, write

$$\frac{1}{\sqrt{\alpha_2^2 - \alpha_3^2 \sec^2\theta}} = \frac{\cos\theta}{\sqrt{\alpha_2^2 \cos^2\theta - \alpha_3^2}} = \frac{\cos\theta}{\sqrt{\alpha_2^2 - \alpha_3^2 - \alpha_2^2 \sin^2\theta}}$$

$$= \left(\alpha_2^2 - \alpha_3^2\right)^{-\frac{1}{2}} \frac{\cos\theta}{\sqrt{1 - \frac{\alpha_2^2}{\alpha_2^2 - \alpha_3^2} \sin^2\theta}}$$

Then

$$\frac{\partial W_2}{\partial \alpha_2} = \frac{\alpha_2}{\sqrt{\alpha_2^2 - \alpha_3^2}} \int_0^\theta \pm \frac{\cos\theta \, \mathrm{d}\theta}{\sqrt{1 - \frac{\alpha_2^2}{\alpha_2^2 - \alpha_3^2} \sin^2\theta}}$$

Define γ, which we shall later identify physically, by

$$\cos\gamma = \frac{\alpha_3}{\alpha_2} \qquad \sin\gamma = \frac{\sqrt{\alpha_2^2 - \alpha_3^2}}{\alpha_2} > 0$$

giving the integral

$$\alpha_3 = \alpha_2 \cos\gamma \qquad \text{for } 0 \leq \gamma \leq \pi$$

The β_2 Integral

Then

$$\frac{\partial W_2}{\partial \alpha_2} = \frac{1}{\sin\gamma} \int_0^\theta \frac{\pm \cos\theta \, \mathrm{d}\theta}{\sqrt{1 - \left(\frac{\sin\theta}{\sin\gamma}\right)^2}}$$

To eliminate the double sign, introduce the variable

$$w = \frac{\sin\theta}{\sin\gamma}$$

so that

$$\mathrm{d}w = \frac{\cos\theta \, \mathrm{d}\theta}{\sin\gamma}$$

Since $\cos\theta \geq |\alpha_3|/\alpha_2 > 0$ always, it follows that $\mathrm{d}w > 0$ is for the upper sign and $\mathrm{d}w < 0$ is for the lower sign. Then

$$\frac{\partial W_2}{\partial \alpha_2} = \int_0^w \frac{\pm \mathrm{d}w}{\sqrt{1 - w^2}}$$

The double sign is still there; so next introduce a uniformizing variable ψ, with $\dot{\psi} > 0$ for all t, such that

$$w = \sin\psi$$

(Note that ψ is defined as the argument of latitude in Chapter 2, Sec. VI.) Then

$$\mathrm{d}w = \cos\psi \, \mathrm{d}\psi$$

and

$$\frac{\pm \mathrm{d}w}{\sqrt{1 - w^2}} = \frac{\pm \cos\psi \, \mathrm{d}\psi}{|\cos\psi|}$$

Since $\dot{\psi} > 0$ always and since $\mathrm{d}w > 0$ for the upper sign and $\mathrm{d}w < 0$ for the lower sign, it follows that $\cos\psi > 0$ for the upper sign and $\cos\psi < 0$ for the lower. Thus

$$\frac{\pm \mathrm{d}w}{\sqrt{1 - w^2}} = \mathrm{d}\psi$$

and

$$\frac{\partial W_2}{\partial \alpha_2} = \psi$$

where $\sin\theta = \sin\psi \sin\gamma$.

Thus

$$\beta_2 = \frac{\partial W_1}{\partial \alpha_2} + \frac{\partial W_2}{\partial \alpha_2}$$

gives the integral

$$\beta_2 = \psi - f \tag{6.15}$$

Integral Involving Only W_2: The β_3 Integral

We have

$$\beta_3 = \frac{\partial W(q, \alpha)}{\partial \alpha_3} = \phi + \frac{\partial W_2}{\partial \alpha_3}$$

where

$$W_2 = \int_0^{\theta} \pm\left(\alpha_2^2 - \alpha_3^2 \sec^2\theta\right)^{\frac{1}{2}} \mathrm{d}\theta$$

Thus

$$\frac{\partial W_2}{\partial \alpha_3} = \int_0^{\theta} \mp \frac{\alpha_3 \sec^2\theta}{\sqrt{\alpha_2^2 - \alpha_3^2 \sec^2\theta}} \mathrm{d}\theta$$

Removing $|\alpha_3|$ from both numerator and denominator, we obtain

$$\frac{\partial W_2}{\partial \alpha_3} = \operatorname{sgn} \alpha_3 \int_0^{\theta} \mp \frac{\sec^2\theta}{\sqrt{\frac{\alpha_2^2}{\alpha_3^2} - \tan^2\theta}} \mathrm{d}\theta$$

$$= \operatorname{sgn} \alpha_3 \int_0^{\theta} \mp \frac{\sec^2\theta}{\sqrt{\frac{\alpha_2^2 - \alpha_3^2}{\alpha_3^2} - \tan^2\theta}} \mathrm{d}\theta$$

However,

$$\tan\gamma = \sqrt{\frac{\alpha_2^2 - \alpha_3^2}{\alpha_3^2}} \tag{6.16}$$

so that

$$\frac{\partial W_2}{\partial \alpha_3} = \operatorname{sgn} \alpha_3 \int_0^{\theta} \mp \frac{\sec^2\theta}{\sqrt{\tan^2\gamma - \tan^2\theta}} \mathrm{d}\theta$$

Introduce the variable

$$u = \frac{\tan\theta}{|\tan\gamma|} \le 1 \tag{6.17}$$

To show the $u \le 1$, note that for real motion

$$\alpha_2^2 - \alpha_3^2 \sec^2\theta \ge 0 \qquad \text{giving} \qquad \sec^2\theta \le \left(\alpha_2^2/\alpha_3^2\right)$$

or

$$\cos^2\theta \ge \left(\alpha_3^2/\alpha_2^2\right) \qquad \text{giving} \qquad \alpha_3^2 \le \alpha_2^2 \cos^2\theta \le \alpha_2^2$$

However,

$$\tan^2\theta = \sec^2\theta - 1 \leq \frac{\alpha_2^2 - \alpha_3^2}{\alpha_3^2} = \tan^2\gamma$$

using Eq. (6.16). Thus

$$|\tan\theta| \leq |\tan\gamma|$$

so that

$$\tan\theta \leq |\tan\gamma|$$

or

$$u \leq 1$$

Return to the integral. We obtain

$$\frac{\partial W_2}{\partial \alpha_2} = \operatorname{sgn}\alpha_3 \int_0^\theta \mp \frac{\mathrm{d}\theta}{\sqrt{1-u^2}}$$

where that $\mathrm{d}u > 0$ is for the upper sign and $\mathrm{d}u < 0$ is for the lower sign. The double sign is still there, but to eliminate it, introduce χ by $\dot{\chi} > 0$ for all t and

$$u = \sin\chi \tag{6.18}$$

Then $\cos\chi > 0$ with the upper sign and < 0 with the lower sign. (Note that $\chi = \phi - \Omega$ gives the physical meaning of the element χ in Chapter 2, Sec. VI.)

We have

$$\mp \frac{\mathrm{d}\theta}{\sqrt{1-u^2}} = \mp \frac{\cos\chi\,\mathrm{d}\chi}{|\cos\chi|} = -\mathrm{d}\chi$$

Thus

$$\frac{\partial W_2}{\partial \alpha_3} = -\chi \operatorname{sgn}\alpha_3$$

where

$$\tan\theta = |\tan\gamma|\sin\chi$$

using Eqs. (6.17) and (6.18). Thus

$$\beta_3 = \frac{\partial W(q,\alpha)}{\partial \alpha_3} = \phi + \frac{\partial W_2}{\partial \alpha_3}$$

gives the integral

$$\beta_3 = \phi - \chi \operatorname{sgn}\alpha_3$$

Summary for β_2 and β_3

$$\beta_2 = \psi - f$$

$$\beta_3 = \phi - \chi \operatorname{sgn} \alpha_3$$

$$\tan\theta = |\tan\gamma| \sin\chi$$

$$\tan\gamma = \sqrt{\frac{\alpha_2^2 - \alpha_3^2}{\alpha_3^2}} = |\tan\gamma| \operatorname{sgn} \alpha_3 \tag{6.19}$$

$$r = \frac{a(1-e^2)}{1 + e\cos f} \qquad (\dot{f} > 0)$$

$$\sin\theta = \sin\psi \sin\gamma \qquad (\dot{\psi} > 0)$$

To understand these equations better, we prove some theorems.

V. Relations Connecting β_2 and β_3 with ω and Ω

Theorem 1: The orbit lies in a plane passing through the origin.
Proof: Use

$$\chi \operatorname{sgn} \alpha_3 = \phi - \beta_3 \qquad \chi = (\phi - \beta_3) \operatorname{sgn} \alpha_3$$

$$\sin\chi = \sin(\phi - \beta_3) \operatorname{sgn} \alpha_3 \qquad \cos\chi = \cos(\phi - \beta_3)$$

$$\tan\theta = |\tan\gamma| \sin\chi = |\tan\gamma| \operatorname{sgn} \alpha_3 \sin(\phi - \beta_3) = \tan\gamma \sin(\phi - \beta_3)$$

using Eq. (6.19). Thus

$$\sin(\phi - \beta_3) - \cot\gamma \tan\theta = 0$$

Multiply this by $r\cos\theta$, to find

$$r\cos\theta[\sin\phi\cos\beta_3 - \cos\phi\sin\beta_3] - r\sin\theta\cot\gamma = 0$$

However, $r\cos\theta\sin\phi = y$, $r\cos\theta\cos\phi = x$, $r\sin\theta = z$. Thus

$$y\cos\beta_3 - x\sin\beta_3 - z\cot\gamma = 0$$

This is the equation of a plane passing through the origin. It follows that the intersection of the orbital plane with the celestial sphere is a great circle, so that we may apply spherical trigonometry.

Theorem 2: $\beta_2 = \omega$, the argument of pericenter.
Proof:

$$\sin\theta = \sin\psi \sin\gamma$$

However, $\beta_2 = \psi - f$, so that

$$\sin\theta = \sin\gamma \sin(\beta_2 + f)$$

At the ascending node $\theta = 0$, so that

$$\sin(\beta_2 + f) = 0$$

$$\cos(\beta_2 + f) = \pm 1$$

To show that the sign is plus, use

$$\cos\theta \frac{\mathrm{d}\theta}{\mathrm{d}f} = \sin\gamma \cos(\beta_2 + f)$$

Here $\sin\gamma > 0$, from the definition $\sin\gamma = \sqrt{\alpha_2^2 - \alpha_3^2}/\alpha_2$.

Now at the ascending node, $\mathrm{d}\theta/\mathrm{d}f > 0$ and $\theta = 0$, so that

$$\cos(\beta_2 + f) > 0$$

Thus, $\cos(\beta_2 + f) = 1$. Since $\sin(\beta_2 + f) = 0$, it follows that $\beta_2 + f = 0$, modulo 2π, at the ascending node. However, $\omega + f = 0$, modulo 2π, at the ascending node. Thus, $\beta = \omega$, as was to be proved.

Theorem 3: $\gamma = I$, the inclination.

Proof: Because $\sin\theta = \sin\gamma \sin(\beta_2 + f)$, we now have

$$\sin\theta = \sin\gamma \sin(\omega + f)$$

From Fig. 2.5

$$\sin\theta = \sin I \sin(\omega + f)$$

Thus

$$\sin\gamma = \sin I$$

and

$$\gamma = I \qquad \text{or} \qquad \gamma = \pi - I$$

By definition

$$\cos\gamma = \alpha_3/\alpha_2$$

However,

$$p_\phi = r^2 \cos^2\theta \dot{\phi}$$

so that α_3 and thus $\cos\gamma$ are positive for direct orbits and negative for retrograde orbits.

We see that $\gamma = I$ satisfies these requirements. The assumption $\gamma = \pi - I$ gives $\cos\gamma = -\cos I$, which would lead to $\alpha_3 < 0$ for a direct orbit and $\alpha_3 > 0$ for a retrograde orbit. Thus, $\gamma = I$, as stated.

Theorem 4: $\beta_3 = \Omega$, the longitude of the ascending node.

Proof: In the proof of Theorem 1, we had

$$\tan\theta = \tan\gamma \sin(\phi - \beta_3)$$

which now becomes

$$\tan\theta = \tan I \sin(\phi - \beta_3)$$

At the ascending node $\theta = 0$, so that

$$\sin(\phi - \beta_3) = 0$$

$$\cos(\phi - \beta_3) = \pm 1$$

To show that $+1$ holds, differentiate the preceding equation to obtain

$$\sec^2\theta \frac{\mathrm{d}\theta}{\mathrm{d}\phi} = \tan I \cos(\phi - \beta_3)$$

so that

$$\frac{\sec^2\theta \ \mathrm{d}\theta/\mathrm{d}\phi}{\tan I} = \cos(\phi - \beta_3)$$

At the ascending node, the numerator and denominator are both plus for direct orbits and both minus for retrograde orbits. Thus

$$\cos(\phi - \beta_3) > 0$$

and thus

$$\sin(\phi - \beta_3) = 0$$

$$\cos(\phi - \beta_3) = 1$$

Thus, at the ascending node

$$\phi = \beta_3 \qquad \text{modulo } 2\pi$$

Also, at the ascending node

$$\phi = \Omega \qquad \text{modulo } 2\pi$$

Thus, $\beta_3 = \Omega$, as stated.

VI. Summary

The results of this chapter show that if (q, p) are the coordinates and momenta for the Kepler problem, in either rectangular or spherical coordinates, we have found new canonical variables (α, β), corresponding to a new Hamiltonian $K = 0$. They are

$$\begin{aligned} \alpha_1 &= -\frac{\mu}{2a} & \beta_1 &= -\tau \\ \alpha_2 &= [\mu a(1 - e^2)]^{\frac{1}{2}} & \beta_2 &= \omega \\ \alpha_3 &= \alpha_2 \cos I & \beta_3 &= \Omega \end{aligned} \tag{6.20}$$

The Kepler problem is defined either by the equation

$$\ddot{\boldsymbol{r}}(t) = -\mu \boldsymbol{r}/r^3$$

or by the Hamiltonian

$$H = \frac{1}{2}\left(p_x^2 + p_y^2 + p_z^2\right) - \frac{\mu}{r}$$

where $V(r) = -\mu/r$, $\mu = G(m_1 + m_2)$.

Its solution, given by

$$
\begin{aligned}
\boldsymbol{r} &= \boldsymbol{A}(\cos E - e) + \boldsymbol{B}\sin E \\
\dot{\boldsymbol{r}} &= \frac{an}{r}(-\boldsymbol{A}\sin E + \boldsymbol{B}\cos E) \\
n &= \sqrt{\mu a^{-3}} \\
r &= a(1 - e\cos E) \\
E - e\sin E &= n(t - \tau)
\end{aligned}
\tag{6.21}
$$

is then a canonical transformation from (q, p) to the new variables (α, β).

In the Chapter 7 we shall consider the effects of adding a perturbing term to the Hamiltonian. A perturbing term $V_1(q)$ added to the Hamiltonian will correspond to a term $-\nabla V_1(q)$ added to the $\ddot{\boldsymbol{r}}$ equation. After adding such a perturbation, we shall treat the α's and β's, or the corresponding Kepler elements, as variables related to the original q's and p's by the same equations [(6.20) and (6.21)] as in the unperturbed problem. If we can find the α's and β's as functions of t, we have simply to use Eqs. (6.20) and (6.21) to find the orbit.

Bibliography

[1]Smart, W. M., *Celestial Mechanics,* Longmans, Green, and Co., London, 1953, pp. 143–148.

Chapter 7

Hamilton–Jacobi Perturbation Theory

SUPPOSE we have a problem characterized by a Hamiltonian

$$H(q, p, t) = H_0(q, p) + H_1(q, p, t)$$

where $H_0(q, p)$ leads to a separable problem and $H_1(q, p, t)$ is a perturbing term.

The separable problem leads to the usual scheme.

1) Solve the HJ equation

$$H_0\left(q, \frac{\partial S}{\partial q}\right) + \frac{\partial S}{\partial t} = 0$$

Then $H_0(q, \partial S/\partial q) =$ const, since it does not depend explicitly on t; call it α_1. Thus

$$\frac{\partial S}{\partial t} = -\alpha_1$$

$$S = -\alpha_1 t + W(q, \alpha_1, \alpha_2, \alpha_3)$$

$$H_0\left(q, \frac{\partial W}{\partial q}\right) = \alpha_1$$

2) Find the q's as functions of t by inverting

$$t + \beta_1 = \frac{\partial W(q, \alpha)}{\partial \alpha_1}$$

$$\beta_2 = \frac{\partial W(q, \alpha)}{\partial \alpha_2}$$

$$\beta_3 = \frac{\partial W(q, \alpha)}{\partial \alpha_3}$$

The q's are then

$$q_k = q_k(\alpha, \beta, t) \qquad k = 1, 2, 3$$

3) Find the p's from

$$p_k = \frac{\partial W(q, \alpha)}{\partial q_k} \qquad k = 1, 2, 3$$

4) Find the $\dot{q}$'s from

$$\dot{q}_k = \frac{\partial H_0(q, p)}{\partial p_k} \qquad k = 1, 2, 3$$

If $H_0(q, p)$ is the Kepler Hamiltonian, this whole procedure is taken care of by Eqs. (6.21). So far, the α's and β's are canonical with respect to $H_0(q, p)$ as Hamiltonian.

To solve the perturbed problem, we introduce the preceding α's and β's as new variables; that is, we use the relations

$$q_k = q_k(\alpha, \beta, t) \qquad p_k = p_k(\alpha, \beta, t) \qquad k = 1, 2, 3$$

as a time-dependent canonical mapping to introduce the new variables α, β into the perturbed problem. If $H_0(q, p)$ is the Kepler Hamiltonian, this mapping is simply Eqs. (6.21); the time dependence is a result of Kepler's equation $E - e \sin E = n(t + \beta_1)$. The α's and β's will no longer be constant but will depend on time.

It is then clear that the perturbed q's and p's will be the same functions of t and the perturbed α's and β's, as the unperturbed q's and p's are of t and the unperturbed α's and β's. It follows that

$$H_0(\text{perturbed}\, q,\ \text{perturbed}\, p) = \alpha_1 \text{ perturbed}$$

To see the meaning of this more clearly, note that, if $\boldsymbol{v}$ is velocity and $H_0(q, p)$ the Kepler Hamiltonian, the equation

$$H_0 = \frac{1}{2}v^2 - \frac{\mu}{r} = -\frac{\mu}{2a} \qquad a = -\frac{\mu}{2\alpha_1}$$

is still exactly true for the perturbed variables.

For the perturbed problem, we have

$$H(q, p, t) = H_0(q, p) + H_1(q, p, t)$$

$$\dot{p}_k = -\frac{\partial H(q, p, t)}{\partial q_k} \qquad p_k = \frac{\partial W(q, \alpha)}{\partial q_k} = \frac{\partial S(q, \alpha, t)}{\partial q_k}$$

$$\dot{q}_k = \frac{\partial H(q, p, t)}{\partial p_k} \qquad \beta_k = \frac{\partial S(q, \alpha, t)}{\partial \alpha_k}$$

Thus, $S(q, \alpha, t)$ is a generating function of the form $S(q, P, t)$ for introducing new canonical variables. It follows that the α's and β's introduced in this way are canonical with respect to

$$K = H + \frac{\partial S}{\partial t}$$

as new Hamiltonian. However,

$$S = -\alpha_1 t + W(q, \alpha)$$

$$H = H_0(q, p) + H_1(q, p, t)$$

so that

$$K = H_0(q, p) + H_1(q, p, t) - \alpha_1$$

Here the q's, p's, and α_1 are all perturbed variables and

$$H_0(q, p) = \alpha_1$$

so that

$$K = H_1(q, p, t)$$

For the perturbed problem,

$$\dot{\alpha}_k = -\frac{\partial K(q, p, t)}{\partial \beta_k} = -\frac{\partial H_1(q, p, t)}{\partial \beta_k}$$

$$\dot{\beta}_k = \frac{\partial K(q, p, t)}{\partial \alpha_k} = \frac{\partial H_1(q, p, t)}{\partial \alpha_k}$$

so that the α's and β's of the perturbed problem are canonical with respect to the perturbing term $H_1(q, p, t)$ as Hamiltonian.

Our problem is now to solve this canonical system for the α's and β's as functions of t. After we do so, the q's and p's that are solutions of

$$\dot{q}_k = \frac{\partial H(q, p, t)}{\partial p_k} = \frac{\partial H_0(q, p)}{\partial p_k} + \frac{\partial H_1(q, p, t)}{\partial p_k}$$

$$\dot{p}_k = -\frac{\partial H(q, p, t)}{\partial q_k} = -\frac{\partial H_0(q, p)}{\partial q_k} - \frac{\partial H_1(q, p, t)}{\partial q_k}$$

will be found from the relations

$$p_k = \frac{\partial S(q, \alpha, t)}{\partial q_k} \qquad \beta_k = \frac{\partial S(q, \alpha, t)}{\partial \alpha_k} \tag{7.1}$$

Here $S(q, \alpha, t)$ has the same functional form in t, the q's, and the α's as it has for the unperturbed problem. If the latter is the Kepler problem, Eqs. (7.1) are equivalent to the Keplerian algorithm for the q's and p's in terms of t and the α's and β's, i.e., to Eqs. (6.21).

Actually, we shall find that $\beta_1 = -\tau$ never appears in $H_1(q, p, t)$ or in the solution except in the combination $\ell = n(t - \tau)$. It should also be remarked that $H_1(q, p, t)$ will not ordinarily contain ℓ explicitly, but rather the true anomaly f; ℓ appears implicitly through the relation connecting f with E and the Kepler equation $E - e \sin E = \ell$.

In Chapter 8 we shall get rid of β_1 as a variable. To understand why, we have to anticipate later developments. A perturbation in orbital mechanics and celestial mechanics ordinarily produces variations that are periodic in t or change monotonically with t, usually linearly. These monotonic variations are called secular variations, and any term in t^2 is called a secular acceleration.

If, however, we use β_1 as a variable, we should find mixed terms of the form t times periodic terms. Authors sometimes call them inconvenient and introduce other variables to get rid of them. How can we do so if they are really there? The answer is that they are not. The element $\beta_1 = -\tau$ never appears except in the combination $nt - n\tau$, and it so happens that nt introduces mixed terms that exactly cancel those of $n\tau$. We shall prove this later in drag-free satellite theory by showing that the variations in $\ell = n(t - \tau)$ are purely linear plus periodic.

To see the compatibility of mixed terms in τ with no mixed terms in ℓ, consider the drag-free case. Here calculation will show that

$$\ell = n(t - \tau) = k_0 + k_1 t + P_1(t)$$

$$n = c[1 + P_2(t)]$$

where k_0, k_1, and c are constants and $P_1(t)$ and $P_2(t)$ are periodic in t. It is clear that nt is mixed because of $t P_2(t)$, so that $n\tau$ must be mixed.

For τ itself

$$\tau = t - \frac{\ell}{n} = t - \frac{k_0 + k_1 t + P_1(t)}{c[1 + P_2(t)]}$$

so that

$$\tau = t - (1/c)[k_0 + k_1 t + P_1(t)]\left[1 - P_2(t) + P_2^2(t) + \cdots\right]$$

Thus, τ is mixed because of the terms $k_1 t P_2(t)$, $k_1 t P_2^2(t)$, etc.

Sometimes ℓ is expressed as

$$\ell = nt + \sigma$$

Here $\sigma = -n\tau$, which is mixed. If, instead of σ, one defines a quantity σ', such that

$$\ell = \int_0^t n\,\mathrm{d}t + \sigma'$$

then σ' will be free of mixed terms. To show this, note that

$$\dot{\ell} = n + \dot{\sigma}' = k_1 + \dot{P}_1$$

and

$$\dot{\sigma}' = k_1 + \dot{P}_1 - c - cP_2$$

which is constant plus periodic. Thus, σ' is linear secular plus periodic, containing no mixed terms.

Bibliography

[1]Garfinkel, B., *Space Mathematics, Part I,* Vol. 5, Lectures in Applied Mathematics, American Mathematical Society, Providence, RI, 1966, pp. 67–68.

Chapter 8

The Vinti Spheroidal Method for Satellite Orbits and Ballistic Trajectories

I. Introduction

THE Earth is approximately an oblate spheroid. The oblate spheroidal system of coordinates is one of the 11 systems in which the motion of a particle in Euclidean space may lead to a separable problem. In this chapter we introduce this system and find a general form for the potential of the Earth that leads to separability of the Hamilton–Jacobi equation. We next introduce this form for the potential into Laplace's equation, solve it, and then expand this solution in spherical harmonics. This solution can be fitted exactly to the zeroth and second zonal harmonics, thereby accounting exactly for the oblateness. Moreover, it makes the first harmonic vanish, as it should for the origin to be at the Earth's center of mass. The fit of the fourth harmonic has the correct sign and about two-thirds of the correct value. The third harmonic is not accounted for in this first approach but has since been incorporated into the potential.[1–4]

II. The Coordinates and the Hamiltonian

Let the origin O be at the Earth's center of mass, the axis Oz along the polar axis, and the axis Ox toward the vernal equinox. We then define the oblate spheroidal coordinates by[5]

$$x + iy = r\cos\theta e^{i\phi} = c[(\xi^2 + 1)(1 - \eta^2)]^{\frac{1}{2}} e^{i\phi} \tag{8.1}$$

$$z = r\sin\theta = c\xi\eta \tag{8.2}$$

Here $e^{i\phi} = \exp i\phi$, r is the geocentric distance of the satellite, θ its latitude or declination, and ϕ its right ascension. The constant c is a parameter to be fitted. As $r \to \infty$, one shows easily that $c\xi \to r$ and $\eta \to \sin\theta$.

The metric $\mathrm{d}s^2$ is given by

$$\mathrm{d}s^2 = h_1^2\,\mathrm{d}\xi^2 + h_2^2\,\mathrm{d}\eta^2 + h_3^2\,\mathrm{d}\phi^2 \tag{8.3}$$

where

$$h_1^2 = c^2(\xi^2 + \eta^2)(\xi^2 + 1)^{-1} \tag{8.4a}$$

$$h_2^2 = c^2(\xi^2 + \eta^2)(1 - \eta^2)^{-1} \tag{8.4b}$$

$$h_3^2 = c^2(\xi^2 + 1)(1 - \eta^2) \tag{8.4c}$$

The level surfaces of ξ are oblate spheroids, those of η are hyperboloids of one

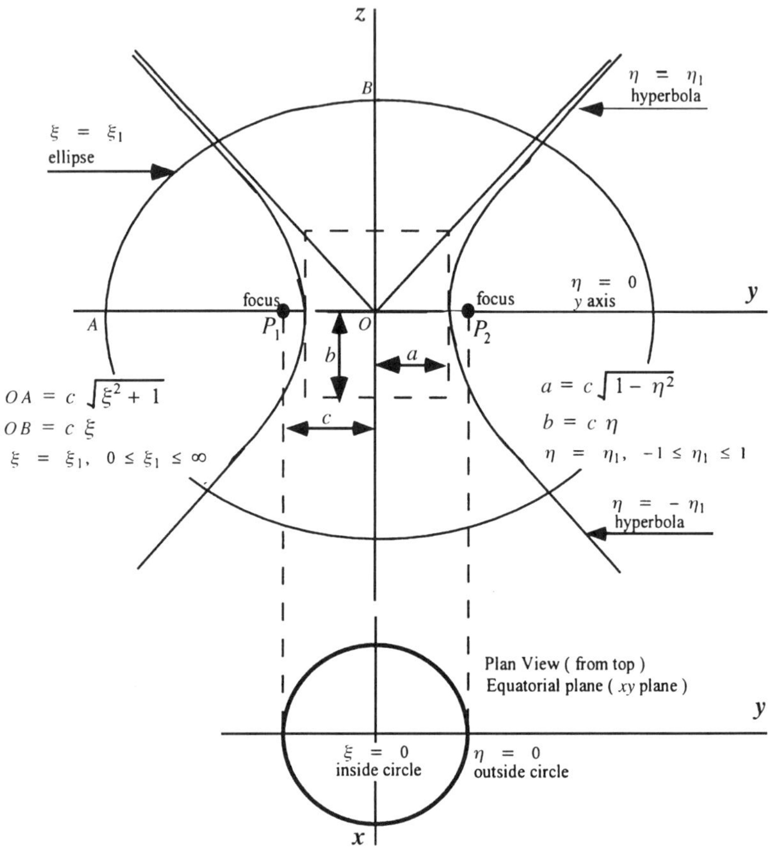

Fig. 8.1 The oblate spheroidal coordinates.

sheet, and those of ϕ are meridian planes. (A section perpendicular to Ox is shown in Fig. 8.1.) The derivations of Eqs. (8.1–8.4), the pertinent analytic geometry, and coordinate transformation are described in Appendix A.

The points P_1 and P_2 are foci both of the ellipsoids ξ = const and of the one-sheet hyperboloids η = const. The positive z axis satisfies $\eta = +1$, the negative z axis $\eta = -1$. The foci lie on a focal circle, of radius c in the equatorial plane. Points in the equatorial plane satisfy $\xi = 0$ inside the circle and $\eta = 0$ outside the circle.

The kinetic energy per unit mass is

$$T = \tfrac{1}{2}\left(h_1^2\dot{\xi}^2 + h_2^2\dot{\eta}^2 + h_3^2\dot{\phi}^2\right) \tag{8.5}$$

The generalized momenta are

$$p_\xi = \frac{\partial T}{\partial \dot{\xi}} = h_1^2\dot{\xi} \tag{8.6a}$$

$$p_\eta = \frac{\partial T}{\partial \dot{\eta}} = h_2^2\dot{\eta} \tag{8.6b}$$

$$p_\phi = \frac{\partial T}{\partial \dot{\phi}} = h_3^2\dot{\phi} \tag{8.6c}$$

If V is the potential, the Lagrangian $L = T - V$, and the Hamiltonian

$$H(q, p, t) = \Sigma_k p_k \dot{q}_k - L = 2T - L = T + V \tag{8.7}$$

Putting $H = H(q, p, t)$, thus

$$H = \tfrac{1}{2}\left(h_1^{-2} p_\xi^2 + h_2^{-2} p_\eta^2 + h_3^{-2} p_\phi^2\right) + V \tag{8.8}$$

Now V is a function of r, θ, and λ, where λ is the geographic longitude.

Since

$$\phi = \lambda + \omega_e t \tag{8.9}$$

where ω_e is the Earth's speed of rotation, we have

$$V = V(\xi, \eta, \phi - \omega_e t) \tag{8.10}$$

The Earth's rotation will spoil separability unless we demand that V depend only on ξ and η:

$$V = V(\xi, \eta) \tag{8.11}$$

This means that we cannot account for tesseral and sectorial harmonics, but at most for zonal harmonics. With such an axially symmetric potential, we obtain from Eqs. (8.8), (8.11), and (8.4):

$$H = \frac{1}{2c^2}\left[\frac{\xi^2 + 1}{\xi^2 + \eta^2} p_\xi^2 + \frac{1 - \eta^2}{\xi^2 + \eta^2} p_\eta^2 + \frac{p_\phi^2}{(\xi^2 + 1)(1 - \eta^2)}\right] + V(\xi, \eta) \tag{8.12}$$

Because Eq. (8.12) is explicitly independent of time, we have

$$H = \alpha_1 \tag{8.13}$$

Here α_1 is the constant energy, with $\alpha_1 < 0$ for a bounded orbit, since V vanishes at infinity. Also, ϕ is not contained in Eq. (8.12), so that it is a cyclic coordinate. Thus

$$p_\phi = \alpha_3 \tag{8.14}$$

a constant. From Eqs. (8.6c), (8.4c), and (8.1)

$$p_\phi = c^2(\xi^2 + 1)(1 - \eta^2)\dot{\phi} = r^2 \cos^2 \theta \dot{\phi} \tag{8.15}$$

Since $r \cos \theta$ is the distance of the satellite from the z axis and $\dot{\phi}$ its angular velocity about that axis, we identify $p_\phi = \alpha_3$ as the z component of angular momentum. This is always conserved with axial symmetry.

III. The Hamilton–Jacobi Equation

Call the coordinates q_k, $k = 1, 2, 3$. If we place

$$p_k = \frac{\partial W}{\partial q_k} \tag{8.16}$$

in Eq. (8.12), we obtain the *HJ* equation. Then

$$p_\phi = \frac{\partial W}{\partial \phi} \tag{8.17}$$

so that

$$W = \alpha_3\phi + \text{a function of } \xi \text{ and } \eta \tag{8.18}$$

If we separate Eq. (8.12), we have

$$W = \alpha_3\phi + W_1(\xi) + W_2(\eta) \tag{8.19}$$

Since we know that $p_\phi = \alpha_3$ in Eq. (8.12), we can write it and apply Eq. (8.16) only to p_ξ and p_η in Eq. (8.12). Then, with use of Eq. (8.19), we find

$$\begin{aligned}(\xi^2+1)W_1'^2 + (1-\eta^2)W_2'^2 &+ \frac{\alpha_3^2(\xi^2+\eta^2)}{(\xi^2+1)(1-\eta^2)} \\ &+ 2c^2(\xi^2+\eta^2)V(\xi,\eta) = 2c^2(\xi^2+\eta^2)\alpha_1\end{aligned} \tag{8.20}$$

Here we have put $H = \alpha_1$ in Eq. (8.12). Now

$$\xi^2 + \eta^2 = (\xi^2+1) - (1-\eta^2) \tag{8.21}$$

so that Eq. (8.20) becomes

$$\begin{aligned}(\xi^2+1)W_1'^2 - \frac{\alpha_3^2}{(\xi^2+1)} - 2c^2\xi^2\alpha_1 + (1-\eta^2)W_2'^2 + \frac{\alpha_3^2}{(1-\eta^2)} \\ - 2c^2\eta^2\alpha_1 + 2c^2(\xi^2+\eta^2)V(\xi,\eta) = 0\end{aligned} \tag{8.22}$$

Inspection of Eq. (8.22) shows that we obtain separability if and only if

$$V = \frac{f(\xi) + g(\eta)}{(\xi^2+\eta^2)} \tag{8.23}$$

This leads us to the problem: What forms must $f(\xi)$ and $g(\eta)$ have to satisfy Laplace's equation?

IV. Laplace's Equation

For axial symmetry $\nabla^2 V = 0$ becomes[5]

$$\frac{\partial}{\partial\xi}\left[(\xi^2+1)\frac{\partial V}{\partial\xi}\right] + \frac{\partial}{\partial\eta}\left[(1-\eta^2)\frac{\partial V}{\partial\eta}\right] = 0 \tag{8.24}$$

We require that V satisfy Eq. (8.23), with the requirement that V have no singularities outside the planet. The solution is long, but the result is simple. It is that V shall be a linear combination of the real and imaginary parts of

$$V = (\xi + i\eta)^{-1} \tag{8.25}$$

The reader can verify that Eq. (8.25) is a solution of Eq. (8.24). Then

$$V = \frac{b_0\xi - b_1\eta}{(\xi^2+\eta^2)} \tag{8.26}$$

which has the correct form to yield separability of the *HJ* equation. The next step is to find how many of the zonal harmonics we can fit with Eq. (8.26).

V. Expansion of Potential in Spherical Harmonics

Begin with

$$(\xi + i\eta)^2 = \xi^2 - \eta^2 + 2i\xi\eta \tag{8.27}$$

From Appendix A

$$(x^2 + y^2)/c^2 = \xi^2 + 1 - \eta^2 - \xi^2\eta^2 \tag{8.28}$$

$$z^2/c^2 = \xi^2\eta^2 \tag{8.29}$$

Thus

$$r^2/c^2 = \xi^2 + 1 - \eta^2 \qquad \text{or} \qquad \xi^2 - \eta^2 = (r^2/c^2) - 1 \tag{8.30}$$

and

$$(\xi + i\eta)^2 = \frac{r^2}{c^2} - 1 + 2i\frac{z}{c} = \frac{r^2}{c^2} - 1 + \frac{2i}{c}r\sin\theta \tag{8.31}$$

Then

$$(\xi + i\eta)^{-1} = \frac{c}{r}\left(1 + \frac{2ic}{r}\sin\theta - \frac{c^2}{r^2}\right)^{-\frac{1}{2}} \tag{8.32}$$

In Eq. (8.32) put

$$h = -(ic/r) \tag{8.33}$$

Then

$$(\xi + i\eta)^{-1} = \frac{c}{r}(1 - 2h\sin\theta + h^2)^{-\frac{1}{2}} \tag{8.34}$$

$$= \frac{c}{r}\sum_{n=1}^{\infty} h^n P_n(\sin\theta) \tag{8.35}$$

if $|h| <$ the smaller of $|\sin\theta \pm (\sin^2\theta - 1)^{1/2}|$ or $|h| <$ the smaller of $|\sin\theta \pm i\cos\theta|$. However, $|\sin\theta \pm i\cos\theta| = 1$. The condition for the validity of the Legendre expansions is thus $|h| \equiv c/r < 1$. We shall see that c will turn out to be small compared to r, so that the Legendre expansion is valid and

$$(\xi + i\eta)^{-1} = \frac{c}{r}\sum_{n=1}^{\infty}\left(-\frac{ic}{r}\right)^n P_n(\sin\theta) \tag{8.36}$$

The real part of this is given by the terms $n = 2k$ and the imaginary part by the terms $n = 2k + 1$. Thus

$$\operatorname{Re}(\xi + i\eta)^{-1} = \frac{c}{r}\sum_{k=0}^{\infty}(-1)^k\left(\frac{c}{r}\right)^{2k} P_{2k}(\sin\theta) \tag{8.37}$$

$$\operatorname{Im}(\xi + i\eta)^{-1} = \frac{c}{r}\sum_{k=0}^{\infty}(-1)^{k+1}\left(\frac{c}{r}\right)^{2k+1} P_{2k+1}(\sin\theta) \tag{8.38}$$

Since $V = b_0 \operatorname{Re}(\xi + i\eta)^{-1} + b_1 \operatorname{Im}(\xi + i\eta)^{-1}$, we find

$$V = \frac{b_0 c}{r} \sum_{k=0}^{\infty} (-1)^k \left(\frac{c}{r}\right)^{2k} P_{2k}(\sin\theta) - \frac{b_1 c}{r} \sum_{k=0}^{\infty} (-1)^k \left(\frac{c}{r}\right)^{2k+1} P_{2k+1}(\sin\theta) \tag{8.39}$$

This is to be compared with the zonal part V_z of the usual spherical harmonic expansion of the potential [Eq. (15.37a)]:

$$V_z = -\frac{\mu}{r}\left[1 - \sum_{n=1}^{\infty} \left(\frac{r_e}{r}\right)^n J_n P_n(\sin\theta)\right] \tag{8.40}$$

$$V_z = -\frac{\mu}{r}\left[1 - \sum_{k=1}^{\infty} \left(\frac{r_e}{r}\right)^{2k} J_{2k} P_{2k}(\sin\theta) - \sum_{k=0}^{\infty} \left(\frac{r_e}{r}\right)^{2k+1} J_{2k+1} P_{2k+1}(\sin\theta)\right] \tag{8.41}$$

Zeroth Harmonic

$$b_0 c = -\mu \tag{8.42}$$

Second Harmonic

$$-b_0 c^3 = \mu r_e^2 J_2 \tag{8.43}$$

From these we obtain

$$c^2 = r_e^2 J_2 \tag{8.44}$$

First Harmonic

$$-b_1 c^2 = \mu r_e J_1 \tag{8.45}$$

With the origin at the center of mass, we have $J_1 = 0$ and thus $b_1 = 0$. That is, with this model, all the odd zonal harmonics drop out.

Even Harmonics in General

$$b_0 c\,(-1)^k (c)^{2k} = \mu r_e^{2k} J_{2k} \tag{8.46}$$

With $b_0 c = -\mu$, this leads to

$$J_{2k} = (-1)^{k+1} \left(\frac{c}{r_e}\right)^{2k} \tag{8.47}$$

However, for this model, $J_2 = c^2/r_e^2$, so that we find

$$J_{2k} = (-1)^{k+1} J_2^k \tag{8.48}$$

In particular

$$J_4 = -J_2^2$$

$$J_6 = J_2^3$$

For the Earth, $J_2 = (1.08263) \times 10^{-3}$ and $r_e = 6378.137$ km, so that $c \approx 209.862$ km, using the World Geodetic System 1984, WGS84 Earth gravity model. The value of J_4 has the correct sign but is only about two-thirds of the correct value. The higher even harmonics of the model are much too small. They diminish rapidly with increasing n, while the actual values diminish slowly with increasing n.

Just the same, the fit is remarkably good, since most of the departure from spherical symmetry comes from the J_2. Since $b_1 = 0$ and $b_0 = -\mu/c$, we find

$$V = \frac{b_0 \xi}{(\xi^2 + \eta^2)} = -\frac{\mu}{c} \frac{\xi}{(\xi^2 + \eta^2)} \tag{8.49}$$

Placing

$$\rho \equiv c\xi \tag{8.50}$$

which approaches r for large r, we find

$$V = -\frac{\mu}{c} \frac{\rho/c}{(\xi^2 + \eta^2)} = -\frac{\mu \rho}{\rho^2 + c^2 \eta^2} \tag{8.51}$$

VI. Return to the *HJ* Equation

In Eq. (8.22) put $V = -\mu\rho(\rho^2 + c^2\eta^2)^{-1}$ and $\xi = \rho/c$. We find

$$(\rho^2 + c^2)\left(\frac{\mathrm{d}W_1}{\mathrm{d}\rho}\right)^2 - \frac{c^2\alpha_3^2}{\rho^2 + c^2} - 2\mu\rho - 2\alpha_1\rho^2 = -(1 - \eta^2)\left(\frac{\mathrm{d}W_2}{\mathrm{d}\eta}\right)^2$$
$$- \frac{\alpha_3^2}{1 - \eta^2} + 2\alpha_1 c^2 \eta^2 = k \tag{8.52}$$

Because the left side depends only on ρ and the right side only on η, each side is equal to a constant k. Now for a bounded orbit we have $\alpha_1 < 0$.

Also, $\eta^2 \leq 1$, so that

$$k < 0 \tag{8.53}$$

Moreover,

$$k + \alpha_3^2 = -(1 - \eta^2)\left(\frac{\mathrm{d}W_2}{\mathrm{d}\eta}\right)^2 + 2\alpha_1 c^2 \eta^2 - \frac{\alpha_3^2 \eta^2}{1 - \eta^2} < 0 \tag{8.54}$$

Thus

$$k < -\alpha_3^2 \tag{8.55}$$

We may put

$$k = -\alpha_2^2 \tag{8.56}$$

where α_2 may be taken as positive without loss of generality. Then

$$\alpha_3^2 < \alpha_2^2 \tag{8.57}$$

On placing $k = -\alpha_2^2$ in Eq. (8.52), we obtain

$$\left(\frac{\mathrm{d}W_1}{\mathrm{d}\rho}\right)^2 = (\rho^2 + c^2)^{-2} F(\rho) \tag{8.58}$$

$$\left(\frac{\mathrm{d}W_2}{\mathrm{d}\eta}\right)^2 = (1 - \eta^2)^{-2} G(\eta) \tag{8.59}$$

where

$$F(\rho) = c^2\alpha_3^2 + (\rho^2 + c^2)\left(-\alpha_2^2 + 2\mu\rho + 2\alpha_1\rho^2\right) \tag{8.60a}$$

$$G(\eta) = -\alpha_3^2 + (1 - \eta^2)\left(\alpha_2^2 + 2\alpha_1 c^2\eta^2\right) \tag{8.60b}$$

Then, by Eq. (8.19),

$$W = \alpha_3\phi + W_1(\rho) + W_2(\eta) \tag{8.61}$$

where

$$W_1(\rho) = \int_{\rho'}^{\rho} \pm(\rho^2 + c^2)^{-1} F(\rho)^{\frac{1}{2}}\,\mathrm{d}\rho \tag{8.62a}$$

$$W_2(\eta) = \int_{0}^{\eta} \pm(1 - \eta^2)^{-1} G(\eta)^{\frac{1}{2}}\,\mathrm{d}\eta \tag{8.62b}$$

It is convenient to let ρ' be the minimum ρ, viz. ρ_1, reached by the satellite. The motivation for this procedure is the same as in the Keplerian case.

VII. The Kinematic Equations

These are

$$t + \beta_1 = \frac{\partial W}{\partial \alpha_1} = \frac{\partial W_1}{\partial \alpha_1} + \frac{\partial W_2}{\partial \alpha_1} \tag{8.63a}$$

$$\beta_2 = \frac{\partial W}{\partial \alpha_2} = \frac{\partial W_1}{\partial \alpha_2} + \frac{\partial W_2}{\partial \alpha_2} \tag{8.63b}$$

$$\beta_3 = \frac{\partial W}{\partial \alpha_3} = \phi + \frac{\partial W_1}{\partial \alpha_3} + \frac{\partial W_2}{\partial \alpha_3} \tag{8.63c}$$

Calculate the $\partial W/\partial\alpha$'s by Eqs. (8.62) and (8.60) and insert the results into Eqs. (8.63), which become

$$t + \beta_1 = R_1 + c^2 N_1 \tag{8.64a}$$

$$\beta_2 = -\alpha_2 R_2 + \alpha_2 N_2 \tag{8.64b}$$

$$\beta_3 = \phi + c^2\alpha_3 R_3 - \alpha_3 N_3 \tag{8.64c}$$

Here

$$R_1 = \int_{\rho_1}^{\rho} \pm \rho^2 F^{-\frac{1}{2}}\, \mathrm{d}\rho \qquad = \frac{\partial W_1}{\partial \alpha_1} \tag{8.65a}$$

$$R_2 = \int_{\rho_1}^{\rho} \pm F^{-\frac{1}{2}}\, \mathrm{d}\rho \qquad = -\frac{1}{\alpha_2}\frac{\partial W_1}{\partial \alpha_2} \tag{8.65b}$$

$$R_3 = \int_{\rho_1}^{\rho} \pm (\rho^2 + c^2)^{-1} F^{-\frac{1}{2}}\, \mathrm{d}\rho \qquad = \frac{1}{c^2\alpha_3}\frac{\partial W_1}{\partial \alpha_3} \tag{8.65c}$$

$$N_1 = \int_{0}^{\eta} \pm \eta^2 G^{-\frac{1}{2}}\, \mathrm{d}\eta \qquad = \frac{1}{c^2}\frac{\partial W_2}{\partial \alpha_1} \tag{8.66a}$$

$$N_2 = \int_{0}^{\eta} \pm G^{-\frac{1}{2}}\, \mathrm{d}\eta \qquad = \frac{1}{\alpha_2}\frac{\partial W_2}{\partial \alpha_2} \tag{8.66b}$$

$$N_3 = \int_{0}^{\eta} \pm (1-\eta^2)^{-1} G(\eta)^{-\frac{1}{2}}\, \mathrm{d}\eta \qquad = -\frac{1}{\alpha_3}\frac{\partial W_2}{\partial \alpha_3} \tag{8.66c}$$

The next steps are to evaluate these six integrals for the R's and N's and then to invert Eqs. (8.64) to find ρ, η, and ϕ as functions of time. Evaluating the integrals requires factoring the functions $F(\rho)$ and $G(\eta)$, and this requires a discussion of possible mean orbital elements. In turn this requires a discussion of initial conditions.

VIII. Orbital Elements

The constant α_1 is the energy per unit mass, with $\alpha_1 < 0$ for a bounded orbit; α_3 is the polar component of angular momentum; and α_2 is a constant closely related to the total angular momentum. It is not exactly equal to it because the latter is not conserved in the noncentral field that we are dealing with. If the subscript i denotes an initial value and u is the speed,

$$\alpha_1 = \frac{1}{2}u_i^2 - \frac{\mu\rho_i}{\rho_i^2 + c^2\eta_i^2} \tag{8.67a}$$

$$\alpha_3 = r_i^2 \cos^2\theta_i \dot{\phi}_i = x_i\dot{y}_i - y_i\dot{x}_i \tag{8.67b}$$

using Eqs. (8.51) and (8.15), respectively.

For α_2, use Eqs. (8.52) and (8.54) and the fact that $\mathrm{d}W_2/\mathrm{d}\eta = p_\eta = h_2^2\dot{\eta}$, where h_2^2 is given by Eq. (8.4b). The result is

$$\alpha_2^2 = -2c^2\eta_i^2\alpha_1 + \left(1-\eta_i^2\right)^{-1}\left[\left(\rho_i^2 + c^2\eta_i^2\right)^2\dot{\eta}_i^2 + \alpha_3^2\right] \tag{8.67c}$$

Thus, a knowledge of the initial coordinates and their initial derivatives (see Appendix A for transformation from the xyz to $\rho\eta\zeta$ system) would provide an estimate of the α's and the orbital elements a_0, e_0, and i_0. By using Keplerian relations,

$$a_0 = -\frac{\mu}{2\alpha_1} \qquad e_0^2 = 1 + \frac{2\alpha_1\alpha_2^2}{\mu^2} \qquad \cos i_0 = \frac{\alpha_3}{\alpha_2} \tag{8.68}$$

If we then define a corresponding semi-latus rectum by

$$p_0 = a_0\left(1 - e_0^2\right) \tag{8.69}$$

we have

$$\alpha_2^2 = \mu p_0 \tag{8.70}$$

The external values ρ_1 and ρ_2 of ρ will then be approximately equal to r_1 and r_2, where

$$r_1 = a_0(1 - e_0) \qquad r_2 = a_0(1 + e_0) \tag{8.71}$$

If one can evaluate the integrals (8.65) and (8.66) in terms of a_0, e_0, and i_0, one can then find the β's by means of Eqs. (8.64) and the initial conditions.

Actually, a knowledge of a_0, e_0, and i_0 does not lead directly to the factoring of $F(\rho)$ that is necessary to evaluate the integrals. At this point, we have to consider the factoring, and this will lead us to another set of orbital elements introduced by Ref. 6.

IX. Factoring the Quartics

If ρ_1 and ρ_2 are the extremal values of ρ actually reached, we need to factor $F(\rho)$ into

$$F(\rho) = -2\alpha_1(\rho - \rho_1)(\rho_2 - \rho)(\rho^2 + A\rho + B) \tag{8.72}$$

where Eq. (8.60a) specifies $F(\rho)$. Expressing Eq. (8.72) as quartic in ρ and comparing it with Eq. (8.60a), we obtain four equations by equating coefficients of ρ^k, $k = 0, 1, 2, 3$. These simultaneous equations express A, B, $\rho_1 + \rho_2$, and $\rho_1\rho_2$ in terms of a_0, e_0, i_0, and c. For convenience, we also bring in $p_0 = a_0(1 - e_0^2)$.

These equations can be solved by successive approximations or by expansion in powers of

$$k_0 = c^2/p_0^2 \tag{8.73}$$

The solution is given in Ref. 2, in terms of

$$x = \left(1 - e_0^2\right)^{\frac{1}{2}} \qquad y = \cos i_0 \tag{8.74}$$

Solved for are A, B, $\rho_1 + \rho_2$, and $\rho_1\rho_2$. From these follow

$$a = \frac{1}{2}(\rho_1 + \rho_2) \qquad e = \frac{\rho_2 - \rho_1}{\rho_2 + \rho_1} \qquad p = a(1 - e^2) \tag{8.74a}$$

in terms of x and y. The a and the e thus introduced are part of another set of orbital elements that is the set actually used, a set directly related to the factoring.

Factoring $G(\eta)$ is easier, because it is a quadratic in η^2. We have

$$G(\eta) = -\alpha_3^2 + (1 - \eta^2)\left(\alpha_2^2 + 2\alpha_1 c^2\eta^2\right) \tag{8.60b}$$

If we write it as $G(\eta) = -2\alpha_1 c^2(\eta_0^2 - \eta^2)(\eta_2^2 - \eta^2)$, the solution for η_0 and η_2

involves the difference of two almost equal quantities. It is better to write it as

$$G(\eta) = \left(\alpha_2^2 - \alpha_3^2\right)\eta^4\left(\eta^{-2} - \eta_0^{-2}\right)\left(\eta^{-2} - \eta_2^{-2}\right) \tag{8.75}$$

Comparison of Eqs. (8.75) and (8.60b) shows that η_0^{-2} are the roots of

$$\left(\alpha_2^2 - \alpha_3^2\right)\eta^{-4} + \left(2\alpha_1 c^2 - \alpha_3^2\right)\eta^{-2} - 2\alpha_1 c^2 = 0 \tag{8.76}$$

These are

$$\left(\eta_0^{-2}, \eta_2^{-2}\right) = \tfrac{1}{2}\left(\alpha_2^2 - 2\alpha_1 c^2\right)\left(\alpha_2^2 - \alpha_3^2\right)^{-1}\left(1 \pm Q^{\frac{1}{2}}\right) \tag{8.77a}$$

$$Q \equiv \left(1 + 8\alpha_1 c^2\right)\left(\alpha_2^2 - \alpha_3^2\right)\left(\alpha_2^2 - 2\alpha_1 c^2\right)^{-2} \tag{8.77b}$$

From these equations it follows that for $\alpha_1 < 0$

$$\eta_0^2 \le \frac{\alpha_2^2 - \alpha_3^2}{\alpha_2^2} \le 1 \tag{8.78}$$

(Note that the eight constants A, B, $\rho_1 + \rho_2$, $\rho_1\rho_2$, $\pm\eta_0$, $\pm\eta_2$ are computed based on the initial set of α's.)

Instead of a_0, e_0, and i_0, it is more convenient to use a, e, and η_0 in setting up the theory. Reference 2 gives the connections in detail and permits one to derive either set from the other. The β's are the same in either case.

We shall write

$$\eta_0 = \sin I \tag{8.79}$$

as the definition of I. The constants A and B are given approximately by

$$\begin{aligned} A &\approx -2k_0 p_0 \cos^2 i_0 \approx -2kp\cos^2 I \\ B &\approx k_0 p_0^2 \sin^2 i_0 \approx kp^2 \sin^2 I \end{aligned} \tag{8.80}$$

where

$$\begin{aligned} k_0 &= c^2/p_0^2 = r_e^2 J_2/p_0^2 \\ k &= c^2/p^2 = r_e^2 J_2/p^2 \end{aligned} \tag{8.81}$$

Thus, A and B are both of order J_2, with $A < 0$ and $B > 0$.

X. The ρ Integrals

Refer back to Eqs. (8.65) and (8.72). From Eq. (8.72)

$$F(\rho)^{-\frac{1}{2}} = (-2\alpha_1)^{-\frac{1}{2}}[(\rho - \rho_1)(\rho_2 - \rho)]^{-\frac{1}{2}}\rho^{-1}\left(1 + \frac{A}{\rho} + \frac{B}{\rho^2}\right)^{-\frac{1}{2}} \tag{8.82}$$

The parentheses in A and B distinguish the present problem from the Kepler problem. To handle it, we define b_1 and b_2 by

$$A = -2b_1 \tag{8.83a}$$

$$B = b_2^2 \tag{8.83b}$$

Then $b_1 > 0$, and b_1 and b_2^2 (or A and B) are both of order J_2. Let us also define

$$\lambda \equiv b_1/b_2 \tag{8.84a}$$

$$h \equiv b_2/\rho \tag{8.84b}$$

Then

$$1 + \frac{A}{\rho} + \frac{B}{\rho^2} = 1 - \frac{2b_1}{\rho} + \frac{b_2^2}{\rho^2} = 1 - 2\lambda h + h^2 \tag{8.85}$$

so that

$$\left(1 + \frac{A}{\rho} + \frac{B}{\rho^2}\right)^{-\frac{1}{2}} = (1 - 2\lambda h + h^2)^{-\frac{1}{2}} = \sum_{n=0}^{\infty} h^n P_n(\lambda) \tag{8.86}$$

provided that the Legendre expansion is valid. From Eqs. (8.80), (8.81), and (8.83), we find

$$b_1 = kp\cos^2 I$$

$$b_2 = k^{\frac{1}{2}} p \sin I \tag{8.87}$$

Reference 2 used the conditions $|h| < 1$ and $|\lambda| < 1$ to put limits on the inclination I. These limits are not correct, however. If one uses the condition

$$|h| < \text{smaller of} \left|\lambda \pm \sqrt{\lambda^2 - 1}\right| \tag{8.88}$$

one can prove that the Legendre expansion is valid for all inclinations, provided that $J_2 < 0.17$ (see Ref. 7). This restriction is easily satisfied for the Earth, for which $J_2 = (1.08263) \times 10^{-3}$.

Thus, we use

$$\left(1 + \frac{A}{\rho} + \frac{B}{\rho^2}\right)^{-\frac{1}{2}} = \sum_{n=0}^{\infty} \left(\frac{b_2}{\rho}\right)^n P_n(b_1/b_2) \tag{8.89}$$

$$= 1 + \frac{b_1}{\rho} + \sum_{n=2}^{\infty} \left(\frac{b_2}{\rho}\right)^n P_n\left(\frac{b_1}{b_2}\right) \tag{8.90}$$

From Eqs. (8.82) and (8.86)

$$F(\rho)^{-\frac{1}{2}} = (-2\alpha_1)^{-\frac{1}{2}} \sum_{n=0}^{\infty} b_2^n \rho^{-1-n} [(\rho - \rho_1)(\rho_2 - \rho)]^{-\frac{1}{2}} P_n(\lambda) \tag{8.91}$$

We now have to insert Eq. (8.91) into Eqs. (8.65) to work out the ρ integrals. To get rid of the double signs in those integrals, we introduce uniformizing variables E and v, defined by

$$\rho = a(1 - e\cos E) = \frac{a(1 - e^2)}{1 + e\cos v} \tag{8.92}$$

where $\dot{E} > 0$ and $\dot{v} > 0$ for all t. Here E and v are analogous to the eccentric and true anomalies in Keplerian motion. Exactly as in that case,

$$\pm[(\rho-\rho_1)(\rho_2-\rho)]^{-\frac{1}{2}}\,\mathrm{d}\rho = \mathrm{d}E = (1-e^2)^{\frac{1}{2}}(1+e\cos v)^{-1}\,\mathrm{d}v \tag{8.93}$$

Insert Eqs. (8.91) and (8.93) into Eqs. (8.65a) and (8.65b). The results are

$$(-2\alpha_1)^{-\frac{1}{2}}R_1 = b_1E + a(E - e\sin E) + (1-e^2)^{\frac{1}{2}}\,p\sum_{n=2}^{\infty}\left(\frac{b_2}{p}\right)^n \times P_n(\lambda)\int_0^v (1+e\cos v)^{n-2}\,\mathrm{d}v \tag{8.94}$$

$$(-2\alpha_1)^{-\frac{1}{2}}R_2 = (1-e^2)^{\frac{1}{2}}p^{-1}\sum_{n=0}^{\infty}\left(\frac{b_2}{p}\right)^n P_n(\lambda)\int_0^v (1+e\cos v)^n\,\mathrm{d}v \tag{8.95}$$

In the limit $J_2 = 0$, the right sides become $a(E - e\sin E)$ and $(1-e^2)^{1/2}\,v/p$, in agreement with Chapter 6.

It is desirable to resolve each result into a secular part proportional to v and a periodic part. To do so, first define

$$f_m(v) = \int_0^v (1+e\cos v)^m\,\mathrm{d}v \tag{8.96}$$

Then $f_m(v) - vf_m(2\pi)/2\pi$ is an odd function of v, of period 2π. However, $f_m(2\pi) = 2f_m(\pi)$, so that

$$f_m(v) = \int_0^v (1+e\cos v)^m\,\mathrm{d}v = \frac{v}{\pi}\int_0^\pi (1+e\cos v)^m\,\mathrm{d}v + \sum_{j=1}^{m} c_{mj}\sin jv \tag{8.97}$$

the periodic part of odd and of period 2π, so that its Fourier expansion contains only terms in $\sin jv$. Also, it is a finite trigonometric polynomial, obtainable as follows: 1) expand $(1+e\cos v)^m$ by the binomial theorem, 2) reject the constant term, and 3) integrate the remaining periodic terms. To obtain a useful form for the secular term, use

$$\int_0^\pi \left(z + \sqrt{z^2-1}\cos v\right)^m \mathrm{d}v = \pi P_m(z) \tag{8.98}$$

(see Ref. 8). In Eq. (8.98) place $z = (1-e^2)^{-1/2}$. Then

$$\int_0^\pi (1+e\cos v)^m\,\mathrm{d}v = \pi(1-e^2)^{\frac{m}{2}}P_m\left[(1-e^2)^{-\frac{1}{2}}\right] = \pi R_m\left(\sqrt{1-e^2}\right) \tag{8.99}$$

where

$$R_m(x) = x^m P_m(1/x) \qquad (0 \le x \le 1) \tag{8.100}$$

a polynomial of degree $[m/2]$ in x^2. Then

$$\int_0^v (1+e\cos v)^m\,\mathrm{d}v = vR_m\left(\sqrt{1-e^2}\right) + \sum_{j=1}^{m} c_{mj}\sin jv \tag{8.101}$$

Thus, R_1 and R_2 are given by

$$R_1 = (-2\alpha_1)^{-\frac{1}{2}}\left[b_1 E + a(E - e\sin E) + vA_1 + \sum_{j=1}^{2} A_{1j}\sin jv\right] \tag{8.102}$$

$$R_2 = (-2\alpha_1)^{-\frac{1}{2}}\left[vA_2 + \sum_{j=1}^{4} A_{2j}\sin jv\right] \tag{8.103}$$

To find R_3, calculate $F^{-1/2}\,\mathrm{d}\rho$ for R_2. From Eq. (8.65c), this has to be multiplied by

$$(\rho^2 + c^2)^{-1} = \rho^{-2}\sum_{j=0}^{\infty}(-1)^j c^{2j}\rho^{-2j} \tag{8.104}$$

On integration, the result is

$$R_3 = (-2\alpha_1)^{-\frac{1}{2}}(1-e^2)^{\frac{1}{2}}p^{-3}\int_0^v \sum_{m=0}^{\infty} D_m(1+e\cos v)^{m+2}\,\mathrm{d}v \tag{8.105}$$

where

$$D_m = \sum_{j=0}^{\infty} d_j\delta_n \tag{8.106}$$

summed over all those nonnegative values of j and n for which

$$2j + n = m \tag{8.107a}$$

and where

$$d_j = (-1)^j(c/p)^{2j} \qquad \delta_n = (b_2/p)^n P_n(\lambda) \tag{8.107b}$$

Then

$$R_3 = (-2\alpha_1)^{-\frac{1}{2}}\left[vA_3 + \sum_{j=1}^{4} A_{3j}\sin jv\right] \tag{8.108}$$

The secular coefficients A_1, A_2, and A_3 and the periodic coefficients A_{1j}, A_{2j}, and A_{3j} of R_1, R_2, and R_3 are listed in the following summary.

Summary: The ρ integrals R_1, R_2, and R_3, which can be computed from Eqs. (8.102), (8.103), and (8.108) are expressed in terms of analytic coefficients. After the factorization process of Sec. IX, the set of orbital elements a, e, $\sin I$, and p and the constants A and B are known. The variables x, b_1, b_2, and λ can also be evaluated, which in turn give the Legendre polynomials $P_n(\lambda)$ and the functions $R_n(x)$. The exact expressions correct through order J_2^2 for the secular coefficients A_1, A_2, and A_3 and the periodic coefficients A_{1j}, A_{2j}, and A_{3j} are also listed as follows:

$$R_1 = (-2\alpha_1)^{-\frac{1}{2}}\left[b_1 E + a(E - e\sin E) + vA_1 + \sum_{j=1}^{2} A_{1j}\sin jv\right]$$

$$R_2 = (-2\alpha_1)^{-\frac{1}{2}}\left[vA_2 + \sum_{j=1}^{4} A_{2j}\sin jv\right]$$

$$R_3 = (-2\alpha_1)^{-\frac{1}{2}}\left[vA_3 + \sum_{j=1}^{4} A_{3j}\sin jv\right]$$

$$A_1 = xp\sum_{n=2}^{\infty}\left(\frac{b_2}{p}\right)^n P_n(\lambda)R_{n-2}(x)$$

$$A_2 = xp^{-1}\sum_{n=0}^{\infty}\left(\frac{b_2}{p}\right)^n P_n(\lambda)R_n(x)$$

$$A_3 = xp^{-3}\sum_{m=0}^{\infty} D_m R_{m+2}(x)$$

$$A_{11} = \tfrac{3}{4}exp^{-3}\left(-2b_1b_2^2p + b_2^4\right)$$

$$A_{12} = (3/32)e^2xp^{-3}b_2^4$$

$$A_{21} = exp^{-1}\left[\frac{b_1}{p} + \frac{3b_1^2 - b_2^2}{p^2} - \frac{9}{2p^3}b_1b_2^2\left(1 + \frac{e^2}{4}\right) + \frac{3}{8p^4}b_2^4(4 + 3e^2)\right]$$

$$A_{22} = e^2xp^{-1}\left[\frac{\left(3b_1^2 - b_2^2\right)}{8p^2} - \frac{9b_1b_2^2}{8p^3} + \frac{3b_2^4(6 + e^2)}{32p^4}\right]$$

$$A_{23} = \frac{e^3}{8}xp^{-1}\left[-\frac{b_1b_2^2}{p^3} + \frac{b_2^4}{p^4}\right]$$

$$A_{24} = \frac{3e^4}{256}xp^{-5}b_2^4$$

$$A_{31} = exp^{-3}\left[2 + \frac{3b_1}{p}\left(1 + \frac{e^2}{4}\right) - \frac{b_2^2 + 2c^2}{2p^2}(4 + 3e^2)\right]$$

$$A_{32} = e^2xp^{-3}\left[\frac{1}{4} + \frac{3b_1}{4p} - \frac{b_2^2 + 2c^2}{8p^2}(e^2 + 6)\right]$$

$$A_{33} = e^3xp^{-3}\left[\frac{b_1}{12p} - \frac{b_2^2 + 2c^2}{6p^2}\right]$$

$$A_{34} = -\frac{1}{64}e^4xp^{-5}\left(b_2^2 + 2c^2\right)$$

where

$$c^2 = r_e^2 J_2 \qquad x = (1-e^2)^{\frac{1}{2}}$$

$$b_1 = -\frac{A}{2} \qquad b_2 = \sqrt{B} \qquad \lambda = \frac{b_1}{b_2} \qquad (A < 0,\, B > 0)$$

$$R_m(x) = x^m P_m(1/x) \qquad (0 \leq x \leq 1)$$

$$D_m = D_{2i} = \sum_{n=0}^{i} (-1)^{i-n} (c/p)^{2i-2n} (b_2/p)^{2n} P_{2n}(\lambda) \qquad (m \text{ is even})$$

$$D_m = D_{2i+1} = \sum_{n=0}^{i} (-1)^{i-n} (c/p)^{2i-2n} (b_2/p)^{2n+1} P_{2n+1}(\lambda) \qquad (m \text{ is odd})$$

$$P_n(\lambda) = \sum_{k=0}^{n/2} \frac{(-1)^k (2n-2k)!\lambda^{n-2k}}{2^n k!(n-2k)!(n-k)!} \qquad \text{Eq. (13.71)}$$

$$P_0(\lambda) = 1$$

$$P_1(\lambda) = \lambda$$

$$P_2(\lambda) = \tfrac{1}{2}(3\lambda^2 - 1)$$

$$P_3(\lambda) = \tfrac{1}{2}(5\lambda^3 - 3\lambda)$$

$$P_4(\lambda) = \tfrac{1}{8}(35\lambda^4 - 30\lambda^2 + 3)$$

XI. The η Integrals

Refer back to Eqs. (8.66) and (8.75). Put

$$\eta = \eta_0 \sin\psi \tag{8.109}$$

where $\dot{\psi}$ is to be positive for all t. Then ψ is analogous to the argument of latitude, since $\eta_0 = \sin I$. (η_0 and η_2 are solutions of factorization.) We obtain

$$\pm G(\eta)^{-\frac{1}{2}}\, \mathrm{d}\eta = \left(\alpha_2^2 - \alpha_3^2\right)^{-\frac{1}{2}} \eta_0 (1 - q^2 \sin^2\psi)^{-\frac{1}{2}}\, \mathrm{d}\psi \tag{8.110}$$

where

$$q^2 (\eta_0/\eta_2)^2 \tag{8.111}$$

of order J_2. We find

$$N_1 = \left(\alpha_2^2 - \alpha_3^2\right)^{-\frac{1}{2}} \eta_0^3 q^{-2} [F(\psi, q) - E(\psi, q)] \tag{8.112}$$

$$N_2 = \left(\alpha_2^2 - \alpha_3^2\right)^{-\frac{1}{2}} \eta_0 F(\psi, q) \tag{8.113}$$

where

$$F(\psi, q) \equiv \int_0^{\psi} (1 - q^2 \sin^2\psi)^{-\frac{1}{2}}\, \mathrm{d}\psi \tag{8.114}$$

$$E(\psi, q) \equiv \int_0^{\psi} (1 - q^2 \sin^2\psi)^{\frac{1}{2}}\, \mathrm{d}\psi \tag{8.115}$$

These functions are, respectively, the incomplete elliptic integrals of the first and second kinds.

We next resolve N_1 and N_2 into secular plus periodic terms. To do so, note that

$$F(\psi + \pi, q) = F(\psi, q) + 2K(q) \tag{8.116}$$

$$K(q) = \int_0^{\pi/2} (1 - q^2 \sin^2 x)^{-\frac{1}{2}}\, \mathrm{d}x \tag{8.117}$$

where $K(q)$ is the complete elliptic integral of the first kind. One readily shows that $F(\psi, q) - (2/\pi)K(q)\psi$ is an odd function of ψ, periodic in ψ with period π. Thus

$$F(\psi, q) = \frac{2}{\pi} K(q)\psi + \sum_{m=1}^{\infty} F_{qm} \sin 2m\psi \tag{8.118}$$

Differentiation of Eq. (8.118) gives

$$(1 - q^2 \sin^2 \psi)^{-\frac{1}{2}} = \frac{2}{\pi} K(q) + 2 \sum_{m=1}^{\infty} m F_{qm} \cos 2m\psi \tag{8.119}$$

The Fourier coefficients F_{qm} are given by

$$F_{qm} = \frac{2}{\pi m} \int_0^{\pi/2} (1 - q^2 \sin^2 x)^{-\frac{1}{2}} \cos 2mx\, \mathrm{d}x \tag{8.120}$$

Expand $(1 - q^2 \sin^2 x)^{-1/2}$ by the binomial theorem

$$(1 - q^2 \sin^2 x)^{-\frac{1}{2}} = 1 + \sum_{n=1}^{\infty} \frac{(2n)! q^{2n} \sin^{2n} x}{2^{2n}(n!)^2} \tag{8.121}$$

Then

$$F_{qm} = \frac{2}{\pi m} \sum_{n=1}^{\infty} \frac{(2n)! q^{2n}}{2^{2n}(n!)^2} \int_0^{\pi/2} \sin^{2n} x \cos 2mx\, \mathrm{d}x \tag{8.122}$$

Express $\sin^{2n} x$ as a trigonometric polynomial. To do so, write $\sin x$ as $(\varepsilon^{ix} - \varepsilon^{-ix})/(2i)$ and expand $\sin^{2n} x$ by the binomial theorem as a sum from $j = 0$ to $j = 2n$. The term $j = n$ will give a constant term. Then group together the terms $j = 0$ to $n - 1$ and the terms $j = n + 1$ to $2n$ to yield cosines.

The result is

$$\sin^{2n} x = \frac{(2n)!}{2^{2n}(n!)^2} + (-1)^n (2)^{1-2n} \sum_{j=0}^{n-1} \frac{(-1)^j (2n)!}{(2n - j)!(j)!} \cos (2n - 2j)x \tag{8.123}$$

Insertion of Eq. (8.123) into Eq. (8.122) gives

$$F_{qm} = (-1)^m m^{-1} \sum_{n=m}^{\infty} \frac{[(2n)!]^2 q^{2n}}{2^{4n}(n + m)!(n - m)!(n!)^2} \tag{8.124}$$

Through order J_2^2 the coefficients are

$$F_{q1} = -\frac{q^2}{8}\left(1 + \frac{3}{4}q^2\right) + \cdots \qquad F_{q2} = \frac{3q^4}{256} + \cdots$$

Thus

$$F(\psi, q) = \frac{2}{\pi} K(q)\psi - \frac{q^2}{8}\left(1 + \frac{3}{4}q^2\right)\sin 2\psi + \frac{3q^4}{256}\sin 4\psi + \cdots \tag{8.125}$$

Similarly, one finds

$$E(\psi, q) = \frac{2}{\pi} K(q)\psi + \frac{q^2}{8}\left(1 + \frac{1}{4}q^2\right)\sin 2\psi + \frac{q^4}{256}\sin 4\psi + \cdots \tag{8.126}$$

where

$$E(q) \equiv \int_0^{\pi/2} (1 - q^2 \sin^2 x)^{\frac{1}{2}}\, dx$$

is the complete elliptic integral of the second kind.

Placing Eqs. (8.125) and (8.126) into Eqs. (8.112) and (8.113) then yields

$$N_1 = \left(\alpha_2^2 - \alpha_3^2\right)^{-\frac{1}{2}} \eta_0^3 \left[B_1\psi - \left(\frac{2+q^2}{8}\right)\sin 2\psi + \frac{q^2}{64}\sin 4\psi + \cdots \right] \tag{8.127}$$

$$N_2 = \left(\alpha_2^2 - \alpha_3^2\right)^{-\frac{1}{2}} \eta_0 \left[B_2\psi - \frac{q^2}{32}(4 + 3q^2)\sin 2\psi + \frac{3q^4}{256}\sin 4\psi + \cdots \right] \tag{8.128}$$

$$B_1 = \frac{2q^{-2}}{\pi}[K(q) - E(q)] = \frac{1}{2} + \frac{3q^2}{16} + \frac{15q^4}{128} + \frac{175q^6}{2048} + \cdots \tag{8.129}$$

$$B_2 = \frac{2}{\pi}K(q) = 1 + \frac{q^2}{4} + \frac{9q^4}{64} + \frac{25q^6}{256} + \cdots \tag{8.130}$$

so that the terms in ψ are exact. In N_2 the periodic terms are correct through order J_2^2, while in N_1 they are correct only through order J_2. This is all the accuracy needed, however, because N_1 is multiplied by $c^2 = r_e^2 J_2$ in the first kinematic equation.

The Integral N_3

From Eqs. (8.66c) and (8.75)

$$N_3 = \left(\alpha_2^2 - \alpha_3^2\right)^{-\frac{1}{2}} \int_0^{\eta} \pm(1 - \eta^2)^{-1}\left(1 - \eta^2/\eta_0^2\right)^{-\frac{1}{2}}\left(1 - \eta^2/\eta_2^2\right)^{-\frac{1}{2}} d\eta \tag{8.131}$$

Insert the binomial expansion

$$\left(1 - \eta^2/\eta_2^2\right)^{-\frac{1}{2}} = \sum_{m=0}^{\infty} \frac{(2m)!}{2^{2m}(m!)^2}(\eta/\eta_2)^{2m} \tag{8.132}$$

into Eq. (8.131) to find

$$N_3 = \left(\alpha_2^2 - \alpha_3^2\right)^{-\frac{1}{2}} \sum_{m=0}^{\infty} \frac{(2m)!}{2^{2m}(m!)^2}\, \eta_2^{-2m} L_m \tag{8.133}$$

Here

$$L_m = \int_0^{\eta} \pm(1-\eta^2)^{-1}\left(1-\eta^2/\eta_0^2\right)^{-\frac{1}{2}}\eta^{2m}\,\mathrm{d}\eta \tag{8.133a}$$

$$L_0 = \int_0^{\eta} \pm(1-\eta^2)^{-1}\left(1-\eta^2/\eta_0^2\right)^{-\frac{1}{2}}\,\mathrm{d}\eta \tag{8.133b}$$

As before, use $\eta = \eta_0 \sin\psi$, where $\dot{\psi} > 0$ for all t. Then

$$\pm\left(1-\eta^2/\eta_0^2\right)^{-\frac{1}{2}}\,\mathrm{d}\eta = \eta_0\,\mathrm{d}\psi \tag{8.134a}$$

so that

$$L_0 = \eta_0 \int_0^{\psi} \left(1-\eta_0^2\sin^2\psi\right)^{-1}\mathrm{d}\psi \tag{8.134b}$$

Here $\eta_0 = \sin I$. Now put

$$\tan\chi = |\cos I|\tan\psi \tag{8.135}$$

In the limiting Keplerian case, the new variable χ is then the projection of the argument of latitude ψ on the equator. With use of Eq. (8.135), we find

$$L_0 = \chi|\tan I| \tag{8.136}$$

To evaluate L_m, write the geometric sum

$$\sum_{n=0}^{m-1}\eta^{2n} = \frac{1-\eta^{2m}}{1-\eta^2} \tag{8.137}$$

Then

$$(1-\eta^2)^{-1}\eta^{2m} = (1-\eta^2)^{-1} - \sum_{n=0}^{m-1}\eta^{2n} \tag{8.138}$$

Put this in Eq. (8.133a). Then

$$L_m = L_0 - \sum_{n=0}^{m-1} L_{1n} \qquad (m \geq 1) \tag{8.139}$$

where

$$L_{1n} = \int_0^{\eta} \pm\left(1-\eta^2/\eta_0^2\right)^{-\frac{1}{2}}\eta^{2n}\,\mathrm{d}\eta \tag{8.140}$$

With use of Eq. (8.134a) we find

$$L_{10} = \eta_0\psi \tag{8.141}$$

and

$$L_{1n} = \eta_0^{2n+1}\int_0^{\psi}\sin^{2n}x\,\mathrm{d}x \qquad (n \geq 1) \tag{8.142}$$

It takes some care to see how to enter L_m into Eq. (8.133). From Eq. (8.139) we have for L_m:

$$\begin{aligned} m = 0: &\quad L_0 \\ m = 1: &\quad L_0 - L_{10} \\ m \geq 2: &\quad L_0 - L_{10} - \sum_{n=1}^{m-1} L_{1n} \end{aligned} \tag{8.143}$$

Entering these quantities into Eq. (8.133), we find

$$N_3 = \left(\alpha_2^2 - \alpha_3^2\right)^{-\frac{1}{2}} \left[L_0 \sum_{m=0}^{\infty} \frac{(2m)!}{2^{2m}(m!)^2} \eta_2^{-2m} - L_{10} \sum_{m=1}^{\infty} \frac{(2m)!}{2^{2m}(m!)^2} \eta_2^{-2m} \right.$$
$$\left. - \sum_{m=2}^{\infty} \frac{(2m)!}{2^{2m}(m!)^2} \eta_2^{-2m} \sum_{n=1}^{m-1} L_{1n} \right] \tag{8.144}$$

Now, from the binomial expansion

$$\left(1 - \eta_2^{-2}\right)^{-\frac{1}{2}} = \sum_{m=0}^{\infty} \frac{(2m)!}{2^{2m}(m!)^2} \eta_2^{-2m} \tag{8.145}$$

we find

$$\sum_{m=1}^{\infty} \frac{(2m)!}{2^{2m}(m!)^2} \eta_2^{-2m} = \left(1 - \eta_2^{-2}\right)^{-\frac{1}{2}} - 1 \tag{8.146}$$

Thus

$$N_3 = \left(\alpha_2^2 - \alpha_3^2\right)^{-\frac{1}{2}} \left[L_0\left(1 - \eta_2^2\right)^{-\frac{1}{2}} - \eta_0 \psi \left\{ \left(1 - \eta_2^{-2}\right)^{-\frac{1}{2}} - 1 \right\} \right.$$
$$\left. - \sum_{m=2}^{\infty} \frac{(2m)!}{2^{2m}(m!)^2} \eta_2^{-2m} \sum_{n=1}^{m-1} \eta_0^{2n+1} \int_0^{\psi} \sin^{2n} x \, \mathrm{d}x \right] \tag{8.147}$$

Here we have used Eq. (8.141) for L_{10} and Eq. (8.142) for L_{1n}.

To write down the secular part of the integrals in Eq. (8.147) use the constant part of $\sin^{2n} x$, viz.,

$$\sin^{2n} x = \frac{(2n)!}{2^{2n}(n!)^2} + \cdots$$

as given by Eq. (8.123). The secular part of the integrals in Eq. (8.147) is then

$$-\psi \sum_{m=2}^{\infty} \frac{(2m)!}{2^{2m}(m!)^2} \eta_2^{-2m} \sum_{n=1}^{m-1} \eta_0^{2n+1} \frac{(2n)!}{2^{2n}(n!)^2} = -\eta_0 \psi \sum_{m=2}^{\infty} \gamma_m \eta_2^{-2m} \tag{8.148}$$

where

$$\gamma_m = \frac{(2m)!}{2^{2m}(m!)^2} \sum_{n=1}^{m-1} \frac{(2n)!}{2^{2n}(n!)^2} \eta_0^{2n} \tag{8.149}$$

We shall use only the term in J_2^2 for the periodic part in Eq. (8.147). It is given by placing $m = 2$ in Eq. (8.147); then $n = 1$. It comes from

$$-\frac{4!}{2^4 2^2}\eta_2^{-4}\eta_0^3 \int_0^{\psi} \sin^2 x \, \mathrm{d}x$$

The periodic part of $\sin^2 x$ is $-(\cos 2x)/2$, so that our whole periodic contribution

$$\frac{4!}{2^6}\eta_2^{-4}\eta_0^3 \frac{1}{4} \sin 2\psi = \frac{3}{32}\eta_2^{-4}\eta_0^3 \sin 2\psi \tag{8.150}$$

Putting everything together, we obtain

$$N_3 = \left(\alpha_2^2 - \alpha_3^2\right)^{-\frac{1}{2}} \left[\chi \eta_0 \left(1 - \eta_0^2\right)^{-\frac{1}{2}} \left(1 - \eta_2^{-2}\right)^{-\frac{1}{2}} + B_3 \psi + \frac{3}{32}\eta_2^{-4}\eta_0^3 \sin 2\psi + \cdots \right] \tag{8.151}$$

where

$$B_3 = \eta_0 \left[1 - \left(1 - \eta_2^{-2}\right)^{-\frac{1}{2}} - \sum_{m=2}^{\infty} \gamma_m \eta_2^{-2m} \right] \tag{8.152}$$

The term in χ comes from $(1-\eta_2^{-2})^{-1/2} L_0$, and $L_0 = \chi |\tan I| = \chi \eta_0 (1-\eta_0^2)^{-1/2}$.

Summary for the η Integrals

The η Integrals N_1, N_2, and N_3 can be computed from Eqs. (8.127), (8.128), and (8.151). After the factorization process of Sec. IX, the constants η_0 and η_2 are known. The given initial position and velocity vectors $\boldsymbol{r}$ and $\dot{\boldsymbol{r}}$ at time t_i can be transformed to give the spheriodal state vector $(\rho_i, \eta_i, \phi_i, \dot{\rho}_i, \dot{\eta}_i, \dot{\phi}_i)$ as shown in Appendix A. At time t_i, the variables ψ, q, B_1, B_2, χ, and γ_m can also be evaluated, which in turn give the η integrals N_1, N_2, and N_3. These integrals of the kinematic equations (8.64a), (8.64b), and (8.64c), which provide expressions correct through order J_2^2, are listed as follows:

$$N_1 = \left(\alpha_2^2 - \alpha_3^2\right)^{-\frac{1}{2}} \eta_0^3 \left[B_1 \psi - \left(\frac{2+q^2}{8}\right) \sin 2\psi + \frac{q^2}{64} \sin 4\psi + \cdots \right]$$

$$N_2 = \left(\alpha_2^2 - \alpha_3^2\right)^{-\frac{1}{2}} \eta_0 \left[B_2 \psi - \frac{q^2}{32}(4 + 3q^2) \sin 2\psi + \frac{3q^4}{256} \sin 4\psi + \cdots \right]$$

$$N_3 = \left(\alpha_2^2 - \alpha_3^2\right)^{-\frac{1}{2}} \left[\chi \eta_0 \left(1 - \eta_0^2\right)^{-\frac{1}{2}} \left(1 - \eta_2^{-2}\right)^{-\frac{1}{2}} + B_3 \psi + \frac{3}{32}\eta_0^3 \eta_2^{-4} \sin 2\psi + \cdots \right]$$

where

$$\sin\psi = \frac{\eta_0}{\eta_i} \qquad q^2 = \left(\frac{\eta_0}{\eta_2}\right)^2 \qquad \text{(at time } t_i\text{)}$$

$$B_1 = \frac{2q^{-2}}{\pi}[K(q) - E(q)] = \frac{1}{2} + \frac{3q^2}{16} + \frac{15q^4}{128} + \frac{175q^6}{2048} + \cdots$$

$$B_2 = \frac{2}{\pi}K(q) = 1 + \frac{q^2}{4} + \frac{9q^4}{64} + \frac{25q^6}{256} + \cdots$$

$$B_3 = \eta_0\left[1 - \left(1 - \eta_2^{-2}\right)^{-\frac{1}{2}} - \sum_{m=2}^{\infty}\gamma_m\eta_2^{-2m}\right]$$

$$\gamma_m = \frac{(2m)!}{2^{2m}(m!)^2}\sum_{n=1}^{m-1}\frac{(2n)!}{2^{2n}(n!)^2}\eta_0^{2n}$$

At this point, the Jacobi constants $(\beta_1, \beta_2, \beta_3)$ can be estimated from the *HJ* equations even though the η integrals are computed at time t_i. As indicated in Chapter 6, the *HJ* solution will yield a canonical transformation of the Cartesian q's and p's ($\boldsymbol{r}$ and $\dot{\boldsymbol{r}}$) or the spheriodal coordinate q's and p's $(\rho, \eta, \phi, \dot{\rho}, \dot{\eta}, \dot{\phi})$ to the α's and β's, which are so closely related to the Keplerian elements. Resubstituting the α's and β's back into the kinematic equations, we can solve the perturbed problem by finding the variable Keplerian elements as functions of the given time t. We can write down the solutions at time t for the position vector $\boldsymbol{r}$ and the velocity $\dot{\boldsymbol{r}}$. The first kinematic equation is, of course, a generalized form of the Kepler's equation for the perturbed problem, and we shall deal with that in the following sections.

XII. The Mean Frequencies

We need to know the mean frequencies to check the secular parts that we shall obtain for the anomalies v and E and for ψ, the argument of latitude.

The action variables are

$$j_1 = \oint p_\rho\,\mathrm{d}\rho = 2\int_{\rho_1}^{\rho_2} p_\rho\,\mathrm{d}\rho$$

$$j_2 = \oint p_\eta\,\mathrm{d}\eta = 4\int_0^{\eta_0} p_\eta\,\mathrm{d}\eta \tag{8.153}$$

$$j_3 = \oint p_\phi\,\mathrm{d}\phi = \int_0^{2\pi} p_\phi\,\mathrm{d}\phi = 2\pi\alpha_3$$

The mean frequencies are[9]

$$\nu_\rho = \nu_1 = \frac{\partial\alpha_1}{\partial j_1}$$

$$\nu_\eta = \nu_2 = \frac{\partial\alpha_1}{\partial j_2} \tag{8.154}$$

$$\nu_\phi = \nu_3 = \frac{\partial\alpha_1}{\partial j_3}$$

To compute them, use

$$\sum_{m=1}^{3} \frac{\partial \alpha_1}{\partial j_m} \frac{\partial j_m}{\partial \alpha_n} = \frac{\partial \alpha_1}{\partial \alpha_n} = \delta_{1n} \tag{8.155}$$

Put

$$j_{mn} \equiv \frac{\partial j_m}{\partial \alpha_n} \tag{8.156}$$

Then

$$\begin{aligned} \nu_1 j_{11} + \nu_2 j_{21} &= 1 \\ \nu_1 j_{12} + \nu_2 j_{22} &= 0 \\ \nu_1 j_{13} + \nu_2 j_{23} + 2\pi \nu_3 &= 0 \end{aligned} \tag{8.157}$$

If

$$\Delta \equiv j_{11} j_{22} - j_{12} j_{21} \tag{8.158}$$

the solution of Eq. (8.157) is

$$\begin{aligned} \nu_1 &= j_{22}/\Delta \\ \nu_2 &= -j_{12}/\Delta \\ 2\pi \nu_3 &= -\nu_1 j_{12} - \nu_2 j_{22} \end{aligned} \tag{8.159}$$

From Eqs. (8.62) and (8.153)

$$\begin{aligned} j_1 &= 2 \int_{\rho_1}^{\rho_2} \pm(\rho^2 + c^2)^{-1} F(\rho)^{\frac{1}{2}} \, \mathrm{d}\rho \\ j_2 &= 4 \int_{0}^{\eta_0} \pm(1 - \eta^2)^{-1} G(\eta)^{\frac{1}{2}} \, \mathrm{d}\eta \end{aligned} \tag{8.160}$$

From Eqs. (8.60)

$$\begin{aligned} \frac{\partial F}{\partial \alpha_1} &= 2\rho^2(\rho^2 + c^2) \\ \frac{\partial F}{\partial \alpha_2} &= -2\alpha_2(\rho^2 + c^2) \\ \frac{\partial F}{\partial \alpha_3} &= 2c^2 \alpha_3 \end{aligned} \tag{8.161}$$

$$\begin{aligned} \frac{\partial G}{\partial \alpha_1} &= 2c^2(1 - \eta^2)\eta^2 \\ \frac{\partial G}{\partial \alpha_2} &= 2\alpha_2(1 - \eta^?) \\ \frac{\partial G}{\partial \alpha_3} &= -2\alpha_3 \end{aligned} \tag{8.162}$$

From

$$\dot{j}_{mn} \equiv \frac{\partial j_m}{\partial \alpha_n} \tag{8.156}$$

and Eqs. (8.160)–(8.162), we find

$$\begin{aligned}
j_{11} &= 2\int_{\rho_1}^{\rho_2} \pm\rho^2 F^{-\frac{1}{2}}\,\mathrm{d}\rho = 2R_1(\rho_2)\\
j_{12} &= -2\alpha_2\int_{\rho_1}^{\rho_2} \pm F^{-\frac{1}{2}}\,\mathrm{d}\rho = -2\alpha_2 R_2(\rho_2)\\
j_{13} &= 2c^2\alpha_3\int_{\rho_1}^{\rho_2} \pm(\rho^2+c^2)^{-1}F^{-\frac{1}{2}}\mathrm{d}\rho = 2c^2\alpha_3 R_3(\rho_2)
\end{aligned} \tag{8.163}$$

The right sides come from Eqs. (8.65). For the others we obtain

$$\begin{aligned}
j_{21} &= 4c^2\int_0^{\eta_0} \pm\eta^2 G^{-\frac{1}{2}}\,\mathrm{d}\eta = 4c^2 N_1(\eta_0)\\
j_{22} &= 4\alpha_2\int_0^{\eta_0} \pm G^{-\frac{1}{2}}\,\mathrm{d}\eta = 4\alpha_2 N_2(\eta_0)\\
j_{23} &= -4\alpha_3\int_0^{\eta_0} \pm(1-\eta^2)^{-1}G^{-\frac{1}{2}}\,\mathrm{d}\eta = -4\alpha_3 N_3(\eta_0)
\end{aligned} \tag{8.164}$$

by means of Eqs. (8.66).

To obtain the first three j_{mn}'s, we use Eqs. (8.102), (8.103), and (8.108), putting $E = v = \pi$. To obtain the next three j_{mn}'s, we use Eqs. (8.127) and (8.128) for N_1 and N_2, putting $\psi = \pi/2$. Then in Eq. (8.151) for N_3, we put $\psi = \chi = \pi/2$. The results are

$$\begin{aligned}
j_{11} &= 2\pi(-2\alpha_1)^{-\frac{1}{2}}(a + b_1 + A_1)\\
j_{12} &= -2\pi\alpha_2(-2\alpha_1)^{-\frac{1}{2}}A_2\\
j_{13} &= 2c^2\pi\alpha_3(-2\alpha_1)^{-\frac{1}{2}}A_3\\
j_{21} &= 2\pi c^2\left(\alpha_2^2 - \alpha_3^2\right)^{-\frac{1}{2}}\eta_0^3\, B_1\\
j_{22} &= 2\pi\alpha_2\left(\alpha_2^2 - \alpha_3^2\right)^{-\frac{1}{2}}\eta_0 B_2\\
j_{23} &= -2\pi\alpha_3\left(\alpha_2^2 - \alpha_3^2\right)^{-\frac{1}{2}}\eta_0\left[B_3 + \left(1-\eta_0^2\right)^{-\frac{1}{2}}\left(1-\eta_2^{-2}\right)^{-\frac{1}{2}}\right]
\end{aligned} \tag{8.165}$$

Insert Eqs. (8.165) into Eqs. (8.159) to find ν_1 and ν_2. These mean frequencies are given by

$$\begin{aligned}
2\pi\nu_1 &= (-2\alpha_1)^{-\frac{1}{2}}\left(a + b_1 + A_1 + c^2\eta_0^2 A_2 B_1 B_2^{-1}\right)^{-1}\\
2\pi\nu_2 &= \left(\alpha_2^2 - \alpha_3^2\right)^{-\frac{1}{2}}\eta_0^{-1}A_2 B_2^{-1}\left(a + b_1 + A_1 + c^2\eta_0^2 A_2 B_1 B_2^{-1}\right)^{-1}
\end{aligned} \tag{8.166}$$

These equations will show that $\bar{\dot{E}} = \bar{\dot{v}} = 2\pi\nu_1$ and $\bar{\dot{\psi}} = 2\pi\nu_2$. Because the variables on the right sides of Eqs. (8.166) are known, the mean frequencies can be computed. From Ref. 2, the mean frequencies can be approximated by

$$2\pi \nu_1 = n_0 + O\left(J_2^2\right)$$

$$2\pi \nu_2 = n_0 + [1 + 3J_2(5\cos i_0 - 1)] + O\left(J_2^2\right)$$

where the Keplerian mean motion n_0 is given by $\mu = n_0^2\, a_0^3$ and $a_0 = -\mu/(2\alpha_1)$.

XIII. Assembly of the Kinematic Equations

We gather together the results that express $t + \beta_1$ and β_2 as functions of the eccentric anomaly E, the true anomaly v, the argument of latitude ψ, constants depending on the orbital elements a, e, $\eta_0 = \sin I$, $p = a(1 - e^2)$, and $c^2 = r_e^2 J_2$. For details see Ref. 2.

Arranged according to their order in J_2, these constants are

$$J_2^0\colon \alpha_1, \alpha_2, \alpha_3, A_2, B_1, B_2, p$$

$$J_2\colon c^2, A_1, q^2, A_{21}, A_{22}, b_1, b_2^2, A, B$$

$$J_2^2\colon A_{11}, A_{12}, A_{23}, A_{24}$$

The equations are

$$t + \beta_1 = (-2\alpha_1)^{-\frac{1}{2}}\left[b_1 E + a(E - e\sin E) + vA_1 + A_{11}\sin v + A_{12}\sin 2v\right]$$
$$+ c^2\left(\alpha_2^2 - \alpha_3^2\right)^{-\frac{1}{2}}\eta_0^3\left[B_1\psi - \left(\frac{2+q^2}{8}\right)\sin 2\psi + \frac{q^2}{64}\sin 4\psi\right]$$
$$+ \text{periodic terms of order } J_2^3 \tag{8.167a}$$

$$\beta_2 = -\alpha_2(-2\alpha_1)^{-\frac{1}{2}}[vA_2 + A_{21}\sin v + A_{22}\sin 2v + A_{23}\sin 3v + A_{24}\sin 4v]$$
$$+ \alpha_2\left(\alpha_2^2 - \alpha_3^2\right)^{-\frac{1}{2}}\eta_0\left[B_2\psi - \frac{q^2}{32}(4 + 3q^2)\sin 2\psi + \frac{3q^4}{256}\sin 4\psi\right]$$
$$+ \text{periodic terms of order } J_2^3 \tag{8.167b}$$

Here $\rho = a(1 - e\cos E) = a(1 - e^2)/(1 + e\cos v)$, $\eta = \eta_0 \sin\psi$.

XIV. Solution of the Kinematic Equations

Before solving the kinematic equations (8.167), it is convenient to have several relations connecting the uniformizing variables E and v. From Chapter 2, Sec. V, we obtain

$$\cos v = \frac{\cos E - e}{1 - e\cos E}$$

$$\sin v = \frac{\sqrt{1 - e^2}\sin E}{1 - e\cos E}$$

The requirements that $\mathrm{d}v/\mathrm{d}t > 0$, $\mathrm{d}E/\mathrm{d}t > 0$ for all t lead to the result that $\mathrm{d}v/\mathrm{d}E > 0$ for all t. Because of this result, the $\sin v$ equation has no ambiguity in sign. For a given value of the eccentric anomaly E, the preceding relations determine the true anomaly υ completely. The three unknowns (E, υ, ψ) of Eqs. (8.167) essentially reduce to two. We assume that the Jacobi constants β_1 and β_2 can be estimated from the application of the initial conditions as discussed in the last paragraph of Sec. XI. Theoretically, we can solve the two equations of (8.167) for the two unknowns (E, ψ) or (υ, ψ), since all the other parameters in Eqs. (8.167) are known.

To solve Eqs. (8.167), place

$$E = E_s + E_p \qquad \upsilon = \upsilon_s + \upsilon_p \qquad \psi = \psi_s + \psi_p \tag{8.168}$$

Here the subscript s means "secular" and the subscript p means "periodic." If ρ goes through N_1 cycles in time T_1 and if η goes through N_2 cycles in time T_2, we have[9]

$$\bar{\dot{E}} = \bar{\dot{\upsilon}} = \dot{E}_s = \dot{\upsilon}_s = \lim_{T_1 \to \infty} \frac{2\pi N_1}{T_1} = 2\pi \nu_1 \tag{8.169a}$$

$$\bar{\dot{\psi}} = \dot{\psi}_s = \lim_{T_2 \to \infty} \frac{2\pi N_2}{T_2} = 2\pi \nu_2 \tag{8.169b}$$

Because we have already obtained exact expressions for ν_1 and ν_2, it is clear that we can obtain the secular terms exactly for the assumed potential. We shall also obtain the periodic terms through order J_2^2.

By Eqs. (8.169) we can write

$$E_s = \upsilon_s = M_s \tag{8.170}$$

where M_s is the secular part of the mean anomaly. Then

$$E = M_s + E_p \qquad \upsilon = M_s + \upsilon_p \qquad \psi = \psi_s + \psi_p \tag{8.171}$$

We may obtain the secular solution of Eqs. (8.167) independently of Sec. XII by dropping all the sines in these equations, placing $E = \upsilon = M_s$, $\psi = \psi_s$, and solving the resulting equations for M_s and ψ_s. These resulting equations are

$$(-2\alpha_1)^{-\frac{1}{2}}(a + b_1 + A_1)M_s + c^2\left(\alpha_2^2 - \alpha_3^2\right)^{-\frac{1}{2}}\eta_0^3 B_1 \psi_s = t + \beta_1 \tag{8.172a}$$

$$-\alpha_2(-2\alpha_1)^{-\frac{1}{2}} A_2 M_s + \alpha_2\left(\alpha_2^2 - \alpha_3^2\right)^{-\frac{1}{2}}\eta_0 B_2 \psi_s = \beta_2 \tag{8.172b}$$

giving

$$M_s = (-2\alpha_1)^{\frac{1}{2}} \frac{B_2(t + \beta_1) - c^2\eta_0^2 B_1 \alpha_2^{-1}\beta_2}{(a + b_1 + A_1)B_2 + c^2\eta_0^2 A_2 B_1} \tag{8.173a}$$

$$\psi_s = \left(\alpha_2^2 - \alpha_3^2\right)^{\frac{1}{2}} \eta_0^{-1} \frac{A_2(t + \beta_1) + (a + b_1 + A_1)\alpha_2^{-1}\beta_2}{(a + b_1 + A_1)B_2 + c^2\eta_0^2 A_2 B_1} \tag{8.173b}$$

Comparison with Eqs. (8.166) verifies that $\dot{M}_s = 2\pi\nu_1, \dot{\psi}_s = 2\pi\nu_2$, as expected. We can rewrite these equations as

$$M_s = 2\pi\nu_1\left[t + \beta_1 - c^2\eta_0^2\alpha_2^{-1}\beta_2 B_1 B_2^{-1}\right] \tag{8.174a}$$

$$\psi_s = 2\pi\nu_2\left[t + \beta_1 + (a + b_1 + A_1)\alpha_2^{-1}\beta_2 A_2^{-1}\right] \tag{8.174b}$$

If one traces through the constants, one finds that $\psi_s = M_s + \beta_2 + O(J_2)$, as expected, with β_2 replacing ω.

XV. The Periodic Terms

To solve the assembled equations (8.167), we put, successively,

$$E_p = E_0 \qquad v_p = v_0 \qquad \psi_p = \psi_0 \qquad \text{(Step 0)}$$

$$E_p = E_0 + E_1 \qquad v_p = v_0 + v_1 \qquad \psi_p = \psi_0 + \psi_1 \qquad \text{(Step 1)}$$

$$E_p = E_0 + E_1 + E_2 \qquad v_p = v_0 + v_1 + v_2 \qquad \psi_p = \psi_0 + \psi_1 + \psi_2 \qquad \text{(Step 2)}$$

In step 0, we retain in Eqs. (8.167) only the periodic term of order J_2^0, viz., $\sin E$. In step 1, we retain all periodic terms of orders J_2^0 and J_2, but none of higher order. In step 2, we retain all periodic terms through order J_2^2, but none higher. In carrying out each step, however, we shall suppose that each quantity involved is calculated to such an accuracy that the error is of order J_2^3.

Step 0

On placing $E = M_s + E_0$, $v = v_s + v_0$, and $\psi = \psi_s + \psi_0$ into Eqs. (8.167) and retaining only the terms $\sin E$ of the periodic terms, we find

$$M_s + E_0 - e' \sin(M_s + E_0) = M_s \tag{8.175a}$$

$$\psi_0 = (-2\alpha_1)^{-\frac{1}{2}} \left(\alpha_2^2 - \alpha_3^2\right)^{\frac{1}{2}} \eta_0^{-1} A_2 B_2^{-1} v_0 \tag{8.175b}$$

$$e' = \frac{ae}{a + b_1} < e \qquad (b_1 > 0) \tag{8.175c}$$

on subtracting Eqs. (8.172a) and (8.172b). Equation (8.175a) is Kepler's equation for $M_s + E_0$, with an effective eccentricity e'. Suppose it is to be solved by the most approximate method, which will depend on the value of e'. We then have $E = M_s + E_0$ and can find $v = v_s + v_0$ by use of $\cos v = (\cos E - e)/(1 - e \cos E)$ and $\sin v = (1 - e^2)^{1/2} \sin E/(1 - e \cos E)$. Substituting v_0 into Eq. (8.175b) gives ψ_0. At this point, we have E_0, v_0, and ψ_0. Note that here e is the orbital eccentricity e and not e'.

Step 1

Knowing M_s, ψ_s, E_0, v_0, and ψ_0, we place $E = M_s + E_0 + E_1$, $v = v_s + v_0 + v_1$, and $\psi = \psi_s + \psi_0 + \psi_1$ into Eqs. (8.167), discarding only periodic terms of order J_2^2. We find

$$M_s + E_0 + E_1 - e' \sin(M_s + E_0 + E_1) = M_s + M_1 \tag{8.176a}$$

$$\psi_1 = (-2\alpha_1)^{-\frac{1}{2}} \left(\alpha_2^2 - \alpha_3^2\right)^{\frac{1}{2}} \eta_0^{-1} B_2^{-1} [A_2 v_1 + A_{21} \sin(M_s + V_0) + A_{22} \sin(2M_s + 2v_0)] + \frac{q^2}{8} B_2^{-1} \sin(2\psi_s + 2\psi_0) \tag{8.176b}$$

$$M_1 = (a + b_1)^{-1} \left[-\left(A_1 + c^2 \eta_0^2 A_2 B_1 B_2^{-1}\right) v_0 + \frac{c^2}{4} (-2\alpha_1)^{-\frac{1}{2}} \left(\alpha_2^2 - \alpha_3^2\right)^{-\frac{1}{2}} \eta_0^3 \sin(2\psi_s + 2\psi_0) \right] \tag{8.176c}$$

on subtracting Eqs. (8.172a) and (8.172b). Equation (8.176a) is Kepler's equation for $M_s + E_0 + E_1$, with the effective eccentricity e' defined in Eq. (8.175c). Using Laguerre's method, Kepler's equation can be efficiently and accurately solved. We have $E = M_s + E_0 + E_1$ and can find $v = v_s + v_0 + v_1$. Substituting v_1 into Eq. (8.176b) gives ψ_1. At this point, we have E_1, v_1, and ψ_1.

Step 2

Finally, knowing M_s, ψ_s, E_0, v_0, ψ_0, E_1, v_1, and ψ_1, we place $E = M_s + E_0 + E_1 + E_2$, $v = v_s + v_0 + v_1 + v_2$, and $\psi = \psi_s + \psi_0 + \psi_1 + \psi_2$ into Eqs. (8.167), discarding only periodic terms of order J_2^3. We find

$$M_s + E_0 + E_1 + E_2 - e' \sin(M_s + E_0 + E_1 + E_2) = M_s + M_1 + M_2 \qquad (8.177a)$$

$$\begin{aligned}\psi_2 = {} & (-2\alpha_1)^{-\frac{1}{2}}\left(\alpha_2^2 - \alpha_3^2\right)^{\frac{1}{2}} \eta_0^{-1} B_2^{-1} [A_2 v_2 + A_{21} v_1 \cos(M_s + v_0) \\ & + 2A_{22} v_1 \cos(2M_s + 2v_0) + A_{23} \sin(3M_s + 3v_0) + A_{24} \sin(4M_s + 4v_0)] \\ & + \frac{q^2}{8} B_2^{-1} \left[\psi_1 \cos(2\psi_s + 2\psi_0) + \frac{3q^2}{8} \sin(2\psi_s + 2\psi_0) \right. \\ & \left. - \frac{3q^2}{64} \sin(4\psi_s + 4\psi_0) \right]\end{aligned} \qquad (8.177b)$$

$$\begin{aligned}M_2 \equiv {} & -(a + b_1)^{-1} \Bigg[-A_1 v_1 + A_{11} \sin(M_s + v_0) + 2A_{12} \sin(2M_s + 2v_0) \\ & + \frac{c^2}{4} (-2\alpha_1)^{-\frac{1}{2}} \left(\alpha_2^2 - \alpha_3^2\right)^{-\frac{1}{2}} \eta_0^3 \Bigg\{ B_1 \psi_1 - \frac{1}{2} \psi_1 \cos(2\psi_s + 2\psi_0) \\ & - \frac{q^2}{8} \sin(2\psi_s + 2\psi_0) + \frac{q^2}{64} \sin(4\psi_s + 4\psi_0) \Bigg\} \Bigg]\end{aligned} \qquad (8.177c)$$

on subtracting Eqs. (8.172a) and (8.172b). Equation (8.177a) is Kepler's equation for $M_s + E_0 + E_1 + E_2$, with the effective eccentricity e' defined in Eq. (8.175c). Again, using Laguerre's method, Kepler's equation can be solved. One could solve the Kepler equation (8.177a) for E_2 directly by using

$$E_2 = \frac{M_2}{1 - e' \cos(M_s + E_0 + E_1)}$$

However, Laguerre's method has been proven to converge for any value of eccentricity. We then have $E = M_s + E_0 + E_1 + E_2$ and can find $v = v_s + v_0 + v_1 + v_2$. Substituting v_2 into Eq. (8.176b) gives ψ_2. At this point, we have E_2, v_2, and ψ_2.

This completes the solution with exact secular terms and periodic terms correct through order J_2^2 for E, v, and ψ and, thus, for the spheroidal coordinates $\rho = a(1 - e \cos E) = a(1 - e^2)/(1 + e \cos v)$, $\eta = \eta_0 \sin \psi$.

XVI. The Right Ascension ϕ

From Eq. (8.64c)

$$\beta_3 = \phi + c^2 \alpha_3 R_3 - \alpha_3 N_3 \qquad (8.64c)$$

In Eq. (8.64c) insert N_3 from Eq. (8.151) and R_3 from Eq. (8.108). The result is

$$\phi = \beta_3 + \alpha_3\left(\alpha_2^2 - \alpha_3^2\right)^{-\frac{1}{2}}\left[\eta_0\left(1-\eta_0^2\right)^{-\frac{1}{2}}\left(1-\eta_2^{-2}\right)^{-\frac{1}{2}}\chi + B_3\chi + \frac{3}{32}\eta_0^3\eta_2^{-4}\sin 2\psi\right] - c^2\alpha_3(-2\alpha_1)^{-\frac{1}{2}}\left[A_3 v + \sum_{n=1}^{4} A_{3n}\sin nv\right] \tag{8.178}$$

Here $\eta_0 = \sin I$ for all J_2. In the limit $J_2 = 0$, this becomes

$$\phi = \beta_3 + \alpha_3\left(\alpha_2^2 - \alpha_3^2\right)^{-\frac{1}{2}}\eta_0\left(1-\eta_0^2\right)^{-\frac{1}{2}}\chi$$

However, in the limit $J_2 = 0$, we also have $\alpha_3 = \alpha_2 \cos I$, so that this reduces to

$$\beta_3 = \phi - \chi \text{ sgn } \alpha_3 \tag{8.179}$$

which is a correct Keplerian equation if $\beta_3 = \Omega$. It is a useful exercise to check that $\bar{\dot{\phi}} = 2\pi\nu_3$.

XVII. Further Developments

See Ref. 10 for a treatment of zonal harmonic perturbations. This article uses the Brouwer–von Zeipel method to handle the effects of J_3 and J_4 on the spheroidal problem as developed in this chapter. For further development of the spheroidal method itself, see Ref. 4. For a summary of the spheroidal method, correcting all previous errata and showing how to avoid troubles with near-polar orbits, see Ref. 11.

References 3 and 4 incorporate J_3 into the separable potential. The history of this topic is as follows. Shortly after the publication of Ref. 1, Brouwer and Pines[12] discovered that the spheroidal potential of this chapter could be found by use of the separable problem of two fixed centers.[13] To see why this is so, imagine a particle of half the mass of the Earth placed on the z axis at a distance c_1 north of the Earth's center of mass and another one of the same mass also placed on the z axis but at a distance c_1 south of the center of mass. If P is a field point at distances r_1 and r_2 from these two masses, the potential produced at P by these masses would be

$$V = -\frac{GM}{2}\left(\frac{1}{r_1} + \frac{1}{r_2}\right)$$

where M is the mass of the Earth. If x, y, z are the coordinates of P, then

$$r_1^2 = x^2 + y^2 + (z - c_1)^2$$

$$r_2^2 = x^2 + y^2 + (z + c_1)^2$$

Now introduce spheroidal coordinates ρ, η, ϕ, defined in this chapter, so that

$$x^2 + y^2 = (\rho^2 + c^2)(1 - \eta^2)$$

$$z = \rho\eta$$

Then

$$r_1^2 = \rho^2 + c^2(1-\eta^2) + c_1^2 - 2c_1\rho\eta$$

$$r_2^2 = \rho^2 + c^2(1-\eta^2) + c_1^2 + 2c_1\rho\eta$$

If we now formally put $c_1 = ic$, where $i = (-1)^{1/2}$, then $c_1^2 + c^2 = 0$ and

$$r_1^2 = \rho^2 - c^2\eta^2 - 2ic_1\rho\eta = (\rho - ic\eta)^2$$

$$r_2^2 = \rho^2 - c^2\eta^2 + 2ic_1\rho\eta = (\rho + ic\eta)^2$$

Thus,

$$r_1 = \rho - ic\eta$$

$$r_2 = \rho + ic\eta$$

and

$$V = -\frac{GM}{2}\left(\frac{1}{\rho - ic\eta} + \frac{1}{\rho + ic\eta}\right) = -\frac{GM\rho}{\rho^2 + c^2\eta^2} = -\frac{\mu\rho}{\rho^2 + c^2\eta^2}$$

This, however, is the separable spheroidal potential.

Aksenov, Grebenikov, and Demin[14] discovered that if the masses and distances are all complex, with $M_1 r_1^{-1}$ and $M_2 r_2^{-1}$ conjugate, the potential

$$V = -G\left(\frac{M_1}{r_1} + \frac{M_2}{r_2}\right)$$

also leads to separability. It enabled them to fit not only μ and J_2, but also J_3, with the origin still located at the center of mass. The author's endeavor to understand this possibility in more physical terms led to Refs. 3 and 4.

References 3 and 4 illustrated by the methods of this chapter that J_3 could be incorporated into the separable potential. This now becomes

$$V = -\frac{\mu(\rho + \eta\delta)}{\rho^2 + c^2\eta^2}$$

where

$$x + iy = [(\rho^2 + c^2)(1-\eta^2)]^{\frac{1}{2}} e^{i\phi}$$

$$z = \rho\eta - \delta$$

$$c^2 = r_e^2 J_2\left(1 - \frac{J_3^2}{J_2^2}\right) \approx r_e^2 J_2$$

$$\delta = -\frac{1}{2} r_e \frac{J_3}{J_2} > 0$$

Here c is again about 210 km for the Earth, and δ is about 7 km.

Unfortunately, there are a number of errors in Ref. 4. They do not change its main conclusions but have to be avoided for applications. Reference 11 eliminates all these errors, except for a final bracket sign for H_3 on page 33 and omission of the e in $\rho = a(1 - e\cos E)$ on page 34.

References

[1]Vinti, J. P., "New Method of Solution for Unretarded Satellite Orbits," *Journal of Research of the National Bureau of Standards,* Vol. 63B, Oct.–Dec. 1959, pp. 105–116.

[2]Vinti, J. P., "Theory of an Accurate Intermediary Orbit for Satellite Astronomy," *Journal of Research of the National Bureau of Standards,* Vol. 65B, July–Sept. 1961, pp. 169–201.

[3]Vinti, J. P., "Invariant Properties of the Spheroidal Potential of an Oblate Planet," *Journal of Research of the National Bureau of Standards,* Vol. 70B, Jan.–March 1966, pp. 1–16.

[4]Vinti, J. P., "Inclusion of the Third Zonal Harmonic in an Accurate Reference Orbit of an Artificial Satellite," *Journal of Research of the National Bureau of Standards,* Vol. 70B, Jan.–March 1966, pp. 17–46.

[5]Morse, P. M., and Feshbach, H., *Methods of Theoretical Physics,* McGraw–Hill, New York, 1953.

[6]Izsak, I., "A Theory of Satellite Motion About an Oblate Planet—I. A Second-Order Solution of Vinti's Dynamical Problem," Smithsonian Institution Astrophysical Observatory, Special Report No. 52, Nov. 1960.

[7]Vinti, J. P., "Intermediary Equatorial Orbits of an Artificial Satellite," *Journal of Research of the National Bureau of Standards,* Vol. 65B, Jan.–March 1962, pp. 5–13.

[8]Whittaker, E. T., and Watson, G. N., *Modern Analysis,* 4th ed., Cambridge Univ. Press, Cambridge, England, 1952, p. 312.

[9]Vinti, J. P., "Mean Motions in Conditionally Periodic Separable Systems," *Journal of Research of the National Bureau of Standards,* Vol. 65B, April–June 1961, pp. 131–135.

[10]Vinti, J. P., "Zonal Harmonic Perturbations of an Accurate Reference Orbit of an Artificial Satellite," *Journal of Research of the National Bureau of Standards,* Vol. 67B, Oct.–Dec. 1963, pp. 191–222.

[11]Vinti, J. P., "Improvement of the Spheroidal Method for Artificial Satellites," *Astronomical Journal,* Vol. 74, Feb. 1969, pp. 25–34.

[12]Brouwer, D., and Pines, S., *Methods of Celestial Mechanics,* edited by D. Brouwer and G. M. Clemence, Academic Press, New York, 1961, pp. 573, 592.

[13]Pars, L. A., *A Treatise on Analytical Dynamics,* Wiley, New York, 1965, p. 309.

[14]Aksenov, E. P., Grebenikov, E. A., and Demin, V. G., "A Qualitative Analysis of the Motion of an Artificial Satellite of the Earth in the Normal Gravitational Field of the Earth," *Soviet Astronomy,* Vol. 7, 1963, pp. 276–282.

Chapter 9

Delaunay Variables

CHAPTER 8 developed Hamilton–Jacobi perturbation theory with the α's and β's as canonical variables and the perturbing Hamiltonian term $H_1(q, p, t)$ as Hamiltonian. The unsuitability of β_1 as a variable now leads us to introduce a new set of canonical variables, called the Delaunay variables. In the case of a general reference Hamiltonian $H_0(q, p)$, they have to be introduced by means of certain other variables called action and angle variables. For the present, we shall not deal with them but bring in the Delaunay variables by a special method applicable only to the Keplerian reference Hamiltonian.

We have

$$\dot{\alpha}_k = -\frac{\partial H_1}{\partial \beta_k} = \frac{\partial F_1}{\partial \beta_k} \qquad \dot{\beta}_k = \frac{\partial H_1}{\partial \alpha_k} = -\frac{\partial F_1}{\partial \alpha_k} \qquad k = 1, 2, 3$$

where we write $F_1 = -H_1$ and $H_1 = H_1(q, p, t)$. This is to follow Delaunay, who reversed the sign of the Hamiltonian; all the literature follows this convention. With F_1 as Hamiltonian governing the behavior of the α's and β's, the α's appear mathematically as coordinates and the β's as momenta.

For the generating function that we need, see Ref. 1. With H_0 as the Kepler $H_0(q, p)$, we introduce a generating function of the form $S(q, P, t)$:

$$S = -\alpha_1 t + \mu(-2\alpha_1)^{-\frac{1}{2}}\beta_1' + \alpha_2\beta_2' + \alpha_3\beta_3'$$

where $\mu = G(m_1 + m_2)$. Note that the β_k' are used because of the new P in S. Here the α's are the "old" coordinates and the β's the "old" momenta; the α_k' are the "new" coordinates and the β_k' the "new" momenta. Then

$$\beta_k = \frac{\partial S}{\partial \alpha_k} \qquad \alpha_k' = \frac{\partial S}{\alpha \beta_k'} \qquad k = 1, 2, 3$$

and the new Hamiltonian will be

$$F = F_1 + \frac{\partial S}{\partial t}$$

Thus

$$\beta_1 = -t + \mu(-2\alpha_1)^{-\frac{3}{2}}\beta_1' \qquad \alpha_1' = \mu(-2\alpha_1)^{-\frac{1}{2}}$$

$$\beta_2 = \beta_2' \qquad \alpha_2' = \alpha_2$$

$$\beta_3 = \beta_3' \qquad \alpha_3' = \alpha_3$$

With use of $\alpha_1 = -\mu/(2a)$, where a is the Keplerian perturbed variable for the semi-major axis and $n = \sqrt{\mu a^{-3}}$, the perturbed mean motion, we find

$$\alpha_1' = \sqrt{\mu a} \qquad \beta_1' = (t + \beta_1)\mu^{-1}(-2\alpha_1)^{\frac{3}{2}} = \ell$$

$$\alpha_2' = \alpha_2 = [\mu a(1 - e^2)]^{\frac{1}{2}} \qquad \beta_2' = \beta_2 = \omega$$

$$\alpha_3' = \alpha_3 = [\mu a(1 - e^2)]^{\frac{1}{2}} \cos I \qquad \beta_3' = \beta_3 = \Omega$$

In Delaunay's notation

$$L = \sqrt{\mu a} \qquad \ell = n(t + \beta_1)$$

$$G = [\mu a(1 - e^2)]^{\frac{1}{2}} \qquad g = \omega$$

$$H = [\mu a(1 - e^2)]^{\frac{1}{2}} \cos I \qquad h = \Omega$$

The Delaunay Hamiltonian is

$$F = F_1 - \alpha_1$$

However, $\alpha_1 = -\mu/(2a)$ and $L^2 = \mu a$, so that

$$\alpha_1 = -(\mu^2/2L^2)$$

and

$$F = F_1 + (\mu^2/2L^2)$$

Note that $F_1 = -H_1$. The canonical equations in Delauney variables are then

$$\frac{dL}{dt} = \frac{\partial F}{\partial \ell} \qquad \frac{d\ell}{dt} = -\frac{\partial F}{\partial L}$$

$$\frac{dG}{dt} = \frac{\partial F}{\partial g} \qquad \frac{dg}{dt} = -\frac{\partial F}{\partial G}$$

$$\frac{dH}{dt} = \frac{\partial F}{\partial h} \qquad \frac{dh}{dt} = -\frac{\partial F}{\partial H}$$

the L, G, H now being "coordinates" and the ℓ, g, h "momenta."

Reference

[1]Garfinkel, B., *Space Mathematics, Part I,* Vol. 5, Lectures in Applied Mathematics, American Mathematical Society, Providence, RI, 1996, p. 65.

Chapter 10

The Lagrange Planetary Equations

LATER on we shall use the Delaunay equations as a canonical system to develop artificial satellite theory, but first it is desirable to use them to derive equations for the variations of the Keplerian elements. These equations are known as the Lagrange planetary equations (not to be confused with the Euler–Lagrange equations of advanced dynamics) or as the V.O.P. equations. The "V.O.P." means variation of parameters after a method called "variation of constants" in books on differential equations. It is not necessary to bring in this latter method, because the variations of the Keplerian elements are an easy by-product of canonical theory that we should have had to develop in any event.

First, let us define two Keplerian sets of variables. There is the "slow" set: a, e, I, Ω, ω, and τ or $\sigma = -n\tau$, and the "fast" set: a, e, I, Ω, ω, and $\ell = nt + \sigma = n(t - \tau)$. It is the presence of nt in ℓ that makes the latter the fast set. Our earlier remarks about $\beta_1 = -\tau$ as a variable should have made it amply clear why we shall consider only the fast set.

If V_1 is the perturbing potential, then $V_1 = H_1 = -F_1$, and the Hamiltonian in Delaunay variables is

$$F = (\mu^2/2L^2) + F_1$$

corresponding to the differential equation

$$\ddot{\boldsymbol{r}} = -\frac{\mu}{r^3}\boldsymbol{r} + \nabla F_1$$

Here $F_1 = -V_1$ is called the disturbing function. It is clear that the Lagrange planetary equations will apply only when the disturbing force is derivable from a potential. Dissipative forces will be taken care of later by the Gaussian equations.

The Delaunay equations

$$\dot{L} = \frac{\partial F}{\partial \ell} \qquad \dot{\ell} = -\frac{\partial F}{\partial L}$$

$$\dot{G} = \frac{\partial F}{\partial g} \qquad \dot{g} = -\frac{\partial F}{\partial G}$$

$$\dot{H} = \frac{\partial F}{\partial h} \qquad \dot{h} = -\frac{\partial F}{\partial H}$$

may be used to calculate $\dot{a}$, $\dot{e}$, $\dot{I}$, $\dot{\Omega}$, $\dot{\omega}$, and $\dot{\ell}$ as functions of a, e, I, Ω, ω, and ℓ. Note that, even though $\ell = n(t - \tau)$ and $n = \sqrt{\mu a^{-3}}$, the ℓ and the a are to be considered as independent variables in these calculations. We begin with

$$F = (\mu^2/2L^2) + F_1(a, e, I, \Omega, \omega, \ell)$$

$$a = L^2/\mu \qquad \omega = g$$

$$1 - e^2 = G^2/L^2 \qquad \Omega = h$$

$$\cos I = H/G \qquad n = \mu^{\frac{1}{2}} a^{-\frac{3}{2}}$$

I. Semi-Major Axis

$$\dot{a} = \frac{2L}{\mu}\dot{L} = \frac{2L}{\mu}\frac{\partial F_1}{\partial \ell} = \frac{2\sqrt{\mu a}}{\mu}\frac{\partial F_1}{\partial \ell}$$

so that

$$\dot{a} = \frac{2}{na}\frac{\partial F_1}{\partial \ell}$$

II. Eccentricity

$$1 - e^2 = G^2/L^2$$

$$\ln(1 - e^2) = 2 \ln G - 2 \ln L$$

$$\frac{-e\dot{e}}{1 - e^2} = \frac{\dot{G}}{G} - \frac{\dot{L}}{L} = \frac{1}{G}\frac{\partial F_1}{\partial \omega} - \frac{1}{L}\frac{\partial F_1}{\partial \ell}$$

so that

$$\dot{e} = \frac{1 - e^2}{eL}\left(\frac{\partial F_1}{\partial \ell} - \frac{L}{G}\frac{\partial F_1}{\partial \omega}\right)$$

However,

$$L = \sqrt{\mu a} = na^2 \qquad L/G = (1 - e^2)^{-\frac{1}{2}}$$

so that

$$\dot{e} = \frac{1 - e^2}{na^2 e}\left[\frac{\partial F_1}{\partial \ell} - (1 - e^2)^{-\frac{1}{2}}\frac{\partial F_1}{\partial \omega}\right]$$

III. Inclination

$$\cos I = H/G$$

$$-\dot{I}\sin I = \frac{\dot{H}}{G} - \frac{H}{G^2}\dot{G} = \frac{1}{G}\frac{\partial F_1}{\partial \Omega} - \frac{\cos I}{G}\frac{\partial F_1}{\partial \omega}$$

However,

$$G = na^2\sqrt{1 - e^2}$$

so that

$$\dot{I} = \frac{1}{na^2\sqrt{(1 - e^2)}}\left(\cot I \frac{\partial F_1}{\partial \omega} - \csc I \frac{\partial F_1}{\partial \Omega}\right)$$

For the other three Keplerian elements, it is necessary to keep in mind which variables are being kept fixed in a partial derivative. The subscript O.D. will mean that other Delaunays are to be kept fixed, and the subscript O.K. will mean that other Keplerians are to be kept fixed.

IV. Mean Anomaly

With other Delaunays fixed

$$\dot{\ell} = -\frac{\partial F}{\partial L} = -\frac{\partial}{\partial L}\left(\frac{\mu^2}{2L^2} + F_1\right) = \frac{\mu^2}{L^3} - \frac{\partial F_1}{\partial L}$$

Since

$$L = \sqrt{\mu a} = na^2$$

$$\mu^2/L^3 = \mu^2(\mu a)^{-\frac{3}{2}} = n$$

Thus

$$\dot{\ell} = n - \left(\frac{\partial F_1}{\partial L}\right)_{\text{O.D.}}$$

However,

$$F_1 = F_1(a, e, I, \Omega, \omega, \ell)$$

Of the Keplerian elements, only a and e depend on L. Thus

$$\left(\frac{\partial F_1}{\partial L}\right)_{\text{O.D.}} = \left(\frac{\partial F_1}{\partial a}\right)_{\text{O.K.}}\left(\frac{\partial a}{\partial L}\right)_{\text{O.D.}} + \left(\frac{\partial F_1}{\partial e}\right)_{\text{O.K.}}\left(\frac{\partial e}{\partial L}\right)_{\text{O.D.}}$$

Here

$$a = L^2/\mu$$

$$\left(\frac{\partial a}{\partial L}\right)_{\text{O.D.}} = \frac{2L}{\mu} = \frac{2(\mu a)^{\frac{1}{2}}}{\mu} = \frac{2}{na}$$

$$1 - e^2 - G^2/L^2$$

$$-e\left(\frac{\partial e}{\partial L}\right)_{\text{O.D.}} = -\frac{G^2}{L^3} = -\frac{1 - e^2}{L} = -\frac{1 - e^2}{na^2}$$

Thus

$$\left(\frac{\partial F_1}{\partial L}\right)_{\text{O.D.}} = \frac{2}{na}\left(\frac{\partial F_1}{\partial a}\right)_{\text{O.K.}} + \frac{1-e^2}{na^2 e}\left(\frac{\partial F_1}{\partial e}\right)_{\text{O.K.}}$$

so that

$$\dot{\ell} = n - \frac{2}{na}\frac{\partial F_1}{\partial a} - \frac{1-e^2}{na^2 e}\frac{\partial F_1}{\partial e}$$

In the final Lagrange planetary equations we do not need the subscript, since it is understood that the variables are the fast Keplerian set.

V. The Argument of Pericenter

$$\dot{\omega} = \dot{g} = -\left(\frac{\partial F}{\partial G}\right)_{\text{O.D.}} = -\left(\frac{\partial F_1}{\partial G}\right)_{\text{O.D.}}$$

In $F_1(a, e, I, \Omega, \omega, \ell)$ only e and I depend on G:

$$1 - e^2 = G^2/L^2 \qquad \cos I = H/G$$

Thus

$$\left(\frac{\partial F_1}{\partial G}\right)_{\text{O.D.}} = \left(\frac{\partial F_1}{\partial e}\right)_{\text{O.K.}}\left(\frac{\partial e}{\partial G}\right)_{\text{O.D.}} + \left(\frac{\partial F_1}{\partial I}\right)_{\text{O.K.}}\left(\frac{\partial I}{\partial G}\right)_{\text{O.D.}}$$

Then

$$-e\left(\frac{\partial e}{\partial G}\right)_{\text{O.D.}} = \frac{G}{L^2} = \frac{(1-e^2)^{\frac{1}{2}}}{na^2}$$

Also

$$-\sin I\left(\frac{\partial I}{\partial G}\right)_{\text{O.D.}} = -\frac{H}{G^2} = -\frac{\cos I}{na^2(1-e^2)^{\frac{1}{2}}}$$

Thus

$$\dot{\omega} = \frac{(1-e^2)^{\frac{1}{2}}}{na^2 e}\frac{\partial F_1}{\partial e} - \frac{\cot I}{na^2(1-e^2)^{\frac{1}{2}}}\frac{\partial F_1}{\partial I}$$

VI. The Longitude of the Node

$$\dot{\Omega} = \dot{h} = -\left(\frac{\partial F}{\partial H}\right)_{\text{O.D.}} = -\left(\frac{\partial F_1}{\partial H}\right)_{\text{O.D.}}$$

In $F_1(a, e, I, \Omega, \omega, \ell)$ only I depends on the Delaunay variable H. Thus

$$\left(\frac{\partial F_1}{\partial H}\right)_{\text{O.D.}} = \left(\frac{\partial F_1}{\partial I}\right)_{\text{O.K.}}\left(\frac{\partial I}{\partial H}\right)_{\text{O.D.}}$$

However,

$$\cos I = H/G$$

$$-\sin I\left(\frac{\partial I}{\partial H}\right)_{\text{O.D.}} = -\frac{1}{G} = -\frac{1}{na^2(1-e^2)^{\frac{1}{2}}}$$

Thus

$$\dot{\Omega} = \frac{\csc I}{na^2(1-e^2)^{\frac{1}{2}}}\frac{\partial F_1}{\partial I}$$

VII. Summary

$$\dot{a} = \frac{2}{na}\frac{\partial F_1}{\partial \ell}$$

$$\dot{e} = \frac{1-e^2}{na^2 e}\left(\frac{\partial F_1}{\partial \ell} - (1-e^2)^{-\frac{1}{2}}\frac{\partial F_1}{\partial \omega}\right)$$

$$\dot{I} = \frac{1}{na^2\sqrt{(1-e^2)}}\left(\cot I\frac{\partial F_1}{\partial \omega} - \csc I\frac{\partial F_1}{\partial \Omega}\right)$$

$$\dot{\omega} = \frac{(1-e^2)^{\frac{1}{2}}}{na^2 e}\frac{\partial F_1}{\partial e} - \frac{\cot I}{na^2(1-e^2)^{\frac{1}{2}}}\frac{\partial F_1}{\partial I}$$

$$\dot{\Omega} = \frac{\csc I}{na^2(1-e^2)^{\frac{1}{2}}}\frac{\partial F_1}{\partial I}$$

$$\dot{\ell} = n - \frac{2}{na}\frac{\partial F_1}{\partial a} - \frac{1-e^2}{na^2 e}\frac{\partial F_1}{\partial e}$$

These are the final Lagrange planetary equations for the variations of the elements of the fast Keplerian set. The partial derivatives are the derivatives of the disturbing function with respect to those elements. Note that ℓ contains an additive term n, the mean motion, which is nonvanishing even in the unperturbed case, which is why this set is called the fast set.

Note that e appears in denominators for $\dot{e}, \dot{\omega}$, and $\dot{\ell}$ and that $\sin I$ appears in denominators for $\dot{I}, \dot{\omega}$, and $\dot{\Omega}$. These appearances mean trouble for circular orbits, $e = 0$, and for orbits in the xy plane, with $\sin I = 0$. The solution of these equations leads to e and $\sin I$ in the denominators of results for most of the Keplerian elements. Actually, they do not produce singularities in the resulting variations of the Cartesian coordinate system of components $x, y, z, \dot{x}, \dot{y}$, and $\dot{z}$ but produce the necessary algebra to show this is heavy.

For this reason, other elements are often used that do not lead to such singularities. One such set is the following[1]: "equinoctial" system—a, h, k, p, q, λ. This is good for all inclinations except $I = 180$.

$$a = a$$

$$h = e\sin(\omega + \Omega)$$

$$k = e\cos(\omega + \Omega)$$

$$p = \tan\frac{I}{2}\sin\Omega$$

$$q = \tan\frac{I}{2}\cos\Omega$$

$$\lambda = \ell + \omega + \Omega$$

The reader will recognize λ as the mean longitude. To handle absolutely all orbits, one may define a "retrograde factor" r, defined by

$$r = 1 \qquad 0 \le I \le 90^\circ$$

$$r = -1 \qquad 90^\circ < I \le 180^\circ$$

Then

$$a = a$$

$$h = e\sin(\omega + r\Omega)$$

$$k = e\cos(\omega + r\Omega)$$

$$p = \tan^r\frac{I}{2}\sin\Omega$$

$$q = \tan^r\frac{I}{2}\cos\Omega$$

$$\lambda = \ell + \omega + r\Omega$$

Reference

[1]Computer Sciences Corporation, *System Description and User's Guide for the GTDS R&D Averaged Orbit Generator,* prepared for NASA Goddard Space Flight Center, Greenbelt, MD, 1978, pp. A1, A2.

Chapter 11

The Planetary Disturbing Function

LET us consider the orbit of a planet about the sun perturbed by other planets whose orbits are known. The orbit to be solved for may be that of a minor planet. The main perturbation will then come from Jupiter, with smaller effects from Mars and Saturn. Such a minor planet would ordinarily be moving in the asteroid belt between Mars and Jupiter.

The same general form of disturbing function arises in the case of an artificial satellite of the Earth. In this latter case, the Earth takes the place of the sun, the satellite the place of the minor planet whose orbit is being solved for, and the sun and the moon the roles of the perturbing planets. The disturbing function is then called the lunar–solar disturbing function.

Return now to the minor planet. Part of the disturbing function will arise from the direct gravitational force of the known perturbing planets, called the direct part, and another part will arise from the nongravitational forces due to the motion of the perturbing planets, called the indirect part.

To carry out the derivation, we introduce two reference systems:

1) A globally inertial system—one in which the universe as a whole is at rest. Operationally, it is one in which no apparent forces appear when we treat the motion of a particle.

2) An inertially oriented system with origin at the center of mass of the sun, z axis perpendicular to the plane of the ecliptic and x axis pointing toward the vernal equinox. (In the case of an artificial satellite, the Earth replaces the sun, and the z axis is along the axis of rotation, i.e., perpendicular to the plane of the equator.)

Let O be the center of mass of the first system and S that of the second system as shown in Fig. 11.1. Also, let M be the mass of the sun, m the mass of the solved-for planet, and m_i, $i = 1, \ldots, N$, the masses of the N perturbing planets. Let $\boldsymbol{\rho}$ denote the position vector of the solved-for planet in the first system, $\boldsymbol{\rho}_s$ that of the sun, and $\boldsymbol{\rho}_i$ that of a perturbing planet.

Also, let $\boldsymbol{r}$ be the position vector of m relative to the sun and $\boldsymbol{r}_i$ that of m_i relative to the sun. Then

$$\boldsymbol{r} = \boldsymbol{\rho} - \boldsymbol{\rho}_s \qquad \boldsymbol{r}_i = \boldsymbol{\rho}_i - \boldsymbol{\rho}_s$$

where $\boldsymbol{r}$ has Cartesian coordinates x, y, z and $\boldsymbol{r}_i$ has Cartesian coordinates x_i, y_i, z_i, both in the second system. Here

$$\boldsymbol{\Delta}_i = \boldsymbol{r}_i - \boldsymbol{r} \qquad \Delta_i = |\boldsymbol{\Delta}_i|$$

$$\boldsymbol{\rho} = \boldsymbol{\rho}_s + \boldsymbol{r}$$

$$\boldsymbol{\rho}_i = \boldsymbol{\rho}_s + \boldsymbol{r}_i$$

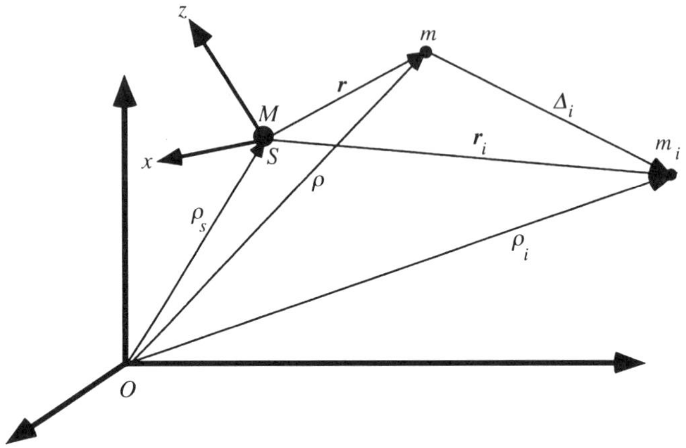

Fig. 11.1 The inertial coordinate systems.

For the sun

$$M\ddot{\boldsymbol{\rho}}_s = \frac{GMm}{r^3}\boldsymbol{r} + \sum_{i=1}^{N}\frac{GMm_i}{r_i^3}\boldsymbol{r}_i \tag{11.1}$$

For the planet to be solved for

$$m\ddot{\boldsymbol{\rho}} = -\frac{GMm}{r^3}\boldsymbol{r} + \sum_{i=1}^{N}\frac{Gmm_i}{\Delta_i^3}\Delta_i \tag{11.2}$$

Since $\boldsymbol{r} = \boldsymbol{\rho} - \boldsymbol{\rho}_s$, we may obtain $\ddot{\boldsymbol{r}}$ by canceling M in Eq. (11.1) and m in Eq. (11.2) and taking the difference of the resulting two equations. The result is

$$\ddot{\boldsymbol{r}} = -\frac{G(M+m)}{r^3}\boldsymbol{r} + \sum_{i=1}^{N}\frac{Gm_i}{\Delta_i^3}(\boldsymbol{r}_i - \boldsymbol{r}) - \sum_{i=1}^{N}\frac{Gm_i}{r_i^3}\boldsymbol{r}_i \tag{11.3}$$

In Eq. (11.3), Δ_i has been replaced by $\boldsymbol{r}_i - \boldsymbol{r}$. To simplify the last two terms in Eq. (11.3), introduce the function

$$U = \sum_{i=1}^{N}\frac{Gm_i}{\Delta_i} - \sum_{i=1}^{N}\frac{Gm_i}{r_i^3}\boldsymbol{r}_i\cdot\boldsymbol{r} \tag{11.4}$$

and differentiate it with respect to x, the x coordinate of $\boldsymbol{r}$ in the system attached to the sun. Then

$$\frac{\partial U}{\partial x} = -\sum_{i=1}^{N}\frac{Gm_i}{\Delta_i^2}\frac{\partial\Delta_i}{\partial x} - \sum_{i=1}^{N}\frac{Gm_i}{r_i^3}x_i \tag{11.5}$$

However,

$$\Delta_i^2 = (x - x_i)^2 - (y - y_i)^2 - (z - z_i)^2$$

so that

$$\frac{\partial\Delta_i}{\partial x} = \frac{x - x_i}{\Delta_i}$$

Then Eq. (11.5) becomes

$$\frac{\partial U}{\partial x} = -\sum_{i=1}^{N} \frac{Gm_i}{\Delta_i^3}(x - x_i) - \sum_{i=1}^{N} \frac{Gm_i}{r_i^3} x_i \tag{11.6a}$$

Similarly

$$\frac{\partial U}{\partial y} = -\sum_{i=1}^{N} \frac{Gm_i}{\Delta_i^3}(y - y_i) - \sum_{i=1}^{N} \frac{Gm_i}{r_i^3} y_i \tag{11.6b}$$

$$\frac{\partial U}{\partial z} = -\sum_{i=1}^{N} \frac{Gm_i}{\Delta_i^3}(z - z_i) - \sum_{i=1}^{N} \frac{Gm_i}{r_i^3} z_i \tag{11.6c}$$

Thus, if ∇_{xyz} is the gradient operator with components $\partial/\partial x$, $\partial/\partial y$, and $\partial/\partial z$, we find

$$\nabla_{xyz} U = \sum_{i=1}^{N} \frac{Gm_i}{\Delta_i^3} \Delta_i - \sum_{i=1}^{N} \frac{Gm_i}{r_i^3} \boldsymbol{r}_i \tag{11.7}$$

Equation (11.3) becomes

$$\ddot{\boldsymbol{r}} = -\frac{G(M+m)}{r^3}\boldsymbol{r} + \nabla_{xyz} U \tag{11.8}$$

The function U is called the disturbing function. Its first term, $\Sigma_i Gm_i/\Delta_i$, is called the direct term, because it is clearly produced by the direct gravitational forces of the perturbing planets. Its second term, $\Sigma_i G(m_i/r_i^3)\boldsymbol{r}_i \cdot \boldsymbol{r}$, is called the indirect term, because it is produced by the acceleration of the perturbing planets. A simple way to verify this is to carry through both the inertial mass of the sun and its gravitational mass, with a separate symbol for each. It will then be found that the indirect term has a factor sun's gravitational mass/sun's inertial mass; thus, if the sun's inertial mass were infinite, it would vanish.

Bibliography

[1]Smart, W. M., *Celestial Mechanics,* Longmans, Green, and Co., London, 1953, pp. 8–10.

Chapter 12

Gaussian Variational Equations for the Jacobi Elements

THE Lagrange planetary equations are applicable when the perturbation is derivable from a potential. If it is given only as a force not so derivable (e.g., if it arises from drag), they are not applicable, and we need another approach. The appropriate variational equations are known as Gaussian, after Gauss, the great German mathematician.

When the perturbation is known only as a force, the motion of an orbiter is

$$\ddot{\boldsymbol{r}} = -\nabla V_0 + \boldsymbol{F} \tag{12.1}$$

where $\boldsymbol{F}$ is not the gradient of any potential. Here V_0 would be $-\mu/r$ in the Keplerian case, but in general V_0 may be any potential function that leads to a solvable Hamilton–Jacobi equation when $\boldsymbol{F} = \boldsymbol{0}$.

If potential V_0 leads to a solvable *HJ* equation, the Hamiltonian is

$$H_0 = T(q, p) + V_0(q) \tag{12.2}$$

and we shall call the corresponding orbit the reference orbit. It is characterized by Jacobi α's and β's, satisfying

$$p_j = \frac{\partial S(q, \alpha, t)}{\partial q_j} \qquad \beta_j = \frac{\partial S(q, \alpha, t)}{\partial \alpha_j} \tag{12.3}$$

where

$$H_0 + \frac{\partial S}{\partial t} = 0 \tag{12.4}$$

Equations (12.3) and (12.4) will also hold for the perturbed problem, because they represent a canonical transformation from the q's and p's to the α's and β's.

Our aim in this chapter is to find equations for the $\dot{\alpha}$'s and $\dot{\beta}$'s in terms of the perturbing force $\boldsymbol{F}$. Before we can proceed with the main derivation, we shall need certain equations, connecting derivatives among the q's and p's, called the Jacobi relations. There are four of these, but we shall need only two of them, viz.,[1–3]

$$\left(\frac{\partial q_i}{\partial \beta_k}\right)_{\alpha,\beta} = \left(\frac{\partial \alpha_k}{\partial p_i}\right)_{q,p} \tag{12.5}$$

$$\left(\frac{\partial q_i}{\partial \alpha_k}\right)_{\alpha,\beta} = -\left(\frac{\partial \beta_k}{\partial p_i}\right)_{q,p} \tag{12.6}$$

In Eqs. (12.5) and (12.6) the subscripts (α, β) mean that all the α's and β's are held constant during the differentiation, except β_k in Eq. (12.5) and α_k in Eq. (12.6). The subscripts (q, p) mean that all the q's and p's are held constant, except p_i in Eq. (12.5) and p_i in Eq. (12.6).

Proof of Eq. (12.5): From Eq. (12.3) we have

$$0 = \left(\frac{\partial \beta_j}{\partial \alpha_k}\right)_{\alpha,\beta} = \frac{\partial^2 S}{\partial \alpha_j \partial \alpha_k} + \frac{\partial^2 S}{\partial \alpha_j \partial q_\ell}\left(\frac{\partial q_\ell}{\partial \alpha_k}\right)_{\alpha,\beta} \tag{12.7}$$

with use of the summation convention. Thus

$$\frac{\partial^2 S}{\partial \alpha_j \partial q_\ell}\left(\frac{\partial q_\ell}{\partial \alpha_k}\right)_{\alpha,\beta} = -\frac{\partial^2 S}{\partial \alpha_j \partial \alpha_k} \tag{12.8}$$

Also, from Eq. (12.3)

$$\delta_{jk} = \left(\frac{\partial \beta_j}{\partial \beta_k}\right)_{\alpha,\beta} = \frac{\partial^2 S}{\partial \alpha_i \partial q_m}\left(\frac{\partial q_m}{\partial \beta_k}\right)_{\alpha,\beta} \tag{12.9}$$

and

$$\delta_{ij} = \left(\frac{\partial p_j}{\partial p_i}\right)_{q,p} = \frac{\partial^2 S}{\partial q_j \partial \alpha_\ell}\left(\frac{\partial \alpha_\ell}{\partial p_i}\right)_{q,p} \tag{12.10}$$

Multiply Eq. (12.10) by $\partial q_j/\partial \beta_k$ and sum on j to obtain

$$\left(\frac{\partial q_i}{\partial \beta_k}\right)_{\alpha,\beta} = \left(\frac{\partial \alpha_\ell}{\partial p_i}\right)_{q,p}\frac{\partial^2 S}{\partial q_j \partial \alpha_\ell}\left(\frac{\partial q_j}{\partial \beta_k}\right)_{\alpha,\beta} \tag{12.11}$$

Because ℓ and j are dummy indices, we may change ℓ to j and j to m. Then

$$\left(\frac{\partial q_i}{\partial \beta_k}\right)_{\alpha,\beta} = \left(\frac{\partial \alpha_j}{\partial p_i}\right)_{q,p}\frac{\partial^2 S}{\partial q_m \partial \alpha_j}\left(\frac{\partial q_m}{\partial \beta_k}\right)_{\alpha,\beta} \tag{12.12}$$

By Eq. (12.9) we can replace

$$\frac{\partial^2 S}{\partial q_m \partial \alpha_j}\left(\frac{\partial q_m}{\partial \beta_k}\right)_{\alpha,\beta}$$

in Eq. (12.12) by $\delta_{j,k}$. Equation (12.12) becomes

$$\left(\frac{\partial q_i}{\partial \beta_k}\right)_{\alpha,\beta} = \left(\frac{\partial \alpha_k}{\partial p_i}\right)_{q,p} \tag{12.5}$$

which is Eq. (12.5) that we wished to prove.

Proof of Eq. (12.6): Multiply Eq. (12.10) by $\partial q_j/\partial \alpha_k$ and sum on j to obtain

$$\left(\frac{\partial q_i}{\partial \alpha_k}\right)_{\alpha,\beta} = \left(\frac{\partial \alpha_\ell}{\partial p_i}\right)_{q,p}\frac{\partial^2 S}{\partial q_j \partial \alpha_\ell}\left(\frac{\partial q_j}{\partial \alpha_k}\right)_{\alpha,\beta} \tag{12.13}$$

Interchange the dummy indices j and ℓ in Eq. (12.13). Then

$$\left(\frac{\partial q_i}{\partial \alpha_k}\right)_{\alpha,\beta} = \left(\frac{\partial \alpha_j}{\partial p_i}\right)_{q,p}\frac{\partial^2 S}{\partial q_\ell \partial \alpha_j}\left(\frac{\partial q_\ell}{\partial \alpha_k}\right)_{\alpha,\beta} \tag{12.14}$$

However, by Eq. (12.8)

$$\frac{\partial^2 S}{\partial q_\ell \partial \alpha_j}\left(\frac{\partial q_\ell}{\partial \alpha_k}\right)_{\alpha,\beta} = -\frac{\partial^2 S}{\partial \alpha_j \partial \alpha_k}$$

Insert this into Eq. (12.14) to obtain

$$\left(\frac{\partial q_i}{\partial \alpha_k}\right)_{\alpha,\beta} = -\frac{\partial}{\partial \alpha_j}\frac{\partial S}{\partial \alpha_k}\left(\frac{\partial \alpha_j}{\partial p_i}\right)_{q,p} = -\left(\frac{\partial}{\partial p_i}\frac{\partial S}{\partial \alpha_k}\right)_{q,p}$$

However, $\partial S/\partial \alpha_k = \beta_k$, so that this becomes

$$\left(\frac{\partial q_i}{\partial \alpha_k}\right)_{\alpha,\beta} = -\left(\frac{\partial \beta_k}{\partial p_i}\right)_{q,p} \tag{12.6}$$

which is the second Jacobi relation. This completes the proof of the Jacobi relations.

We now return to Eqs. (12.3). They can be inverted to give the α's and β's as functions of the q's and p's. The q's and p's can then be expressed in terms of the rectangular coordinates x_k and rectangular velocities $\dot{x}_k$. In this way, we can express the α's and β's as functions of the x's and $\dot{x}$'s. (Parenthetically, let it be remarked that we essentially did this in Chapters 6–8 for the Keplerian H_0 when we expressed the Keplerian elements in terms of the x's and $\dot{x}$'s; the further step of expressing the Keplerian α's and β's in terms of the x's and $\dot{x}$'s is simple.)

Thus, we may write

$$\alpha_j = \alpha_j(x, \dot{x}) \tag{12.15}$$

$$\beta_j = f_j(x, \dot{x}) - t\delta_{j1} \tag{12.16}$$

With the summation convention,

$$\dot{\alpha}_j = \left(\frac{\partial \alpha_j}{\partial x_k}\right)_{x,\dot{x}} \dot{x}_k + \left(\frac{\partial \alpha_j}{\partial \dot{x}_k}\right)_{x,\dot{x}} \ddot{x}_k \tag{12.17}$$

$$\dot{\beta}_j = -\delta_{j1} + \left(\frac{\partial f_j}{\partial x_k}\right)_{x,\dot{x}} \dot{x}_k + \left(\frac{\partial f_j}{\partial \dot{x}_k}\right)_{x,\dot{x}} \ddot{x}_k \tag{12.18}$$

However, by Eq. (12.1)

$$\ddot{\boldsymbol{r}} = -\nabla V_0 + \boldsymbol{F} \tag{12.1}$$

so that

$$\ddot{x}_k = -\frac{\partial V_0}{\partial x_k} + F_k \tag{12.19}$$

On inserting Eq. (12.19) into Eqs. (12.17) and (12.18), we find

$$\dot{\alpha}_j = \frac{\partial \alpha_j}{\partial x_k}\dot{x}_k - \frac{\partial \alpha_j}{\partial \dot{x}_k}\frac{\partial V_0}{\partial x_k} + F_k\frac{\partial \alpha_j}{\partial \dot{x}_k} \tag{12.20}$$

$$\dot{\beta}_j = -\delta_{j1} + \frac{\partial f_j}{\partial x_k}\dot{x}_k - \frac{\partial f_j}{\partial \dot{x}_k}\frac{\partial V_0}{\partial x_k} + F_k\frac{\partial f_j}{\partial \dot{x}_k} \tag{12.21}$$

or

$$\dot{\alpha}_j = \Phi_j + F_k \frac{\partial \alpha_j}{\partial \dot{x}_k} \tag{12.22}$$

$$\dot{\beta}_j = \Psi_j + F_k \frac{\partial \beta_j}{\partial \dot{x}_k} \tag{12.23}$$

We have used here $\partial f_j / \partial \dot{x}_k = \partial \beta_j / \partial \dot{x}_k$ and have denoted by Φ_j and Ψ_j the terms that do not involve F_k.

If we were to turn off the force $\boldsymbol{F}$ at time $t = t_0$, we should have for $t > t_0$

$$\Phi_j(t > t_0) = 0 \qquad \Psi_j(t > t_0) = 0 \tag{12.24}$$

since we would then be back to the unperturbed problem, where the α's and β's are constants. Now

$$\Phi_j = \Phi_j(x, \dot{x}) \qquad \Psi_j = \Psi_j(x, \dot{x})$$

depending explicitly only on the x's and $\dot{x}$'s and not explicitly on t. At the moment t_0 when we turn off the perturbing force $\boldsymbol{F}$, the acceleration $\ddot{\boldsymbol{r}}$ changes instantaneously, but the x's and $\dot{x}$'s do not. This means that Φ_j and Ψ_j do not change value at time t_0. Because they vanish for $t > t_0$, however, they must also vanish at time t_0. However, t_0 is *any* time. Thus

$$\Phi_j = 0 \qquad \Psi_j = 0 \tag{12.25}$$

Insertion of Eq. (12.25) into Eqs. (12.22) and (12.23) yields

$$\dot{\alpha}_j = F_k \frac{\partial \alpha_j}{\partial \dot{x}_k} \tag{12.26}$$

$$\dot{\beta}_j = F_k \frac{\partial \beta_j}{\partial \dot{x}_k} \tag{12.27}$$

Equations (12.26) and (12.27) express one possible form for the desired variational equations, but not the most convenient one.

To express them in the most convenient form, we need a lemma, viz.,

$$\left(\frac{\partial \alpha_j}{\partial \dot{x}_i}\right)_{x,\dot{x}} = \left(\frac{\partial x_k}{\partial \beta_j}\right)_{\alpha,\beta} \tag{12.28}$$

$$\left(\frac{\partial \beta_j}{\partial \dot{x}_k}\right)_{x,\dot{x}} = -\left(\frac{\partial x_k}{\partial \alpha_j}\right)_{\alpha,\beta} \tag{12.29}$$

Proof of Lemma: Begin with

$$x_k = x_k(q) \tag{12.30}$$

$$\dot{x}_k = \Sigma_r \frac{\partial x_k}{\partial q_r} \dot{q}_r \tag{12.31}$$

(It is best to avoid the summation convention in this proof.)

Thus

$$\left(\frac{\partial \dot{x}_k}{\partial \dot{q}_j}\right)_{q,\dot{q}} = \frac{\partial x_k}{\partial q_j} \tag{12.32}$$

The kinetic energy per unit mass is

$$T = \tfrac{1}{2}\Sigma_r \dot{x}_k^2 \tag{12.33}$$

so that

$$p_j = \left(\frac{\partial T}{\partial \dot{q}_k}\right)_{x,\dot{x}} = \Sigma_k \dot{x}_k \frac{\partial \dot{x}_k}{\partial \dot{q}_j} = \Sigma_k \dot{x}_k \frac{\partial x_k}{\partial \dot{q}_j} \tag{12.34}$$

by Eq. (12.32). Now p_j also satisfies

$$p_j = \frac{\partial S(q,\alpha)}{\partial q_j} \tag{12.35}$$

Let us seek a similar equation for $\dot{x}_r$. To do so, multiply Eq. (12.34) by $(\partial q_j/\partial x_r)_x$ and sum on j. We obtain

$$\Sigma_k \dot{x}_k \Sigma_j \frac{\partial x_k}{\partial q_j}\frac{\partial q_j}{\partial x_r} = \Sigma_j p_j \frac{\partial q_j}{\partial x_r} \tag{12.36}$$

$$= \Sigma_j \frac{\partial S(q,\alpha)}{\partial q_j}\frac{\partial q_j}{\partial x_r} \tag{12.37}$$

with use of Eq. (12.35). Then

$$\Sigma_k \dot{x}_k \Sigma_j \frac{\partial x_k}{\partial q_j}\frac{\partial q_j}{\partial x_r} = \left(\frac{\partial S(q,\alpha)}{\partial q_j}\right)_\alpha \tag{12.38}$$

Now, because the q's are functions of the x's and the x's functions of the q's, we have

$$\mathrm{d}x_k = \Sigma_j \frac{\partial x_k}{\partial q_j}\mathrm{d}q_j = \Sigma_j \frac{\partial x_k}{\partial q_j}\Sigma_r \frac{\partial q_j}{\partial x_r}\mathrm{d}x_r$$

$$= \Sigma_r \Sigma_j \left(\frac{\partial x_k}{\partial q_j}\frac{\partial q_j}{\partial x_r}\right)\mathrm{d}x_r$$

Thus

$$\Sigma_j \left(\frac{\partial x_k}{\partial q_j}\frac{\partial q_j}{\partial x_r}\right) = \delta_{kr} \tag{12.39}$$

Inserting Eq. (12.39) into Eq. (12.38), we find

$$\dot{x}_r = \left(\frac{\partial S(q,\alpha)}{\partial x_r}\right)_\alpha = \left(\frac{\partial G(x,\alpha)}{\partial x_r}\right)_\alpha \tag{12.40}$$

where we have put

$$S(q,\alpha) = G(x,\alpha) \tag{12.41}$$

Also

$$\beta_r = \frac{\partial S(q,\alpha)}{\partial \alpha_r} = \frac{\partial G(x,\alpha)}{\partial x_r} \tag{12.42}$$

Equations (12.40) and (12.42) have the same form as Eqs. (12.3), with x_r replacing q_r, $\dot{x}_r$ replacing p_r, and $G(x,\alpha)$ replacing $S(q,\alpha)$. In terms of the x's and $\dot{x}$'s [from Eqs. (12.5) and (12.6)], the first Jacobi relation becomes

$$\left(\frac{\partial x_i}{\partial \beta_k}\right)_{\alpha,\beta} = \left(\frac{\partial \alpha_k}{\partial \dot{x}_i}\right)_{x,\dot{x}} \tag{12.43}$$

and the second Jacobi relation becomes

$$\left(\frac{\partial x_i}{\partial \alpha_k}\right)_{\alpha,\beta} = -\left(\frac{\partial \beta_k}{\partial \dot{x}_i}\right)_{x,\dot{x}} \tag{12.44}$$

Now, in Eqs. (12.43) and (12.44), change k to j and i to k to obtain

$$\left(\frac{\partial \alpha_j}{\partial \dot{x}_i}\right)_{x,\dot{x}} = \left(\frac{\partial x_k}{\partial \beta_j}\right)_{\alpha,\beta}$$

$$\left(\frac{\partial \beta_j}{\partial \dot{x}_k}\right)_{x,\dot{x}} = -\left(\frac{\partial x_k}{\partial \alpha_j}\right)_{\alpha,\beta}$$

The results constitute the lemma we set out to prove.

Now insert the lemma equations into Eqs. (12.26) and (12.27). The results are

$$\dot{\alpha}_j = F_k\left(\frac{\partial x_k}{\partial \beta_j}\right)_{\alpha,\beta}$$

$$\dot{\beta}_j = -F_k\left(\frac{\partial x_k}{\partial \alpha_j}\right)_{\alpha,\beta}$$

In vector form, these become

$$\dot{\alpha}_j = \boldsymbol{F}\cdot\left(\frac{\partial \boldsymbol{r}}{\partial \beta_j}\right)_{\alpha,\beta} \tag{12.45}$$

$$\dot{\beta}_j = -\boldsymbol{F}\cdot\left(\frac{\partial \boldsymbol{r}}{\partial \alpha_j}\right)_{\alpha,\beta} \tag{12.46}$$

These are the Gaussian variational equations for the Jacobi elements. In the special case that the perturbing force $\boldsymbol{F}$ is derivable from a potential V_1

$$\boldsymbol{F} = -\nabla V_1 \tag{12.47}$$

The Gaussian equations then become

$$\dot{\alpha}_j = -\nabla V_1\cdot\frac{\partial \boldsymbol{r}}{\partial \beta_j}$$

$$\dot{\beta}_j = \nabla V_1\cdot\frac{\partial \boldsymbol{r}}{\partial \alpha_j}$$

However,

$$dV_1 = \frac{\partial V_1}{\partial x}\,dx + \frac{\partial V_1}{\partial y}\,dy + \frac{\partial V_1}{\partial z}\,dz = \nabla V_1 \cdot d\boldsymbol{r}$$

so that the equations become

$$\dot{\alpha}_j = -\frac{\partial V_1}{\partial \beta_j}$$

$$\dot{\beta}_j = \frac{\partial V_1}{\partial \alpha_j}$$

the same as we found in Chapter 7, where $H_1 = V_1$.

References

[1]Smart, W. M., *Celestial Mechanics,* Longman, Green, and Co., London, 1953, pp. 140–142.

[2]Tisserand, F., *Mecanique Celeste,* Gauthier–Villars, Paris, 1889, pp. 20–23.

[3]Vinti, J. P., "Gauss Variational Equations for Osculating Elements of an Arbitrary Separable Reference Orbit," *Celestial Mechanics 7,* 1973, pp. 367–375.

Chapter 13

Gaussian Variational Equations for the Keplerian Elements

I. Preliminaries

CHAPTER 12 derived Gaussian variational equations for the Jacobi α's and β's:

$$\dot{\alpha}_k = \boldsymbol{F} \cdot \frac{\partial \boldsymbol{r}}{\partial \beta_k} \qquad \dot{\beta}_k = -\boldsymbol{F} \cdot \frac{\partial \boldsymbol{r}}{\partial \alpha_k} \tag{13.1}$$

where F is the perturbing force. The present chapter will be devoted to obtaining Gaussian equations that tell how the Keplerian elements a, e, I, Ω, ω, and ℓ vary with time because of such a perturbing force. It is convenient to begin with two lemmas that will be needed in the derivations.

Lemma 1: If a vector $\boldsymbol{r}$ rotates around a fixed axis pointing along a unit vector $\boldsymbol{J}$, so that the angle $(\boldsymbol{r}, \boldsymbol{J})$ remains constant (Fig. 13.1), then if η is the angle of rotation

$$\frac{\partial \boldsymbol{r}}{\partial \eta} = \boldsymbol{J} \times \boldsymbol{r} \tag{13.2}$$

The proof is simple. As η increases by $\mathrm{d}\eta$

$$|\mathrm{d}\boldsymbol{r}| = r \sin \xi \, \mathrm{d}\zeta \tag{13.3}$$

The direction of $\mathrm{d}\boldsymbol{r}$ as shown in Fig. 13.1 is along the tangent to the circle in the plan view (Fig. 13.2). Since rotation of a right-handed screw through $\mathrm{d}\eta$ would produce screw translation along $\boldsymbol{J}$, the direction of $\mathrm{d}\boldsymbol{r}$ or of $\partial \boldsymbol{r}/\partial \eta$ is along $\boldsymbol{J} \times \boldsymbol{r}$.

Now

$$|\boldsymbol{J} \times \boldsymbol{r}| = r \sin \zeta \tag{13.4}$$

so that by Eqs. (13.3) and (13.4)

$$\left| \frac{\partial \boldsymbol{r}}{\partial \eta} \right| = |\boldsymbol{J} \times \boldsymbol{r}|$$

Because $\partial \boldsymbol{r}/\partial \eta$ is along $\boldsymbol{J} \times \boldsymbol{r}$ and has the magnitude $|\boldsymbol{J} \times \boldsymbol{r}|$, it follows that

$$\frac{\partial \boldsymbol{r}}{\partial \eta} = \boldsymbol{J} \times \boldsymbol{r} \tag{13.2}$$

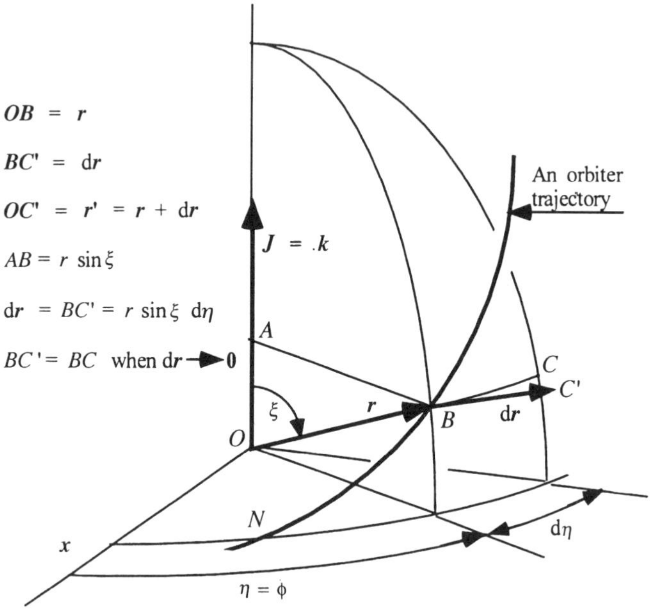

Fig. 13.1 An example of $\eta = \phi$ and $J = k$.

Lemma 2: If, in the osculating elliptic orbit shown in Fig. 13.3, $\boldsymbol{r}$ is the position vector of the orbiter, $r = |\boldsymbol{r}|$, f the true anomaly, $\boldsymbol{l}_A$ a unit vector pointing from the force center toward pericenter, and $\boldsymbol{l}_B$ a unit vector perpendicular to $\boldsymbol{l}_A$, so that f has to increase by 90° to rotate $\boldsymbol{r}$ from $\boldsymbol{l}_A$ to $\boldsymbol{l}_B$, then

$$\boldsymbol{r} = \mathrm{Re}\left[(\boldsymbol{l}_A + i\boldsymbol{l}_B)r\varepsilon^{-if}\right] \tag{13.5}$$

where $\varepsilon^{-if} = \exp(-if)$.

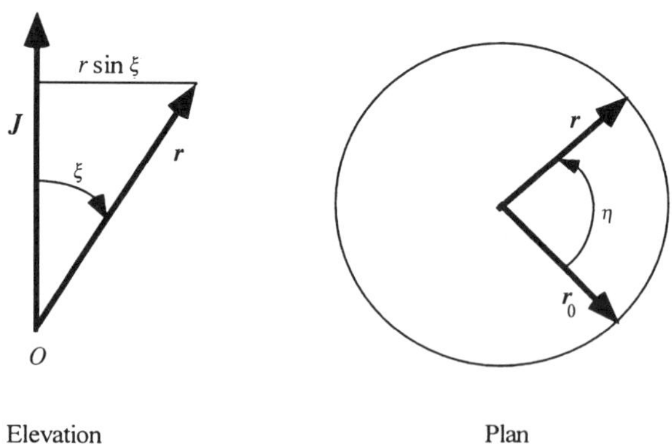

Fig. 13.2 A vector $\boldsymbol{r}$ rotating around a fixed axis $\boldsymbol{J}$.

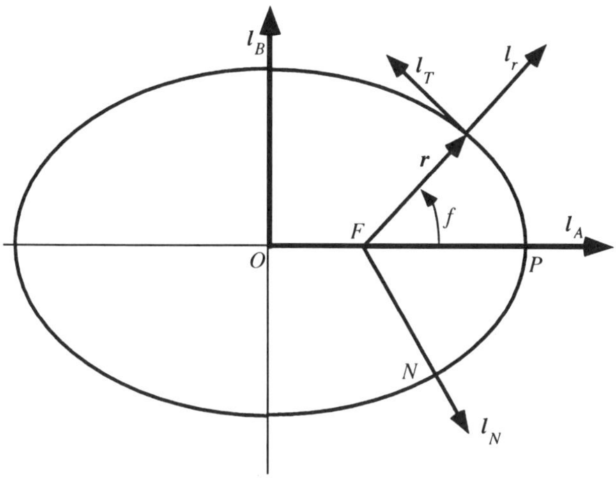

Fig. 13.3 An osculating elliptic orbit.

Proof:

$$\boldsymbol{r} = \boldsymbol{l}_A r \cos f + \boldsymbol{l}_B r \sin f = \mathrm{Re}\left[(\boldsymbol{l}_A + i\boldsymbol{l}_B) r \varepsilon^{-if}\right]$$

The perturbing force $\boldsymbol{F}$ may be expressed as

$$\boldsymbol{F} = \boldsymbol{l}_r R + \boldsymbol{l}_T T + \boldsymbol{l}_w W \tag{13.6}$$

Here $\boldsymbol{l}_r$ is a unit vector along $\boldsymbol{r}$; $\boldsymbol{l}_T$ is a unit vector along the transverse direction in the plane of the orbit; and $\boldsymbol{l}_W$ is a unit vector along the angular momentum, i.e., perpendicular to the orbital plane. Then $\boldsymbol{l}_r, \boldsymbol{l}_T, \boldsymbol{l}_W$ form a cyclic orthonormal triad of vectors, satisfying

$$\boldsymbol{l}_r \times \boldsymbol{l}_T = \boldsymbol{l}_w \qquad \boldsymbol{l}_T \times \boldsymbol{l}_W = \boldsymbol{l}_r \qquad \boldsymbol{l}_W \times \boldsymbol{l}_r = \boldsymbol{l}_T \tag{13.7}$$

Equation for $\boldsymbol{l}_W$

Let $\boldsymbol{l}_N$ be a unit vector pointing along the line of nodes toward the ascending node. From Fig. 13.4 (the octant figure) we have

$$\boldsymbol{l}_N = \boldsymbol{i} \cos \Omega + \boldsymbol{j} \sin \Omega \tag{13.8}$$

$$\boldsymbol{k} \times \boldsymbol{l}_W = \boldsymbol{l}_N \sin I \tag{13.9}$$

where $\boldsymbol{i}, \boldsymbol{j}, \boldsymbol{k}$ are unit vectors along the inertial axes.

Equation (13.9) follows from these facts: $\boldsymbol{l}_N$ lies in both the orbital plane and the equatorial plane, so that it is perpendicular to their respective normals $\boldsymbol{l}_W$ and $\boldsymbol{k}$; the angle between these planes is I, so that the angle between $\boldsymbol{k}$ and $\boldsymbol{l}_W$ is I. If we form the vector product of $\boldsymbol{k}$ with Eq. (13.9), we obtain

$$\boldsymbol{k} \times (\boldsymbol{k} \times \boldsymbol{l}_W) - \boldsymbol{k}(\boldsymbol{k} \cdot \boldsymbol{l}_W) - \boldsymbol{l}_W = \boldsymbol{k} \times \boldsymbol{l}_N \sin I \tag{13.10}$$

From Eq. (13.8)

$$\boldsymbol{k} \times \boldsymbol{l}_N = \boldsymbol{j} \cos \Omega - \boldsymbol{i} \sin \Omega \tag{13.11}$$

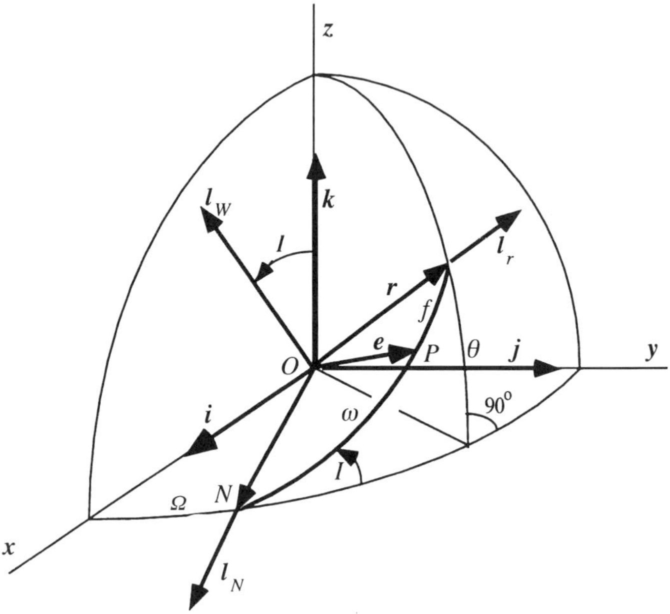

Fig. 13.4 Octant figure.

On inserting Eq. (13.11) into Eq. (13.10) and using $\boldsymbol{k}\cdot\boldsymbol{l}_W = \cos I$, we find

$$\boldsymbol{l}_W = \boldsymbol{i}\sin\Omega\sin I - \boldsymbol{j}\cos\Omega\sin I + \boldsymbol{k}\cos I \tag{13.12}$$

Equations for $\boldsymbol{l}_r, \boldsymbol{l}_T, \boldsymbol{l}_W$

From Fig. 13.3, we have

$$\boldsymbol{l}_r = \boldsymbol{l}_A\cos f + \boldsymbol{l}_B\sin f \tag{13.13}$$

$$\begin{aligned}\boldsymbol{l}_T &= \boldsymbol{l}_A\cos(f+\pi/2) + \boldsymbol{l}_B\sin(f+\pi/2)\\ &= -\boldsymbol{l}_A\sin f + \boldsymbol{l}_B\cos f\end{aligned} \tag{13.14}$$

$$\boldsymbol{l}_r + i\boldsymbol{l}_T = (\boldsymbol{l}_A + i\boldsymbol{l}_B)\varepsilon^{-if} \tag{13.15}$$

$$\boldsymbol{l}_A + i\boldsymbol{l}_B = (\boldsymbol{l}_r + i\boldsymbol{l}_T)\varepsilon^{if} \tag{13.16}$$

$$\boldsymbol{l}_A \times \boldsymbol{l}_B = \boldsymbol{l}_r \times \boldsymbol{l}_T = \boldsymbol{l}_W \tag{13.17}$$

II. Equations for $\dot{\alpha}_1$ and $\dot{a}$

Use Eqs. (13.1) and (13.5). Then

$$\dot{\alpha}_1 = \boldsymbol{F}\cdot\frac{\partial\boldsymbol{r}}{\partial\beta_1} = \mathrm{Re}\left[\boldsymbol{F}\cdot\frac{\partial}{\partial\beta_1}\{(\boldsymbol{l}_A + i\boldsymbol{l}_B)r\varepsilon^{-if}\}\right] \tag{13.18}$$

Consult Chapter 2 for

$$\boldsymbol{l}_A = \boldsymbol{i}[\cos\Omega\cos\omega - \sin\Omega\cos I\sin\omega] + \boldsymbol{j}[\sin\Omega\cos\omega + \cos\Omega\cos I\sin\omega] + \boldsymbol{k}\sin I\sin\omega \tag{13.19a}$$

$$\boldsymbol{l}_B = -\boldsymbol{i}[\cos\Omega\sin\omega + \sin\Omega\cos I\sin\omega] + \boldsymbol{j}[-\sin\Omega\sin\omega + \cos\Omega\cos I\cos\omega] + \boldsymbol{k}\sin I\cos\omega \tag{13.19b}$$

Recall that $\beta_3 = \Omega$, $\beta_2 = \omega$, $\beta_1 = -\tau$, $\cos I = (\alpha_3/\alpha_2)$.
Thus, $\boldsymbol{l}_A$ and $i\boldsymbol{l}_B$ do not depend on β_1, and therefore

$$\dot{\alpha}_1 = \boldsymbol{F}\cdot\frac{\partial \boldsymbol{r}}{\partial\beta_1} = \mathrm{Re}\left[\boldsymbol{F}\cdot(\boldsymbol{l}_A + i\boldsymbol{l}_B)\frac{\partial}{\partial\beta_1}(r\varepsilon^{-if})\right] \tag{13.20}$$

By Eq. (13.16)

$$\boldsymbol{F}\cdot(\boldsymbol{l}_A + i\boldsymbol{l}_B) = \boldsymbol{F}\cdot(\boldsymbol{l}_r + i\boldsymbol{l}_T)\varepsilon^{if} = (R + iT)\varepsilon^{if} \tag{13.21}$$

so that

$$\dot{\alpha}_1 = \boldsymbol{F}\cdot\frac{\partial \boldsymbol{r}}{\partial\beta_1} = \mathrm{Re}\left[(R + iT)\varepsilon^{if}\frac{\partial}{\partial\beta_1}(r\varepsilon^{-if})\right] \tag{13.22}$$

Now

$$\begin{gathered} r\cos f = a(\cos E - e) \\ r\sin f = b\sin E \qquad b = a\sqrt{1-e^2} \\ \boldsymbol{F}\cdot(\boldsymbol{l}_A + i\boldsymbol{l}_B) = \boldsymbol{F}\cdot(\boldsymbol{l}_r + i\boldsymbol{l}_T)\varepsilon^{if} = (R + iT)\varepsilon^{if} \\ r\varepsilon^{-if} = a(\cos E - e) - ib\sin E \end{gathered} \tag{13.23}$$

In Eqs. (13.23), only E depends on β_1. However,

$$E - e\sin E = n\,(t + \beta_1)$$

so that

$$(1 - e\cos E)\frac{\partial E}{\partial\beta_1} = n \qquad \frac{\partial E}{\partial\beta_1} = \frac{na}{r} \tag{13.24}$$

From Eqs. (13.23)

$$\begin{aligned} \frac{\partial}{\partial\beta_1}(r\varepsilon^{-if}) &= (-a\sin E - ib\cos E)\frac{\partial E}{\partial\beta_1} \\ &= -\frac{na}{r}(a\sin E + ib\cos E) \end{aligned} \tag{13.25}$$

Use the anomaly connections

$$\frac{\sin E}{r} = \frac{\sqrt{1-e^2}\sin f}{p} \qquad \frac{\cos E}{r} = \frac{e + \cos f}{p}$$

so that

$$\frac{\partial}{\partial \beta_1}(r\varepsilon^{-if}) = -\frac{na^2\sqrt{1-e^2}}{p}[\sin f + i(e + \cos f)]$$

$$= -\frac{na^2\sqrt{1-e^2}}{p}[ie + i\varepsilon^{-if}] \tag{13.26}$$

Insert Eqs. (13.26) into (13.22) to find

$$\dot{\alpha}_1 = -\frac{na^2\sqrt{1-e^2}}{p}\mathrm{Re}[(R + iT)(i + ie\varepsilon^{if})]$$

$$= \frac{na^2\sqrt{1-e^2}}{p}[eR\sin f + T(1 + e\cos f)] \tag{13.27}$$

With use of $p = a(1 - e^2)$, this becomes

$$\dot{\alpha}_1 = \frac{na}{\sqrt{1-e^2}}[eR\sin f + T(1 + e\cos f)] \tag{13.28}$$

and the semi-major axis is

$$a = -\frac{\mu}{2\alpha_1}$$

$$\dot{a} = \frac{\mu}{2\alpha_1^2}\dot{\alpha}_1 = \frac{2a^2}{\mu}\dot{\alpha}_1$$

$$= \frac{2}{n\sqrt{1-e^2}}[eR\sin f + T(1 + e\cos f)] \tag{13.29}$$

III. Equations for $\dot{\alpha}_2$ and $\dot{e}$

Here

$$\dot{\alpha}_2 = \boldsymbol{F}\cdot\frac{\partial \boldsymbol{r}}{\partial \beta_2} \tag{13.30}$$

Now

$$\psi = \beta_2 + f$$

where the true anomaly f depends on E and e. However, $e = e(\alpha_1, \alpha_2)$, and $E = E(e, \beta_1, a)$. Thus, f depends only on α_1, α_2, and β_1, so that it is independent of β_2. Thus, the argument ψ of latitude has no dependence on β_2 through f. It follows that changing β_2 only leads to $\mathrm{d}\psi = \mathrm{d}\beta_2$, so that

$$\frac{\partial \boldsymbol{r}}{\partial \beta_2} = \frac{\partial \boldsymbol{r}}{\partial \psi} \tag{13.31}$$

Thus

$$\dot{\alpha}_2 = \boldsymbol{F}\cdot\frac{\partial \boldsymbol{r}}{\partial \psi} \tag{13.32}$$

If $\boldsymbol{r}$ is changed by a change in ψ only, it gets rotated about an axis perpendicular to the orbital plane. Here $\boldsymbol{l}_W$ is a unit vector along this axis, and $(\boldsymbol{r}, \boldsymbol{l}_W)$ remains 90° during such rotation. Thus, in Eq. (13.2), η becomes ψ, so that

$$\frac{\partial \boldsymbol{r}}{\partial \beta_2} = \frac{\partial \boldsymbol{r}}{\partial \psi} = \boldsymbol{l}_W \times r \tag{13.33}$$

Insertion of Eq. (13.33) into Eq. (13.30) yields

$$\dot{\alpha}_2 = \boldsymbol{l}_W \times \boldsymbol{r} \cdot \boldsymbol{F} = \boldsymbol{l}_W \cdot \boldsymbol{r} \times \boldsymbol{F} \tag{13.34}$$

Since $\boldsymbol{F} = \boldsymbol{l}_r R + \boldsymbol{l}_T T + \boldsymbol{l}_w W$, then

$$\boldsymbol{r} \times \boldsymbol{F} = rT\boldsymbol{l}_W - rW\boldsymbol{l}_T \tag{13.35}$$

$$\boldsymbol{l}_W \cdot \boldsymbol{r} \times \boldsymbol{F} = rT \tag{13.36}$$

It follows that

$$\dot{\alpha}_2 = rT \tag{13.37}$$

Now to find $\dot{e}$, use

$$\begin{aligned} \alpha_2^2 &= \mu a(1 - e^2) \\ 2 \ln \alpha_2 &= \ln \mu + \ln a + 2 \ln (1 - e^2) \\ \frac{2\dot{\alpha}_2}{\alpha_2} &= \frac{\dot{a}}{a} - \frac{2\, e\dot{e}}{1 - e^2} \end{aligned} \tag{13.38}$$

In Eq. (13.38), insert Eqs. (13.29) and (13.37), so that

$$\frac{2e\dot{e}}{1 - e^2} = \frac{2}{na\sqrt{1 - e^2}}[Re \sin f + T(1 + e \cos f)] - \frac{2rT}{na^2\sqrt{1 - e^2}} \tag{13.39}$$

Thus

$$e\dot{e} = \frac{\sqrt{1 - e^2}}{na}[Re \sin f + T(1 + e \cos f) - rT] \tag{13.40}$$

In Eq. (13.40), insert $r = a(1 - e \cos E)$ to find

$$e\dot{e} = \frac{\sqrt{1 - e^2}}{na}[Re \sin f + T(1 + e \cos f) - T(1 - e \cos E)] \tag{13.41}$$

so that

$$\dot{e} = \frac{\sqrt{1 - e^2}}{na}[R \sin f + T(\cos E + \cos f)] \tag{13.42}$$

IV. Equations for $\dot{\alpha}_3$ and $\dot{I}$

Here

$$\dot{\alpha}_3 = \boldsymbol{F} \cdot \frac{\partial \boldsymbol{r}}{\partial \beta_3} \tag{13.43}$$

Now $\beta_3 = \Omega$ and the longitude

$$\phi = \beta_3 + \chi \tag{13.44}$$

where, by Sec. VII of Chapter 2,

$$\tan\chi = \cos I \tan\psi \tag{13.45}$$

Here, according to Sec. III of this chapter, ψ depends only on α_1, α_2, β_1, and β_2. Also

$$\cos I = \alpha_3/\alpha_2$$

Thus, χ is independent of $\beta_3 = \Omega$. It follows that

$$\frac{\partial\phi}{\partial\beta_3} = 1$$

Then

$$\frac{\partial \boldsymbol{r}}{\partial\beta_3} = \frac{\partial \boldsymbol{r}}{\partial\phi}\frac{\partial\phi}{\partial\beta_3} = \frac{\partial \boldsymbol{r}}{\partial\phi} \tag{13.46}$$

Thus

$$\dot{\alpha}_3 = \boldsymbol{F}\cdot\frac{\partial \boldsymbol{r}}{\partial\phi} \tag{13.47}$$

Here $\partial\boldsymbol{r}/\partial\phi$ is the rate of change of $\boldsymbol{r}$ as $\boldsymbol{r}$ is rotated around the inertial z axis with constant θ. That is, in Eq. (13.2), η becomes ϕ, and $\boldsymbol{J}$ is $\boldsymbol{k}$, the unit vector along the inertial axis Oz. Thus

$$\frac{\partial \boldsymbol{r}}{\partial\phi} = \boldsymbol{k}\times\boldsymbol{r} \tag{13.48}$$

so that

$$\begin{aligned}\dot{\alpha}_3 &= \boldsymbol{k}\times\boldsymbol{r}\cdot\boldsymbol{F}\\ &= \boldsymbol{k}\cdot\boldsymbol{r}\times\boldsymbol{F} = r\boldsymbol{k}\cdot(\boldsymbol{l}_r\times\boldsymbol{F})\end{aligned} \tag{13.49}$$

With use of

$$\begin{aligned}\boldsymbol{F} &= \boldsymbol{l}_r R + \boldsymbol{l}_T T + \boldsymbol{l}_w W\\ \boldsymbol{l}_r\times\boldsymbol{F} &= T\boldsymbol{l}_W - W\boldsymbol{l}_T\end{aligned} \tag{13.50}$$

so that

$$\dot{\alpha}_3 = r\boldsymbol{k}\cdot(T\boldsymbol{l}_W - W\boldsymbol{l}_T) \tag{13.51}$$

However,

$$\boldsymbol{k}\cdot\boldsymbol{l}_W = \cos I \tag{13.52}$$

and

$$\boldsymbol{l}_T = -\boldsymbol{l}_A\sin f + \boldsymbol{l}_B\cos f \tag{13.53}$$

by Sec. I of this chapter. Also, by Eq. (13.19)

$$\boldsymbol{k} \cdot \boldsymbol{l}_A = \sin I \sin \omega$$

$$\boldsymbol{k} \cdot \boldsymbol{l}_B = \sin I \cos \omega$$

Thus

$$\boldsymbol{k} \cdot \boldsymbol{l}_T = -\sin I \sin \omega \sin f + \sin I \cos \omega \cos f \tag{13.54}$$

Insertion of Eqs. (13.52) and (13.54) into Eq. (13.51) thus yields

$$\dot{\alpha}_3 = r[T \cos I - W \sin I \cos(\omega + f)] \tag{13.55}$$

To find $\dot{I}$, use

$$\cos I = \alpha_3/\alpha_2 \tag{13.56}$$

$$-\dot{I} \sin I = \frac{\dot{\alpha}_3}{\alpha_2} - \frac{\alpha_3}{\alpha_2^2}\dot{\alpha}_2 \tag{13.57}$$

Insertion of $\dot{\alpha}_2 = rT$ and of Eqs. (13.55) and (13.56) into Eq. (13.57) yields

$$-\dot{I} \sin I = (r/\alpha_2)[T \cos I - W \sin I \cos(\omega + f) - T \cos I]$$

so that, with use of $\alpha_2 = \sqrt{\mu a(1-e^2)} = na^2\sqrt{1-e^2}$, we find

$$\dot{I} = \frac{rW \cos(\omega + f)}{na^2\sqrt{1-e^2}} \tag{13.58}$$

V. Equations for $\dot{\beta}_3 = \dot{\Omega}$

Here

$$\dot{\beta}_3 = -\boldsymbol{F} \cdot \frac{\partial \boldsymbol{r}}{\partial \alpha_3} \tag{13.59}$$

Because of the six Keplerian elements only I depends on α_3, we may proceed as follows. Use

$$\cos I = \frac{\alpha_3}{\alpha_2} \qquad -\sin I \frac{\partial I}{\partial \alpha_3} = \frac{1}{\alpha_2}$$

$$\frac{\partial \boldsymbol{r}}{\partial \alpha_3} = \frac{\partial \boldsymbol{r}}{\partial I}\frac{\partial I}{\partial \alpha_3} = -\frac{\csc I}{\alpha_2}\frac{\partial \boldsymbol{r}}{\partial I} \tag{13.60}$$

Then

$$\dot{\beta}_3 = \frac{\csc I}{\alpha_2}\boldsymbol{F} \cdot \frac{\partial \boldsymbol{r}}{\partial I} \tag{13.61}$$

Now $\partial \boldsymbol{r}/\partial I$ corresponds to a rotation about $\boldsymbol{l}_N$, the unit vector along the line of nodes. In Eq. (13.2) η becomes I, and $\boldsymbol{J}$ is $\boldsymbol{l}_N$. Then

$$\frac{\partial \boldsymbol{r}}{\partial I} = \boldsymbol{l}_N \times \boldsymbol{r} \tag{13.62}$$

and

$$\dot{\beta}_3 = \frac{\csc I}{\alpha_2} \boldsymbol{F} \cdot \boldsymbol{l}_N \times \boldsymbol{r} \tag{13.63}$$

$$= \frac{\csc I}{\alpha_2} \boldsymbol{l}_N \cdot \boldsymbol{r} \times \boldsymbol{F}$$

$$= \frac{r \csc I}{\alpha_2} \boldsymbol{l}_N \cdot (\boldsymbol{l}_r \times \boldsymbol{F}) \tag{13.64}$$

$$= \frac{r \csc I}{\alpha_2} \boldsymbol{l}_N \cdot (T\boldsymbol{l}_W - W\boldsymbol{l}_T) \tag{13.65}$$

with use of Eqs. (13.50). However, $\boldsymbol{l}_N \cdot \boldsymbol{l}_W = 0$, then

$$\dot{\beta}_3 = -\frac{r \csc I}{\alpha_2} W(\boldsymbol{l}_N \cdot \boldsymbol{l}_T) \tag{13.66}$$

Now

$$\boldsymbol{l}_N \cdot \boldsymbol{l}_r = \csc \psi \tag{13.67}$$

and

$$\boldsymbol{l}_N \cdot \boldsymbol{l}_T = \csc(\psi + \pi/2) = -\sin \psi \tag{13.68}$$

from Fig. 13.4. Thus

$$\dot{\beta}_3 = \dot{\Omega} = \frac{rW \csc I \sin \psi}{\alpha_2} \tag{13.69}$$

or

$$\dot{\Omega} = \frac{rW \csc I \sin \psi}{na^2\sqrt{1 - e^2}} \tag{13.70}$$

VI. Equations for $\dot{\beta}_2 = \dot{\omega}$

We begin by proving two lemmas.

Lemma 1: With Keplerian elements a, e, I, Ω, ω, and ℓ as independent variables

$$\frac{\partial}{\partial e}\left(\frac{a}{r}\right) = \left(\frac{a}{r}\right)^2 \cos f \tag{13.71}$$

Proof:

$$\frac{\partial}{\partial e}\left(\frac{a}{r}\right) = -\frac{a}{r^2}\frac{\partial r}{\partial e}$$

$$r = a(1 - e \cos E) \tag{13.71a}$$

$$\frac{\partial r}{\partial e} = -a \cos E + ae \sin E \frac{\partial E}{\partial e}$$

However,

$$E - e \sin E = \ell$$

so that

$$(1 - e\cos E)\frac{\partial E}{\partial e} - \sin E = 0$$

$$\frac{\partial E}{\partial e} = \frac{\sin E}{1 - e\cos E}$$

Thus

$$\frac{\partial r}{\partial e} = -a\cos E + \frac{ae\sin^2 E}{1 - e\cos E} = \frac{a(e - \cos E)}{1 - e\cos E} = -a\cos f \qquad (13.71\text{b})$$

Then

$$\frac{\partial}{\partial e}\left(\frac{a}{r}\right) = \left(\frac{a}{r}\right)^2 \cos f \qquad (13.71\text{c})$$

This completes the proof of Lemma 1.

Lemma 2:

$$\frac{\partial f}{\partial e} = \left(\frac{a}{r} + \frac{1}{1 - e^2}\right)\sin f \qquad (13.72)$$

Proof:

$$\cos f = \frac{a}{r}(\cos E - e)$$

With use of Lemma 1 we find

$$-\sin f\frac{\partial f}{\partial e} = \left(\frac{a}{r}\right)^2 \cos f(\cos E - e) + \left(\frac{a}{r}\right)\left(-\sin E\frac{\partial E}{\partial e} - 1\right)$$

However,

$$\frac{\partial E}{\partial e} = \frac{\sin E}{1 - e\cos E} = \frac{a}{r}\sin E$$

so that

$$-\sin f\frac{\partial f}{\partial e} = \left(\frac{a}{r}\right)\cos^2 f - \frac{a^2\sin^2 E}{r^2} - \frac{a}{r} = \left(\frac{a}{r}\right)\sin^2 f - \frac{\sin^2 f}{1 - e^2}$$

since

$$\sin f = \frac{a}{r}\sqrt{(1 - e^2)}\sin E$$

Thus

$$\frac{\partial f}{\partial e} = \left(\frac{a}{r} + \frac{1}{1 - e^2}\right)\sin f \qquad (13.72)$$

which is Lemma 2.

Now

$$\dot{\beta}_2 = -\boldsymbol{F} \cdot \frac{\partial \boldsymbol{r}}{\partial \alpha_2} \tag{13.72a}$$

where $\boldsymbol{r} = \boldsymbol{r}(a, e, I, \Omega, \omega, \ell)$. Of the six Keplerian variables, only e and I depend on α_2. Thus

$$\left(\frac{\partial \boldsymbol{r}}{\partial \alpha_2}\right)_{\alpha,\beta} = \left(\frac{\partial \boldsymbol{r}}{\partial e}\right)_K \left(\frac{\partial e}{\partial \alpha_2}\right)_{\alpha,\beta} + \left(\frac{\partial \boldsymbol{r}}{\partial I}\right)_K \left(\frac{\partial I}{\partial \alpha_2}\right)_{\alpha,\beta} \tag{13.73}$$

Here

$$\begin{gathered} e^2 = 1 + \frac{2\alpha_1\alpha_2^2}{\mu^2} \qquad e\frac{\partial e}{\partial \alpha_2} = \frac{2\alpha_1\alpha_2}{\mu^2} \\ \frac{\partial e}{\partial \alpha_2} = \frac{2\alpha_1\alpha_2}{e\mu^2} \end{gathered} \tag{13.74}$$

$$\begin{gathered} \cos I = \frac{\alpha_3}{\alpha_2} \qquad -\sin I \frac{\partial I}{\partial \alpha_2} = -\frac{\cos I}{\alpha_2} \\ \frac{\partial I}{\partial \alpha_2} = \frac{\cot I}{\alpha_2} \end{gathered} \tag{13.75}$$

Insert Eqs. (13.74) and (13.75) into Eq. (13.73) to find

$$\frac{\partial \boldsymbol{r}}{\partial \alpha_2} = \frac{\cot I}{\alpha_2}\frac{\partial \boldsymbol{r}}{\partial I} + \frac{2\alpha_1\alpha_2}{e\mu^2}\frac{\partial \boldsymbol{r}}{\partial e} \tag{13.76}$$

From Eqs. (13.76) and (13.72a)

$$\dot{\beta}_2 = -\frac{\cot I}{\alpha_2}\boldsymbol{F} \cdot \frac{\partial \boldsymbol{r}}{\partial I} - \frac{2\alpha_1\alpha_2}{e\mu^2}\boldsymbol{F} \cdot \frac{\partial \boldsymbol{r}}{\partial e} \tag{13.77}$$

or

$$\dot{\beta}_2 = N_1 + N_2 \tag{13.78}$$

$$N_1 = -\frac{\cot I}{\alpha_2}\boldsymbol{F} \cdot \frac{\partial \boldsymbol{r}}{\partial I} \qquad N_2 = -\frac{2\alpha_1\alpha_2}{e\mu^2}\boldsymbol{F} \cdot \frac{\partial \boldsymbol{r}}{\partial e} \tag{13.79}$$

From Eq. (13.61)

$$\dot{\Omega} = \frac{\csc I}{\alpha_2}\boldsymbol{F} \cdot \frac{\partial \boldsymbol{r}}{\partial I} \tag{13.61}$$

so that

$$N_1 = -\dot{\Omega}\cos I \tag{13.80}$$

We now have to calculate N_2. For this we need $(\partial \boldsymbol{r}/\partial e)_K$. Use

$$\boldsymbol{r} = \mathrm{Re}[(\boldsymbol{l}_A + i\boldsymbol{l}_B)r\varepsilon^{-if}]$$

Because $\boldsymbol{l}_A + i\boldsymbol{l}_B$ depends only on ω, Ω, and I,

$$\left(\frac{\partial \boldsymbol{r}}{\partial e}\right)_K = \mathrm{Re}\left[(\boldsymbol{l}_A + i\boldsymbol{l}_B)\frac{\partial}{\partial e}(r\varepsilon^{-if})\right] \tag{13.81}$$

$$\frac{\partial}{\partial e}(r\varepsilon^{-if}) = \varepsilon^{-if}\frac{\partial r}{\partial e} - ir\,\varepsilon^{-if}\frac{\partial f}{\partial e} = \varepsilon^{-if}\left(\frac{\partial r}{\partial e} - ir\frac{\partial f}{\partial e}\right)$$

By Eq. (13.16), $(\boldsymbol{l}_A + i\boldsymbol{l}_B)\varepsilon^{-if} = \boldsymbol{l}_r + i\boldsymbol{l}_T$, so that

$$\left(\frac{\partial \boldsymbol{r}}{\partial e}\right)_K = \mathrm{Re}\left[(\boldsymbol{l}_r + i\boldsymbol{l}_T)\left(\frac{\partial r}{\partial e} - ir\frac{\partial f}{\partial e}\right)\right] \tag{13.82}$$

By Eq. (13.71b)

$$\frac{\partial r}{\partial e} = -a\cos f$$

Use this and Lemma 2 to find

$$\frac{\partial r}{\partial e} - ir\frac{\partial f}{\partial e} = -a\cos f - i\left(a + \frac{r}{1-e^2}\right)\sin f \tag{13.83}$$

Then

$$\left(\frac{\partial \boldsymbol{r}}{\partial e}\right)_K = \mathrm{Re}\left[(\boldsymbol{l}_r + i\boldsymbol{l}_T)[-a\cos f - i\left(a + \frac{r}{1-e^2}\right)\sin f]\right] \tag{13.84}$$

$$= -\boldsymbol{l}_r a\cos f + \boldsymbol{l}_T\left(a + \frac{r}{1-e^2}\right)\sin f \tag{13.85}$$

With

$$\boldsymbol{F} = \boldsymbol{l}_r R + \boldsymbol{l}_T T + \boldsymbol{l}_w W$$

this gives

$$\boldsymbol{F}\cdot\frac{\partial \boldsymbol{r}}{\partial e} = a[-R\cos f + T(1 + r/p)\sin f] \tag{13.86}$$

because $p = a(1-e^2)$. Place this in Eq. (13.79) and use $-2\alpha_1 = \mu/a$. We find

$$N_2 = (\alpha_2/e\mu)[-R\cos f + T(1 + r/p)\sin f] \tag{13.87}$$

However,

$$\frac{\alpha_2}{e\mu} = \frac{\sqrt{\mu a(1-e^2)}}{e\mu} = \frac{\sqrt{1-e^2}}{ena} \tag{13.88}$$

so that

$$N_2 = -\frac{\sqrt{1-e^2}}{ena}\left[R\cos f - T\left(1 + \frac{r}{p}\right)\sin f\right] \tag{13.89}$$

Then

$$\dot{\beta}_2 = \dot{\omega} = -\dot{\Omega}\cos I - \frac{\sqrt{1-e^2}}{ena}\left[R\cos f - T\left(1+\frac{r}{p}\right)\sin f\right] \tag{13.90}$$

VII. Equations for $\dot{\beta}_1$ and $\dot{\ell}$

$$\dot{\beta}_1 = -\boldsymbol{F}\cdot\frac{\partial \boldsymbol{r}}{\partial \alpha_1} \tag{13.91}$$

Of the Keplerian variables, a, e, and ℓ depend on α_1. Thus

$$\frac{\partial \boldsymbol{r}}{\partial \alpha_1} = \left(\frac{\partial \boldsymbol{r}}{\partial a}\right)_K \frac{\partial a}{\partial \alpha_1} + \left(\frac{\partial \boldsymbol{r}}{\partial e}\right)_K \frac{\partial e}{\partial \alpha_1} + \left(\frac{\partial \boldsymbol{r}}{\partial \ell}\right)_K \frac{\partial \ell}{\partial \alpha_1} \tag{13.92}$$

$$-\dot{\beta}_1 = \frac{\partial a}{\partial \alpha_1}\boldsymbol{F}\cdot\left(\frac{\partial \boldsymbol{r}}{\partial a}\right)_K + \frac{\partial e}{\partial \alpha_1}\boldsymbol{F}\cdot\left(\frac{\partial \boldsymbol{r}}{\partial e}\right)_K + \frac{\partial \ell}{\partial \alpha_1}\boldsymbol{F}\cdot\left(\frac{\partial \boldsymbol{r}}{\partial \ell}\right)_K \tag{13.93}$$

From Sec. VI

$$-\frac{2\alpha_1\alpha_2}{e\mu^2}\boldsymbol{F}\cdot\left(\frac{\partial \boldsymbol{r}}{\partial e}\right)_K = \dot{\omega} + \dot{\Omega}\cos I \tag{13.94}$$

Thus

$$\boldsymbol{F}\cdot\left(\frac{\partial \boldsymbol{r}}{\partial e}\right)_K = -\frac{e\mu^2}{2\alpha_1\alpha_2}(\dot{\omega} + \dot{\Omega}\cos I)$$

Because

$$e^2 = 1 + \frac{2\alpha_1\alpha_2^2}{\mu^2} \qquad \frac{\partial e}{\partial \alpha_1} = \frac{\alpha_2^2}{e\mu^2}$$

then

$$\frac{\partial e}{\partial \alpha_1}\boldsymbol{F}\cdot\left(\frac{\partial \boldsymbol{r}}{\partial e}\right)_K = -\frac{\alpha_2^2}{e\mu^2}\frac{e\mu^2}{2\alpha_1\alpha_2}(\dot{\omega} + \dot{\Omega}\cos I)$$

However,

$$-\frac{\alpha_2^2}{e\mu^2}\frac{e\mu^2}{2\alpha_1\alpha_2} = \frac{\sqrt{1-e^2}}{n}$$

Thus

$$\frac{\partial e}{\partial \alpha_1}\boldsymbol{F}\cdot\left(\frac{\partial \boldsymbol{r}}{\partial e}\right)_K = \frac{\sqrt{1-e^2}}{n}(\dot{\omega} + \dot{\Omega}\cos I) \tag{13.95}$$

The term in $\partial \boldsymbol{r}/\partial a$:

$$\left(\frac{\partial \boldsymbol{r}}{\partial a}\right)_K = \mathrm{Re}\left[(\boldsymbol{l}_A + i\boldsymbol{l}_B)\frac{\partial}{\partial a}(r\varepsilon^{-if})\right] \tag{13.96}$$

$$r = \frac{a(1-e^2)}{1+e\cos f} \qquad \frac{\partial r}{\partial a} = \frac{r}{a} \tag{13.97}$$

Now f depends only on e and ℓ and not on a. Thus

$$\frac{\partial}{\partial a}(r\varepsilon^{-if}) = \frac{r}{a}\varepsilon^{-if} \tag{13.98}$$

and thus

$$\boldsymbol{F}\cdot\left(\frac{\partial \boldsymbol{r}}{\partial a}\right)_K = \frac{r}{a}\mathrm{Re}[\boldsymbol{F}\cdot(\boldsymbol{l}_r + i\boldsymbol{l}_T)] = \frac{r}{a}R \tag{13.99}$$

Because

$$a = -\frac{\mu}{2\alpha_1} \qquad \frac{\partial a}{\partial \alpha_1} = \frac{\mu}{2\alpha_1^2} = \frac{2a^2}{\mu}$$

then

$$\frac{\partial a}{\partial \alpha_1}\boldsymbol{F}\cdot\left(\frac{\partial \boldsymbol{r}}{\partial a}\right)_K = \frac{2a^2}{\mu}\frac{r}{a}R = \frac{2rR}{n^2a^2} \tag{13.100}$$

The term in $\partial \boldsymbol{r}/\partial \ell$:

$$\begin{aligned}\left(\frac{\partial \boldsymbol{r}}{\partial \ell}\right)_K &= \mathrm{Re}\left[(\boldsymbol{l}_A + i\boldsymbol{l}_B)\frac{\partial}{\partial \ell}(r\varepsilon^{-if})\right]\\ &= \mathrm{Re}\left[(\boldsymbol{l}_A + i\boldsymbol{l}_B)\varepsilon^{-if}\left(\frac{\partial r}{\partial \ell} - ir\frac{\partial f}{\partial \ell}\right)\right]\\ &= \mathrm{Re}\left[(\boldsymbol{l}_r + i\boldsymbol{l}_T)\left(\frac{\partial r}{\partial \ell} - ir\frac{\partial f}{\partial \ell}\right)\right]\end{aligned} \tag{13.101}$$

Then

$$\begin{aligned}\boldsymbol{F}\cdot\left(\frac{\partial \boldsymbol{r}}{\partial \ell}\right)_K &= \mathrm{Re}\left[(R + iT)\left(\frac{\partial r}{\partial \ell} - ir\frac{\partial f}{\partial \ell}\right)\right]\\ &= R\frac{\partial r}{\partial \ell} + Tr\frac{\partial f}{\partial \ell}\end{aligned} \tag{13.102}$$

For $\partial r/\partial \ell$, use

$$\begin{gathered} r = a(1 - e\cos E) \qquad \frac{\partial r}{\partial \ell} = ae\sin E\frac{\partial E}{\partial \ell}\\ E - e\sin E = \ell \qquad (1 - e\cos E)\frac{\partial E}{\partial \ell} = 1 \qquad \frac{\partial E}{\partial \ell} = \frac{a}{r}\\ \frac{\partial r}{\partial \ell} = \frac{a^2e}{r}\sin E = \frac{ae\sin f}{\sqrt{1-e^2}}\end{gathered} \tag{13.103}$$

For $\partial f/\partial \ell$, use

$$\tan\frac{f}{2} = \sqrt{\frac{1+e}{1-e}}\tan\frac{E}{2}$$

$$\frac{1}{2}\sec^2\frac{f}{2}\frac{\partial f}{\partial \ell} = \frac{1}{2}\sqrt{\frac{1+e}{1-e}}\sec^2\frac{E}{2}\frac{\partial E}{\partial \ell}$$

$$\frac{\partial f}{\partial \ell} = \sqrt{\frac{1+e}{1-e}}\frac{\cos^2(f/2)}{\cos^2(E/2)}\left(\frac{a}{r}\right)$$

$$= \frac{a}{r}\sqrt{\frac{1+e}{1-e}}\frac{1+\cos f}{1+\cos E}$$

However,

$$\cos E = \frac{e+\cos f}{1+e\cos f}$$

$$1+\cos E = \frac{r(1+e)}{p}(1+\cos f)$$

$$\frac{1+\cos f}{1+\cos E} = \frac{p}{r(1+e)}$$

Then

$$\frac{\partial f}{\partial \ell} = \frac{a}{r}\sqrt{\frac{1+e}{1-e}}\frac{p}{r(1+e)} = \frac{a^2}{r^2}\sqrt{1-e^2} \tag{13.104}$$

This proof could perhaps be shortened by using the expressions for an unperturbed orbit, viz.,

$$r^2\dot{f} = nab = na^2\sqrt{1-e^2}$$

$$\mathrm{d}\ell = n\,\mathrm{d}t$$

Then

$$r^2\frac{\mathrm{d}f}{\mathrm{d}t} = a^2\sqrt{1-e^2}$$

However, since we wish to be sure that the equation for $\partial f/\partial \ell$ holds generally for Keplerian variables, the first proof is perhaps more convincing. The expression that we shall soon obtain for $\dot{\ell}$ is not n. [See Eq. (13.109).]

Next, insert Eqs. (13.103) and (13.104) into Eqs. (13.102) to obtain

$$F \cdot \left(\frac{\partial \boldsymbol{r}}{\partial \ell}\right)_K = \frac{Rae\sin f}{\sqrt{1-e^2}} + \frac{Ta^2}{2}\sqrt{1-e^2} \tag{13.105}$$

For $\partial\ell/\partial\alpha_1$, use

$$\ell = n(t+\beta_1) \qquad \frac{\partial \ell}{\partial \alpha_1} = (t+\beta_1)\frac{\partial n}{\partial \alpha_1} = \frac{\ell}{n}\frac{\partial n}{\partial \alpha_1}$$

$$n = \mu^{\frac{1}{2}}a^{-\frac{3}{2}} \qquad a = -\frac{\mu}{2\alpha_1}$$

$$\frac{\partial n}{\partial \alpha_1} = -\frac{3}{2}\mu^{\frac{1}{2}}a^{-\frac{5}{2}}\frac{\partial a}{\partial \alpha_1} \qquad \frac{\partial a}{\partial \alpha_1} = \frac{\mu}{2\alpha_1^2} = \frac{2a^2}{\mu}$$

$$\frac{\partial n}{\partial \alpha_1} = -\frac{3}{2}\mu^{\frac{1}{2}}a^{-\frac{5}{2}}\frac{2a^2}{\mu} = -\frac{3}{na^2}$$

Then

$$\frac{\partial \ell}{\partial \alpha_1} = -\frac{3\ell}{n^2a^2} \tag{13.106}$$

Thus

$$\frac{\partial \ell}{\partial \alpha_1}\boldsymbol{F}\cdot\left(\frac{\partial \boldsymbol{r}}{\partial \ell}\right)_K = -\frac{3\ell}{n^2a^2}\left(\frac{Rae\sin f}{\sqrt{1-e^2}} + \frac{Ta^2}{2}\sqrt{1-e^2}\right)$$
$$= -\frac{3\ell}{n^2a\sqrt{1-e^2}}\left(eR\sin f + \frac{pT}{r}\right) \tag{13.107}$$

Inserting Eqs. (13.95), (13.100), and (13.107) into Eq. (13.93), we find

$$\dot\beta_1 = -\frac{2rR}{n^2a^2} - \frac{\sqrt{1-e^2}}{n}(\dot\omega + \dot\Omega\cos I) + \frac{3\ell}{n^2a\sqrt{1-e^2}}\left(eR\sin f + \frac{pT}{r}\right) \tag{13.108}$$

Now to find $\dot\ell$, use $\ell = n(t+\beta_1)$

$$\dot\ell = \dot n(t+\beta_1) + n(1+\dot\beta_1) = n + n\dot\beta_1 + (\dot n\ell/n) \tag{13.109}$$

We need $\dot n$

$$n = \mu^{\frac{1}{2}}a^{-\frac{3}{2}} \qquad \dot n = -\frac{3}{2}\mu^{\frac{1}{2}}a^{-\frac{5}{2}}\dot a = -\frac{3n\dot a}{2a}$$

From Sec. II,

$$\dot a = \frac{2}{n\sqrt{1-e^2}}[eR\sin f + T(1+e\cos f)]$$

Thus

$$\dot n = -\frac{3}{a\sqrt{1-e^2}}[eR\sin f + T(1+e\cos f)]$$

$$\frac{\dot n\ell}{n} = -\frac{3\ell}{na\sqrt{1-e^2}}[eR\sin f + T(1+e\cos f)]$$

$$n\dot\beta_1 = -\frac{2rR}{na^2} - \sqrt{1-e^2}(\dot\omega + \dot\Omega\cos I) + \frac{3\ell}{na\sqrt{1-e^2}}\left(eR\sin f + \frac{pT}{r}\right)$$

The $\dot{n}\ell/n$ cancels one of the terms in $n\dot{\beta}_1$, since $p/r = 1 + e\cos f$. Thus

$$\dot{\ell} = n - \frac{2rR}{na^2} - \sqrt{1-e^2}(\dot{\omega} + \dot{\Omega}\cos I) \tag{13.110}$$

VIII. Summary

$$\dot{a} = \frac{2}{n\sqrt{1-e^2}}[eR\sin f + T(1 + e\cos f)]$$

$$\dot{e} = \frac{\sqrt{1-e^2}}{na}[R\sin f + T(\cos E + \cos f)]$$

$$\dot{I} = \frac{rW\cos(\omega + f)}{na^2\sqrt{1-e^2}}$$

$$\dot{\Omega} = \frac{rW\csc I\sin(\omega + f)}{na^2\sqrt{1-e^2}}$$

$$\dot{\omega} = -\dot{\Omega}\cos I - \frac{\sqrt{1-e^2}}{ena}\left[R\cos f - T\left(1 + \frac{r}{p}\right)\sin f\right]$$

$$\dot{\ell} = n - \frac{2rR}{na^2} - \sqrt{1-e^2}\,(\dot{\omega} + \dot{\Omega}\cos I)$$

Chapter 14

Potential Theory

I. Introduction

IN SOLVING for the orbit of an artificial satellite around a planet, it is necessary to take into account the nonspherical figure of the planet. We shall first derive an approximate formula (MacCullagh's) for its gravitational potential and then derive the full expansion in spherical harmonics.

Let us consider the planet to be made up of particles, the ith one having mass m_i. Such a particle at Q_i will have a colatitude θ_i and a longitude λ_i, relative to axes fixed in the planet, with origin at the center of mass. Also, consider a field point P outside the planet, with colatitude θ and longitude λ as shown in Fig. 14.1. Then, for a source point i,

$$x_i + iy_i = r_i \sin\theta_i \; \varepsilon^{i\lambda_i}$$

$$z_i = r_i \cos\theta_i$$

and for the field point

$$x + iy = r \sin\theta \; \varepsilon^{i\lambda}$$

$$z = r \cos\theta$$

Assume that $r_i < r$ for every source point. There is a difficulty here, because a field point close to a pole of an oblate planet may be nearer the center of mass than a source point close to the surface in an equatorial plane. We shall not dwell on this difficulty now.

If $\boldsymbol{R}_i$ is the vector from a source point to the field point and $\boldsymbol{r}$ and $\boldsymbol{r}_i$ are the position vectors of the field point and the source point, then

$$\boldsymbol{R}_i = \boldsymbol{r} - \boldsymbol{r}_i$$

$$R_i^2 = r^2 + r_i^2 - 2rr_i \cos\psi_i$$

where ψ_i is the angle $(\boldsymbol{r}_i, \boldsymbol{r})$. Then

$$R_i^2 = r^2\left(1 - 2\frac{r_i}{r}\cos\psi_i + \frac{r_i^2}{r^2}\right)$$

$$\frac{1}{R_i} = \frac{1}{r}\left(1 - 2\frac{r_i}{r}\cos\psi_i + \frac{r_i^2}{r^2}\right)^{-\frac{1}{2}}$$

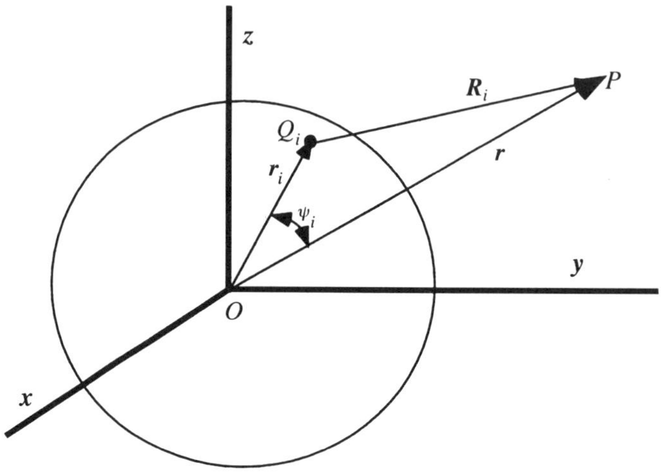

Fig. 14.1 Field point *P* outside a planet with a nonspherical figure.

The potential V at the field point is given by

$$V = -G\Sigma_i \frac{m_i}{R_i} = -GU$$

where

$$U = \frac{1}{r}\Sigma_i m_i \left(1 - 2\frac{r_i}{r}\cos\psi_i + \frac{r_i^2}{r^2}\right)^{-\frac{1}{2}} \tag{14.1}$$

For a field point sufficiently far from the planet that

$$\left|2\frac{r_i}{r}\cos\psi_i - \frac{r_i^2}{r^2}\right| < 1$$

for every source point, we may expand Eq. (14.1) by the binomial theorem

$$(1+\varepsilon)^{-\frac{1}{2}} = 1 - \frac{1}{2}\varepsilon + \frac{3}{8}\varepsilon^2 - \frac{15}{48}\varepsilon^3 + \cdots \tag{14.2}$$

so that

$$\left(1 - 2\frac{r_i}{r}\cos\psi_i + \frac{r_i^2}{r^2}\right)^{-\frac{1}{2}} = 1 + \frac{r_i}{r}\cos\psi_i - \frac{1}{2}\frac{r_i^2}{r^2} + \frac{3}{2}\frac{r_i^2}{r^2}\cos^2\psi_i + \cdots$$

Then

$$rU = \Sigma_i m_i + \frac{1}{r}\Sigma_i m_i r_i \cos\psi_i + \frac{1}{2r^2}\Sigma_i m_i\left(3r_i^2\cos^2\psi_i - r_i^2\right) + O\left(\frac{1}{r^3}\right) \tag{14.3}$$

If we choose a new Z axis along OP, then

$$Z_i = r_i \cos\psi_i$$

$$\Sigma_i m_i r_i \cos\psi_i = \Sigma_i m_i Z_i \cos\psi_i = M\bar{Z} = 0 \tag{14.4}$$

where $M = \Sigma_i m_i$ and $\bar{Z} = 0$ with the origin at the center of mass. Placing $\cos^2 = 1 - \sin^2$ in Eq. (14.3), we find

$$rU = M + \frac{1}{2r^2}(2\Sigma_i m_i r_i^2 - 3\Sigma_i m_i r_i^2 \sin^2 \psi_i) + O\left(\frac{1}{r^3}\right) \tag{14.5}$$

where M is the total mass of the planet. However, $r_i \sin \psi_i$ is the distance from Q_i to the OZ axis, so that

$$\Sigma_i m_i r_i^2 \sin^2 \psi_i = I \tag{14.6}$$

the moment of inertia about OP. Then

$$rU = M + \frac{1}{2r^2}\left(2\Sigma_i m_i r_i^2 - 3I\right) + \cdots \tag{14.7}$$

If the moments of inertia relative to the principal axes $O\xi$, $O\eta$, $O\zeta$ are A, B, C, then

$$A = \Sigma_i m_i\left(\eta_i^2 + \zeta_i^2\right) \qquad B = \Sigma_i m_i\left(\zeta_i^2 + \xi_i^2\right) \qquad C = \Sigma_i m_i\left(\xi_i^2 + \eta_i^2\right) \tag{14.8}$$

and

$$A + B + C = 2\Sigma_i m_i\left(\xi_i^2 + \eta_i^2 + \zeta_i^2\right) = 2\Sigma_i m_i r_i^2 \tag{14.9}$$

Thus, from Eqs. (14.7) and (14.9),

$$rU = M + \frac{1}{2r^2}(A + B + C - 3I) + \cdots \tag{14.10}$$

and

$$V = -\frac{GM}{r} - \frac{G}{2r^3}(A + B + C - 3I) + O\left(\frac{1}{r^4}\right) \tag{14.11}$$

This is MacCullagh's formula, which is good for many problems such as the theory of the precession and nutation of the Earth's axis, but not for the theory of satellite orbits.

II. Laplace's Equation

From

$$R_i = \left[(x - x_i)^2 - (y - y_i)^2 - (z - z_i)^2\right]^{\frac{1}{2}}$$

one deduces readily that $\nabla^2(1/R_i) = 0$ outside the planet, so that

$$\nabla^2 V = 0 \tag{14.12}$$

outside the planet. The spherical harmonic expansion of the potential that we wish to derive is an orthogonal expansion in separated solutions of this equation in spherical coordinates. With

$$x = r \sin\theta \cos\phi \qquad y = r \sin\theta \sin\phi \qquad z = r\cos\theta \tag{14.13}$$

the Laplace equation $\nabla^2 V = 0$ becomes

$$\frac{\partial^2 V}{\partial r^2} + \frac{2}{r}\frac{\partial V}{\partial r} + \frac{1}{r^2}\frac{\partial^2 V}{\partial \theta^2} + \frac{\cot\theta}{r^2}\frac{\partial V}{\partial \theta} + \frac{1}{r^2 \sin^2\theta}\frac{\partial^2 V}{\partial \phi^2} = 0 \tag{14.14}$$

After some manipulation, one finds

$$\frac{1}{r^2}\frac{\partial}{\partial r}\left(r^2 \frac{\partial V}{\partial r}\right) + \frac{1}{r^2 \sin\theta}\frac{\partial}{\partial \theta}\left(\sin\theta \frac{\partial V}{\partial \theta}\right) + \frac{1}{r^2 \sin^2\theta}\frac{\partial^2 V}{\partial \phi^2} = 0 \tag{14.15}$$

Here $\phi = \lambda$, the longitude, and θ is the colatitude.

To separate this equation, put

$$V = R(r)\Theta(\theta)\Phi(\phi) \tag{14.16}$$

Then

$$\frac{\Theta\Phi}{r^2}\frac{d}{\mathrm{d}r}(r^2 R') + \frac{R\Phi}{r^2 \sin\theta}\frac{d}{\mathrm{d}\theta}(\sin\theta\Theta') + \frac{R\Theta}{r^2 \sin^2\theta}\Phi'' = 0 \tag{14.17}$$

(The primed values of R, Θ, and Φ denote total derivatives.) Multiply this by $r^2 \sin^2\theta/(R\Theta\Phi)$ to obtain

$$\frac{\sin^2\theta}{R}\frac{d}{\mathrm{d}r}(r^2 R') + \frac{\sin\theta}{\Theta}\frac{d}{\mathrm{d}\theta}(\sin\theta\, \Theta') = -\frac{\Phi''}{\Phi} = m^2 \tag{14.18}$$

The left side depends only on r and θ and the right side only on ϕ, so that both are constant. The constant is chosen positive as m^2, since Φ would otherwise vary like $\exp\phi$ and would not be a single-valued function of position. Moreover, it is necessary that $m = 0, 1, 2, 3, \ldots$. Thus

$$\Phi = \text{linear combination of } \cos m\phi \text{ and } \sin m\phi \tag{14.19}$$

Next, divide Eq. (14.18) by $\sin^2\theta$ and transpose to obtain

$$\frac{1}{\Theta \sin\theta}\frac{d}{\mathrm{d}\theta}(\Theta' \sin\theta) - \frac{m^2}{\sin^2\theta} = -\frac{1}{R}\frac{d}{\mathrm{d}r}(r^2 R') \tag{14.20}$$

In Eq. (14.20), put $\Theta' \sin\theta = \Theta' \sin^2\theta/\sin\theta$ and denote $\cos\theta$ by x and Θ by y, where x and y are not to be confused with rectangular coordinates. Equation (14.20) becomes

$$\frac{1}{y}\frac{d}{\mathrm{d}x}\left[(1 - x^2)\frac{\mathrm{d}y}{\mathrm{d}x}\right] - \frac{m^2}{1 - x^2} = -\frac{1}{R}\frac{d}{\mathrm{d}r}(r^2 R') = -\lambda \tag{14.21}$$

where λ is a constant. With

$$\frac{\mathrm{d}y}{\mathrm{d}x} = y'$$

Eq. (14.21) becomes

$$(1 - x^2)y'' - 2xy' + \left(\lambda - \frac{m^2}{1 - x^2}\right)y = 0 \tag{14.22}$$

The solutions of Eq. (14.22) turn out to be finite for all θ between $-\pi$ and π only if λ is equal to an eigenvalue

$$\lambda = n(n+1) \qquad n = 0, 1, 2, 3, \ldots \tag{14.23}$$

Otherwise, y would become infinite at the poles ($x = \pm 1$). We next try to give some indication that this statement is true.

Rewrite Eq. (14.22) as

$$y'' - \frac{2xy'}{1-x^2} + \left(\lambda - \frac{m^2}{1-x^2}\right)\frac{y}{1-x^2} = 0 \tag{14.24}$$

This equation has singularities at $x = \pm 1$. At $x = +1$, we put $z = 1 - x$ and seek a solution by series in the form

$$y = z^\alpha \sum_{k=0}^{\infty} a_k z^k \tag{14.25}$$

Then

$$\frac{\mathrm{d}^2 y}{\mathrm{d}z^2} + \frac{2(1-z)}{z(2-z)}\frac{\mathrm{d}y}{\mathrm{d}z} + \left(\frac{\lambda}{z(2-z)} - \frac{m^2}{z^2(2-z)^2}\right) y = 0 \tag{14.26}$$

By Eq. (14.25)

$$y = a_0 z^\alpha + a_1 z^{\alpha+1} + a_2 z^{\alpha+2} + \cdots$$

$$\frac{\mathrm{d}y}{\mathrm{d}z} = a_0 \alpha z^{\alpha-1} + \cdots$$

$$\frac{\mathrm{d}^2 y}{\mathrm{d}z^2} = a_0 \alpha(\alpha-1) z^{\alpha-2} + \cdots$$

$$\frac{2(1-z)}{z(2-z)}\frac{\mathrm{d}y}{\mathrm{d}z} = a_0 \alpha z^{\alpha-2} + \cdots$$

$$\frac{\lambda y}{z(2-z)} = \frac{\lambda}{2} a_0 \alpha z^{\alpha-1} + \cdots$$

$$-\frac{m^2 y}{z^2(2-z)^2} = -\frac{m^2}{4} a_0 z^{\alpha-2} + \cdots$$

The first term in the series for Eq. (14.26) is

$$\left(a_0\alpha(\alpha-1) + a_0\alpha - \frac{m^2}{4}a_0\right) z^{\alpha-2}$$

Because a power series is unique and because the right side of Eq. (14.26) vanishes, we obtain the so-called indicial equation for α:

$$a_0\alpha(\alpha-1) + a_0\alpha - \frac{m^2}{4}a_0 = 0$$

or

$$a_0\left(\alpha^2-\frac{m^2}{4}\right)=0 \tag{14.27}$$

Thus

$$\alpha=\pm\frac{m}{2} \tag{14.28}$$

To obtain a solution finite at $x=+1$ (i.e., at $z=0$), we choose $\alpha=m/2$, since $m\geq 0$.

We may handle the situation at the other pole (i.e., at $x=-1$) by using $z=1+x$. It follows in the same way that $\alpha=m/2$ near $x=-1$. To put the two results together, we may then write

$$y=(1-x)^{m/2}(1-x)^{m/2}v(x)=(1-x^2)^{m/2}v(x) \tag{14.29}$$

Next insert Eq. (14.29) into Eq. (14.22) to obtain the differential equation for $v(x)$. The result is

$$(1-x^2)v''-2(m+1)xv'+(\lambda-m-m^2)v=0 \tag{14.30}$$

Since we have now taken care of indicial effects, we may now expand $v(x)$ as

$$v=\sum_{k=0}^{\infty}b_kx^k \tag{14.31}$$

It is known that there exists a regular solution for $v(x)$ over the whole interval $-1\leq x\leq 1$. This follows from Fuchs's theorem.[1,2] We shall show that for this to be true the series must terminate. Now insert Eq. (14.31) into Eq. (14.30). The result is

$$2b_2+6b_3x-2(m+1)b_1x+(\lambda-m-m^2)(b_0+b_1x)$$
$$+\sum_{k=2}^{\infty}x^k\big[(k+1)(k+2)b_{k+2}+(\lambda-m-m^2-2mk-k-k^2)b_k\big]=0 \tag{14.32}$$

Because the coefficient of x^k must vanish,

$$\frac{b_{k+2}}{b_k}=\frac{N}{D} \tag{14.33}$$

where

$$\begin{aligned}N&=k^2+k+2mk+m+m^2-\lambda\\&=(k+1)(k+2)+(2m-2)k+m+m^2-\lambda-2\end{aligned} \tag{14.34a}$$

$$D=(k+1)(k+2) \tag{14.34b}$$

Thus

$$\frac{b_{k+2}}{b_k}=1+\frac{(2m-2)k}{(k+1)(k+2)}+\frac{m+m^2-\lambda-2}{(k+1)(k+2)} \tag{14.35}$$

The series (14.31) for v breaks up into two series, a series of even powers and a series of odd powers.

III. The Eigenvalue Problem

We shall show next that both of these series diverge at $x = \pm 1$, unless the constant λ has certain characteristic values called eigenvalues. To do so, write

$$v(x) = u(x) + w(x) \tag{14.36}$$

where

$$u(x) = \sum_{j=0}^{\infty} b_{2j} x^{2j} = \sum_{j=0}^{\infty} a_j x^{2j} \tag{14.37a}$$

$$w(x) = \sum_{j=0}^{\infty} b_{2j+1} x^{2j} = \sum_{j=0}^{\infty} c_j x^{2j} \tag{14.37b}$$

Even Series

Here $k = 2j$ and Eq. (14.35) becomes

$$\frac{a_{j+1}}{a_j} = \frac{b_{2j+2}}{b_{2j}} = 1 + \frac{(2m-2)2j}{(2j+1)(2j+2)} + \frac{m + m^2 - \lambda - 2}{(2j+1)(2j+2)} \tag{14.38a}$$

Odd Series

$$\frac{c_{j+1}}{c_j} = \frac{b_{2j+3}}{b_{2j+1}} = 1 + \frac{(2m-2)(2j+1)}{(2j+2)(2j+3)} + \frac{m + m^2 - \lambda - 2}{(2j+2)(2j+3)} \tag{14.38b}$$

After some manipulations, these equations become

$$\frac{a_{j+1}}{a_j} = 1 + \frac{1-m}{j} + \frac{\theta_1}{j^2} \tag{14.39a}$$

$$\frac{c_{j+1}}{c_j} = 1 + \frac{1-m}{j} + \frac{\theta_2}{j^2} \tag{14.39b}$$

where

$$\theta_1 = \frac{(4 + m^2 - 5m - \lambda)j^3 + (2 - 2m)j^2}{j(2j+1)(2j+2)} \tag{14.40a}$$

$$\theta_2 = \frac{(6 + m^2 - 7m - \lambda)j^3 + (6 - 6m)j^2}{j(2j+2)(2j+3)} \tag{14.40b}$$

The ratio test for convergence or divergence of these series fails, because the ratio of successive terms approaches unity as $j \to \infty$.

There is a test due to Raabe, however, that works.[3] "If, at an endpoint, the successive terms of the series are of constant sign and if the ratio of the $(j+1)^{\text{th}}$

term to the jth can be expressed as $1 - q/j + \theta(j)/j^2$, where q is independent of k and $\theta(j)$ is bounded as $j \to \infty$, then the series converges if $q > 1$ and diverges if $q \leq 1$."

It is clear from Eqs. (14.40) that the θ's are bounded as $j \to \infty$. Also, in either case $q = 1 - m \leq 1$, because $m \geq 0$. Both series, the even and the odd, diverge unless they terminate. This means the series (14.31) for $v(x)$ diverges unless it terminates. By Eqs. (14.33) and (14.34) the series for v can terminate at some value $k = k_f$ if and only if

$$\lambda = k_f^2 + (2m + 1)k_f + m(m + 1) \tag{14.41}$$

This can be factored

$$\lambda = (k_f + m)(k_f + m + 1) \tag{14.42}$$

Put

$$n = k_f + m \tag{14.43}$$

The eigenvalues of λ are thus

$$\lambda = n(n + 1) \qquad n = 0, 1, 2, 3, \ldots \tag{14.44}$$

The factoring is unique. To show this, suppose $\lambda = \ell(\ell + 1)$, where ℓ is an integer. Then $\ell(\ell + 1) - n(n + 1) = 0$, a quadratic equation for ℓ with solutions $\ell = n$ or $\ell = -n - 1$. However, ℓ must be a positive integer, so that $\ell = n$.

Now consider the case $m = 0$; $v(x)$ becomes

$$(1 - x^2)v'' - 2xv' + n(n + 1)v = 0 \tag{14.45}$$

on putting $m = 0$ and $\lambda = n(n + 1)$ in Eq. (14.30). As we have seen, the solution takes the form

$$v(x) = u(x) + w(x) \tag{14.46}$$

where $u(x)$ is an even series and $w(x)$ an odd series. Here $u(x)$ begins with b_0 and $w(x)$ with $b_1 x$. We may write

$$v(x) = b_0 U(x) + b_1 W(x) \tag{14.47}$$

Since $v(x)$ is to be finite at $x = \pm 1$, either b_0 or b_1 must vanish because, for $m = 0$, $n = k_f$ and $\lambda = k_f(k_f + 2)$. Here k_f is either even or odd. If it is even, there is no odd k_f that can satisfy $\lambda = k_f(k_f + 2)$. That is, if the $U(x)$ series terminates, the $W(x)$ series cannot terminate. Similarly, if the $W(x)$ series terminates, the $U(x)$ series cannot terminate.

For $m > 0$, we have $k_f = n - m$, by Eq. (14.43). If $n - m$ is even, only the even series terminates, so that $b_1 = 0$ and v is an even polynomial in x, of degree $n - m$. If $n - m$ is odd, only the odd series terminates, so that $b_0 = 0$ and v is an odd polynomial in x of degree $n - m$.

Summary of the Θ Equation

$$(1 - x^2)\Theta'' - 2x\Theta' + \left(\lambda - \frac{m^2}{1 - x^2}\right)\Theta = 0$$

The solutions are finite at $x = \pm 1$ if and only if $\lambda = n(n + 1)$, $n = 0, 1, 2, 3, \ldots$.

Then

$$\Theta(x) = (1 - x^2)^{m/2} P_{nm}(x) = \sin^m \theta \, P_{nm}(\cos\theta) \tag{14.48}$$

where $P_{nm}(x)$ is a polynomial of degree $n - m$, containing only even powers or only odd powers.

IV. The $R(r)$ Equation

From Eq. (14.21)

$$\frac{1}{R}\frac{d}{\mathrm{d}r}(r^2 R') = \lambda = n(n+1) \tag{14.49}$$

To solve this, place $r = r_0 \varepsilon^z$. Then

$$\frac{\mathrm{d}^2 R}{\mathrm{d}z^2} + \frac{\mathrm{d}R}{\mathrm{d}z} - n(n+1)R = 0 \tag{14.50}$$

Here

$$R = \varepsilon^{pz} \tag{14.51}$$

is a solution where

$$p^2 + p - n(n+1) = 0 \tag{14.52}$$

so that

$$p = n \quad \text{or} \quad -n - 1 \tag{14.53}$$

and, therefore, ε^{nz} and $\varepsilon^{-(n+1)z}$ are solutions. That is, the solutions are $(r/r_0)^n$ and $(r_0/r)^{n+1}$. Thus

$$R = c_1 r^n + c_2 r^{-n-1} \tag{14.54}$$

Outside a planet, the potential becomes zero at $r = \infty$; so we may reject the r^n.

V. The Assembled Solution

The total solution of Laplace's equation for V is thus a linear combination of products of

$$r^{-n-1} \sin^m \theta \, P_{nm}(\cos\theta) \cos m\phi$$

$$r^{-n-1} \sin^m \theta \, P_{nm}(\cos\theta) \sin m\phi$$

It can be written

$$V = \sum_{n=0}^{\infty} r^{-n-1} \sum_{m=0}^{n} \big[C_{nm} \sin^m \theta \, P_{nm}(\cos\theta) \cos m\phi + S_{nm} \sin^m \theta \, P_{nm}(\cos\theta) \sin m\phi \big] \tag{14.55}$$

Our next task is to find $P_{nm}(\cos\theta)$, so that $\sin^m \theta \, P_{nm}(\cos\theta)$ will be the appropriate solution of the Θ equation. To do so, we approach the problem indirectly, by first considering certain Legendre polynomials $P_n(x)$. With $x = \cos\theta$, we shall show that $(1 - x^2)^{m/2}(d^m/\mathrm{d}x^m)[P_n(x)]$ satisfies the Θ equation.

VI. Legendre Polynomials

Consider the function

$$f(h, x) = (1 - 2xh + h^2)^{-\frac{1}{2}} \tag{14.56}$$

where x is a complex number and h a complex variable. The function f has singularities at those values of h that satisfy

$$h^2 - 2xh + 1 = 0 \tag{14.57}$$

viz.,

$$h_1 = x + \sqrt{x^2 - 1}$$

$$h_2 = x - \sqrt{x^2 - 1}$$

We define the Legendre polynomials by the expansion

$$f(h, x) = (1 - 2xh + h^2)^{-\frac{1}{2}} = \sum_{n=0}^{\infty} h^n P_n(x) \tag{14.58}$$

Here $f(h)$ is called the generating function for the Legendre polynomials $P_n(x)$. Note that by Eq. (14.58), $P_n(1) = 1$.

One could find the $P_n(x)$ by expanding Eq. (14.56) by the binomial theorem and collecting together the powers of h, but it would seem necessary that $|h^2 - 2xh| < 1$ for the validity of the expansion. In due time we shall develop another method for handling Eq. (14.56). By Eq. (14.57) the series that we find for $f(h, x)$ will then be valid for

$$|h| = \text{smaller of } \left|x \pm \sqrt{x^2 - 1}\right| \tag{14.59}$$

That is, it will be valid within any circle in the complex plane that does not include the nearest singularity. Such a power series expansion is unique, so that it must agree with that given by the binomial expansion.

From Eq. (14.58) one can develop various recursion formulas for the $P_n(x)$ by means of which one can prove that $P_n(x)$ satisfies Eq. (14.45).

$$(1 - x^2)v'' - 2xv' + n(n+1)v = 0 \tag{14.45}$$

is known as Legendre's equation. Proof that $P_n(x)$ satisfies Legendre's equation can be found in Refs. 3 and 4. [Certain other functions $Q_n(x)$ also satisfy Eq. (14.45), but they are not regular at $x = \pm 1$.]

VII. The Results for $P_n(x)$

Lagrange's expansion theorem, for which the proof can be found in Refs. 5 and 6, states that if

$$y = x + \alpha\phi(y) \tag{14.60}$$

then

$$F(y) = F(x) + \sum_{n=1}^{\infty} \frac{\alpha^n}{n!} \frac{d^{n-1}}{dx^{n-1}} (\phi^n(x) F'(x)) \tag{14.61}$$

α being "small." We shall apply this theorem to derive Rodrigue's formula,[7] which is

$$P_n(x) = \frac{1}{2^n n!} \frac{d^n}{dx^n}(x^2 - 1)^n \tag{14.62}$$

To do so, in Eq. (14.60) put $F(x) = x, \alpha = t/2$, and $\phi(y) = y^2 - 1$. Then, by Eq. (14.60)

$$y = x + \frac{t}{2}(y^2 - 1) \tag{14.63}$$

By Eq. (14.61)

$$y = x + \sum_{n=1}^{\infty} \frac{t^n}{2^n n!} \frac{d^{n-1}}{dx^{n-1}}[(x^2 - 1)^n] \tag{14.64}$$

Solve Eq. (14.63) for y:

$$y = \frac{1}{t}\left(1 \pm \sqrt{1 - 2xt + t^2}\right) \tag{14.65}$$

For small t, $y \approx x$, by Eq. (14.63), which tells us to choose the minus sign in Eq. (14.65):

$$y = \frac{1}{t}\left(1 - \sqrt{1 - 2xt + t^2}\right) \tag{14.66}$$

From Eq. (14.66)

$$\frac{\partial y}{\partial x} = (1 - 2xt + t^2)^{-\frac{1}{2}} \tag{14.67}$$

By Eqs. (14.67) and (14.58)

$$\frac{\partial y}{\partial x} = \sum_{n=0}^{\infty} t^n P_n(x) \tag{14.68}$$

However, by Eq. (14.64)

$$\frac{\partial y}{\partial x} = \sum_{n=0}^{\infty} \frac{t^n}{2^n n!} \frac{d^n}{dx^n}[(x^2 - 1)^n] \tag{14.69}$$

Comparison of Eqs. (14.68) and (14.69) yields Rodrigue's formula (14.62).

In Eq. (14.62), if one expands $(x^2 - 1)^n$ by the binomial theorem and differentiates n times, one obtains a polynomial expansion for $P_n(x)$. The calculation has to be done for n even and for n odd, but one can put the results together as

$$P_n(x) = \sum_{k=0}^{[n/2]} \frac{(-1)^k (2n - 2k)! x^{n-2k}}{2^n k!(n - 2k)!(n - k)!} \tag{14.70}$$

where $[n/2] = n$ if n is even and $(n - 1)/2$ if n is odd. The first few P_n's are

$$P_0 = 1$$

$$P_1 = x$$

$$P_2 = \tfrac{1}{2}(3x^2 - 1)$$

$$P_3 = \tfrac{1}{2}(5x^3 - 3x)$$

$$P_4 = \tfrac{1}{8}(35x^4 - 30x^2 + 3)$$

VIII. The Θ Solution for $m \geq 0$

For $m = 0$, we have $\lambda = n(n+1)$ by Eq. (14.22)

$$(1 - x^2)y'' - 2xy' + \left[n(n+1) - \frac{m^2}{1 - x^2}\right] y = 0 \tag{14.71}$$

Here $x = \cos\theta$ and $y = \Theta$.

Define

$$P_n^{(m)}(x) = \frac{d^m}{dx^m} P_n(x) \tag{14.72}$$

$$P_n^m(x) = (1 - x^2)^{m/2} P_n^{(m)}(x) \tag{14.73}$$

Consult Refs. 3 or 4 for a proof that $P_n^m(x)$ is a solution of the Θ equation (14.71). Note that for $m = 0$, $P_n^m(x)$ reduces to $P_n(x)$.

By Eq. (14.73) the quantity $\sin^m\theta\, P_{nm}(\cos\theta)$ of Eq. (14.55) is now $P_n^m(\cos\theta)$, so that the potential is expressible as

$$V = \sum_{n=0}^{\infty} r^{-n-1} \sum_{m=0}^{n} \left[C_{nm} P_n^m(\cos\theta) \cos m\phi + S_{nm} P_n^m(\cos\theta) \sin m\phi\right] \tag{14.74}$$

in place of Eq. (14.55). This may also be written as

$$V = \sum_{n=0}^{\infty} r^{-n-1} Y_n(\theta, \phi) \tag{14.75}$$

where

$$Y_n(\theta, \phi) = \sum_{m=0}^{n} P_n^m(\cos\theta)[C_{nm} \cos m\phi + S_{nm} \sin m\phi] \tag{14.76}$$

References

[1]Goursat, D., *Differential Equations,* translated by Hedrick and Dunnel, Dover Publications, 1945, pp. 134–136.

[2]Whittaker, B. T., and Watson, G. N., *A Course of Modern Analysis,* Cambridge Univ. Press, 1952, pp. 194–201.

[3]Woods, F. S., *Advanced Calculus,* Ginn and Co., Boston, 1926, pp. 268–272.

[4]Bethe, H., *Handbuch der Physik XXIVI,* Springer, Berlin, 1933, pp. 551–554.

[5]Smart, W. M., *Celestial Mechanics,* Longmans, Green, and Co., London, 1953, pp. 27–28.

[6]Brown, E. W., and Shook, C. A., *Planetary Theory,* Dover Publications, New York, 1964, pp. 37–39.

[7]Morse, P. M., and Feshbach, H., *Methods of Theoretical Physics,* McGraw–Hill, New York, 1953, Vol. I, p. 751.

Chapter 15

The Gravitational Potential of a Planet

I. The Addition Theorem for Spherical Harmonics

TO MAKE the series in Eq. (14.75) of the preceding chapter more definite, we need to obtain expressions for the coefficients C_{nm} and S_{nm}. To do this, we next develop an addition theorem for the Legendre polynomials and the associated functions $P_n^m(x)$.

In Fig. 15.1 let OQ' and OQ be unit vectors pointing, respectively, to a source point Q' and a field point Q. Let Q' have colatitude θ' and longitude ϕ' and Q have the values θ and ϕ. Also, let $(OQ, OQ') = \psi$. The addition theorem states that

$$P_n(\cos\psi) = P_n(\cos\theta)P_n(\cos\theta') + 2\sum_{m=1}^{n}\frac{(n-m)!}{(n+m)!}P_n^m(\cos\theta)P_n^m(\cos\theta')\cos(m\phi - m\phi') \tag{15.1}$$

(See Refs. 1 and 2.)

To prove Eq. (15.1), first write the Laplace equation in spherical coordinates

$$\nabla^2 V = \frac{1}{r^2}\frac{\partial}{\partial r}\left(r^2\frac{\partial V}{\partial r}\right) + \frac{1}{r^2\sin\theta}\frac{\partial}{\partial\theta}\left(\sin\theta\frac{\partial V}{\partial\theta}\right) + \frac{1}{r^2\sin^2\theta}\frac{\partial^2 V}{\partial\phi^2} = 0 \tag{15.2}$$

Then

$$\nabla^2 V = \frac{1}{r^2}\frac{\partial}{\partial r}\left(r^2\frac{\partial V}{\partial r}\right) + \frac{1}{r^2}M^2 V = 0 \tag{15.3}$$

where M^2 is the operator

$$M^2 = \frac{1}{\sin\theta}\frac{\partial}{\partial\theta}\left(\sin\theta\frac{\partial}{\partial\theta}\right) + \frac{1}{\sin^2\theta}\frac{\partial^2}{\partial\phi^2} \tag{15.4}$$

Because r and ∇^2 are both invariant to a rotation of the coordinate system, it follows that M^2 is also invariant. If we go from $Oxyz$ to a rotated system, $Ox'y'z'$, where

$$\begin{aligned} x + iy &= r\sin\theta\varepsilon^{i\lambda} & x' + iy' &= r\sin\psi\varepsilon^{i\beta} \\ z &= r\cos\theta & z' &= r\cos\psi \end{aligned} \tag{15.5}$$

then

$$M'^2 = M^2 \tag{15.6}$$

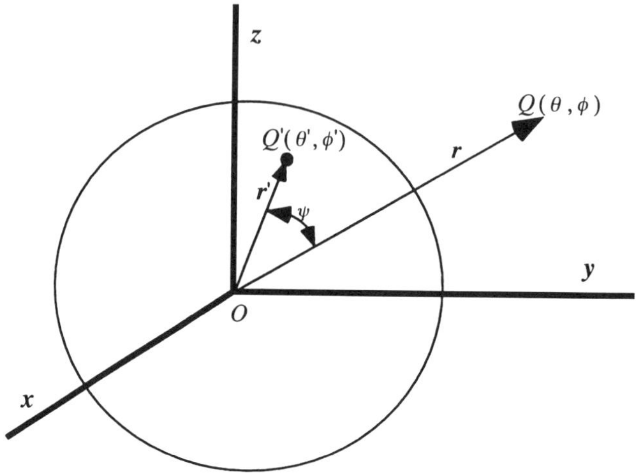

Fig. 15.1 Diagram of unit vectors OQ' and OQ.

That is,

$$\frac{1}{\sin\psi}\frac{\partial}{\partial\psi}\left(\sin\psi\frac{\partial}{\partial\psi}\right)+\frac{1}{\sin^2\psi}\frac{\partial^2}{\partial\beta^2}=\frac{1}{\sin\theta}\frac{\partial}{\partial\theta}\left(\sin\theta\frac{\partial}{\partial\theta}\right)+\frac{1}{\sin^2\theta}\frac{\partial^2}{\partial\phi^2} \tag{15.7}$$

Now a separated solution of $\nabla^2 V = 0$, viz.,

$$V = R(r)\,\Theta\,(\theta)\Phi(\phi) = R(r)Y(\theta,\phi) \tag{15.8}$$

satisfies, by Eq. (15.2)

$$\frac{Y}{r^2}\frac{\partial}{\partial r}\left(r^2\frac{\partial R}{\partial r}\right)+\frac{R}{r^2}M^2Y = 0 \tag{15.9}$$

However, by Eq. (14.49)

$$\frac{1}{R}\frac{d}{\mathrm{d}r}(r^2R') = n(n+1) \tag{15.10}$$

By Eqs. (15.9) and (15.10) Y satisfies

$$M^2Y_n + n(n+1)Y_n = 0 \tag{15.11}$$

where the subscript n on Y means that it corresponds to the eigenvalue n.

Since M^2 is invariant to a rotation, Y_n also satisfies

$$M'^2Y_n + n(n+1)Y_n = 0 \tag{15.12}$$

By Eq. (14.77),

$$Y_n(\theta,\phi) = \sum_{m=0}^{n} P_n^m(\cos\theta)[C_{nm}\cos m\phi + S_{nm}\sin m\phi] \tag{15.13}$$

which is the complete solution of Eq. (15.11) and thus of Eq. (15.12).

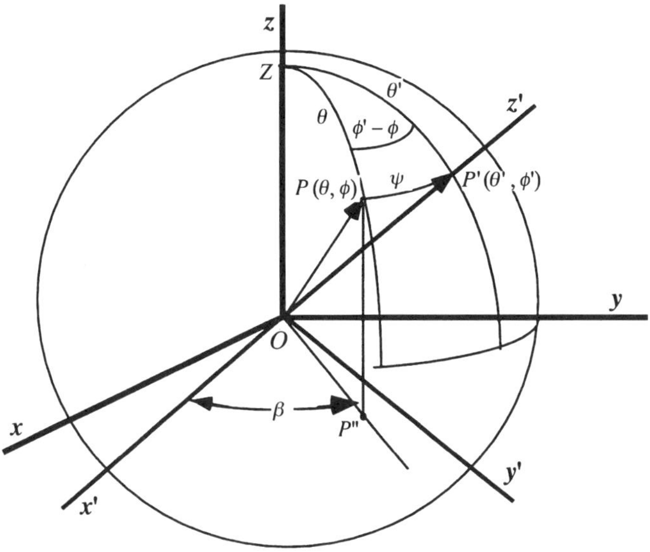

Fig. 15.2 Unit sphere containing points *P* and *P′*.

Next draw a unit sphere with points P and P' on it as shown in Fig. 15.2. Because $\phi' - \phi = PZP', \theta = (OP, OZ)$, and $\psi = (OP, OP')$,

$$\cos\psi = \cos\theta\cos\theta' + \sin\theta\sin\theta'\cos(\phi' - \phi) \tag{15.14}$$

Here the angular coordinates θ and ϕ are relative to $Oxyz$. Let us also use a rotated system $Ox'y'z'$, where Oz' is along OP'. Then, P has angular coordinates ψ and β in $Ox'y'z'$. The angle β is the angle from Ox' to the line OP'', where P'' is the foot of the perpendicular from P to the plane $Ox'y'$.

Now, Y_n satisfies both Eqs. (15.11) and (15.12). Because $P_n(\cos\psi)$ is a solution of Eq. (15.12), it also satisfies Eq. (15.11), so that

$$P_n(\cos\psi) = \sum_{m=0}^{n} P_n^m(\cos\theta)[a_{nm}\cos m\phi + b_{nm}\sin m\phi] \tag{15.15}$$

By Eq. (15.14), $\cos\psi$ is symmetric in θ and θ' and in ϕ and ϕ'. We can thus rewrite Eq. (15.15) as

$$P_n(\cos\psi) = c_{n0}P_n(\cos\theta)P_n(\cos\theta') + \sum_{m=1}^{n} P_n^m(\cos\theta)P_n^m(\cos\theta')$$
$$\times\ [c_{nm}\cos m\phi\cos m\phi' + d_{nm}\sin m\phi\sin m\phi'] \tag{15.16}$$

This equation must hold when P and P' are both coincident on Oz, in which case $\psi = 0$ and $\theta = \theta' = 0$. In this case, the P_n^m vanish, since they contain a factor $\sin^m\theta$. Also

$$P_n(\cos\psi) = P_n(\cos\theta) = P_n(\cos\theta') = P_n(1) = 1$$

Thus

$$c_{n0} = 1 \tag{15.17}$$

Next, specialize only to $\phi = \phi'$. Then $\psi = \theta - \theta'$ and

$$P_n[\cos(\theta - \theta')] = P_n(\cos\theta)P_n(\cos\theta') + \left[\sum_{m=1}^{n} P_n^m(\cos\theta)P_n^m(\cos\theta')\big(c_{nm}\cos^2 m\phi + d_{nm}\sin^2 m\phi\big)\right]_1 \tag{15.18}$$

Because the left side is independent of ϕ, so is the first term on the right side. This means that $[\ ]_1$ is independent of ϕ, and this can happen only if $d_{nm} = c_{nm}$, as may be shown by differentiation. Placing $c_{n0} = 1$ and $d_{nm} = c_{nm}$ in Eq. (15.16), we find

$$P_n(\cos\psi) = P_n(\cos\theta)P_n(\cos\theta') + \sum_{m=1}^{n} c_{nm}\, P_n^m(\cos\theta)P_n^m(\cos\theta')\cos(m\phi - m\phi') \tag{15.19}$$

To evaluate c_{nm}, multiply this equation by $P_n^p(\cos\theta)\cos p\phi$ and integrate over the unit sphere. On the left side, use for the surface element $\mathrm{d}S = \sin\psi\,\mathrm{d}\psi\,\mathrm{d}\beta$ and on the right $\mathrm{d}S = \sin\theta\,\mathrm{d}\theta\,\mathrm{d}\phi$. The ϕ integral on the right is

$$\int_0^{2\pi} \cos p\phi\cos(m\phi - m\phi')\,\mathrm{d}\phi = \pi\delta_{pm}\cos m\phi' \tag{15.20}$$

The right side becomes

$$\text{R.S.} = \pi c_{np} P_n^p(\cos\theta')\cos p\phi' \int_0^{\pi} \big(P_n^p(\cos\theta)\big)^2 \sin\theta\,\mathrm{d}\theta \tag{15.21}$$

However,

$$\int_0^{\pi} \big(P_n^p(\cos\theta)\big)^2 \sin\theta\,\mathrm{d}\theta = \frac{2(n+p)!}{(2n+1)(n-p)!} \tag{15.22}$$

(Ref. 3), so that the right side becomes

$$\text{R.S.} = \frac{2\pi c_{np}(n+p)!}{(2n+1)(n-p)!} P_n^p(\cos\theta')\cos m\phi' \tag{15.23}$$

The left side becomes

$$\text{L.S.} = \int_0^{\pi} \mathrm{d}\beta \int_0^{\pi} P_n^p(\cos\theta)\cos p\phi P_n(\cos\psi)\sin\psi\,\mathrm{d}\psi \tag{15.24}$$

The coefficient c_{np} is given by equating L.S. to R.S.

To evaluate L.S., note that $P_n^p(\cos\theta)\cos p\phi$ is a solution of Eq. (15.11) and thus of Eq. (15.12), so that

$$P_n^p(\cos\theta)\cos p\phi = f_{n0}P_n(\cos\psi) + \sum_{m=1}^{n} P_n^m(\cos\psi)[f_{nm}\cos m\beta + g_{nm}\sin m\beta] \tag{15.25}$$

To evaluate f_{n0}, note that if $\theta = \theta'$ and $\phi = \phi'$, then $\psi = 0$, so that $P_n(\cos\psi) = P_n(1) = 1$ and $P_n^m(\cos\psi) = 0$ for $m > 0$. Thus

$$f_{n0} = P_n^p(\cos\theta')\cos p\phi' \tag{15.26}$$

Also, note that the terms in $\cos m\beta$ and $\sin m\beta$ do not contribute anything to the integral in Eq. (15.24). Thus, Eq. (15.24) becomes

$$\begin{aligned} \text{L.S.} &= 2\pi P_n^p(\cos\theta')\cos p\phi' \int_0^\pi [P_n(\cos\psi)]^2 \sin\psi \,\mathrm{d}\psi \\ &= \frac{4\pi}{2n+1} P_n^p(\cos\theta')\cos p\phi' \end{aligned} \tag{15.27}$$

On equating Eq. (15.27) to Eq. (15.23), we find

$$c_{np} = \frac{2(n-p)!}{(n+p)!} \tag{15.28}$$

$$c_{nm} = \frac{2(n-m)!}{(n+m)!} \tag{15.29}$$

Insertion of Eq. (15.29) into Eq. (15.19) leads to Eq. (15.1), the desired addition theorem for spherical harmonics.

II. The Standard Series

From Eq. (14.1), we have

$$V = -\frac{G}{r}\Sigma_i m_i \left(1 - 2\frac{r_i}{r}\cos\psi_i + \frac{r_i^2}{r^2}\right)^{-\frac{1}{2}} \tag{15.30}$$

By the generating function for Legendre polynomials,

$$\left(1 - 2\frac{r_i}{r}\cos\psi_i + \frac{r_i^2}{r^2}\right)^{-\frac{1}{2}} = \sum_{n=0}^{\infty}\left(\frac{r_i}{r}\right)^n P_n(\cos\psi_i) \tag{15.31}$$

Thus

$$V = -\frac{G}{r}\Sigma_i m_i \sum_{n=0}^{\infty}\left(\frac{r_i}{r}\right)^n P_n(\cos\psi_i) \tag{15.32}$$

By the addition theorem of Sec. I,

$$\begin{aligned} P_n(\cos\psi_i) &= P_n(\cos\theta)P_n(\cos\theta_i) \\ &\quad + 2\sum_{m=0}^{n}\frac{(n-m)!}{(n+m)!}P_n^m(\cos\theta)P_n^m(\cos\theta_i)\cos(m\phi - m\phi_i) \end{aligned} \tag{15.33}$$

Here, r_i, θ_i, and ϕ_i are the spherical coordinates of a source point, and r, θ, and ϕ are those of the field point. Then

$$\begin{aligned} V = -\frac{G}{r}\sum_{n=0}^{\infty} r^{-n}\Sigma_i m_i r_i^n \Bigg[& P_n(\cos\theta)P_n(\cos\theta_i) \\ & + 2\sum_{m=0}^{n}\frac{(n-m)!}{(n+m)!}P_n^m(\cos\theta)P_n^m(\cos\theta_i)\cos(m\phi - m\phi_i)\Bigg] \end{aligned} \tag{15.34}$$

We next resolve the potential into zonal harmonics ($m = 0$) and into tesseral ($m > 0, m \neq n$) and sectorial harmonics ($m > 0, m = n$). Thus

$$V = V_Z + V_{TS} \tag{15.35}$$

Here the zonal part is given by

$$V_Z = -\frac{G}{r}\sum_{n=0}^{\infty} r^{-n}[\Sigma_i m_i r_i^n P_n(\cos\theta)P_n(\cos\theta_i)] \tag{15.36a}$$

and the tesseral-sectorial part by

$$V_{TS} = -\frac{G}{r}\sum_{n=1}^{\infty} 2r^{-n}\Sigma_i m_i r_i^n \sum_{m=1}^{n}\frac{(n-m)!}{(n+m)!}P_n^m(\cos\theta)P_n^m(\cos\theta_i)\cos(m\phi - m\phi_i) \tag{15.36b}$$

Note that each term in V, for a given n, is a solution of Laplace's equation. The zeros of the zonal harmonics divide the unit sphere into zones bounded by parallels of latitude. The zeros of the sectorial harmonics divide the unit sphere into lunes, bounded by meridians. The zeros of the tesseral harmonics divide the unit sphere into curved rectangles (tesserae) bounded both by parallels of latitude and by meridians. (For a graphical description, see Ref. 4.)

In the case of the Earth, the standard notation adopted here is the following. Let r_e be the equatorial radius of the Earth and $\mu = GM$, where M is the mass of the Earth. Then

$$V_Z = -\frac{\mu}{r}\left[1 - \sum_{n=1}^{\infty}\left(\frac{r_e}{r}\right)^n J_n P_n(\cos\theta)\right] \tag{15.37a}$$

$$V_{TS} = -\frac{\mu}{r}\sum_{n=1}^{\infty}\left(\frac{r_e}{r}\right)^n \sum_{m=1}^{n} P_n^m(\cos\theta)[C_{nm}\cos m\phi + S_{nm}\sin m\phi] \tag{15.37b}$$

Let us now compare Eq. (15.37a) with Eq. (15.36a) to obtain expressions for the J_n.

$n = 0$:

$$\mu = G\Sigma_i m_i = GM$$

$n = 1$:

$$\mu r_e J_1 = -G\Sigma_i m_i r_i P_1(\cos\theta_i) = -G\Sigma_i m_i r_i \cos\theta_i = -\mu\bar{z}$$

where $\bar{z}$ is the z coordinate of the Earth's center of mass. Thus

$$J_1 = -\frac{\bar{z}}{r_e} \tag{15.38}$$

Thus, J_1 vanishes if the origin is at the center of mass. This condition is ordinarily imposed in the reduction of satellite observations to determine the coefficient of potential. If one adopts standard values for station positions, there would of course be small errors in the J_n's unless one determined a corresponding nonvanishing J_1. Ideally, in reducing such observations, one should solve for station positions as well as J_n's while imposing the condition $J_1 = 0$.

General n for the Zonals

Comparison of Eqs. (15.37a) and (15.36a) yields

$$\mu r_e^n J_n = -G\Sigma_i m_i r_i^n P_n(\cos\theta_i) \tag{15.39}$$

Changing this from a sum to an integral makes this

$$J_n = -\frac{1}{M r_e^n}\int_{\text{Earth}} r^n P_n(\cos\theta)\rho\,\mathrm{d}\tau \tag{15.40}$$

where ρ is the density and $d\tau$ the volume element.

$n = 2$:

$$P_n(\cos\theta) = \frac{1}{2}(3\cos^2\theta - 1) = \frac{1}{2}\left(3\frac{z^2}{r^2} - 1\right)$$

Insertion of this into Eq. (15.40) gives, with $n = 2$

$$J_2 = -\frac{1}{2Mr_e^2}\int_{\text{Earth}} \rho(3z^2 - r^2)\,\mathrm{d}\tau \tag{15.41}$$

The integral is related to the moments of inertia

$$I_x = \int \rho(y^2 + z^2)\,\mathrm{d}\tau \qquad I_y = \int \rho(z^2 + x^2)\,\mathrm{d}\tau \qquad I_z = \int \rho(x^2 + y^2)\,\mathrm{d}\tau$$

Indeed

$$I_z - \frac{1}{2}(I_x + I_y) = \frac{1}{2}\int \rho(x^2 + y^2 - 2z^2)\,\mathrm{d}\tau = -\frac{1}{2}\int \rho(3z^2 - r^2)\,\mathrm{d}\tau \tag{15.42}$$

Comparison of Eqs. (15.41) and (15.42) shows that

$$J_2 = \frac{I_z - \frac{1}{2}(I_x + I_y)}{Mr_e^2} \tag{15.43}$$

Next, let the moments of inertia about the three principal axes be $A < B < C$. It is known for the Earth that the polar axis Oz lies very close to the principal axis of greatest moment of inertia C. Now, if we rotate the x and y axes, so that they coincide with the principal axes corresponding to A and B, the integrand in Eqs. (15.40) and (15.41) does not change. This means that J_2 is invariant to such a rotation, so that in Eq. (15.43) we can replace I_x, I_y, and I_z by A, B, and C, respectively. Thus

$$J_2 = \frac{C - \frac{1}{2}(A + B)}{Mr_e^2} \tag{15.44}$$

Note that, if a planet were a flat disk of uniform density, the value of J_2 would be only one-fourth, so that for small oblateness it is clear that $J_2 \ll 1/4$. Actually, for the Earth

$$\begin{aligned} J_2 &\approx 1082.63 \times 10^{-6} \\ J_3 &\approx -2.53 \times 10^{-6} \\ J_4 &\approx -1.61 \times 10^{-6} \end{aligned} \tag{15.45}$$

are World Geodetic System 1984 (WGS84) constants. Thus, J_3 and J_4 are of order J_2^2, and this behavior persists up to rather large values of n. What is the physical implication of this fact? We can write Eq. (15.40) as

$$J_n = -\frac{1}{M}\int_{\text{Earth}}\left(\frac{r}{r_e}\right)^n P_n(\cos\theta)\rho\,\mathrm{d}\tau \tag{15.46}$$

Because $r_e \geq r \geq 0$ and by Eq. (15.46), $(r/r_e)^n$ becomes very small for large values of n, unless $r \approx r_e$. The slow diminution of J_n as n increases implies that the higher coefficients arise mostly from matter near the surface, probably in the Earth's crust. Furthermore, if the density there were constant, the integral in Eq. (15.46) would vanish since, for $n > 0$,

$$\int_0^{2\pi}\mathrm{d}\phi\int_0^{\pi} P_n(\cos\theta)\sin\theta\,\mathrm{d}\theta = 2\pi\int_{-1}^{1} P_n(\lambda)\,\mathrm{d}\lambda = 0$$

Thus, there must be important density anomalies in the Earth's crust.

Tesseral-Sectorial Terms

On equating Eqs. (15.36b) and (15.37b), we find

$$\frac{M}{2}r_e^n(C_{nm} + iS_{nm}) = \frac{(n-m)!}{(n+m)!}\Sigma_i m_i r_i^n P_n^m(\cos\theta_i)\varepsilon^{im\phi_i} \tag{15.47}$$

$n = 1, m = 1$:

$$Mr_e(C_{11} + iS_{11}) = \Sigma_i m_i r_i P_1^1(\cos\theta_i)\varepsilon^{i\phi_i}$$

Now, with $\lambda = \cos\theta_i$

$$P_1(\lambda) = \lambda \qquad P_1^1(\lambda) = (1-\lambda^2)^{\frac{1}{2}} \qquad P_1^{(1)}(\lambda) = 1$$

Thus

$$Mr_e(C_{11} + iS_{11}) = \Sigma_i m_i r_i \sin\theta_i\varepsilon^{i\phi_i} = \Sigma_i m_i(x_i + iy_i) = M(\bar{x} + i\bar{y})$$

where $\bar{x}$ and $\bar{y}$ are coordinates of the center of mass. Thus

$$C_{11} = \bar{x}/r_e \qquad S_{11} = \bar{y}/r_e \tag{15.48}$$

With origin at the center of mass, C_{11} and S_{11} both vanish.

$n = 2, m = 1$: Eq. (15.47) yields

$$\frac{M}{2}r_e^2(C_{21} + iS_{21}) = \frac{1}{6}\Sigma_i m_i r_i^2 P_2^1(\cos\theta_i)\varepsilon^{i\phi_i}$$

Now

$$P_2(\lambda) = \tfrac{1}{2}(3\lambda^2 - 1) \qquad P_2^1(\lambda) = 3(1-\lambda^2)^{\frac{1}{2}}\lambda^2 \qquad P_2^{(1)}(\lambda) = 3\lambda$$

Thus

$$Mr_e^2(C_{21} + iS_{21}) = \Sigma_i m_i(r_i\cos\theta_i)(r_i\sin\theta_i)\varepsilon^{i\phi_i} = \Sigma_i m_i z_i(x_i + iy_i)$$

and therefore

$$C_{21} = \frac{\Sigma_i m_i z_i x_i}{Mr_e^2} \qquad S_{21} = \frac{\Sigma_i m_i z_i y_i}{Mr_e^2} \tag{15.49}$$

Both these coefficients vanish if the polar axis Oz is a principal axis. To show this, let Ox', Oy' be the principal axes. If α is the angle from Ox to Ox'

$$x = x'\cos\alpha - y'\sin\alpha \qquad y = x'\sin\alpha + y'\cos\alpha \tag{15.50}$$

By Eqs. (15.49) and Eqs. (15.50)

$$Mr_e^2 C_{21} = \Sigma_i m_i z_i (x_i' \cos\alpha - y_i' \sin\alpha)$$

$$Mr_e^2 S_{21} = \Sigma_i m_i z_i (x_i' \sin\alpha + y_i' \cos\alpha)$$

However, relative to the principal axes, all the products of inertia vanish, including $\Sigma_i m_i z_i x_i'$ and $\Sigma_i m_i z_i y_i'$. Therefore, if Oz is a principal axis, C_{21} and S_{21} vanish.

For Earth, the pole of rotation wanders by a small amount, very roughly over a circle of about 6-m radius, corresponding to an angle of about 0.2 arcsec between the pole and the mean pole. The mean pole is close to the axis of greatest moment of inertia, so that the wandering about the principal axis is small. It is, therefore, customary to put $C_{21} = S_{21} = 0$ in calculating orbits or in reducing satellite observations. For the moon, C_{21} and S_{21} are larger.

$n = 2$, $m = 2$: Eq. (15.47) yields

$$\frac{M}{2} r_e^2 (C_{22} + iS_{22}) = \frac{1}{24} \Sigma_i m_i r_i^2 P_2^2(\cos\theta_i)\varepsilon^{i2\phi_i} \tag{15.51}$$

With the use of

$$P_2^2(\lambda) = 3(1 - \lambda^2)$$

$$\cos 2\phi = \cos^2\phi - \sin^2\phi \qquad \sin 2\phi = 2\cos\phi\sin\phi$$

we obtain

$$\begin{aligned} Mr_e^2 C_{22} &= \tfrac{1}{4}\Sigma_i m_i \left(x_i^2 - y_i^2\right) \\ Mr_e^2 S_{22} &= \tfrac{1}{2}\Sigma_i m_i x_i y_i \end{aligned} \tag{15.52}$$

If all the axes were principal axes, we should have $S_{22} = 0$. This is not the case, however, because Ox passes through the Greenwich meridian and is not a principal axis. To find C_{22} and S_{22} in terms of moments of inertia, rewrite Eqs. (15.50) as

$$x + iy = (x' + iy')\varepsilon^{i\alpha} \tag{15.53}$$

Then

$$(x + iy) = (x' + iy')^2 \varepsilon^{i2\alpha}$$

$$x^2 - y^2 = (x'^2 - y'^2)\cos 2\alpha - 2x'y'\sin 2\alpha$$

$$2xy = (x'^2 - y'^2)\sin 2\alpha + 2x'y'\cos 2\alpha$$

Thus

$$Mr_e^2 C_{22} = \tfrac{1}{4}\left[\Sigma_i m_i \left(x_i'^2 - y_i'^2\right)\cos 2\alpha - 2\Sigma_i m_i x_i' y_i \sin 2\alpha\right]$$

$$Mr_e^2 S_{22} = \tfrac{1}{4}\left[\Sigma_i m_i \left(x_{i'}'^2 - y_i'^2\right)\sin 2\alpha + 2\Sigma_i m_i x_i' y_i' \cos 2\alpha\right]$$

Here

$$\Sigma_i m_i x_i' y_i' = 0$$

Because $A = \Sigma_i m_i (y_i'^2 + z_i'^2)$ and $B = \Sigma_i m_i (z_i'^2 + x_i'^2)$, we have

$$B - A = \Sigma_i m_i \left(x_i'^2 - y_i'^2\right)$$

so that

$$M r_e^2 (C_{22} + i S_{22}) = \tfrac{1}{4}(B - A)(\cos 2\alpha + i \sin 2\alpha)$$

or

$$C_{22} = \frac{(B - A)\cos 2\alpha}{4 M r_e^2} \qquad S_{22} = \frac{(B - A)\sin 2\alpha}{4 M r_e^2}$$

From G and $\mu = GM$, one can determine M and, thus, $B - A$, and α can be calculated from C_{22} and S_{22}. From J_2, one can determine $C - (B + A)/2$. It turns out that one can determine $C[C - (B + A)/2]^{-1}$ from data on precession and nutation of the polar axis; these data serve to determine A, B, and C.

III. Orthogonality of Spherical Harmonics

From Eqs. (15.37a) and (15.37b), the potential V can be expressed as

$$V = -\frac{\mu}{r} \sum_{n=0}^{\infty} \left(\frac{r_e}{r}\right)^n \sum_{m=0}^{n} P_n^m(\cos\theta)[C_{nm} \cos m\phi + S_{nm} \sin m\phi] \qquad (15.54)$$

To make this agree with Eqs. (15.37), we must put

$$C_{n0} = -J_n$$

$$S_{n0} = 0$$

$$J_0 = -1$$

A term $P_n^m(\cos\theta)\cos m\phi$ or $P_n^m(\cos\theta)\sin m\phi$ is called a surface spherical harmonic. Two such terms are distinct 1) if one has the cosine for ϕ and the other the sine; or 2) if both have cosines for ϕ or sines for ϕ, $m_1 \neq m_2$; or 3) if both have cosines for ϕ or both sines for ϕ and $m_1 = m_2$, then $n_1 \neq n_2$.

Two such functions ψ_1 and ψ_2 are said to be orthogonal over the unit sphere if

$$\int_0^{2\pi} \mathrm{d}\phi \int_0^{\pi} \psi_1 \psi_2 \sin\theta \, \mathrm{d}\theta = 0$$

Here, $\mathrm{d}S = \sin\theta \, \mathrm{d}\theta \, \mathrm{d}\phi$, the surface element on the unit sphere. The reader can easily verify that

$$P_{n1}^{m1}(\cos\theta)\begin{pmatrix}\cos m_1\phi \\ \sin m_1\phi\end{pmatrix} \qquad \text{and} \qquad P_{n2}^{m2}(\cos\theta)\begin{pmatrix}\cos m_2\phi \\ \sin m_2\phi\end{pmatrix}$$

are orthogonal if either case1 or 2 holds. A simple integration over ϕ from 0 to 2π shows this.

To show that any two distinct spherical harmonics are orthogonal, it remains only to consider case 3. The functions are orthogonal if

$$\int_0^{\pi} P_{n1}^m(\cos\theta)\, P_{n2}^m(\cos\,\theta)\sin\theta\,\mathrm{d}\theta = 0 \qquad (n_1 \neq n_2) \tag{15.55}$$

To prove Eq. (15.55), note that $P_n^m(\cos\theta)$ is simply the θ_{nm} of Sec. II. With $\theta = y$ and $\lambda = n(n+1)$, it satisfies Eq. (14.22), which can be written

$$\frac{d}{\mathrm{d}x}[(1-x^2)y'] + \left[n(n+1) - \frac{m^2}{1-x^2}\right]y = 0 \tag{15.56}$$

Now, let $y_1 = P_{n1}^m(\cos\theta)$, $y_2 = P_{n2}^m(\cos\theta)$ and recall that in Eq. (15.56), we have $x = \cos\theta$. The orthogonality condition (15.55) becomes

$$\int_{-1}^{1} y_1 y_2\,\mathrm{d}x = 0 \tag{15.57}$$

Proving Eq. (15.57) proves Eq. (15.55). To prove Eq. (15.57), write Eq. (15.56) once for y_1 with $n = n_1$ and once for y_2 with $n = n_2$, as follows

$$\frac{d}{\mathrm{d}x}[(1-x^2)y_1'] = -\left[n_1(n_1+1) - \frac{m^2}{1-x^2}\right]y_1 \tag{15.58a}$$

$$\frac{d}{\mathrm{d}x}[(1-x^2)y_2'] = -\left[n_2(n_2+1) - \frac{m^2}{1-x^2}\right]y_2 \tag{15.58b}$$

Multiply Eq. (15.58a) by y_2 and integrate over x from -1 to $+1$. Multiply Eq. (15.58b) by y_1 and integrate over x from -1 to $+1$. Take the difference of the two results. The reader should do this as an exercise; note that the integrals on the left must be evaluated by integrating by parts. The difference of the resulting right sides is zero. We obtain

$$[n_2(n_2+1) - n_1(n_1+1)]\int_{-1}^{1} y_1 y_2\,\mathrm{d}x = 0 \tag{15.59}$$

Thus, if $n_1 \neq n_2$, we obtain Eq. (15.57) and the orthogonality is proved.

Suppose a function is developed in an infinite series of orthogonal polynomials and the coefficients are $b_0, b_1, b_2, \ldots$. If we try to approximate the series by a finite sum of these functions, with coefficients $c_0, c_1, c_2, \ldots$, the integrated square of the error is a minimum if $c_k = b_k$, $k = 0, 1, 2, \ldots$. This is a well-known theorem, and its meaning for the development of the potential is clear. Once a certain number of spherical harmonic coefficients have been correctly determined for the Earth's potential field, the fit of the potential cannot be improved, in the least-square sense, by changing the coefficients.

For the case $m = 0$, the orthogonality of the spherical harmonics $P_n^m(\cos\theta) \times \cos m\phi$ leads to the orthogonality of the Legendre polynomials $P_n(x)$. That is

$$\int_{-1}^{1} P_n(x)P_k(x)\,\mathrm{d}x = 0 \qquad (n \neq k) \tag{15.60}$$

IV. The Normalized Coefficients and Harmonics

For large values of n, the coefficients of potential are small, and the corresponding spherical harmonics have large values. Because it is inconvenient in a long computation to do many multiplications of small numbers by large numbers, it has become customary to normalize the tesseral-sectorial harmonics and sometimes the zonal harmonics.

Denote quantities in the normalized system by superscript bars. Then

$$\bar{J}_n \bar{P}_n = J_n P_n \qquad \bar{C}_{nm} \bar{P}_n^m = C_{nm} P_n^m \qquad \bar{S}_{nm} \bar{P}_n^m = S_{nm} P_n^m$$

For this purpose, the harmonics are customarily normalized to 4π. That is

$$\int_0^{2\pi} \mathrm{d}\phi \int_0^{\pi} [\bar{P}_n^m(\cos\theta)]^2 \sin\theta \,\mathrm{d}\theta = 4\pi \tag{15.61}$$

or

$$\int_0^{\pi} \bar{P}_n^2 \sin\theta \,\mathrm{d}\phi = 2$$
$$\int_0^{2\pi} \mathrm{d}\theta \int_0^{\pi} [\bar{P}_n^m]^2 \begin{pmatrix} \cos^2 m\phi \\ \sin^2 m\phi \end{pmatrix} \sin\theta \,\mathrm{d}\theta = 4\pi \tag{15.62}$$

In dealing with Eq. (15.61), we need the integral

$$\int_{-1}^{1} [P_k(x)]^2 \,\mathrm{d}x = \frac{2}{2n+1} \tag{15.63}$$

It is easy to derive Eq. (15.63) by means of the generating function

$$(1 - 2xh + h^2)^{-\frac{1}{2}} = \sum_{n=0}^{\infty} h^n P_n(x) \tag{15.64}$$

Rewrite this as

$$(1 - 2xh + h^2)^{-\frac{1}{2}} = \sum_{k=0}^{\infty} h^k P_k(x) \tag{15.65}$$

Multiply Eq. (15.65) by Eq. (15.64) and integrate from -1 to $+1$, using the orthogonality. The result is

$$L \equiv \sum_{n=0}^{\infty} h^{2n} \int_{-1}^{1} [P_n(x)]^2 \,\mathrm{d}x = \int_{-1}^{1} (1 - 2xh + h^2)^{-1} \,\mathrm{d}x \tag{15.66}$$

If the integral is evaluated on the right side, by putting $u = 1 - 2xh + h^2$, it can be shown that

$$\int_{-1}^{1} (1 - 2xh + h^2)^{-1} \,\mathrm{d}x = h^{-1}[\ln(1+h) - \ln(1-h)]$$

Expansion by McLaurin's theorem reduces this to

$$L = 2\left(1 + \frac{h^2}{3} + \frac{h^4}{5} + \cdots + \frac{h^{2n}}{2n+1} + \cdots\right) \tag{15.67}$$

Equating coefficients of h^{2n} on both sides then yields

$$\int_{-1}^{1} [P_k(x)]^2 \, \mathrm{d}x = \frac{2}{2n+1} \tag{15.63}$$

as stated previously. We also need the integral

$$\int_{-1}^{1} \left[P_n^m(x)\right]^2 \mathrm{d}x = \frac{2}{2n+1}\frac{(n+m)!}{(n-m)!} \tag{15.68}$$

For $m > 0$, this cannot be evaluated so easily. It is done in Ref. 3 by repeated integration by parts.

With the use of Eqs. (15.61–15.63) and (15.68), we obtain for

Zonals:

$$\frac{\bar{P}_n}{P_n} = \frac{J_n}{\bar{J}_n} = (2n+1)^{\frac{1}{2}} \tag{15.69}$$

Tesseral-Sectorials:

$$\frac{\bar{P}_n^m}{P_n^m} = \frac{C_{nm}}{\bar{C}_{nm}} = \frac{S_{nm}}{\bar{S}_{nm}} = \left[\frac{2(2n+1)(n-m)!}{(n+m)!}\right]^{\frac{1}{2}} \qquad (m > 0) \tag{15.70}$$

Note that Eq. (15.69) does not follow from Eq. (15.70) by placing $m = 0$.

V. The Figure of the Earth

From the gravitational potential that we have deduced and from the apparent forces acting on a particle of water in the open sea, we can deduce the figure of the open sea, with disregard of tides, waves, and ocean currents. This figure is called the geoid, i.e., the Figure of the Earth (Fig. 15.3).

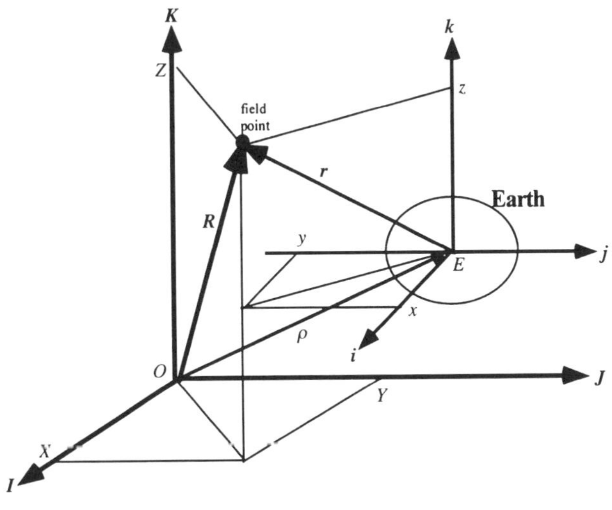

Fig. 15.3 Figure of the Earth.

If we let $Exyz$ be an Earth-fixed system and $OXYZ$ a truly inertial system, then a field point will have corresponding position vectors

$$\boldsymbol{r} = x\boldsymbol{i} + y\boldsymbol{j} + z\boldsymbol{k}$$

$$\boldsymbol{R} = X\boldsymbol{I} + Y\boldsymbol{J} + Z\boldsymbol{K}$$

The position vector of E relative to O will then be $\boldsymbol{\rho}$, where

$$\boldsymbol{R} = \boldsymbol{r} + \boldsymbol{\rho}$$

and

$$\ddot{\boldsymbol{R}} = \ddot{\boldsymbol{r}} + \ddot{\boldsymbol{\rho}} = \boldsymbol{f} + \boldsymbol{f}_S + \boldsymbol{f}_M + \boldsymbol{f}_D \tag{15.71}$$

Here

$\boldsymbol{f} \;\; = -\nabla V =$ gravitational field of the Earth
$\boldsymbol{f}_S =$ gravitational field of the sun
$\boldsymbol{f}_M =$ gravitational field of the moon
$\boldsymbol{f}_D =$ nongravitational force per unit mass

Now

$$\dot{\boldsymbol{r}} = \dot{x}\boldsymbol{i} + \dot{y}\boldsymbol{j} + \dot{z}\boldsymbol{k} + x\dot{\boldsymbol{i}} + y\dot{\boldsymbol{j}} + z\dot{\boldsymbol{k}}$$

Here

$$\boldsymbol{v} = \dot{x}\boldsymbol{i} + \dot{y}\boldsymbol{j} + \dot{z}\boldsymbol{k}$$

is the velocity of a particle relative to the Earth. The other term in $\dot{\boldsymbol{r}}$ can be found from

$$\dot{\boldsymbol{i}} = \boldsymbol{\omega} \times \boldsymbol{i} \qquad \dot{\boldsymbol{j}} = \boldsymbol{\omega} \times \boldsymbol{j} \qquad \dot{\boldsymbol{k}} = \boldsymbol{\omega} \times \boldsymbol{k}$$

where ω is the angular velocity of the Earth. Thus

$$x\dot{\boldsymbol{i}} + y\dot{\boldsymbol{j}} + z\dot{\boldsymbol{k}} = \boldsymbol{\omega} \times (x\boldsymbol{i} + y\boldsymbol{j} + z\boldsymbol{k})$$

so that

$$\dot{\boldsymbol{r}} = \boldsymbol{v} + \boldsymbol{\omega} \times \boldsymbol{r}$$

A second differentiation gives

$$\ddot{\boldsymbol{r}} = \boldsymbol{a} + 2\boldsymbol{\omega} \times \boldsymbol{v} + \boldsymbol{\omega} \times (\boldsymbol{\omega} \times \boldsymbol{r}) + \dot{\boldsymbol{\omega}} \times \boldsymbol{r} \tag{15.72}$$

as can be readily verified. Here $\boldsymbol{a}$ is the acceleration of the particle relative to the Earth.

Insert Eq. (15.72) into Eq. (15.71). The result is

$$\boldsymbol{a} = \boldsymbol{f} - \boldsymbol{\omega} \times (\boldsymbol{\omega} \times \boldsymbol{r}) - 2\boldsymbol{\omega} \times \boldsymbol{v} - \dot{\boldsymbol{\omega}} \times \boldsymbol{r} + \boldsymbol{f}_S + \boldsymbol{f}_M + \boldsymbol{f}_D - \ddot{\boldsymbol{\rho}} \tag{15.73}$$

In Eq. (15.73), the sum

$$\boldsymbol{f}_S + \boldsymbol{f}_M + \boldsymbol{f}_D = \boldsymbol{f}_{LS} \tag{15.74}$$

goes by various names—the lunar–solar perturbation, the tidal force, or the gravity-gradient force. It is small and would vanish for a particle at the center of the Earth. The force $-\dot{\boldsymbol{\omega}} \times \boldsymbol{r}$ is also small. The term $-2\boldsymbol{\omega} \times \boldsymbol{v}$ is the Coriolis force, which would vanish for a particle at rest in the Earth system. The term $-\boldsymbol{\omega} \times (\boldsymbol{\omega} \times \boldsymbol{r})$ is the centrifugal force. The term $\boldsymbol{f}_D$ is ordinarily a drag.

In defining the acceleration of gravity $\boldsymbol{g}$ only the first two terms in Eq. (15.73) are taken into account. Measurements of $\boldsymbol{g}$ must primarily correct for drag and Coriolis force. Thus, we define

$$\boldsymbol{g} = \boldsymbol{f} - \boldsymbol{\omega} \times (\boldsymbol{\omega} \times \boldsymbol{r}) \tag{15.75}$$

With disregard of any time change in $\boldsymbol{\omega}$, we have

$$\boldsymbol{\omega} = \omega_e \boldsymbol{k} \tag{15.76}$$

where $\boldsymbol{k}$ is along the polar axis and ω_e is the sidereal angular velocity of the Earth, approximately (366/365) times $2\pi/86{,}400$ radians per s.

It is easy to show that insertion of Eq. (15.76) into Eq. (15.75) yields

$$\boldsymbol{g} = \boldsymbol{f} + \omega_e^2 (x\boldsymbol{i} + y\boldsymbol{j}) \tag{15.77}$$

or

$$\boldsymbol{g} = \boldsymbol{f} + \nabla\left[\frac{\omega_e^2}{2}(x^2 + y^2)\right] \tag{15.78}$$

Because $\boldsymbol{f} = -\nabla V$, we have

$$\boldsymbol{g} = -\nabla \Omega \tag{15.79}$$

where

$$\Omega = V - \frac{\omega_e^2}{2}(x^2 + y^2) \tag{15.80}$$

is called the gravity potential.

The geoid is now defined as the level surface of Ω that includes mean sea level. Mean sea level is defined as the surface of the sea with tides, waves, and ocean currents averaged out. It must be a level surface of Ω, or else water would flow to make it so.

To find an equation for the geoid, we equate the gravity potential in Eq. (15.80) to its value at the equator. We shall neglect terms of order J_2^2 in V, so we may write Ω as

$$\Omega = -\frac{\mu}{r}\left[1 - \left(\frac{r_e}{r}\right)^2 J_2 P_2(\sin\theta)\right] - \frac{\omega_e^2}{2}(x^2 + y^2) \tag{15.81}$$

Here we take θ to be the latitude, rather than the colatitude. The equation of the geoid is thus

$$-\frac{\mu}{r}\left[1 - \left(\frac{r_e}{r}\right)^2 J_2\left(\frac{3}{2}\sin^2\theta - \frac{1}{2}\right)\right] - \frac{\omega_e^2}{2} r^2 \cos^2\theta = \Omega_0 \tag{15.82}$$

where

$$\Omega_0 = -\frac{\mu}{r_e}\left(1 - \frac{1}{2}J_2\right) - \frac{\omega_e^2 r_e^2}{2}$$

To simplify Eq. (15.82) put

$$r = r_e(1 + Q) \tag{15.83}$$

The value of Q at the poles is called the flattening F. Thus

$$r_p = r_e(1 - F) \tag{15.84a}$$

$$F = \frac{r_e - r_p}{r_e} \tag{15.84b}$$

Next, insert Eq. (15.83) into Eq. (15.82), neglecting J_2, Q, Q^2, and $\omega_e^2 r_e^3 Q/\mu$. The term $\omega_e^2 r_e^2/\mu$ is roughly the ratio of the centrifugal force at the equator to the gravitational force. It is easy to show that

$$Q = F \sin^2\theta \tag{15.85a}$$

where

$$F = \frac{3}{2}J_2 + \frac{\omega_e^2 r_e^3}{2\mu} \tag{15.85b}$$

Here $3J_2/2 = 0.00162$ and $\omega_e^2 r_e^3/(2\mu) = 0.00173$, so that the flattening $F = 1/298.5$. This corresponds to $r_e - r_p \approx 22$ km.

VI. Geoid as an Oblate Spheroid

We can now show that the Earth, as represented by the geoid, is approximately an oblate spheroid. This is an ellipsoid of revolution obtained by rotating an ellipse about its minor axis. To show this, note that by Eqs. (15.83) and (15.85a)

$$\frac{r}{r_e} = 1 - F\sin^2\theta \tag{15.86}$$

$$\frac{r^2}{r_e^2} = 1 - 2F\sin^2\theta + O(F^2) \tag{15.87}$$

Thus, approximately,

$$\frac{x^2 + y^2 + z^2}{r_e^2} = 1 - 2F\frac{z^2}{r_e^2} \tag{15.88}$$

or

$$\frac{x^2 + y^2}{r_e^2} + \frac{(1 + 2F)z^2}{r_e^2} = 1$$

or

$$\frac{x^2 + y^2}{r_e^2} + \left(\frac{z}{r_e/\sqrt{1 + 2F}}\right)^2 = 1 \tag{15.89}$$

This is the equation of an ellipsoid of revolution. A section through the z axis is an ellipse of semi-major axis r_e and semi-minor axis $r_p = r_e(1 + 2F)^{-\frac{1}{2}}$. If e is the eccentricity of such a meridian ellipse, we find

$$\sqrt{1-e^2} = \frac{r_p}{r_e} = (1+2F)^{-1/2}$$

$$1-e^2 = (1+2F)^{-1} = 1-2F+O(F^2)$$

$$e = \sqrt{2F}$$

For a flattening $F = 1/298.5$, the eccentricity is found to be about one-twelfth.

References

[1]Stratton, J. A., *Electromagnetic Theory,* McGraw–Hill, New York, 1941, pp. 406–408.

[2]Jackson, J. D., *Classical Electrodynamics,* 2nd ed., Wiley, New York, 1975, p. 101.

[3]Whittaker, E. T., and Watson, G. N., *Modern Analysis,* 4th ed., Cambridge Univ. Press, Cambridge, 1952, p. 325.

[4]Chobotov, V. A., *Orbital Mechanics,* AIAA Education Series, AIAA, Washington, DC, 1991, p. 233.

Chapter 16

Elementary Theory of Satellite Orbits with Use of the Mean Anomaly

I. A Few Numbers

FOR a drag-free close circular equatorial orbit, the period is $P = 2\pi/n$, where n is the mean motion $\mu^{1/2}r_e^{-3/2}$. Using the WGS84 constants, $\mu = 3.986005 \times 10^5$ km^3/s^2 and the equatorial radius $r_e = 6378.137$ km, we find $P = 5069$ s or about 84 min. Since the acceleration of gravity is approximately $g_e = \mu/r_e^2$, we can also express P as

$$P = 2\pi\left(r_e^3/\mu\right)^{\frac{1}{2}} = 2\pi(r_e/g_e)^{\frac{1}{2}}$$

This is the period of a Schuler pendulum, of length r_e, in a gravitational field equal to that at the Earth's surface; it occurs in the theory of inertial guidance.

The velocity in such an orbit is

$$v_{\text{close}} = nr_e = (\mu/r_e)^{\frac{1}{2}} = 7.905\,\text{km/s}$$

The escape velocity corresponds to zero energy, for which

$$\frac{1}{2}v_{\text{esc}}^2 - \frac{\mu}{r_e} = 0$$

so that

$$v_{\text{esc}} = (2\mu/r_e)^{\frac{1}{2}} = \sqrt{2}\,v_{\text{close}} = 11.18\,\text{km/s}$$

II. The Disturbing Function

In Chapter 10, the time derivations of the Keplerian elements were given in terms of the derivatives of the disturbing function, with respect to the six Keplerian elements a, e, I, Ω, ω, and $\ell = n(t-\tau)$. Here, the disturbing function is $F_1 = -V_1$, where V_1 is the part of the potential beyond $-\mu/r$ in its expansion in spherical harmonics. Since the oblateness term in J_2 is by far the largest term beyond $-\mu/r$, we shall deal only with it in a first look at the elementary solution for a drag-free satellite orbit.

Thus

$$V = -\frac{\mu}{r}\left[1 - \left(\frac{r_e}{r}\right)^2 J_2 P_2(\sin\theta) + \cdots\right] \tag{16.1}$$

leading to

$$F_1 = -V_1 = -\frac{\mu}{r}\left(\frac{r_e}{r}\right)^2 J_2 P_2(\sin\theta) \tag{16.2}$$

To obtain the first-order solution, we insert unperturbed values of the Keplerian elements on the right side of the Lagrange variational equations and integrate each one with respect to time. Because $\ell = n(t - \tau)$, such a procedure is equivalent to integrating with respect to ℓ, the mean anomaly. With the mean anomaly as an independent variable, we shall need to express the disturbing function F_1 as a Fourier series in ℓ. Before we do so, however, it is desirable to express F_1 as a function of f, the true anomaly; so we write

$$P_2(\sin\theta) = \tfrac{3}{2}\sin^2\theta - \tfrac{1}{2} \tag{16.3}$$

$$\sin\theta = \sin I \sin(\omega + f) \tag{16.4}$$

where θ is the latitude, I is the inclination, and ω is the argument of perigee. Then

$$\begin{aligned} P_2(\sin\theta) &= \tfrac{3}{2}\sin^2 I \sin^2(\omega + f) - \tfrac{1}{2} \\ &= \tfrac{3}{4}\sin^2 I[1 - \cos(2\omega + 2f)] - \tfrac{1}{2} \\ &= \tfrac{1}{4} - \tfrac{3}{4}\cos^2 I - \left[\tfrac{3}{4} - \tfrac{3}{4}\cos^2 I\right]\cos(2\omega + 2f) \end{aligned} \tag{16.5}$$

Insertion of Eqs. (16.5) into Eq. (16.2) gives the result

$$\begin{aligned} F_1 &= \frac{\mu}{r}\left(\frac{r_e}{r}\right)^2 J_2\left[-\frac{1}{4} + \frac{3}{4}\cos^2 I\right] \\ &+ \frac{\mu}{r}\left(\frac{r_e}{r}\right)^2 J_2\left[\frac{3}{4} - \frac{3}{4}\cos^2 I\right]\cos(2\omega + 2f) \end{aligned} \tag{16.6}$$

Here, the Keplerian elements ω and I are evident. The elements a, e, and ℓ are hidden, coming from r and f through the relations

$$r = a(1 - e\cos E) = \frac{a(1 - e^2)}{1 + e\cos f}$$

$$E - e\sin E = \ell$$

$$\cos f = \frac{\cos E - e}{1 - e\cos E} = \frac{a}{r}(\cos E - e)$$

$$\sin f = \frac{\sqrt{1 - e^2}\sin E}{1 - e\cos E} = \frac{a}{r}\sqrt{1 - e^2}\sin E$$

The element Ω does not appear in this F_1 that arises only from the second harmonic, because the latter is axially symmetric.

It is convenient to rewrite Eq. (16.6) as

$$F_1 = \frac{\mu r_e^2 J_2}{a^3}\left\{\left[-\frac{1}{4}+\frac{3}{4}\cos^2 I\right]\left(\frac{a}{r}\right)^3 + \left[\frac{3}{4}-\frac{3}{4}\cos^2 I\right]\left(\frac{a}{r}\right)^3 \cos(2\omega+2f)\right\} \tag{16.7}$$

We could work entirely with f, the true anomaly, as an independent variable, rather than with ℓ. We should be able to treat orbits with e approaching 1, but we shall defer such an approach to Chapter 17.

Using ℓ as an independent variable provides a parallel to the first approach to planetary theory and lunar theory. It will give practice in obtaining Fourier expansions of the Keplerian elements, necessary in so much of celestial mechanics. To obtain the Fourier series in ℓ for F_1, we must first build up to it by deriving Fourier series—or elliptic expansions as they are called—for $(a/r)^3$, $(a/r)^3\cos f$, and $(a/r)^3 \sin f$.

III. Elliptic Expansions[1]

cos E as a Fourier Series in ℓ

We have

$$E - e\sin E = \ell \qquad d\ell = (1-e\cos E)\,dE$$

Here, E is an odd function of ℓ, so that $E(-\ell) = -E(\ell)$ and

$$\cos E = \cos[-E(\ell)] = \cos[E(-\ell)]$$

Thus, $\cos E$ is even in ℓ.

Lemma: If any function of E is periodic in E with period 2π, it is also periodic in ℓ with period 2π.

To prove this, note that $1 - e\cos E > 0$ for $e < 1$, so that by $d\ell = (1 - e\cos E)\,dE$, we see that ℓ is monotonic in E. Hence, E is monotonic in ℓ. Thus, Kepler's equation makes either ℓ or E a single-valued function of the other. Also, let $f(E) = g(\ell)$ be periodic in E with period 2π. If $\Delta E = 2\pi$, it follows that $\Delta\ell = 2\pi$. Thus

$$g(\ell) = f(E) = f(E+2\pi) = g(\ell+2\pi)$$

This proves the lemma.

Thus, $\cos E$ is not only even in ℓ, but also periodic in ℓ with period 2π. It can be expanded as a cosine series in ℓ with period 2π:

$$\cos E = \frac{A_0}{2} + \sum_{k=1}^{\infty} A_k \cos k\ell \tag{16.8}$$

Integrate Eq. (16.8) from 0 to π to obtain

$$\frac{\pi}{2}A_0 = \int_0^{\pi} \cos E(1-e\cos E)\,dE = -e\frac{\pi}{2} \tag{16.8a}$$

so that

$$A_0 = -e$$

Next, multiply Eq. (16.8) by $\cos n\ell$ and integrate from 0 to π, with $n = 1, 2, 3, \ldots$. From the orthogonality of the functions $\cos k\ell$

$$A_n \int_0^\pi \cos^2 n\ell \, \mathrm{d}\ell = \int_0^\pi \cos E \cos n\ell \, \mathrm{d}\ell$$

so that

$$\frac{\pi}{2} A_n = \int_0^\pi \frac{\cos E}{n} \, \mathrm{d}(\sin n\ell) = \left[\frac{\cos E}{n} \sin n\ell \right]_0^\pi + \frac{1}{n} \int_0^\pi \sin n\ell \sin E \, \mathrm{d}E$$

or

$$A_n = \frac{2}{\pi n} \int_0^\pi \left[\frac{1}{2} \cos(n\ell - E) - \frac{1}{2} \cos(n\ell + E) \right] \mathrm{d}E$$

However

$$n\ell - E = n(E - e \sin E) - E = (n-1)E - ne \sin E$$

$$n\ell + E = n(E - e \sin E) + E = (n+1)E - ne \sin E$$

Thus

$$A_n = \frac{1}{\pi n} \int_0^\pi \{\cos[(n-1)E - ne \sin E] - \cos[(n+1)E - ne \sin E]\} \, \mathrm{d}E$$

The Bessel function $J_n(x)$ is defined by

$$\pi J_n(x) = \int_0^\pi \cos(n\theta - x \sin\theta) \, \mathrm{d}\theta$$

so that

$$A_n = \frac{1}{n} [J_{n-1}(ne) - J_{n+1}(ne)] \tag{16.9}$$

There is a recurrence relation

$$J_{n-1}(x) - J_{n+1}(x) = 2J_n'(x)$$

so that Eq. (16.9) can also be expressed as

$$A_n = \frac{2}{n} \frac{d}{\mathrm{d}(ne)} J_n(ne) = \frac{2}{n^2} \frac{d}{\mathrm{d}e} J_n(ne) \tag{16.10}$$

Use of Eqs. (16.8–16.10) yields for $\cos E$:

$$\cos E = -\frac{e}{2} + \sum_{n=1}^{\infty} \frac{1}{n} [J_{n-1}(ne) - J_{n+1}(ne)] \cos n\ell \tag{16.11}$$

or

$$\cos E = -\frac{e}{2} + \sum_{n=1}^{\infty} \frac{2}{n^2} \frac{d}{de} [J_n(ne)] \cos n\ell \tag{16.12}$$

We also have

$$\frac{r}{a} = 1 - \dot{e} \cos E = 1 + \frac{e^2}{2} - \sum_{n=1}^{\infty} \frac{2e}{n^2} \frac{d}{de} [J_n(ne)] \cos n\ell \tag{16.13}$$

sin E as a Fourier Series in ℓ

From $\ell = E - e \sin E$, it follows that ℓ is odd in E and E and sin E are odd in ℓ. Since sin E is periodic in E with period 2π, it follows from the lemma in Sec. III that sin E is periodic in ℓ with period 2π. Thus, sin E can be expanded as a Fourier sine series in ℓ:

$$\sin E = \sum_{k=1}^{\infty} B_k \sin k\ell$$

Multiply by $\sin n\ell \, \mathrm{d}\ell$ and integrate from 0 to π. From the orthogonality, it follows that

$$\begin{aligned}
\frac{\pi}{2} B_n &= \int_0^{\pi} \sin E \sin n\, \ell \, \mathrm{d}\ell = -\frac{1}{n} \int_0^{\pi} \sin E \, \mathrm{d}(\cos n\ell) \\
&= -\left[\frac{\sin E}{n} \cos n\ell \right]_0^{\pi} + \frac{1}{n} \int_0^{\pi} \cos n\ell \cos E \, \mathrm{d}E \\
&= \frac{1}{n} \int_0^{\pi} \cos n\ell \cos E \, \mathrm{d}E
\end{aligned}$$

Thus

$$B_n = \frac{1}{\pi n} \int_0^{\pi} [\cos(n\ell + E) + \cos(n\ell - E)] \, \mathrm{d}E$$

As before

$$n\ell \pm E = (n \pm 1)E - ne \sin E$$

Thus

$$\begin{aligned}
B_n &= \frac{1}{\pi n} \int_0^{\pi} \big[\cos[(n+1)E - ne \sin E] + \cos[(n-1)E - ne \sin E] \big] \, \mathrm{d}E \\
&= \frac{1}{n} [J_{n+1}(ne) + J_{n-1}(ne)]
\end{aligned}$$

and

$$\sin E = \sum_{n=1}^{\infty} \frac{1}{n} [J_{n+1}(ne) + J_{n-1}(ne)] \sin n\ell \tag{16.14}$$

However,

$$J_k(x) = \frac{x}{2k} [J_{k+1}(x) + J_{k-1}(x)]$$

a well-known recursion formula for the Bessel function. Thus

$$J_{n+1}(ne) + J_{n-1}(ne) = \frac{2}{e} J_n(ne)$$

so that

$$\sin E = \frac{2}{e} \sum_{n=1}^{\infty} \frac{J_n(ne)}{n} \sin n\ell \tag{16.15}$$

Then

$$E = \ell + e \sin E = 2 \sum_{n=1}^{\infty} \frac{J_n(ne)}{n} \sin n\ell \tag{16.16}$$

***a/r* as a Fourier Series in ℓ**

Because

$$\ell = E - e \sin E$$

it follows that

$$\frac{d\ell}{dE} = 1 - e \cos E = \frac{r}{a}$$

so that

$$\frac{a}{r} = \frac{dE}{d\ell}$$

Thus

$$\frac{a}{r} = 1 + 2 \sum_{n=1}^{\infty} J_n(ne) \cos n\ell \tag{16.17}$$

cos *f* as a Fourier Series in ℓ

Because

$$\frac{a}{r} = \frac{1 + e \cos f}{1 - e^2}$$

$$\cos f = -\frac{1}{e} + \frac{1 - e^2}{e} \frac{a}{r}$$

On applying Eq. (16.17), we find

$$\cos f = -\frac{1}{e} + \frac{1 - e^2}{e} + \frac{2(1 - e^2)}{e} \sum_{n=1}^{\infty} J_n(ne) \cos n\ell$$

or

$$\cos f = -e + \frac{2(1 - e^2)}{e} \sum_{n=1}^{\infty} J_n(ne) \cos n\ell \tag{16.18}$$

$\sin f$ as a Fourier Series in ℓ

From

$$\frac{r}{a} = \frac{1-e^2}{1+e\cos f}$$

$$\frac{d}{d\ell}\left(\frac{r}{a}\right) = \frac{e(1-e^2)}{(1+e\cos f)^2}\sin f\frac{df}{d\ell} = \frac{e(r/a)^2}{1-e^2}\sin f\frac{df}{d\ell}$$

However,

$$r^2\dot{f} = na^2\sqrt{1-e^2}$$

so that

$$r^2\frac{df}{d\ell} = a^2\sqrt{1-e^2}$$

Thus

$$\frac{d}{d\ell}\left(\frac{r}{a}\right) = \frac{e}{\sqrt{1-e^2}}\sin f$$

Then

$$\sin f = \frac{\sqrt{1-e^2}}{e}\frac{d}{d\ell}\left(\frac{r}{a}\right)$$

Using Eq. (16.13) to evaluate $[(d/d\ell)(r/a)]$, we find

$$\sin f = \sqrt{1-e^2}\sum_{n=1}^{\infty}\frac{2}{n}\frac{d}{de}[J_n(ne)]\sin n\ell \tag{16.19}$$

$r^{-2}\cos f$ and $r^{-2}\sin f$ as a Fourier Series in ℓ

The differential equation

$$\ddot{\boldsymbol{r}} = -\frac{\mu}{r^3}\boldsymbol{r}$$

can be split into

$$\ddot{\xi} = -\frac{\mu}{r^3}\xi \qquad \ddot{y} = -\frac{\mu}{r^3}y$$

where

$$\xi = r\cos f \qquad y = r\sin f$$

Thus

$$\frac{\cos f}{r^2} = \frac{\xi}{r^3} = -\frac{\ddot{\xi}}{\mu} = -\frac{n^2}{\mu}\frac{d^2\xi}{d\ell^2} = -\frac{1}{a^3}\frac{d^2\xi}{d\ell^2}$$

$$\frac{\sin f}{r^2} = \frac{y}{r^3} = -\frac{\ddot{y}}{\mu} = -\frac{n^2}{\mu}\frac{d^2y}{d\ell^2} = -\frac{1}{a^3}\frac{d^2y}{d\ell^2}$$

where $n = \sqrt{\mu a^{-3}}$, the mean motion. Now, in terms of the eccentric anomaly

$$\xi = a(\cos E - e) \qquad y = b \sin E = a\sqrt{1-e^2}\sin E$$

Thus

$$\frac{\mathrm{d}^2\xi}{\mathrm{d}\ell^2} = a\frac{d^2}{\mathrm{d}\ell^2}\cos E \qquad \frac{\mathrm{d}^2 y}{\mathrm{d}\ell^2} = b\frac{d^2}{\mathrm{d}\ell^2}\sin E$$

so that

$$\frac{\cos f}{r^2} = -\frac{1}{a^2}\frac{d^2}{\mathrm{d}\ell^2}\cos E \qquad \frac{\sin f}{r^2} = -\frac{b}{a^2}\frac{d^2}{\mathrm{d}\ell^2}\sin E$$

On inserting Eqs. (16.12) and (16.15) into these equations, we find

$$\frac{\cos f}{r^2} = \frac{2}{a^2}\sum_{p=1}^{\infty}\frac{d}{\mathrm{d}e}[J_p(pe)]\cos p\ell \tag{16.20}$$

$$\frac{\sin f}{r^2} = \frac{2\sqrt{1-e^2}}{ea^2}\sum_{p=1}^{\infty} pJ_p(pe)\sin p\ell \tag{16.21}$$

Fourier Series for the Disturbing Function F_1

In Eq. (16.7) for F_1, we need Fourier series for $(a/r)^3$, $(a/r)^3\cos 2f$, and $(a/r)^3\sin 2f$. We already have Fourier series for a number of functions and could develop one for $(a/r)^3$ by similar methods, but each coefficient would be an infinite series of Bessel functions. Instead, we proceed as follows.

Write down a/r as a Fourier series in ℓ and cube it. From Eq. (16.17) and Table 16.1, we find

$$a/r = 1 + e\cos\ell + e^2\cos 2\ell + O(e^3)$$

Then

$$(a/r)^3 = 1 + \frac{3}{2}e^2 + 3e\cos\ell + \frac{9}{2}e^2\cos 2\ell + O(e^3) \tag{16.22}$$

Table 16.1 Table of Bessel Functions[1]

p	$J_p(pe)$	$\frac{d}{\mathrm{d}e}J_p(pe)$
0	$1 - \frac{e^2}{4} + O(e^4)$	$-\frac{e}{2} + O(e^3)$
1	$\frac{e}{2} - \frac{e^3}{16} + O(e^5)$	$\frac{1}{2} - \frac{3e^2}{16} + O(e^4)$
2	$\frac{e^2}{2} + O(e^4)$	$e + O(e^3)$
3	$\frac{9e^3}{16} + O(e^5)$	$\frac{27e^2}{16} + O(e^4)$
4	$\frac{2e^4}{3} + O(e^6)$	$\frac{8e^3}{3} + O(e^5)$

To find the series for $(a/r)^3\cos(2\omega+2f)$, we build it up in the following way. From the series for $(a/r)^2\cos 2f$ and $(a/r)^2\sin 2f$, we find the series for $(a/r)^2\varepsilon^{if}$, then square this series to obtain the series for $(a/r)^4\varepsilon^{i2f}$. Multiply the result by the series for r/a to obtain $(a/r)^3\varepsilon^{i2f}$. Multiply this result by $\varepsilon^{i2\omega}$ to find the series for $(a/r)^3\varepsilon^{i(2\omega+2f)}$ and then take the real part $(a/r)^3\cos(2\omega+2f)$.

Next, verify that the intermediate results are

$$\frac{1}{2}\left(\frac{a}{r}\right)^2\cos f = \left(\frac{1}{2}-\frac{3}{16}e^2\right)\cos\ell + e\cos 2\ell + \frac{27}{16}e^2\cos 3\ell + O(e^3)$$

$$\frac{1}{2}\left(\frac{a}{r}\right)^2\sin f = \left(\frac{1}{2}-\frac{5}{16}e^2\right)\sin\ell + e\sin 2\ell + \frac{27}{16}e^2\sin 3\ell + O(e^3)$$

$$\frac{1}{2}\left(\frac{a}{r}\right)^2\varepsilon^{if} = \left(\frac{1}{2}-\frac{5}{16}e^2\right)\varepsilon^{i\ell} - \frac{i}{8}e^2\sin\ell + e\varepsilon^{i2f} + \frac{27}{16}e^2\varepsilon^{i3f} + O(e^3)$$

Square this last line to obtain

$$\frac{1}{4}(a/r)^4\varepsilon^{i2f} = \frac{1}{16}e^2 + \left[\frac{1}{4}-\frac{e^2}{4}\right]\varepsilon^{i2\ell} + e\varepsilon^{i3\ell} + \frac{43}{16}e^2\varepsilon^{i4\ell} + \cdots$$

Multiply this by

$$r/a = 1 + \frac{e^2}{2} - e\cos\ell - \frac{e^2}{2}\cos 2\ell + \cdots$$

to find

$$(a/r)^3\varepsilon^{i2f} = -\frac{e}{2}\varepsilon^{i\ell} + \left[1-\frac{5e^2}{2}\right]\varepsilon^{i2\ell} + \frac{7}{2}e\varepsilon^{i3\ell} + \frac{17}{2}e^2\varepsilon^{i4\ell} + \cdots$$

On multiplying this by $\varepsilon^{i2\omega}$ and taking the real part, we obtain

$$(a/r)^3\cos(2\omega+2f) = -\frac{e}{2}\cos(2\omega+\ell) + \left[1-\frac{5e^2}{2}\right]\cos(2\omega+2\ell) + \frac{7}{2}e\cos(2\omega+3\ell) + \frac{17}{2}e^2\cos(2\omega+4\ell) + O(e^3) \tag{16.23}$$

Next insert Eq. (16.22) for $(a/r)^3$ and Eq. (16.23) into Eq. (16.7) for F_1. The result is

$$F_1 = \frac{\mu r_e^2 J_2}{a^3}\Bigg[\left(-\frac{1}{4}+\frac{3}{4}\cos^2 I\right)\left(1+\frac{3}{2}e^2+3e\cos\ell+\frac{9}{2}e^2\cos 2\ell\right) + \frac{3}{4}\sin^2 I\Bigg\{-\frac{e}{2}\cos(2\omega+\ell)+\left[1-\frac{5e^2}{2}\right]\cos(2\omega+2\ell) + \frac{7}{2}e\cos(2\omega+3\ell)+\frac{17}{2}e^2\cos(2\omega+4\ell)\Bigg\}\Bigg]_1 + O(e^3) \tag{16.24}$$

IV. Solution of the Lagrange Variational Equations

In solving the Lagrange variational equations, we begin by placing unperturbed values of the Keplerian elements on the right sides of the equations. If a subscript zero denotes an initial value, the unperturbed values of the first five of these elements will be a_0, e_0, I_0, ω_0, and Ω_0. The unperturbed value for ℓ will be $n_0(t - \tau_0)$, where

$$n_0 = \mu^{\frac{1}{2}} a_0^{-\frac{3}{2}} \tag{16.25}$$

the initial mean motion. We shall call it simply ℓ, where

$$\ell = n_0 t + \ell_0 \tag{16.26}$$

$$\ell_0 = -n_0 \tau_0 \tag{16.26a}$$

Since the symbol ℓ is thus preempted for the unperturbed value of the mean anomaly, we use M for the perturbed mean anomaly. In integrating these variational equations with respect to t, we use, from Eq. (16.26)

$$\mathrm{d}t = \mathrm{d}\ell / n_0 \tag{16.27}$$

V. Motion of Perigee, First Approximation

For this purpose we have to integrate the equation derived in Chapter 10:

$$\frac{\mathrm{d}\omega}{\mathrm{d}t} = \frac{(1 - e^2)^{\frac{1}{2}}}{na^2 e} \frac{\partial F_1}{\partial e} - \frac{\cot I}{na^2(1 - e^2)^{\frac{1}{2}}} \frac{\partial F_1}{\partial I} \tag{16.28}$$

From Eq. (16.24)

$$\begin{aligned} \frac{\partial F_1}{\partial e} = \frac{\mu r_e^2 J_2}{a^3} \Bigg[& \left(-\frac{1}{4} + \frac{3}{4} \cos^2 I \right) (3e + 3\cos\ell + 9e\cos 2\ell) \\ & + \frac{3}{4} \sin^2 I \Bigg\{ -\frac{1}{2} \cos(2\omega + \ell) - 5e\cos(2\omega + 2\ell) \\ & + \frac{7}{2} \cos(2\omega + 3\ell) + 17e\cos(2\omega + 4\ell) \Bigg\} \Bigg]_2 + O(e^2) \end{aligned} \tag{16.29}$$

When we enter $\partial F_1 / \partial e$ into Eq. (16.28), we have to divide it by e, and this division increases the error to $0(e)$. To this order of accuracy, we can disregard the $(1 - e^2)^{1/2}$ in Eq. (16.28) and use

$$\frac{(1 - e^2)^{\frac{1}{2}}}{na^2 e} \frac{\partial F_1}{\partial e} = \frac{\mu r_e^2 J_2}{ena^5} [\ ]_2 + O(e) \tag{16.30}$$

Also

$$\frac{\partial F_1}{\partial I} = \frac{\mu r_e^2 J_2}{a^3} \left[-\frac{3}{2} \cos I \sin I + \frac{3}{2} \cos I \sin I \cos(2\omega + 2\ell) \right]_2 + O(e) \tag{16.31}$$

We have omitted here terms of order e and e^2, to correspond to the accuracy of Eq. (16.30). Then

$$-\frac{\cot I}{na^2(1-e^2)^{\frac{1}{2}}}\frac{\partial F_1}{\partial I} = \frac{\mu r_e^2 J_2}{na^5}\left[\frac{3}{2}\cos^2 I - \frac{3}{2}\cos^2 I\cos(2\omega+2\ell)\right]_2 + O(e) \tag{16.32}$$

Addition of Eqs. (16.30) and (16.32), with use of $\mu = n^2a^3$, gives

$$\begin{aligned}\frac{d\omega}{dt} = \frac{nr_e^2J_2}{a^2}\Bigg[&\left(-\frac{3}{4}+\frac{15}{4}\cos^2 I\right)+\left(-\frac{1}{4}+\frac{3}{4}\cos^2 I\right)\\ &\times\left(\frac{3}{e}\cos\ell + 9\cos 2\ell\right) - \frac{3}{2}\cos^2 I\cos(2\omega+2\ell) + \frac{3}{4}\sin^2 I\\ &\times\bigg\{-\frac{1}{2e}\cos(2\omega+\ell) - 5\cos(2\omega+2\ell) + \frac{7}{2e}\cos(2\omega+3\ell)\\ &+17\cos(2\omega+4\ell)\bigg\}\Bigg] + O(e)\end{aligned} \tag{16.33}$$

Place quantities on the right side of Eq. (16.33) equal to their unperturbed values and integrate it with use of Eq. (16.27). The result is

$$\begin{aligned}\omega = k_\omega + \frac{n_0r_e^2J_2}{a_0^2}\Bigg[&\left(-\frac{3}{4}+\frac{15}{4}\cos^2 I_0\right)t\\ &+\left(-\frac{1}{4}+\frac{3}{4}\cos^2 I_0\right)\left(\frac{3}{e_0}\sin\ell + \frac{9}{2}\sin 2\ell\right) - \frac{3}{4}\cos^2 I_0\sin(2\omega_0+2\ell)\\ &+\frac{3}{4}\sin^2 I_0\bigg\{-\frac{1}{2e_0}\sin(2\omega_0+\ell) - \frac{5}{2}\sin(2\omega_0+2\ell)\\ &+\frac{7}{6e_0}\sin(2\omega_0+3\ell) + \frac{17}{4}\sin(2\omega_0+4\ell)\bigg\}\Bigg] + O(e)\end{aligned} \tag{16.34}$$

If the perturbation is turned off at time t, the values of the Keplerian elements at that time are called the osculating elements. Thus, a_0, e_0, I_0, ω_0, Ω_0, and ℓ_0 are the osculating elements at $t = 0$. Equation (16.34) gives an approximate value for the osculating element at time t. The integration constant k_ω can be found by placing $\omega = \omega_0$ on the left and placing $t = 0$ and $\ell = \ell_0$ on the right.

Note that in Eq. (16.34) only the term in t has a nonvanishing time average. We find

$$\bar{\dot{\omega}} = \frac{3n_0r_e^2J_2}{4a_0^2}(5\cos^2 I_0 - 1) \tag{16.35}$$

This is the secular rate of change of ω, and the term from which it arises is called the secular variation, the other terms are short periodic. Note that $\bar{\dot{\omega}}$ vanishes if $\cos^2 I_0 = 1/5$; this corresponds to $I_0 = 63.4°$ or $116.6°$, the "critical inclinations."

Long periodic terms of order J_2 arise only when one carries the calculation through order J_2^2. They are terms like $\cos\omega$ or $\cos 2\omega$. To see what their period is like, we have to examine Eq. (16.35) numerically. Clearly, their period becomes infinite at a critical inclination.

At the inclination $I_0 = 0$, we find

$$\dot{\bar{\omega}} = \frac{3n_0 r_e^2 J_2}{a_0^2}$$

For a close orbit, this becomes

$$\dot{\bar{\omega}} \approx 3n_0 J_2 \approx (3n_0)10^{-3}$$

From the numbers in Sec. I, we see that for a close orbit $n_0 \approx 16$ revolutions per day, so that

$$\dot{\bar{\omega}} \approx 3n_0 J_2 \approx (48)10^{-3} \text{ revolutions per day}$$

or about 1/20 revolution per day. This means that, for an equatorial close orbit, the long periodic terms will have periods of about 20 days. The short periodic terms have periods of about 90 min.

Some of the preceding short periodic terms contain the eccentricity e_0 in denominators. This occurrence has already been noted in Chapter 10 and the cure for it mentioned, viz., use of the so-called "equinoctial elements."

VI. Motion of the Node, First Approximation

The appropriate variational equation is

$$\frac{d\Omega}{dt} = \frac{\csc I}{na^2(1-e^2)^{\frac{1}{2}}} \frac{\partial F_1}{\partial I} \tag{16.36}$$

On inserting Eq. (16.31) for $\partial F_1/\partial I$, we find, on replacing the Keplerian elements on the right by their unperturbed values,

$$\frac{d\Omega}{dt} = \frac{n_0 r_e^2 J_2}{a_0^2\left(1-e_0^2\right)^{\frac{1}{2}}}\left[-\frac{3}{2}\cos I_0 + \frac{3}{2}\cos I_0 \cos(2\omega_0 + 2\ell)\right] + O(e_0) \tag{16.37}$$

To order e_0, we can drop the $(1-e_0^2)^{1/2}$. Integration then yields

$$\Omega = k_\Omega + \frac{n_0 r_e^2 J_2}{a_0^2}\left[-\frac{3t}{2}\cos I_0 + \frac{3}{4n_0}\cos I_0 \sin(2\omega_0 + 2\ell)\right] + O(e_0) \tag{16.38}$$

Again, there is a secular term and a short periodic term. From the secular term, we find

$$\dot{\bar{\Omega}} = -\frac{3n_0 r_e^2 J_2}{2a_0^2}\cos I_0 \tag{16.39}$$

For a polar orbit, the first-order secular motion of the node vanishes. For a direct orbit, $\cos I_0 > 0$, and the node moves in a westerly direction. For a retrograde orbit, $\cos I_0 < 0$, and the node moves in an easterly direction. The maximum secular rate for the node occurs for an equatorial orbit, viz., just at the inclination for which the node ceases to have a meaning.

For a close-equatorial orbit, this becomes

$$\dot{\bar{\Omega}} = \frac{3}{2} n_0 J_2 \approx \left(\frac{3}{2} n_0\right) 10^{-3} \approx (24) 10^{-3} \text{ revolutions per day}$$

or one revolution in about 6 weeks.

VII. The Semi-Major Axis

The Lagrange variational equation is

$$\frac{\mathrm{d}a}{\mathrm{d}t} = \frac{2}{na} \frac{\partial F_1}{\partial \ell} \tag{16.40}$$

Using $\mathrm{d}t = \mathrm{d}\ell / n$, we find

$$a = \frac{2}{n^2 a} \int \frac{\partial F_1}{\partial \ell} \, \mathrm{d}\ell$$

or

$$a = k_a + \frac{2}{n_0^2 a_0} F_1 \tag{16.41}$$

From Eq. (16.24), the disturbing function F_1 is constant plus short periodic, so that δa has no secular part in the first-order approximation.

VIII. The Inclination

The variational equation is

$$\frac{\mathrm{d}I}{\mathrm{d}t} = \frac{1}{na^2 \sqrt{(1 - e^2)}} \left(\cot I \frac{\partial F_1}{\partial \omega} - \csc I \frac{\partial F_1}{\partial \Omega} \right) \tag{16.42}$$

Here, $\partial F_1 / \partial \Omega = 0$, and from Eq. (16.24), we find

$$\frac{\partial F_1}{\partial \omega} = \frac{\mu r_e^2 J_2}{a^3} \frac{3}{4} \sin^2 I \big[e \sin(2\omega + \ell) - (2 - 5e^2) \sin(2\omega + 2\ell) - 7e \sin(2\omega + 3\ell) - 17e^2 \sin(2\omega + 4\ell) \big] + O(e^3) \tag{16.43}$$

Place Eq. (16.43) in Eq. (16.42), use unperturbed quantities in the result, and integrate with use of $\mathrm{d}t = \mathrm{d}\ell / n$. We find

$$I = k_I + \frac{3}{4} \frac{r_e^2 J_2}{a_0^2 (1 - e_0^2)^{\frac{1}{2}}} \sin I_0 \cos I_0 \Big[-e_0 \cos(2\omega + \ell) + \left(1 - \frac{5}{2} e_0^2\right) \cos(2\omega_0 + 2\ell) + \frac{7}{2} e_0 \cos(2\omega_0 + 3\ell) + \frac{17}{4} e_0^2 \cos(2\omega_0 + 4\ell) \Big] + O\left(e_0^3\right) \tag{16.44}$$

There are no secular terms, only a constant plus short periodic terms.

IX. The Eccentricity

The variational equation is

$$\frac{\mathrm{d}e}{\mathrm{d}t} = \frac{1-e^2}{na^2 e}\left[\frac{\partial F_1}{\partial \ell} - (1-e^2)^{-\frac{1}{2}}\frac{\partial F_1}{\partial \omega}\right] \tag{16.45}$$

Because

$$\int \frac{\partial F_1}{\partial \ell}\mathrm{d}t = \frac{F_1}{n} + \text{const}$$

we have

$$e = k_e' + \frac{1-e_0^2}{n_0 a_0^2 e_0}\left[\frac{F_1}{n_0} - \frac{\left(1-e_0^2\right)^{-\frac{1}{2}}}{n_0}\int \frac{\partial F_1}{\partial \omega}\,\mathrm{d}\ell\right] \tag{16.46}$$

From Eq. (16.43)

$$\int \frac{\partial F_1}{\partial \omega}\,\mathrm{d}\ell = \frac{3\mu r_e^2 J_2}{4a_0^3}\sin^2 I_0\Bigg[-e_0\cos(2\omega_0+\ell) + \left(1-\frac{5}{2}e_0^2\right)\cos(2\omega_0+2\ell)$$
$$+\frac{7}{3}e_0\cos(2\omega_0+3\ell) + \frac{17}{4}e_0^2\cos(2\omega_0+4\ell)\Bigg]_2 + O\left(e_0^3\right) \tag{16.47}$$

$$F_1 = \frac{\mu r_e^2 J_2}{a^3}\Bigg[\left(-\frac{1}{4}+\frac{3}{4}\cos^2 I\right)\left(1+\frac{3}{2}e^2 + 3e\cos\ell + \frac{9}{2}e^2\cos 2\ell\right)$$
$$+\frac{3}{4}\sin^2 I\Bigg\{-\frac{e}{2}\cos(2\omega+\ell) + \left[1-\frac{5e^2}{2}\right]\cos(2\omega+2\ell)$$
$$+\frac{7}{2}e\cos(2\omega+3\ell) + \frac{17}{2}e^2\cos(2\omega+4\ell)\Bigg\}\Bigg]_1 + O(e^3)$$

Placing Eq. (16.47) in Eq. (16.46), we find

$$e = k_e' + \frac{\left(1-e_0^2\right)r_e^2 J_2}{a_0^2 e_0}\left([\,]_1 - \left(1-e_0^2\right)^{-\frac{1}{2}}\frac{3}{4}\sin^2 I_0[\,]_2\right) + O\left(e_0^3\right) \tag{16.48}$$

There is a term $1+3e_0^2/2$ that may be absorbed into k_e', so that we obtain

$$e = k_e + \frac{\left(1-e_0^2\right)r_e^2 J_2}{a_0^2 e_0}\Bigg\{\Bigg[\left(-\frac{1}{4}+\frac{3}{4}\cos^2 I_0\right)\left(3e_0\cos\ell + \frac{9}{2}e_0^2\cos 2\ell\right)$$
$$+\frac{3}{4}\sin^2 I_0\Bigg(-\frac{e_0}{2}\cos(2\omega_0+\ell) + \left(1-\frac{5}{2}e_0^2\right)\cos(2\omega_0+2\ell)$$
$$+\frac{7}{2}e_0\cos(2\omega_0+3\ell) + \frac{17}{2}e_0^2\cos(2\omega_0+4\ell)\Bigg)\Bigg]$$
$$-\left(1-e_0^2\right)^{-\frac{1}{2}}\frac{3}{4}\sin^2 I_0\Bigg[-e_0\cos(2\omega_0+\ell) + \left(1-\frac{5}{2}e_0^2\right)\cos(2\omega_0+2\ell)$$
$$+\frac{7}{3}e_0\cos(2\omega_0+3\ell) + \frac{17}{4}e_0^2\cos(2\omega_0+4\ell)\Bigg]\Bigg\} + O\left(e_0^3\right) \tag{16.49}$$

The variation of e is entirely short periodic in this first approximation.

X. Variation of the Mean Motion

Because

$$n = \mu^{\frac{1}{2}} a^{-\frac{3}{2}}$$

$$\delta n = -\frac{3}{2}\mu^{\frac{1}{2}} a^{-\frac{5}{2}} \delta a = -\frac{3n_0}{2a_0}\delta a \tag{16.50}$$

$$n = n_0\left[1 - \frac{3}{2}\frac{\delta a}{a_0}\right]$$

Now, by Eq. (16.41)

$$a = k_a + \frac{2}{n_0^2 a_0} F_1 \tag{16.51}$$

where F_1 is given in Eq. (16.24). In Eq. (16.24), the contribution to a of the term with $1 + 3e_0^2/2$ as a factor can be absorbed into the constant k_a in Eq. (16.51), so that we may write

$$a = k_a + J_2 Q \tag{16.52}$$

where Q is the product of $2/(n_0^2 a_0)$, $\mu r_e^2/(a_0^3)$, and that part of $[\ ]_1$ in Eq. (16.24) that does not contain $1 + 3e_0^2/2$. The product of the first two of these factors is $2r_e^2/a_0$, so that

$$\begin{aligned} Q = \frac{2r_e^2}{a_0}\Bigg[&\left(-\frac{1}{4} + \frac{3}{4}\cos^2 I_0\right)\left(3e_0\cos\ell + \frac{9}{2}e_0^2\cos 2\ell\right) \\ &+ \frac{3}{4}\sin^2 I_0\left(-\frac{e_0}{2}\cos(2\omega_0 + \ell) + \left(1 - \frac{5}{2}e_0^2\right)\cos(2\omega_0 + 2\ell)\right. \\ &\left.+ \frac{7}{2}e_0\cos(2\omega_0 + 3\ell) + \frac{17}{2}e_0^2\cos(2\omega_0 + 4\ell)\right)\Bigg] \end{aligned} \tag{16.53}$$

Denote by Q_0 the value of Q for $\ell = \ell_0$. Then

$$a_0 = k_a + J_2 Q_0 \tag{16.54}$$

so that

$$\delta a \equiv a - a_0 = J_2(Q - Q_0) \tag{16.55}$$

Insert Eq. (16.55) into Eq. (16.50) to find

$$n = n_0\left[1 + \frac{3}{2}\frac{J_2}{a_0}Q_0\right] - \frac{3}{2}\frac{n_0 J_2}{a_0}Q \tag{16.56}$$

This is the varied mean motion. Because the time average of Q vanishes, we find, for the average perturbed mean motion,

$$\bar{n} = n_0\left[1 + \frac{3}{2}\frac{J_2}{a_0}Q_0\right] \tag{16.57}$$

XI. Variation of the Mean Anomaly

With M as the perturbed mean anomaly, the variational equation is

$$\dot{M} = n - \frac{2}{na}\frac{\partial F_1}{\partial a} - \frac{1 - e^2}{na^2 e}\frac{\partial F_1}{\partial e} \tag{16.58}$$

Then

$$M - M_0 = \int_0^t n\,\mathrm{d}t - \frac{2}{n_0 a_0}\int_0^t \frac{\partial F_1}{\partial a}\,\mathrm{d}t - \frac{1-e_0^2}{n_0 a_0^2 e_0}\int_0^t \frac{\partial F_1}{\partial e}\,\mathrm{d}t \tag{16.59}$$

To find $\int_0^t n\,\mathrm{d}t$, use Eqs. (16.53) and (16.56). Before doing so, however, note that F_1 from Eq. (16.24) has an error of order e_0^3, so that F_1/e will have an error of order e_0^2 and the last integral in Eq. (16.59) an error of order e_0. The first two integrals in Eq. (16.59) should not be carried beyond an error of order e_0.

For evaluating $\int_0^t n\,\mathrm{d}t$, we can abbreviate Q from Eq. (16.53) to

$$Q = \frac{2r_e^2}{a_0}\frac{3}{4}\sin^2 I_0 \cos(2\omega_0 + 2\ell) \tag{16.60}$$

From Eqs. (16.56) and (16.60)

$$\int_0^t n\,\mathrm{d}t = n_0\left[1 + \frac{3}{2}\frac{J_2}{a_0}Q_0\right]t - \frac{9}{8}\frac{r_e^2 J_2}{a_0^2}\sin^2 I_0 \sin(2\omega_0 + 2\ell) + \text{const} \tag{16.61}$$

From Eq. (16.24)

$$-\frac{2}{na}\frac{\partial F_1}{\partial a} = -\frac{2}{na}\left(-\frac{3\mu r_e^2 J_2}{a^4}\right)\left[-\frac{1}{4} + \frac{3}{4}\cos^2 I + \frac{3}{4}\sin^2 I\cos(2\omega + 2\ell)\right] + O(e) \tag{16.62}$$

$$-\frac{2}{na}\int_0^t \frac{\partial F_1}{\partial a}\,\mathrm{d}t = \frac{3r_e^2 J_2}{2a_0^2}\left[n_0(-1 + 3\cos^2 I_0)\,t + \frac{3}{2}\sin^2 I_0 \sin(2\omega_0 + 2\ell)\right] + O(e_0) \tag{16.63}$$

$$\begin{aligned}
-\frac{1-e^2}{na^2 e}\frac{\partial F_1}{\partial e} &= -\frac{1-e^2}{na^2 e}\frac{\mu r_e^2 J_2}{a^3}\left[\left(-\frac{1}{4} + \frac{3}{4}\cos^2 I\right)(3e + 3\cos\ell + 9e\cos 2\ell)\right. \\
&\quad + \frac{3}{4}\sin^2 I\left\{-\frac{1}{2}\cos(2\omega + \ell) - 5e\cos(2\omega + 2\ell)\right. \\
&\quad \left.\left. + \frac{7}{2}\cos(2\omega + 3\ell) + 17e\cos(2\omega + 4\ell)\right\}\right] + O(e^2)
\end{aligned} \tag{16.64}$$

$$\begin{aligned}
-\frac{1-e^2}{na^2 e}\int_0^t \frac{\partial F_1}{\partial e}\,\mathrm{d}t &= \frac{3r_e^2 J_2}{4a_0^2}n_0(1 - 3\cos^2 I_0)t \\
&\quad + \frac{3r_e^2 J_2}{4a_0^2}(1 - 3\cos^2 I_0)\left(\frac{1}{e_0}\sin\ell + \frac{3}{2}\sin 2\ell\right) \\
&\quad + \frac{3r_e^2 J_2}{8a_0^2}\sin^2 I_0\left[\frac{1}{e_0}\sin(2\omega_0 + \ell) + 5\sin(2\omega_0 + 2\ell)\right. \\
&\quad \left. - \frac{7}{3e_0}\sin(2\omega_0 + 3\ell) - \frac{17}{2}\sin(2\omega_0 + 4\ell)\right] + O(e_0)
\end{aligned} \tag{16.65}$$

On adding the secular parts of Eqs. (16.61), (16.63), and (16.65), we find

$$M_s = n_0 t\left[1 + \frac{3r_e^2 J_2}{2a_0^2}\left\{1 - \frac{3}{2}\sin^2 I_0 + \frac{3}{2}\sin^2 I_0 \cos(2\omega_0 + 2\ell)\right\}\right] + O(e_0) \quad (16.66)$$

The short periodic part is

$$\begin{aligned} M_\ell = {} & \frac{3r_e^2 J_2}{4a_0^2}(1 - 3\cos^2 I_0)\left[\frac{1}{e_0}\sin\ell + \frac{3}{2}\sin 2\ell\right] \\ & + \frac{3r_e^2 J_2}{8a_0^2}\sin^2 I_0\left[\frac{1}{e_0}\sin(2\omega_0 + \ell) + 8\sin(2\omega_0 + 2\ell)\right. \\ & \left. - \frac{7}{3e_0}\sin(2\omega_0 + 3\ell) - \frac{17}{2}\sin(2\omega_0 + 4\ell)\right] + O(e_0) \end{aligned} \quad (16.67)$$

The average rate of change of the mean anomaly is then

$$\bar{\dot{M}} = \dot{M}_s = n_0\left[1 + \frac{3r_e^2 J_2}{2a_0^2}\left\{1 - \frac{3}{2}\sin^2 I_0 + \frac{3}{2}\sin^2 I_0 \cos(2\omega_0 + 2\ell_0)\right\}\right] + O(e_0)$$
$$(16.68)$$

This is not the same as the average value of the mean motion $\bar{n}$, which is

$$\bar{n} = n_0\left[1 + \frac{9}{4}\frac{r_e^2 J_2}{a_0^2}\sin^2 I_0 \cos(2\omega_0 + 2\ell_0)\right]$$

There is another form for $\bar{\dot{M}}$, when e is small, that can be obtained easily from initial conditions. Because

$$\sin\theta_0 = \sin I_0 \sin(\omega_0 + f_0)$$

it follows, when e is small, that

$$\sin\theta_0 \approx \sin I_0 \sin(\omega_0 + \ell_0) + O(e_0)$$

Then

$$\sin^2\theta_0 = \sin^2 I_0 \sin^2(\omega_0 + \ell_0) = \tfrac{1}{2}\sin^2 I_0 [1 - \cos(2\omega_0 + 2\ell_0)]$$

and

$$1 - 3\sin^2\theta_0 = 1 - \tfrac{3}{2}\sin^2 I_0 + \tfrac{3}{2}\sin^2 I_0 \cos(2\omega_0 + 2\ell_0)$$

so that

$$\bar{\dot{M}} = n_0\left[1 + \frac{3r_e^2 J_2}{2a_0^2}(1 - 3\sin^2\theta_0)\right] + O(e_0)$$

Use of $\bar{\dot{M}}$ in place of n_0 helps to improve the accuracy of first-order calculations. A second-order solution is barely possible with the use of the preceding methods. It can be carried far enough to show that it leads to long periodic terms of the first order in J_2 (see Ref. 2).

References

[1]Smart, W. M., *Celestial Mechanics,* Longmans, Green, and Co., London, 1953.

[2]Kovalevsky, J., *Introduction to Celestial Mechanics,* Springer–Verlag, New York, 1963, pp. 88–90.

Chapter 17

Elementary Theory of Satellite Orbits with Use of the True Anomaly

I. Introduction

CHAPTER 16 used the mean anomaly as an independent variable in treating satellite orbits to give some rough idea of the treatment of a planetary orbit. This method is valid only for small eccentricities. In this chapter, we shall use the true anomaly f as an independent variable, and this procedure will enable us to treat the case of eccentricities close to unity.

The Lagrange variational equations are the same, but we handle the disturbing function F_1 differently. Instead of expanding it in a Fourier series, we separate it into a part F_s that is constant in the first approximation and a part F_p that is short periodic. Then F_s gives rise to secular variations proportional to the time t and F_p to short periodic variations.

Again, we consider only the oblateness term in J_2, which is

$$F_1 = -\frac{\mu r_e^2}{r^3} J_2 P_2(\sin\theta) = -\frac{\mu r_e^2}{2a^3} J_2 \left(\frac{a}{r}\right)^3 (3\sin^2\theta - 1) \tag{17.1}$$

Then

$$F_s = P^{-1}\int_0^P F_1\, \mathrm{d}t = \frac{1}{2\pi}\int_0^{2\pi} F_1\, \mathrm{d}\ell \tag{17.2}$$

where P is the period of the unperturbed orbit and ℓ is the mean anomaly. In calculation of the first-order perturbations, we begin with unperturbed quantities on the right sides of the variational equations, so that among these quantities we have the relations

$$\mathrm{d}\ell = n\, \mathrm{d}t \tag{17.3}$$

$$P = 2\pi/n \tag{17.4}$$

$$n = \mu^{\frac{1}{2}} a^{-\frac{3}{2}} \tag{17.5}$$

$$r^2 \dot{f} = na^2\sqrt{1-e^2} = nab \tag{17.6}$$

Equations (17.3) and (17.4) were used to obtain Eq. (17.2) as an integral over ℓ. From Eqs. (17.3) and (17.6) we obtain

$$\mathrm{d}f = \frac{a^2}{r^2}\sqrt{1-e^2}\, \mathrm{d}\ell \tag{17.7}$$

or

$$d\ell = \frac{r^2}{a^2\sqrt{1-e^2}}\,df \tag{17.8}$$

From Eqs. (17.1), (17.2), and (17.8), we find

$$F_s = -\frac{\mu r_e^2(1-e^2)^{-\frac{1}{2}}J_2}{4\pi a^3}\int_0^{2\pi}\frac{a}{r}(3\sin^2\theta - 1)\,df \tag{17.9}$$

With use of the formulas for the unperturbed motion

$$\frac{a}{r} = \frac{1+e\cos f}{1-e^2} \qquad \sin\theta = \sin I\sin(\omega + f)$$

we find

$$3\sin^2\theta - 1 = \tfrac{3}{2}\sin^2 I[1-\cos(2\omega+2f)] - 1 \tag{17.9a}$$

and

$$F_s = -\frac{\mu r_e^2(1-e^2)^{-\frac{3}{2}}J_2}{4\pi a^3}\int_0^{2\pi}(1+e\cos f)\left[\frac{3}{2}\sin^2 I - 1 - \frac{3}{2}\sin^2 I\cos(2\omega+2f)\right]df \tag{17.10}$$

Because

$$\cos f\cos(2\omega+2f) = \tfrac{1}{2}\cos(2\omega+f) + \tfrac{1}{2}\cos(2\omega+3f)$$

the integrand Q becomes

$$Q = \tfrac{3}{2}\sin^2 I - 1 + \left(\tfrac{3}{2}\sin^2 I - 1\right)e\cos f - \tfrac{3}{2}\sin^2 I\cos(2\omega+2f) - \tfrac{3}{4}e\sin^2 I\cos(2\omega+f) - \tfrac{3}{4}e\sin^2 I\cos(2\omega+3f)$$

Of the terms in Q, only $(3\sin^2 I/2) - 1$ contributes to the integral, so that

$$F_s = -\frac{\mu r_e^2 J_2(1-e^2)^{-\frac{3}{2}}}{a^3}\left(\frac{3}{4}\sin^2 I - \frac{1}{2}\right) \tag{17.11}$$

To find F_p, first insert Eq. (17.9a) into Eq. (17.1) to obtain

$$F_1 = -\frac{\mu r_e^2 J_2}{a^3}\left(\frac{a}{r}\right)^3\left[\frac{3}{4}\sin^2 I - \frac{1}{2} - \frac{3}{4}\sin^2 I\cos(2\omega+2f)\right] \tag{17.12}$$

and then subtract Eq. (17.11) from Eq. (17.12). The result is

$$F_p = -\frac{\mu r_e^2 J_2}{a^3}\left(\frac{a}{r}\right)^3\left[\left\{\frac{3}{4}\sin^2 I - \frac{1}{2}\right\}\left\{1 - \left(\frac{r}{a}\right)^3(1-e^2)^{-\frac{3}{2}}\right\} - \frac{3}{4}\sin^2 I\cos(2\omega+2f)\right] \tag{17.13}$$

Thus

$$F_1 = F_s(a, e, I) + F_p(a, e, I, \omega, \ell)$$

II. Derivatives with Respect to *e*

To calculate short periodic parts of the derivatives of the Keplerian elements, we shall need Keplerian formulas for $\partial/\partial e(a/r)$ and $\partial f/\partial e$. In these differentiations, all other Keplerian elements a, I, ω, Ω, and ℓ are to be kept fixed. These results were derived in Chapter 13, and we simply quote them here.

$$\frac{\partial}{\partial e}\left(\frac{a}{r}\right) = -\frac{a}{r^2}\frac{\partial r}{\partial e} = \left(\frac{a}{r}\right)^2 \cos f \tag{17.14}$$

$$\frac{\partial f}{\partial e} = \left(\frac{a}{r} + \frac{1}{1-e^2}\right)\sin f \tag{17.15}$$

III. The Semi-Major Axis *a*

The Lagrange variational equation is

$$\frac{\mathrm{d}a}{\mathrm{d}t} = \frac{2}{na}\frac{\partial F_1}{\partial \ell} = \frac{2}{na}\frac{\partial F_p}{\partial \ell} \tag{17.16}$$

because $F_1 = F_s + F_p$ and F_s does not depend on ℓ. With the use of unperturbed quantities on the right side, we have $\mathrm{d}\ell = n\,\mathrm{d}t$, so that

$$a - a_0 = \frac{2}{n_0 a_0}\int_0^t \frac{\partial F_p}{\partial \ell}\,\mathrm{d}t = \frac{2}{n_0^2 a_0}\int_{\ell_0}^{\ell} \frac{\partial F_p}{\partial \ell}\,\mathrm{d}\ell \tag{17.17}$$

or

$$a - a_0 = \frac{2}{n_0^2 a_0}[F_p(\ell) - F_p(\ell_0)] \tag{17.18}$$

where $F_p(\ell)$ is given by Eq. (17.13). We see that $F_p(\ell_0)$ is given by putting zero as a subscript on a, r, e, ω, ℓ, and I in $F_p(\ell)$.

It is of some interest to derive this equation in another way. If a is the osculating semi-major axis, we have as an exact equation

$$\frac{1}{2}v^2 - \frac{\mu}{r} = -\frac{\mu}{2a} \tag{17.19}$$

where v is the velocity. The perturbed energy is

$$W = \frac{1}{2}v^2 - \frac{\mu}{r} - F_1 = \text{const} \tag{17.20}$$

From Eqs. (17.19) and (17.20)

$$\frac{\mu}{2a} = -W - F_1(\ell) \tag{17.21}$$

$$\frac{\mu}{2a_0} = -W - F_1(\ell_0) \tag{17.22}$$

On taking the difference of Eqs. (17.21) and (17.22), we find

$$\frac{\mu}{2a} = \frac{\mu}{2a_0} + F_1(\ell_0) - F_1(\ell)$$

or

$$\frac{1}{a} = \frac{1}{a_0}\left[1 + \frac{2a_0}{\mu}\{F_1(\ell_0) - F_1(\ell)\}\right] \tag{17.23}$$

Now, the term involving $F_1(\ell_0) - F_1(\ell)$ has a factor J_2. If we call it $J_2\varepsilon$, we have

$$\frac{1}{a} = \frac{1}{a_0}[1 + J_2\varepsilon]$$

so that

$$a = a_0[1 - J_2\varepsilon] + O\left(J_2^2\right)$$

This becomes

$$a = a_0\left[1 + \frac{2a_0}{\mu}\{F_1(\ell) - F_1(\ell_0)\}\right] + O\left(J_2^2\right)$$

or

$$a = a_0\left[1 + \frac{2}{n_0^2 a_0^2}\{F_1(\ell) - F_1(\ell_0)\}\right] + O\left(J_2^2\right) \tag{17.24}$$

the same as Eq. (17.18).

IV. The Eccentricity *e*

The Lagrange variational equation contains derivatives of the disturbing function F_1 with respect to ℓ and ω. Since F_s does not depend on these Keplerian elements, we may replace F_1 by F_p, so that

$$\frac{\mathrm{d}e}{\mathrm{d}t} = \frac{1-e^2}{na^2e}\left[\frac{\partial F_p}{\partial \ell} - (1-e^2)^{-\frac{1}{2}}\frac{\partial F_p}{\partial \omega}\right] \tag{17.25}$$

Then

$$e - e_0 = \frac{1-e_0^2}{n_0 a_0^2 e_0}\left(\frac{1}{n_0}[F_p(\ell) - F_p(\ell_0)] - \frac{(1-e^2)^{-\frac{1}{2}}}{n_0}\int_{\ell_0}^{\ell}\frac{\partial F_p}{\partial \omega}\,\mathrm{d}\ell\right) \tag{17.26}$$

on using $\mathrm{d}\ell = n_0\,\mathrm{d}t$.

From Eq. (17.13)

$$\frac{\partial F_p}{\partial \omega} = -\frac{3\mu r_e^2 J_2}{2a^3}\left(\frac{a}{r}\right)^3 \sin^2 I \sin(2\omega + 2\ell) \tag{17.27}$$

since f depends only on ℓ and e, and not on ω. To calculate the integral of this with respect to ℓ, we use Eq. (17.8) and insert the usual zeros as subscripts. We obtain

$$\int_{\ell_0}^{\ell}\frac{\partial F_p}{\partial \omega}\,\mathrm{d}\ell = -\frac{3\mu r_e^2 J_2}{2a_0^3}\sin^2 I_0\left(1-e_0^2\right)^{-\frac{1}{2}}\int_{f_0}^{f}\frac{a_0}{r}\sin(2\omega_0 + 2f)\,\mathrm{d}f \tag{17.28}$$

However

$$\frac{a_0}{r} = \frac{1 + e_0 \cos f}{1 - e_0^2}$$

so that

$$\frac{a_0}{r}\sin(2\omega_0 + 2f) = \left(1 - e_0^2\right)^{-1}\left[\sin(2\omega_0 + 2f) + \frac{e_0}{2}\sin(2\omega_0 + f) + \frac{e_0}{2}\sin(2\omega_0 + 3f)\right]$$

because

$$\sin\alpha \sin\beta = \tfrac{1}{2}\sin(\alpha + \beta) + \tfrac{1}{2}\sin(\alpha - \beta)$$

Thus

$$\int_{\ell_0}^{\ell}\frac{\partial F_p}{\partial \omega}\,\mathrm{d}\ell = -\frac{3\mu r_e^2 J_2}{2a_0^3}\sin^2 I_0\left(1 - e_0^2\right)^{-\frac{3}{2}}\int_{f_0}^{f}\left[\sin(2\omega_0 + 2f) + \frac{e_0}{2}\sin(2\omega_0 + f) + \frac{e_0}{2}\sin(2\omega_0 + 3f)\right]\mathrm{d}f$$

$$= -\frac{3\mu r_e^2 J_2}{2a_0^3}\sin^2 I_0\left(1 - e_0^2\right)^{-\frac{3}{2}}\left[\frac{1}{2}\cos(2\omega_0 + 2f) + \frac{e_0}{2}\cos(2\omega_0 + f) + \frac{e_0}{6}\cos(2\omega_0 + 3f)\right] \tag{17.29}$$

On inserting Eq. (17.29) into Eq. (17.26), we find

$$e - e_0 = \frac{1 - e_0^2}{n_0^2 a_0^2 e_0}[F_p(\ell) - F_p(\ell_0)] - \frac{3r_e^2 J_2 \sin^2 I_0}{4a_0^2\left(1 - e_0^2\right)} \times\left[\cos(2\omega_0 + 2f) + e_0\cos(2\omega_0 + f) + \frac{e_0}{3}\cos(2\omega_0 + 3f)\right] \tag{17.30}$$

V. The Inclination I

Since the disturbing function does not contain Ω, the Lagrange variational equation for I is

$$\frac{\mathrm{d}I}{\mathrm{d}t} = \frac{1}{na^2\sqrt{(1 - e^2)}}\cot I\frac{\partial F_p}{\partial \omega} \tag{17.31}$$

With use of $\mathrm{d}t = \mathrm{d}\ell/n_0$, we find

$$I - I_0 = \frac{1}{n_0^2 a_0^2\sqrt{\left(1 - e_0^2\right)}}\cot I_0\int_{\ell_0}^{\ell}\frac{\partial F_p}{\partial \omega}\,\mathrm{d}\ell \tag{17.32}$$

Next, insert Eq. (17.29) for the integral over ℓ, placing $\mu = n_0^2 a_0^3$ and $p_0 = a_0(1 - e_0^2)$, into Eq. (17.32). The result is

$$I - I_0 = \frac{3r_e^2 J_2 \sin 2I_0}{8p_0}\left[\cos(2\omega_0 + 2f) + e_0 \cos(2\omega_0 + f) + \frac{e_0}{3}\cos(2\omega_0 + 3f)\right]_{f_0}^{f} \tag{17.33}$$

Note that, if the unperturbed orbit is equatorial or polar, the factor $\sin 2I_0$ vanishes, so that the inclination of such an orbit does not get changed by the J_2 perturbation.

VI. The Motion of the Node

Here

$$\frac{d\Omega}{dt} = \frac{\csc I}{na^2(1-e^2)^{\frac{1}{2}}}\frac{\partial F_1}{\partial I} \tag{17.34}$$

$$F_1 = F_s(a, e, I) + F_p(a, e, I, \omega, \ell)$$

There are both secular and short periodic variations.

The Secular Variation $\dot{\Omega}_s$

By Eq. (17.11)

$$F_s = -\frac{\mu r_e^2 (1-e^2)^{-\frac{3}{2}} J_2}{a^3}\left(\frac{3}{4}\sin^2 I - \frac{1}{2}\right)$$

Thus

$$\frac{\partial F_s}{\partial I} = -\frac{3\mu r_e^2 J_2 (1-e^2)^{-\frac{3}{2}}}{2a^3}\sin I \cos I \tag{17.35}$$

With use of $\mu = n^2 a^3$ and $p = a(1 - e^2)$, we find from Eqs. (17.34) and (17.35), on using zero subscripts,

$$\dot{\Omega}_s = -\frac{3n_0 r_e^2 J_2}{2p_0^2}\cos I_0 \tag{17.36}$$

Thus

$$\delta\Omega_s = -\frac{3n_0 r_e^2 J_2}{2p_0^2}(\cos I_0)t \tag{17.37}$$

Again, we find the same results for the secular motion of the node as in Chapter 16: no motion for polar orbits, westward motion for direct orbits, eastward for retrograde orbits, and a minimum rate for equatorial orbits. For a close equatorial orbit, the rate is $3n_0 J_2/2$, so that with $J_2 \approx 10^{-3}$ and $n_0 \approx 16$ revolutions per day, the rate is $(24)10^{-3}$ revolutions per day, leading to a period of about six weeks.

The Short Periodic Motion $\dot{\Omega}_p$

From Eq. (17.34)

$$\delta\Omega_p = \frac{\csc I_0}{n_0^2 a_0^2 (1 - e_0^2)^{\frac{1}{2}}} \int_{\ell_0}^{\ell} \frac{\partial F_p}{\partial I} \,\mathrm{d}\ell \tag{17.38}$$

From Eq. (17.13)

$$\frac{\partial F_p}{\partial I} = -\frac{3\mu r_e^2 J_2}{2a_0^3} \sin I_0 \cos I_0 \left[\left(\frac{a_0}{r}\right)^3 - \left(1 - e_0^2\right)^{-\frac{3}{2}} - \left(\frac{a_0}{r}\right)^3 \cos(2\omega_0 + 2f)\right] \tag{17.39}$$

Then

$$\delta\Omega_p = -\frac{3r_e^2 J_2}{2p_0^2} \cos I_0[-(\ell - \ell_0)] - \frac{3r_e^2 J_2}{2a_0^2}\left(1 - e_0^2\right)^{-\frac{1}{2}}$$
$$\times \cos I_0 \int_{\ell_0}^{\ell} \left(\frac{a_0}{r}\right)^3 [1 - \cos(2\omega_0 + 2f)] \,\mathrm{d}\ell \tag{17.40}$$

With use of

$$\mathrm{d}\ell = \frac{r^2}{a_0^2\sqrt{1 - e_0^2}} \,\mathrm{d}f \qquad \frac{a_0}{r} = \frac{1 + e_0 \cos f}{1 - e_0^2}$$

we find

$$\delta\Omega_p = \frac{3r_e^2 J_2}{2p_0^2} \cos I_0 \left[(\ell - \ell_0) - \int_{f_0}^{f} (1 + e_0 \cos f)[1 - \cos(2\omega_0 + 2f)] \,\mathrm{d}f\right] \tag{17.41}$$

$$\delta\Omega_p = \frac{3r_e^2 J_2}{2p_0^2} \cos I_0(\ell - f) + \frac{3r_e^2 J_2}{2p_0^2} \cos I_0 \left[-e_0 \sin f + \frac{1}{2} \sin(2\omega_0 + 2f)\right.$$
$$\left. + \frac{e_0}{2} \sin(2\omega_0 + f) + \frac{e_0}{6} \sin(2\omega_0 + 3f)\right]_{f_0}^{f} \tag{17.42}$$

The term involving $f - \ell$ is short periodic, since it can be expressed as a sine Fourier series in ℓ. The agreement of f and ℓ at all multiples of 2π implies that $\ell_0 = f_0$.[1] Both $\delta\Omega_s$ and $\delta\Omega_p$ vanish if the orbit is polar.

VII. The Motion of Perigee

The Lagrange variational equation is

$$\frac{\mathrm{d}\omega}{\mathrm{d}t} = \frac{(1 - e^2)^{\frac{1}{2}}}{na^2 e} \frac{\partial F_1}{\partial e} - \frac{\cot I}{na^2(1 - e^2)^{\frac{1}{2}}} \frac{\partial F_1}{\partial I} \tag{17.43}$$

By Eq. (17.34)

$$-\frac{\cot I}{na^2(1-e^2)^{\frac{1}{2}}}\frac{\partial F_1}{\partial I} = -\dot{\Omega}\cos I \tag{17.44}$$

By Eqs. (17.43) and (17.44)

$$\delta\omega = -\cos I\,\delta\Omega + \frac{(1-e^2)^{\frac{1}{2}}}{n\,a^2 e}\int_0^t \frac{\partial F_1}{\partial e}\,\mathrm{d}t \tag{17.45}$$

Here

$$F_1 = F_s + F_p$$

$$F_s = \frac{\mu r_e^2 J_2 (1-e^2)^{-\frac{3}{2}}}{2a^3}\left(1 - \frac{3}{2}\sin^2 I\right) \tag{17.11}$$

$$F_p = \frac{\mu r_e^2 J_2}{2a^3}\left(\frac{a}{r}\right)^3\left[\left\{1-\frac{3}{2}\sin^2 I\right\}\left\{1-\left(\frac{r}{a}\right)^3(1-e^2)^{-\frac{3}{2}}\right\}\right.$$
$$\left. + \frac{3}{2}\sin^2 I\cos(2\omega+2f)\right] \tag{17.13}$$

Then

$$\frac{\partial F_s}{\partial e} = \frac{3\mu r_e^2 J_2}{2a^3}e(1-e^2)^{-\frac{5}{2}}\left(1-\frac{3}{2}\sin^2 I\right) \tag{17.46}$$

so that

$$\delta\omega_s = -\cos I_0\,\delta\Omega_s + \frac{3r_e^2 J_2}{2p_0^2}n_0\left(1-\frac{3}{2}\sin^2 I_0\right)t \tag{17.47}$$

Insert Eq. (17.37) into Eq. (17.47). The result is

$$\delta\omega_s = \frac{3r_e^2 J_2}{4p_0^2}(5\cos^2 I_0 - 1)n_0 t \tag{17.48}$$

agreeing with the value found in Chapter 16.

To calculate $\delta\omega_p$, we need $\partial F_p/\partial e$. With use of Eqs. (17.13–17.15), we find

$$\frac{\partial F_p}{\partial e} = \frac{\mu r_e^2 J_2}{2a^3}\left[-3e(1-e^2)^{-\frac{5}{2}}\left(1-\frac{3}{2}\sin^2 I\right) + 3\left(\frac{a}{r}\right)^4\right.$$
$$\times\left(1-\frac{3}{2}\sin^2 I\right)\cos f + \frac{9}{2}\sin^2 I\left(\frac{a}{r}\right)^4\cos f\cos(2\omega+2f)$$
$$\left. - 3\sin^2 I\left(\frac{a}{r}\right)^3\sin(2\omega+2f)\left(\frac{a}{r}+\frac{1}{1-e^2}\right)\sin f\right]_3 \tag{17.49}$$

From Eqs. (17.45) and (17.49)

$$\delta\omega_p + \cos I_0 \delta\Omega_p = \frac{r_e^2 J_2 (1-e_0^2)^{\frac{1}{2}}}{2a_0^2 e_0} \int_{\ell_0}^{\ell} \Big[\;\Big]_3 \, \mathrm{d}\ell$$

$$= -\frac{3r_e^2 J_2}{2p_0^2}\left(1 - \frac{3}{2}\sin^2 I_0\right)(\ell - \ell_0) + \frac{r_e^2 J_2}{2a_0^2 e_0}(1-e_0^2)^{\frac{1}{2}}$$

$$\times \int_{f_0}^{f} Q \frac{r^2}{a^2}(1-e_0^2)^{-\frac{1}{2}} \, \mathrm{d}f \tag{17.50}$$

where Q = sum of terms in $\cos f$, $\cos f \cos(2\omega + 2f)$, $\sin(2\omega + 2f)$ inside the bracket in Eq. (17.49). Then

$$\int_{f_0}^{f} Q \frac{r^2}{a^2} \, \mathrm{d}f = \int_{f_0}^{f} \left[3\left(\frac{a}{r}\right)^2 \left(1 - \frac{3}{2}\sin^2 I\right)\cos f \right.$$

$$+ \frac{9}{2}\sin^2 I \left(\frac{a}{r}\right)^2 \cos f \cos(2\omega + 2f)$$

$$\left. - 3\sin^2 I \left(\frac{a}{r}\right) \sin(2\omega + 2f)\left(\frac{a}{r} + \frac{1}{1-e^2}\right)\sin f \right] \mathrm{d}f \tag{17.51}$$

$$\int_{f_0}^{f} Q \frac{r^2}{a^2} \, \mathrm{d}f = \left(1 - e_0^2\right)^{-2} \int_{f_0}^{f} \left[3(1 + e_0 \cos f)^2 \left(1 - \frac{3}{2}\sin^2 I\right)\cos f \right.$$

$$+ \frac{9}{2}\sin^2 I (1 + e\cos f)^2 \cos f \cos(2\omega + 2f)$$

$$\left. - 3\sin^2 I (1 + e\cos f)\sin(2\omega + 2f)(2 + e\cos f)\sin f \right] \mathrm{d}f \tag{17.52}$$

$$\int_{f_0}^{f} Q \frac{r^2}{a^2} \, \mathrm{d}f = \left(1 - e_0^2\right)^{-2} P \tag{17.53}$$

where P is the integral in Eq. (17.52). By Eqs. (17.50) and (17.53)

$$\delta\omega_p + \cos I_0 \delta\Omega_p = -\frac{3r_e^2 J_2}{2p_0^2}\left(1 - \frac{3}{2}\sin^2 I_0\right)(\ell - \ell_0) + \frac{r_e^2 J_2}{2p_0^2 e_0} P \tag{17.54}$$

To evaluate P, we need expressions as trigonometric polynomials of $(1 + e\cos f)^2 \cos f$, $(1 + e\cos f)^2 \cos f \cos(2\omega + 2f)$, and $(1 + e\cos f)(2 + e\cos f) \times \sin(2\omega + 2f)\sin f$. From

$$\cos(\alpha \pm \beta) = \cos\alpha\cos\beta \mp \sin\alpha\sin\beta$$

we obtain

$$\cos\alpha\cos\beta = \tfrac{1}{2}\cos(\alpha-\beta) + \tfrac{1}{2}\cos(\alpha+\beta)$$

$$\sin\alpha\sin\beta = \tfrac{1}{2}\cos(\alpha-\beta) - \tfrac{1}{2}\cos(\alpha+\beta)$$

$$(1+e\cos f)^2 = 1 + \frac{e^2}{2} + 2e\cos f + \frac{e^2}{2}\cos 2f$$

Then

$$(1+e\cos f)^2\cos f = e + \left(1+\frac{3e^2}{4}\right)\cos f + e\cos 2f + \frac{e^2}{4}\cos 3f$$

$$\begin{aligned}(1+e\cos f)^2\cos f\cos(2\omega+2f) &= \frac{e}{2}\cos 2\omega + \frac{e^2}{8}\cos(2\omega-2f)\\ &+ \left(\frac{1}{2}+\frac{3e^2}{8}\right)\cos(2\omega+f) + e\cos(2\omega+2f)\\ &+ \left(\frac{1}{2}+\frac{3e^2}{8}\right)\cos(2\omega+3f) + \frac{e}{2}\cos(2\omega+4f)\\ &+ \frac{e^2}{8}\cos(2\omega+5f)\end{aligned}$$

$$\begin{aligned}(1+e\cos f)(2+e\cos f)\sin(2\omega+2f)\sin f &= \frac{3e}{4}\cos 2\omega + \frac{e^2}{8}\cos(2\omega-f)\\ &+ \left(1+\frac{e^2}{8}\right)\cos(2\omega+f) - \left(1+\frac{e^2}{8}\right)\cos(2\omega+3f)\\ &- \frac{3e}{4}\cos(2\omega+4f) - \frac{e^2}{8}\cos(2\omega+5f)\end{aligned}$$

If N is the integrand of P, we then obtain

$$\begin{aligned}N &= 3\left(1-\frac{3}{2}\sin^2 I\right)\left[e + \left(1+\frac{3e^2}{4}\right)\cos f + e\cos 2f + \frac{e^2}{4}\cos 3f\right]\\ &+ \frac{3}{2}\sin^2 I\left[\frac{e^2}{8}\cos(2\omega-f) + \left(-\frac{1}{2}+\frac{7e^2}{8}\right)\cos(2\omega+f)\right.\\ &+ 3e\cos(2\omega+2f) + \left(\frac{7}{2}+\frac{11e^2}{8}\right)\cos(2\omega+3f) + 3e\cos(2\omega+4f)\\ &\left.+ \frac{5e^2}{8}\cos(2\omega+5f)\right]\end{aligned} \tag{17.55}$$

Using

$$P = \int_{f_0}^{f} N\,\mathrm{d}f \tag{17.56}$$

and inserting the result into Eq. (17.54), we find

$$
\begin{aligned}
\delta\omega_p + \cos I_0 \delta\Omega_p = & -\frac{3r_e^2 J_2}{2p_0^2}\left(1 - \frac{3}{2}\sin^2 I_0\right)(\ell - f) + \frac{3r_e^2 J_2}{2p_0^2}\left[\left(1 - \frac{3}{2}\sin^2 I_0\right)\right. \\
& \times \left\{\left(\frac{1}{e_0} + \frac{3e_0}{4}\right)\sin f + \frac{1}{2}\sin 2f + \frac{e_0}{12}\sin 3f\right\} + \frac{1}{2}\sin^2 I_0 \\
& \times \left\{-\frac{e_0}{8}\sin(2\omega_0 - f) + \left(-\frac{1}{2e_0} + \frac{7e_0}{8}\right)\sin(2\omega_0 + f)\right. \\
& + \frac{3}{2}\sin(2\omega_0 + 2f) + \left(\frac{7}{6e_0} + \frac{11e_0}{24}\right)\sin(2\omega_0 + 3f) + \frac{3}{4}\sin(2\omega_0 + 4f) \\
& \left.\left. + \frac{e_0}{8}\sin(2\omega_0 + 5f)\right\}\right]_{f_0}^{f}
\end{aligned}
\tag{17.57}
$$

From Eq. (17.42)

$$
\begin{aligned}
\cos I_0 \delta\Omega_p = & \frac{3r_e^2 J_2}{2p_0^2}\cos^2 I_0(\ell - f) + \frac{3r_e^2 J_2}{2p_0^2}\cos^2 I_0\left[-e_0 \sin f\right. \\
& \left. + \frac{1}{2}\sin(2\omega_0 + 2f) + \frac{e_0}{2}\sin(2\omega_0 + f) + \frac{e_0}{6}\sin(2\omega_0 + 3f)\right]_{f_0}^{f}
\end{aligned}
\tag{17.58}
$$

Subtract Eq. (17.58) from Eq. (17.57). The result is

$$
\begin{aligned}
\delta\omega_p = & \frac{3r_e^2 J_2}{4p_0^3}(1 - 5\cos^2 I_0)(\ell - f) \\
& + \frac{3r_e^2 J_2}{2p_0^2}\left[\left\{\left(1 - \frac{3}{2}\sin^2 I_0\right)\left(\frac{1}{e_0} + \frac{3e_0}{4}\right) + e_0\cos^2 I_0\right\}\sin f\right. \\
& + \left(1 - \frac{3}{2}\sin^2 I_0\right)\left\{\frac{1}{2}\sin 2f + \frac{e_0}{12}\sin 3f\right\} - \frac{e_0}{16}\sin^2 I_0 \sin(2\omega_0 - f) \\
& + \left\{\left(\frac{15e_0}{16} - \frac{1}{4e_0}\right)\sin^2 I_0 - \frac{e_0}{2}\right\}\sin(2\omega_0 + f) + \left(\frac{5}{4}\sin^2 I_0 - \frac{1}{2}\right) \\
& \times \sin(2\omega_0 + 2f) + \left\{\left(\frac{19e_0}{48} - \frac{7}{12e_0}\right)\sin^2 I_0 - \frac{e_0}{6}\right\}\sin(2\omega_0 + 3f) \\
& \left. + \frac{3}{4}\sin(2\omega_0 + 4f) + \frac{e_0}{8}\sin(2\omega_0 + 5f)\right]_{f_0}^{f}
\end{aligned}
\tag{17.59}
$$

The term involving $f - \ell$ is short periodic, since it can be expressed as a sine Fourier series in ℓ.[1] The agreement of f and ℓ at all multiples of 2π ($\ell_0 = f_0$) also shows this, since it leads to $\bar{f} = \bar{\ell}$.

VIII. Variation of the Mean Anomaly

Call the perturbed mean anomaly M. By Chapter 10,

$$\dot{M} = n - \frac{2}{na}\frac{\partial F_1}{\partial a} - \frac{1-e^2}{na^2 e}\frac{\partial F_1}{\partial e} \tag{17.60}$$

By Eq. (17.45)

$$\frac{(1-e^2)^{\frac{1}{2}}}{na^2 e}\frac{\partial F_1}{\partial e} = \dot{\omega} + \dot{\Omega}\cos I \tag{17.61}$$

By Eq. (17.12)

$$F_1 = \frac{1}{a^3}U \tag{17.62}$$

where U does not depend on a. Thus

$$\frac{\partial F_1}{\partial a} = -\frac{3}{a}F_1 \tag{17.63}$$

By Eqs. (17.60), (17.61), and (17.63)

$$\dot{M} = n + \frac{6}{na^2}F_1 - (1-e^2)^{\frac{1}{2}}(\dot{\omega} + \dot{\Omega}\cos I) \tag{17.64}$$

By Eq. (17.18)

$$a = a_0 + \frac{2}{n_0^2 a_0}[F_1(\ell) - F_1(\ell_0)] \tag{17.65}$$

or

$$a = a_0\left(1 + \frac{2}{n_0^2 a_0^2}[F_1(\ell) - F_1(\ell_0)]\right) \tag{17.66}$$

where the term $F_1(\ell) - F_1(\ell_0)$ is of order J_2. Thus

$$n = \mu^{\frac{1}{2}}a^{-\frac{3}{2}} = \mu^{\frac{1}{2}}a_0^{-\frac{3}{2}}\left(1 - \frac{3}{n_0^2 a_0^2}[F_1(\ell) - F_1(\ell_0)]\right) + O\left(J_2^2\right) \tag{17.67}$$

or

$$n = n_0\left(1 - \frac{3}{n_0^2 a_0^2}[F_1(\ell) - F_1(\ell_0)]\right) \tag{17.68}$$

By Eqs. (17.64) and (17.68), it follows that

$$\dot{M} = n_0 + \frac{3}{n_0 a_0^2}F_1(\ell_0) + \frac{3}{n_0 a_0^2}F_1(\ell) - \left(1 - e_0^2\right)^{\frac{1}{2}}(\dot{\omega} + \dot{\Omega}\cos I) \tag{17.69}$$

Put

$$n' = n_0 + \frac{3}{n_0 a_0^2}F_1(\ell_0) \tag{17.70}$$

a constant. Then

$$\dot{M} = n' + \frac{3}{n_0 a_0^2}F_1(\ell) - \left(1 - e_0^2\right)^{\frac{1}{2}}(\dot{\omega} + \dot{\Omega}\cos I) \tag{17.71}$$

so that

$$\delta M = n't + \frac{3}{n_0 a_0^2}\int_0^t F_1\, dt - \left(1 - e_0^2\right)^{\frac{1}{2}}(\delta\omega + \delta\Omega \cos I) \qquad (17.72)$$

On integrating Eq. (17.72) for F_1, we find

$$\frac{3}{n_0 a_0^2}\int_0^t F_1\, dt = -\frac{\mu r_e^2 J_2}{2a_0^3}\frac{3}{n_0 a_0^2}\int_0^t \left(\frac{a_0}{r}\right)^3$$
$$\times \left[\frac{3}{4}\sin^2 I - \frac{1}{2} - \frac{3}{4}\sin^2 I \cos(2\omega_0 + 2f)\right] dt \qquad (17.73)$$

The coefficient is $3n_0 r_e^2 J_2/a_0^2$. Placing $dt = d\ell/n_0$, $d\ell = (1-e_0^2)^{-1/2} r^2/a_0^2\, df$, $a_0/r = (1-e_0^2)^{-1}(1 + e_0\cos f)$, we find

$$\frac{3}{n_0 a_0^2}\int_0^t F_1\, dt = -\frac{3r_e^2 J_2\left(1-e_0^2\right)^{\frac{1}{2}}}{p_0^2}\int_{f_0}^f (1 + e_0\cos f)\left[\frac{3}{4}\sin^2 I_0 - \frac{1}{2}\right.$$
$$\left. - \frac{3}{4}\sin^2 I_0 \cos(2\omega_0 + 2f)\right] df \qquad (17.74)$$

$$\frac{3}{n_0 a_0^2}\int_0^t F_1\, dt = -\frac{3r_e^2 J_2\left(1-e_0^2\right)^{\frac{1}{2}}}{2p_0^2}\left[\left(1 - \frac{3}{2}\sin^2 I_0\right)(f + e_0 \sin f)\right.$$
$$\left. + \frac{3}{4}\sin^2 I_0\left\{\sin(2\omega_0 + 2f) + e_0\sin(2\omega_0 + f) + \frac{e_0}{3}\sin(2\omega_0 + 3f)\right\}\right]_{f_0}^{f}$$
$$(17.75)$$

Then

$$\delta M = n't + (75) - (1 - e^2)^{\frac{1}{2}}(\delta\omega + \delta\Omega\cos I) \qquad (17.76)$$

From Eq. (17.47) and Eq. (17.57), we have

$$(1-e^2)^{\frac{1}{2}}(\delta\omega + \delta\Omega\cos I) = \frac{3r_e^2 J_2}{2p_0^2}(1-e^2)^{\frac{1}{2}}\left[\left(1 - \frac{3}{2}\sin^2 I_0\right)\right.$$
$$\times\left\{f + \left(\frac{1}{e_0} + \frac{3e_0}{4}\right)\sin f + \frac{1}{2}\sin 2f + \frac{e_0}{12}\sin 3f\right\}$$
$$+ \frac{1}{2}\sin^2 I_0\left\{-\frac{e_0}{8}\sin(2\omega_0 - f) + \left(-\frac{1}{2e_0} + \frac{7e_0}{8}\right)\sin(2\omega_0 + f)\right.$$
$$+ \frac{3}{2}\sin(2\omega_0 + 2f) + \left(\frac{7}{6e_0} + \frac{11e_0}{24}\right)\sin(2\omega_0 + 3f)$$
$$\left.\left. + \frac{3}{4}\sin(2\omega_0 + 4f) + \frac{e_0}{8}\sin(2\omega_0 + 5f)\right\}\right]_{f_0}^{f} \qquad (17.77)$$

From Eqs. (17.74), (17.75), and (17.77), we obtain

$$\delta M = M - \ell_0 = n't + \frac{3r_e^2 J_2}{8p_0^2}(1-e^2)^{\frac{1}{2}}\Bigg[\left(1-\frac{3}{2}\sin^2 I_0\right)\left(e_0-\frac{3}{e_0}\right)\sin f$$

$$+ (3\sin^2 I_0 - 2)\sin 2f - \frac{e_0}{3}\left(1-\frac{3}{2}\sin^2 I_0\right)\sin 3f$$

$$+ \frac{1}{2}\sin^2 I_0\Bigg\{\frac{e_0}{2}\sin(2\omega_0 - f) + \left(\frac{2}{e_0}+\frac{5e_0}{2}\right)\sin(2\omega_0+f)$$

$$+ \left(-\frac{14}{3e_0}+\frac{e_0}{6}\right)\sin(2\omega_0+3f) - 3\sin(2\omega_0+4f)$$

$$- \frac{e_0}{2}\sin(2\omega_0+5f)\Bigg\}\Bigg]_{f_0}^{f} \tag{17.78}$$

The terms in f and $\sin(2\omega_0 + 2f)$ canceled out.

The secular part $n't$ is of some interest. From Eq. (17.70)

$$n' = n_0\left[1 + \frac{3}{n_0^2 a_0^2}F_1(\ell_0)\right] \tag{17.79}$$

From Eq. (17.1)

$$F_1(\ell_0) = -\frac{\mu r_e^2}{2r_0^3}J_2\left(3\sin^2\theta_0 - 1\right) \tag{17.80}$$

Thus

$$n' = n_0\left[1 + \frac{3a_0 r_e^2 J_2}{2r_0^3}\left(1 - 3\sin^2\theta_0\right)\right] \tag{17.81}$$

For the case of vanishing eccentricity, this becomes

$$n' = n_0\left[1 + \frac{3r_e^2 J_2}{2a_0^2}\left(1 - 3\sin^2\theta_0\right)\right] \tag{17.82}$$

in agreement with the value for $\dot{\bar{M}}$ found in Chapter 16 for the case of small eccentricity.

Reference

[1]Smart, W. M., *Celestial Mechanics,* Longmans, Green, and Co., London, 1953, p. 38.

Chapter 18

The Effects of Drag on Satellite Orbits

I. Introduction

WE SHALL consider the drag on a satellite orbiting around a spherical Earth. The interaction of the oblateness and the drag is too difficult a problem for an elementary treatment. We leave open the question as to what accuracy can be obtained when the two effects are superposed: that of oblateness without drag and that of drag without oblateness.

Because the drag is not derivable from a potential, we need to use the Gaussian equations for the Keplerian elements. For convenience, we list them here.

$$\dot{a} = \frac{2}{n\sqrt{1-e^2}}[eR\sin f + T(1+e\cos f)] \tag{18.1}$$

$$\dot{e} = \frac{\sqrt{1-e^2}}{na}[R\sin f + T(\cos E + \cos f)] \tag{18.2}$$

$$\dot{I} = \frac{rW\cos(\omega+f)}{na^2\sqrt{1-e^2}} \tag{18.3}$$

$$\dot{\Omega} = \frac{rW\csc I\sin(\omega+f)}{na^2\sqrt{1-e^2}} \tag{18.4}$$

$$\dot{\omega} = -\dot{\Omega}\cos I - \frac{\sqrt{1-e^2}}{ena}\left[R\cos f - T\left(1+\frac{r}{p}\right)\sin f\right] \tag{18.5}$$

$$\dot{n} = -\frac{3n\dot{a}}{2a} = -\frac{3}{a\sqrt{1-e^2}}[eR\sin f + T(1+e\cos f)] \tag{18.6}$$

$$\dot{\ell} = n - \frac{2rR}{na^2} - \sqrt{1-e^2}(\dot{\omega}+\dot{\Omega}\cos I) \tag{18.7}$$

If $\boldsymbol{v}_a$ is the velocity of the satellite relative to the atmosphere, the usual expression for the force of drag is

$$\boldsymbol{F}_D = -\tfrac{1}{2}AC_D\rho v_a\boldsymbol{v}_a \tag{18.8}$$

where

$$\boldsymbol{v}_a = v_a\boldsymbol{l}_a \tag{18.9}$$

$\boldsymbol{l}_a$ being a unit vector along $\boldsymbol{v}_a$. Here A is the projected area of the satellite perpendicular to the flow, ρ is the atmospheric density, and C_D is a dimensionless constant of order of magnitude 2.2. For an accurate calculation, one should know A as a function of time or of $\boldsymbol{r}$, the position vector of the satellite's center of mass. If the satellite is spherical, A is known and A can be estimated if the satellite is oriented by a gravity-gradient method. If a nonspherical satellite is tumbling, A could be known accurately only by simultaneous solution of the rotational and orbital problems.

For a mean value of A, consider a convex satellite. "Convex" means that a straight line intersects the satellite in only two points. If such a convex satellite is tumbling at random, its mean projected area is one-fourth of the total surface area.[1] The factor 1/4 can be remembered by thinking of a sphere of radius b, for which the total and projected areas are, respectively, $4\pi b^2$ and πb^2.

If we assume that the atmosphere rotates rigidly with the Earth, then

$$\boldsymbol{v}_a = \dot{\boldsymbol{r}} - \boldsymbol{w} \tag{18.10}$$

where $\boldsymbol{r}$ is the position vector of the center of mass and where the rotational velocity $\boldsymbol{w}$ is given by

$$\boldsymbol{w} = \omega_e \boldsymbol{k} \times \boldsymbol{r} \tag{18.11}$$

Here $\boldsymbol{k}$ is a unit vector along the Earth's polar axis and

$$\omega_e = 2\pi/86{,}164.2 \text{ rad/s} \tag{18.12}$$

the sidereal speed of rotation of the Earth.

Now, R and T lie in the plane of the orbit, and W is perpendicular to it. If we neglect the rotation of the atmosphere, $\boldsymbol{v}_a$, and thus $\boldsymbol{F}_D$, would lie in the plane of the orbit by Eqs. (18.8) and (18.10). Then W would vanish and so would $\dot{I}$ and $\dot{\Omega}$ by Eqs. (18.3) and (18.4). Thus

$$\dot{I} = \dot{\Omega} = 0 \tag{18.13}$$

if we neglect the rotation of the atmosphere. In such a case

$$\boldsymbol{v}_a = \dot{\boldsymbol{r}} = \boldsymbol{v} \tag{18.14}$$

$$v_a \boldsymbol{v}_a = v \boldsymbol{v} = v^2 \boldsymbol{t} \tag{18.14a}$$

where $\boldsymbol{t}$ is a unit vector along the tangent to the orbit in the direction of motion. Insertion of Eq. (18.14a) into Eq. (18.8) yields

$$\boldsymbol{f}_D = \frac{\boldsymbol{F}_D}{m} = -\frac{1}{2} k \rho v^2 \boldsymbol{t} \tag{18.15}$$

$$k = \frac{A C_D}{m} \tag{18.15a}$$

If we let ϕ be the angle from $\boldsymbol{r}$ to $\boldsymbol{t}$, then

$$R = f_D \cos\phi \tag{18.16}$$

$$T = f_D \sin\phi \tag{18.17}$$

where

$$\boldsymbol{f}_D = f_D \boldsymbol{t} \tag{18.17a}$$

and where R and T are the components of the drag per unit mass along the radial and transverse directions.

II. Components of the Drag in Terms of the Anomalies E and f

To find $\cos\phi$ and $\sin\phi$, we first do some spade work. We have

$$\boldsymbol{r}\cdot\dot{\boldsymbol{r}} = rv\cos\phi \tag{18.18}$$

$$\boldsymbol{r}\times\dot{\boldsymbol{r}} = (\mu p)^{\frac{1}{2}}\boldsymbol{l}_w \tag{18.19}$$

Here, $v = |\dot{\boldsymbol{r}}|$, μ is the product of G and the mass of the Earth, p is the osculating semi-latus rectum, and $\boldsymbol{l}_w$ is a unit vector perpendicular to the plane of the orbit along the angular momentum vector. The angular momentum per unit mass is $(\mu p)^{1/2}$.

As in Chapter 2, let $\boldsymbol{A} = \boldsymbol{l}_A a$, $\boldsymbol{B} = \boldsymbol{l}_B b$, where a and b are, respectively, the osculating semi-major axis and semi-minor axis; $\boldsymbol{l}_A$ a unit vector pointing from the Earth's center toward perigee; and $\boldsymbol{l}_B$ a unit vector parallel to the semi-minor axis, so that $\boldsymbol{l}_A\times\boldsymbol{l}_B = \boldsymbol{l}_w$. If e is the osculating eccentricity and E the eccentric anomaly, then

$$\boldsymbol{r} = \boldsymbol{A}(\cos E - e) + \boldsymbol{B}\sin E \tag{18.20}$$

$$\dot{\boldsymbol{r}} = (na/r)(-\boldsymbol{A}\sin E + \boldsymbol{B}\cos E) \tag{18.21}$$

as before. Then

$$\begin{aligned} rv\cos\phi = \boldsymbol{r}\cdot\dot{\boldsymbol{r}} &= \frac{na}{r}[-\boldsymbol{A}\sin E + \boldsymbol{B}\cos E]\cdot[\boldsymbol{A}(\cos E - e) + \boldsymbol{B}\sin E] \\ &= \frac{na}{r}[-a^2\sin E(\cos E - e) + a^2(1-e^2)\sin E\cos E] \\ &= na^2 e\sin E \end{aligned} \tag{18.22}$$

To find rv, use the equation for the osculating a,

$$\frac{1}{2}v^2 - \frac{\mu}{r} = -\frac{\mu}{2a} \tag{18.23}$$

Then

$$r^2v^2 = 2\mu r - \frac{\mu}{a}r^2 = 2\mu a(1 - e\cos E) - \mu a(1 - e\cos E)^2$$

$$r^2v^2 = \mu a(1 - e\cos E)(1 + e\cos E)$$

$$r^2v^2 = n^2a^4(1 - e^2\cos^2 E)$$

so that

$$rv = na^2(1 - e^2\cos^2 E)^{\frac{1}{2}} \tag{18.24}$$

Then from Eqs. (18.22) and (18.24)

$$\cos\phi = \frac{e\sin E}{(1 - e^2\cos^2 E)^{\frac{1}{2}}} \tag{18.25}$$

From Eq. (18.19)

$$rv\sin\phi = |\boldsymbol{r}\times\dot{\boldsymbol{r}}| = (\mu p)^{\frac{1}{2}} = na^2(1 - e^2)^{\frac{1}{2}} \tag{18.26}$$

From Eqs. (18.26) and (18.24)

$$\sin\phi = \left(\frac{1-e^2}{1-e^2\cos^2 E}\right)^{\frac{1}{2}} \tag{18.27}$$

Equations (18.25) and (18.27) check $\sin^2\phi + \cos^2\phi = 1$.

To obtain $\cos\phi$ and $\sin\phi$ in terms of the true anomaly f, use the anomaly connections

$$\cos E = \frac{e+\cos f}{1+e\cos f} \qquad \sin E = \frac{\sqrt{1-e^2}\sin f}{1+e\cos f}$$

It can be shown that

$$\cos\phi = \frac{e\sin f}{(1+e^2+2e\cos f)^{\frac{1}{2}}} \tag{18.28}$$

$$\sin\phi = \frac{1+e\cos f}{(1+e^2+2e\cos f)^{\frac{1}{2}}} \tag{18.29}$$

From Eqs. (18.16), (18.17), and (18.25–18.29), it follows that

$$R = \frac{f_D e\sin E}{(1-e^2\cos^2 E)^{\frac{1}{2}}} = \frac{f_D e\sin f}{(1+e^2+2e\cos f)^{\frac{1}{2}}} \tag{18.30}$$

$$T = \frac{f_D(1-e^2)^{\frac{1}{2}}}{(1-e^2\cos^2 E)^{\frac{1}{2}}} = \frac{f_D(1+e\cos f)}{(1+e^2+2e\cos f)^{\frac{1}{2}}} \tag{18.31}$$

III. Equations for $\dot{a}$ and $\dot{e}$ in Terms of the True Anomaly

From Eq. (18.15)

$$f_D = -\tfrac{1}{2}k\rho v^2 \tag{18.15}$$

On inserting Eq. (18.15) and the f forms of Eqs. (18.30) and (18.31) into Eq. (18.1), we find

$$\dot{a} = -\frac{k\rho v^2}{n\sqrt{1-e^2}}\left[\frac{e^2\sin^2 f + (1+e\cos f)^2}{(1+e^2+2e\cos f)^{\frac{1}{2}}}\right] \tag{18.32}$$

or

$$\dot{a} = -\frac{k\rho v^2}{n}(1-e^2)^{-\frac{1}{2}}(1+e^2+2e\cos f)^{\frac{1}{2}} \tag{18.33}$$

Similarly

$$\dot{e} = -\frac{k\rho v^2}{2na}(1-e^2)^{\frac{1}{2}}[\cos\phi\sin f + \sin\phi(\cos E+\cos f)] \tag{18.34}$$

Inserting $\cos\phi$ and $\sin\phi$ from Eqs. (18.28) and (18.29) and using $\cos E = (e + \cos f)/(1 + e\cos f)$, we find

$$\dot{e} = -\frac{k\rho v^2}{na}(1-e^2)^{\frac{1}{2}}(e+\cos f)(1+e^2+2e\cos f)^{-\frac{1}{2}} \tag{18.35}$$

We can also show that

$$\dot{\omega} = -\frac{k\rho v^2}{nae}(1-e^2)^{\frac{1}{2}}\sin f(1+e^2+2e\cos f)^{-\frac{1}{2}} \tag{18.36}$$

$$\dot{\ell} = n + \frac{k\rho v^2}{na}\left[\frac{(1-e^2)\sin f}{(1+e^2+2e\cos f)^{\frac{1}{2}}}\right]\left[\frac{1}{e} + \frac{e}{1+e\cos f}\right] \tag{18.37}$$

IV. Secular Behavior of a, e, ω, and ℓ

If we use a spherical model for the atmosphere, ρ is a function only of r and, thus, only of $\cos f$. Also v^2 depends only on r and, thus, only on $\cos f$. It follows from the preceding equations that $\dot{a}$ and $\dot{e}$ are functions of $\cos f$ only and that $\dot{\omega}$ and $\dot{\ell} - n$ are products of $\sin f$ and a function of $\cos f$.

Let e_k be any of the Keplerian elements. If P is the period of the unperturbed motion,

$$\bar{\dot{e}}_k = \frac{1}{P}\int_0^P \dot{e}_k \,\mathrm{d}t$$

On the right side of Eqs. (18.33–18.37), we may put

$$\mathrm{d}t = \frac{\mathrm{d}\ell}{n} = \frac{r^2}{na^2}(1-e^2)^{-\frac{1}{2}}\,\mathrm{d}f$$

Then

$$\bar{\dot{e}}_k = \frac{1}{2\pi}\int_0^{2\pi} \psi(f)\,\mathrm{d}f = \frac{1}{2\pi}\int_{-\pi}^{\pi} \psi(f)\,\mathrm{d}f \tag{18.38}$$

where $\psi(f) = \dot{e}_k(r^2/na^2)(1-e^2)^{-1/2}$.

If $\dot{e}_k$ is $\dot{a}$ or $\dot{e}$, $\psi(f)$ is an even function of f; if it is $\dot{\omega}$ or $\dot{\ell} - n$, it is an odd function. It follows from Eq. (18.38) that

$$\bar{\dot{\omega}} = 0 \qquad \bar{\dot{\ell}} - \bar{n} = 0 \tag{18.39}$$

Thus, $\bar{\dot{\omega}}$ and $\dot{\ell} - n$ have no secular parts. If $\dot{e}_k$ is $\dot{a}$ or $\dot{e}$

$$\bar{\dot{e}}_k = \frac{1}{\pi}\int_0^{\pi} \psi(f)\,\mathrm{d}f \tag{18.40}$$

By Eq. (18.33), a always diminishes. By Eq. (18.35), e diminishes when $1 + e\cos f > 0$. If the orbit has initially a large eccentricity, ρ is appreciable only when the orbiter is close to perigee. As it moves toward apogee, ρ diminishes, so that the important changes in e occur when $\cos f \approx 1$. On the average, $\dot{e} < 0$. Qualitatively, this is easy to see. As the satellite comes in from the distant apogee, it loses a good deal of energy going through the denser atmosphere near perigee, so that it then lacks the energy to reach as distant an apogee the next time. The orbit thus becomes more nearly circular.

V. Equations for *a* and *e* in Terms of the Eccentric Anomaly

To find $\dot{a}$ in terms of E, first put $\cos E = (e + \cos f)/(1 + e\cos f)$ in Eq. (18.33). Then

$$(1 + e^2 + 2e\cos f)^{\frac{1}{2}} = (1 - e^2)^{\frac{1}{2}} \left(\frac{1 + e\cos E}{1 - e\cos E} \right)^{\frac{1}{2}} \tag{18.41}$$

From Eq. (18.24) and $r = a(1 - e\cos E)$, we next find

$$v = na \left(\frac{1 + e\cos E}{1 - e\cos E} \right)^{\frac{1}{2}} \tag{18.42}$$

Then

$$(1 + e^2 + 2e\cos f)^{\frac{1}{2}} = \frac{v}{na}(1 - e^2)^{\frac{1}{2}} \tag{18.43}$$

From Eqs. (18.33) and (18.43)

$$\dot{a} = -\frac{k\rho v^3}{n^2 a} \tag{18.44}$$

To find $\dot{e}$, use Eqs. (18.35) and (18.43)

$$e + e\cos f = e + \frac{\cos E - e}{1 - e\cos E} = (1 - e^2)\frac{a}{r}\cos E \tag{18.45}$$

Then

$$\dot{e} = -\frac{k\rho v a}{r}(1 - e^2)\cos E \tag{18.46}$$

or

$$\dot{e} = -\frac{k\rho n a}{r}(1 - e^2)\cos E \frac{(1 + e\cos E)^{\frac{1}{2}}}{(1 - e\cos E)^{\frac{3}{2}}} \tag{18.47}$$

We can also show that

$$\dot{\omega} = -\frac{k\rho v}{e}(1 - e^2)^{\frac{1}{2}} \frac{\sin E}{1 - e\cos E} \tag{18.48}$$

$$\dot{\ell} = n + k\rho v e \sin E + k\rho v \frac{(1 - e)^2 \sin E}{e(1 - e\cos E)} \tag{18.49}$$

VI. An Equation for *E*

From Kepler's equation

$$E - e\sin E = \ell \tag{18.50}$$

we obtain

$$(1 - e\cos E)\dot{E} - \dot{e}\sin E = \dot{\ell} \tag{18.51}$$

Now, insert Eqs. (18.46) and (18.49) into Eq. (18.51), and we obtain

$$\frac{r}{a}\dot{E} = -\frac{k\rho v a}{r}(1-e^2)\cos E \sin E + n + k\rho v e \sin E + k\rho v \frac{(1-e^2)a \sin E}{er} \tag{18.52}$$

This becomes

$$\frac{r}{a}\dot{E} = n + k\rho v \sin E\left[e + \frac{(1-e)^2}{e}\right] \tag{18.53}$$

$$= n + \frac{k\rho v}{e}\sin E \tag{18.54}$$

Thus

$$\dot{E} = \frac{na}{r}\left(1 + \frac{k\rho v}{ne}\sin E\right) \tag{18.55}$$

VII. Equations for the Integration

If we treat the atmosphere as spherical, it is customary to represent the density by the expression

$$\rho = \rho_0 \exp[-(r-r_0)/\lambda] \tag{18.56}$$

where ρ_0 is the density at radius r_0, which we take to be the radius at perigee. Here, λ is called the scale height. If we put $r = a(1 - e\cos E)$, with $r = r_0$ at perigee, we obtain

$$r_0 = a(1-e)$$

$$r - r_0 = ae(1-\cos E) \tag{18.57}$$

Then

$$\rho(r) = \rho_0 \exp\left[-\frac{ae}{\lambda}(1-\cos E)\right] = \rho_0 \varepsilon^{-c}\varepsilon^{c\cos E} \tag{18.58}$$

where

$$c \equiv ae/\lambda \tag{18.59}$$

The simplicity of this function has led various authors to use E as an independent variable in doing the integration. Then

$$\frac{\mathrm{d}a}{\mathrm{d}E} = \frac{\dot{a}}{\dot{E}} \qquad \frac{\mathrm{d}e}{\mathrm{d}E} = \frac{\dot{e}}{\dot{E}} \tag{18.60}$$

If there were no drag, we should have

$$\dot{E} = na/r \tag{18.61}$$

by Eq. (18.55). Jupp[2] has pointed out that $\dot{E} = na/r$ may be a poor approximation for nearly circular orbits, where $k\rho v/(ne)$ may approach unity. King-Hele[3] has suggested that, for actual cases that arise, the resulting error is likely to be serious only during the final day of the satellite's lifetime.

Having stated this warning, we now proceed with the usual treatment, using $\dot{E} = na/r$. From Eqs. (18.44), (18.60), and (18.61), we obtain

$$\frac{\mathrm{d}a}{\mathrm{d}E} = -\frac{k\rho v^3}{n^2 a}\frac{r}{na} = -\frac{k\rho(rv)^3}{n^2a^2r^2} \tag{18.62}$$

With $r = a(1 - e\cos E)$ and $rv = na^2(1 - e^2\cos^2 E)^{\frac{1}{2}}$, this becomes

$$\frac{\mathrm{d}a}{\mathrm{d}E} = -\frac{k\rho a^2(1 + e\cos E)^{\frac{3}{2}}}{(1 - e\cos E)^{\frac{1}{2}}} \tag{18.63}$$

Similarly, from Eqs. (18.47), (18.60), and (18.61), we obtain

$$\frac{\mathrm{d}e}{\mathrm{d}E} = -k\rho(1 - e^2)a\cos E\left(\frac{1 + e\cos E}{1 - e\cos E}\right)^{\frac{1}{2}} \tag{18.64}$$

In finding Δa and Δe for one revolution, it is customary to treat k, a, and e as constant on the right sides of Eqs. (18.63) and (18.64). The results are

$$\Delta a = -ka^2\int_0^{2\pi}\frac{\rho(1 + e\cos E)^{\frac{3}{2}}}{(1 - e\cos E)^{\frac{1}{2}}}\,\mathrm{d}E \tag{18.65}$$

$$\Delta e = -ka(1 - e^2)\int_0^{2\pi}\rho\left(\frac{1 + e\cos E}{1 - e\cos E}\right)^{\frac{1}{2}}\cos E\,\mathrm{d}E \tag{18.66}$$

For ω and ℓ, the corresponding results are

$$\Delta\omega = 0 \tag{18.67}$$

$$\Delta\ell = \int_0^P n\,\mathrm{d}t \tag{18.68}$$

The integrands in Eqs. (18.65) and (18.66) are even functions of E of period 2π. Thus

$$\Delta a = -2ka^2\int_0^{\pi}\frac{\rho(1 + e\cos E)^{\frac{3}{2}}}{(1 - e\cos E)^{\frac{1}{2}}}\,\mathrm{d}E \tag{18.69}$$

$$\Delta e = -2ka(1 - e^2)\int_0^{\pi}\rho\left(\frac{1 + e\cos E}{1 - e\cos E}\right)^{\frac{1}{2}}\cos E\,\mathrm{d}E \tag{18.70}$$

Before integrating these expressions, it is well to discuss the scale height λ in $\rho = \rho_0\exp[-(r - r_0)/\lambda]$. It actually varies with altitude and may be defined by

$$\lambda = -\rho\Big/\frac{\mathrm{d}\rho}{\mathrm{d}r} \tag{18.71}$$

At this point, we refer the reader to Refs. 4 and 5.

The *exosphere* is said to begin at the altitude at which the scale height equals the mean free path. Above this altitude, the temperature is considered to have a constant value, the exospheric temperature T_{ex}.

Values of the density ρ may be found in the *U.S. Standard Atmosphere.*[6] The exospheric temperature is the key to entering the tables. It depends on altitude, time of day, the phase of sunspot activity, and the season of the year. It also depends on unusual solar activity. From reports on solar activity (i.e., of the 10.7 cm solar flux) published regularly, there is a procedure given in the *U.S. Standard Atmosphere* for correcting for such activity. Obviously, this is no good for predictions but can be useful for analyzing orbital data already obtained.

After integrating Eqs. (18.63) and (18.64) analytically, we shall have

$$\frac{\mathrm{d}a}{\mathrm{d}E} = \frac{\Delta a}{2\pi} = \psi_1(a, e) \tag{18.72}$$

$$\frac{\mathrm{d}e}{\mathrm{d}E} = \frac{\Delta e}{2\pi} = \psi_2(a, e) \tag{18.73}$$

One can then integrate Eqs. (18.72) and (18.73) numerically, with large steps, to find $a(E)$ and $e(E)$.

To do the analytical integrations for one revolution, we return to Eqs. (18.69) and (18.70). With

$$\rho = \rho_0 \exp[-(r - r_0)/\lambda] \tag{18.56}$$

we have

$$\rho(r) = \rho_0 \exp\left[-\frac{ae}{\lambda}(1 - \cos E)\right] = \rho_0 \varepsilon^{-c} \varepsilon^{c \cos E} \tag{18.58}$$

$$c \equiv ae/\lambda \tag{18.59}$$

Insert Eq. (18.58) into Eqs. (18.69) and (18.70). The results are

$$\Delta a = -2ka^2 \rho_0 \varepsilon^{-c} \int_0^{\pi} \frac{\varepsilon^{c \cos E}(1 + e \cos E)^{\frac{3}{2}}}{(1 - e \cos E)^{\frac{1}{2}}} \,\mathrm{d}E \tag{18.74}$$

$$\Delta e = -2ka(1 - e^2)\rho_0 \varepsilon^{-c} \int_0^{\pi} \varepsilon^{c \cos E} \left(\frac{1 + e \cos E}{1 - e \cos E}\right)^{\frac{1}{2}} \cos E \,\mathrm{d}E \tag{18.75}$$

We shall evaluate only Δa to find the rate of change of the period, viz., $\dot{P}$, where $P = 2\pi/n$. From

$$n^2 a^3 = \mu$$

we have

$$\begin{aligned} \frac{4\pi^2}{P^2} a^3 &= \mu \\ P^2 &= \frac{4\pi^2}{\mu} a^3 \end{aligned} \tag{18.76}$$

$$2 \ln P = \ln 4\pi^2 + 3 \ln a - \ln \mu$$

$$\frac{2\dot{P}}{P} = \frac{3\dot{a}}{a} \qquad \dot{P} = \frac{3\dot{a}}{2a} P$$

With $\dot{a} = \Delta a/P$, this becomes

$$\dot{P} = \frac{3}{2}\frac{\Delta a}{a} \tag{18.77}$$

From Eqs. (18.74) and (18.77)

$$\dot{P} = -3ka\rho_0\varepsilon^{-c}\int_0^{\pi} \frac{\varepsilon^{c\cos E}(1+e\cos E)^{\frac{3}{2}}}{(1-e\cos E)^{\frac{1}{2}}}\,\mathrm{d}E \tag{18.78}$$

Now

$$\frac{(1+e\cos E)^{\frac{3}{2}}}{(1-e\cos E)^{\frac{1}{2}}} = \frac{(1+e\cos E)^2}{(1-e^2\cos^2 E)^{\frac{1}{2}}} \tag{18.79}$$

$$= (1+2e\cos E + e^2\cos^2 E)(1-e^2\cos^2 E)^{-\frac{1}{2}}$$

$$= (1+2e\cos E+\cdots) \tag{18.80}$$

Thus

$$\dot{P} = -3ka\rho_0\varepsilon^{-c}\int_0^{\pi}\varepsilon^{c\cos E}(1+2e\cos E+\cdots)\,\mathrm{d}E \tag{18.81}$$

This integral can be evaluated in terms of Bessel functions of imaginary argument, which are tabulated. The evaluation proceeds as follows.

From

$$J_n(x) = \frac{1}{\pi}\int_0^{\pi}\cos(n\phi - x\sin\phi)\,\mathrm{d}\phi \tag{18.82}$$

and

$$I_n(x) \equiv i^{-1}J_n(ix) \tag{18.83}$$

we obtain

$$I_0(c) = \frac{1}{\pi}\int_0^{\pi}\cos(-ic\sin\phi)\,\mathrm{d}\phi$$

$$I_0(c) = \frac{1}{2\pi}\left(\int_0^{\pi}\varepsilon^{c\sin\phi}\,\mathrm{d}\phi + \int_0^{\pi}\varepsilon^{-c\sin\phi}\,\mathrm{d}\phi\right)$$

Putting $\phi = \pi/2 - E$ gives

$$\int_0^{\pi}\varepsilon^{c\sin\phi}\,\mathrm{d}\phi = -\int_{\pi/2}^{-\pi/2}\varepsilon^{c\cos E}\,\mathrm{d}E = \int_{-\pi/2}^{\pi/2}\varepsilon^{c\cos E}\,\mathrm{d}E$$

Putting $\phi = E - \pi/2$ gives

$$\int_0^{\pi}\varepsilon^{-c\sin\phi}\,\mathrm{d}\phi = \int_{\pi/2}^{3\pi/2}\varepsilon^{c\cos E}\,\mathrm{d}E$$

Thus

$$I_0(c) = \frac{1}{2\pi}\int_{-\pi/2}^{3\pi/2}\varepsilon^{c\cos E}\,\mathrm{d}E = \frac{1}{2\pi}\int_{-\pi}^{\pi}\varepsilon^{c\cos E}\,\mathrm{d}E = \frac{1}{\pi}\int_0^{\pi}\varepsilon^{c\cos E}\,\mathrm{d}E \tag{18.84}$$

Then

$$I_0'(c) = \frac{1}{\pi}\int_0^{\pi} \varepsilon^{c\cos E}\cos E\,\mathrm{d}E \tag{18.85}$$

Lemma:

$$I_0'(c) = I_1(c) \tag{18.86}$$

Proof: Use the recurrence relations

$$J_{n-1}(x) - J_{n+1}(x) = 2J_n'(x) \tag{18.87}$$

$$J_{n-1}(x) + J_{n+1}(x) = \frac{2n}{x}J_n(x) \tag{18.88}$$

These give

$$J_n'(x) = \frac{n}{x}J_n(x) - J_{n+1}(x) \tag{18.89}$$

However,

$$\frac{d}{\mathrm{d}x}\left(x^{-n}J_n\right) = -nx^{-n-1}J_n + x^{-n}J_n' \tag{18.90}$$

On inserting Eq. (18.89) into Eq. (18.90), we find

$$\frac{d}{\mathrm{d}x}\left(x^{-n}J_n\right) = -x^{-n}J_{n+1} \tag{18.91}$$

or

$$\frac{d}{\mathrm{d}y}\left[y^{-n}J_n(x)\right] = -y^{-n}J_{n+1}(x) \tag{18.92}$$

Now in Eq. (18.92), put $y = ix$. Then

$$\frac{1}{i}\frac{d}{\mathrm{d}x}\left[(ix)^{-n}J_n(ix)\right] = -\,i^{-n}x^{-n}J_{n+1}(ix) \tag{18.93}$$

Here

$$J_n(ix) = i^n I_n(x) \tag{18.94}$$

from Eq. (18.83). Insert Eq. (18.94) into (18.93) to obtain

$$\frac{1}{i}\frac{d}{\mathrm{d}x}\left[x^{-n}I_n(x)\right] = -\,i^{-n}x^{-n}i^{n+1}I_{n+1}(x)$$

or

$$\frac{d}{\mathrm{d}x}\left[x^{-n}I_n(x)\right] = x^{-n}I_{n+1}(x) \tag{18.95}$$

If $n = 0$ and $x = c$, this becomes

$$I_0'(c) = I_1(c) \tag{18.86}$$

which is the lemma to be proved.

Now insert Eqs. (18.84–18.86) into Eq. (18.81). The result is

$$\dot{P} = -3\pi k a \rho_0 \varepsilon^{-c}\left[I_0(c) + 2eI_1(c) + O(e^2)\right] \tag{18.96}$$

where $c = ae/\lambda$. This is the result from Eq. (18.81).

Observations of $\dot{P}$ for two or more satellites at different perigee heights, or of one satellite at different dates, will suffice to determine ρ_0 and λ. The heights must not be too different, or ρ_0 and λ will be too different at the various heights.[7]

References

[1]Sterne, T. E., *An Introduction to Celestial Mechanics,* Interscience, New York, 1960.

[2]Jupp, A. H., *Celestial Mechanics,* Vol. 14, 1971, pp. 335–339.

[3]King-Hele, D. G., *Celestial Mechanics,* Vol. 18, 1978, pp. 163–164.

[4]Jacchia, L. G., Vol. 5, Paper 17, in *Solar Terrestrial Physics, Terrestrial Aspects,* MIT Press, Cambridge, MA, 1969.

[5]Jacchia, L. G., *Geophysical Research,* Vol. 76, 1971, pp. 4602–4606.

[6]*U.S. Standard Atmosphere,* U.S. Government Printing Office, Washington, DC.

[7]King-Hele, D. G., *Theory of Satellite Orbits in an Atmosphere,* Butterworths, London, 1964, p. 138.

Chapter 19

The Brouwer–von Zeipel Method I

I. Introduction

IN THE Brouwer–von Zeipel method for calculating orbits of artificial satellites, one uses the Delaunay form of the canonical equations and eliminates the lower case variables from the Hamiltonian by means of successive canonical transformations. (See Refs. 1 and 2.) From Chapter 9 the Delaunay variables are

$$\begin{aligned}
L &= (\mu a)^{\frac{1}{2}} & \ell &= n(t-\tau) = \text{mean anomaly} \\
G &= [\mu a(1-e^2)]^{\frac{1}{2}} = L(1-e^2)^{\frac{1}{2}} & g &= \omega \\
H &= [\mu a(1-e^2)]^{\frac{1}{2}} \cos I = G \cos I & h &= \Omega
\end{aligned} \tag{19.1}$$

with the Hamiltonian

$$F = (\mu^2/2L^2) + F_1 \tag{19.2}$$

where $F_1 = -V_1$ and V_1 is the Earth's potential beyond $-\mu/r$. The Delaunay canonical equations are

$$\begin{aligned}
\frac{\mathrm{d}L}{\mathrm{d}t} &= \frac{\partial F}{\partial \ell} = \frac{\partial F_1}{\partial \ell} & \frac{\mathrm{d}\ell}{\mathrm{d}t} &= -\frac{\partial F}{\partial L} = \frac{\mu^2}{L^3} - \frac{\partial F_1}{\partial L} = n - \frac{\partial F_1}{\partial L} \\
\frac{\mathrm{d}G}{\mathrm{d}t} &= \frac{\partial F}{\partial g} = \frac{\partial F_1}{\partial g} & \frac{\mathrm{d}g}{\mathrm{d}t} &= -\frac{\partial F}{\partial G} = -\frac{\partial F_1}{\partial G} \\
\frac{\mathrm{d}H}{\mathrm{d}t} &= \frac{\partial F}{\partial h} = \frac{\partial F_1}{\partial h} & \frac{\mathrm{d}h}{\mathrm{d}t} &= -\frac{\partial F}{\partial H} = -\frac{\partial F_1}{\partial H}
\end{aligned} \tag{19.3}$$

Here n is the mean motion.

To begin, take V_1 through the second zonal harmonic only:

$$F_1 = -V_1 = -\frac{\mu r_e^2}{r^3} J_2 P_2(\sin\theta) \tag{19.4}$$

as in Eq. (16.1). Here θ is the latitude. In this first approach with zonal harmonics only, $h = \Omega$ does not appear in F_1, so that $H = \text{const}$. From Eq. (16.7)

$$F_1 = \frac{\mu r_e^2 J_2}{a^3}\left\{\left[-\frac{1}{4} + \frac{3}{4}\cos^2 I\right]\left(\frac{a}{r}\right)^3 + \left[\frac{3}{4} - \frac{3}{4}\cos^2 I\right]\left(\frac{a}{r}\right)^3 \cos(2\omega + 2f)\right\} \tag{19.5}$$

where a depends only on L, $\cos I = H/G$, and f depends on ℓ and e or ultimately on ℓ, L, and G. Altogether, F is a function of L, G, H, ℓ, and g, but not of $h = \Omega$.

II. Splitting F_1 into Two Parts

We may average F_1 over the osculating orbit to find a quantity $\bar{F}_1$ independent of ℓ and of short periodic parts. The remainder, however,

$$F_{1\ell} = F_1 - \bar{F}_1 \tag{19.6}$$

will be short periodic. Here

$$\bar{F}_1 = \frac{1}{2\pi}\int_0^{2\pi} F_1 \, \mathrm{d}\ell \tag{19.7}$$

From Eqs. (17.2–17.11), we have

$$\bar{F}_1 = \frac{\mu r_e^2 J_2 (1-e^2)^{-\frac{3}{2}}}{2a^3}\left(-\frac{1}{2}+\frac{3}{2}\cos^2 I\right) \tag{19.8}$$

or

$$\bar{F}_1 = \frac{\mu r_e^2 J_2}{2a^3}\frac{L^3}{G^3}\left(-\frac{1}{2}+\frac{3}{2}\frac{H^2}{G^2}\right) \tag{19.9}$$

Rewriting Eq. (19.5) as

$$F_1 = \frac{\mu r_e^2 J_2}{2a^3}\left\{-\frac{1}{2}\left(\frac{a}{r}\right)^3 + \frac{3}{2}\frac{H^2}{G^2}\left(\frac{a}{r}\right)^3 + \frac{3}{2}\left[1-\frac{H^2}{G^2}\right]\left(\frac{a}{r}\right)^3\cos(2g+2f)\right\} \tag{19.10}$$

and subtracting Eq. (19.9) from Eq. (19.10), we obtain

$$F_{1\ell} = \frac{\mu r_e^2 J_2}{2a^3}\left\{\left[-\frac{1}{2}+\frac{3}{2}\frac{H^2}{G^2}\right]\left[\frac{a^3}{r^3}-\frac{L^3}{G^3}\right] + \frac{3}{2}\left[1-\frac{H^2}{G^2}\right]\left(\frac{a}{r}\right)^3\cos(2g+2f)\right\} \tag{19.11}$$

Here L, G, H are the q's, and ℓ, g, h are the p's. Remember that $a = L^2/\mu$, that ℓ enters through r and f, and that h is missing.

The Delaunay equations become

$$\begin{aligned} \frac{\mathrm{d}L}{\mathrm{d}t} &= \frac{\partial F_1}{\partial \ell} & \frac{\mathrm{d}\ell}{\mathrm{d}t} &= n - \frac{\partial F_1}{\partial L} \\ \frac{\mathrm{d}G}{\mathrm{d}t} &= \frac{\partial F_1}{\partial g} & \frac{\mathrm{d}g}{\mathrm{d}t} &= -\frac{\partial F_1}{\partial G} \\ H &= \text{const} & \frac{\mathrm{d}h}{\mathrm{d}t} &= -\frac{\partial F_1}{\partial H} \end{aligned} \tag{19.12}$$

III. Elimination of ℓ

To solve Eq. (19.12), we make a canonical transformation to new, primed variables $L', G', H', \ell', g', h'$ by means of a generating function of the form $S(p, Q)$.

Here the p's are ℓ, g, h; the q's are L, G, H; the P's are ℓ', g', h'; and the Q's are L', G', H'. For short, denote L, G, H by L_k $(k = 1, 2, 3)$ and ℓ, g, h by ℓ_k.

Then

$$q_k = \frac{\partial S(p, Q)}{\partial p_k} \qquad P_k = \frac{\partial S(p, Q)}{\partial Q_k} \tag{19.13}$$

become

$$L_k = \frac{\partial S(\ell_k, L'_k)}{\partial \ell_k} \qquad \ell'_k = \frac{\partial S(\ell_k, L'_k)}{\partial L'_k} \tag{19.14}$$

If F^* is the new Hamiltonian, the new variables will satisfy

$$\frac{\mathrm{d}L'_k}{\mathrm{d}t} = \frac{\partial F^*}{\partial \ell'_k} \qquad \frac{\mathrm{d}\ell'_k}{\mathrm{d}t} = -\frac{\partial F^*}{\partial L'_k} \tag{19.15}$$

The primed variables will not differ greatly from the unprimed variables, because it is only the J_2 perturbation that makes them change. Thus, S must start off with the identity transformation function (see Chapter 5, Sec. II, case d, $S = \Sigma_k Q_k p_k$)

$$S_0 = \Sigma_k L'_k \ell_k = L'\ell + G'g + H'h \tag{19.16}$$

Inserted into Eqs. (19.14), this would give $L_k = L'_k$ and $\ell_k = \ell'_k$. We then write

$$S = S_0 + S_1(L', G', H', \ell, g, -) + S_2(L', G', H', \ell, g, -) \tag{19.17}$$

Here, it is understood that S_1 contains a factor J_2 and S_2 a factor J_2^2. The variable $h = \Omega$ is not indicated in S_1 and S_2 because it is not present in the Hamiltonian. Insertion of Eq. (19.17) into Eqs. (19.14) gives

$$\begin{aligned} L &= L' + \frac{\partial S_1}{\partial \ell} + \frac{\partial S_2}{\partial \ell} \\ G &= G' + \frac{\partial S_1}{\partial g} + \frac{\partial S_2}{\partial g} \\ H &= H' \end{aligned} \tag{19.18}$$

$$\begin{aligned} \ell' &= \ell + \frac{\partial S_1}{\partial L'} + \frac{\partial S_2}{\partial L'} \\ g' &= g + \frac{\partial S_1}{\partial G'} + \frac{\partial S_2}{\partial G'} \\ h' &= h + \frac{\partial S_1}{\partial H'} + \frac{\partial S_2}{\partial H'} \end{aligned} \tag{19.19}$$

The old Hamiltonian

$$F = \frac{\mu^2}{2L^2} + F_1 = F_0(L) + F_1 \tag{19.20}$$

The new Hamiltonian F^* will be equal to F, because the generating function S is independent of t, but will have a different functional form in the new variables:

$$F^* = F_0^*(L'_k) + F_1^*(L'_k, \ell'_k) + F_2^*(L'_k, \ell'_k) \tag{19.21}$$

Here, it is understood that F_1^* is of order J_2, and F_2^* is of order J_2^2. The old Hamiltonian is

$$F = F(L, G, H, \ell, g, -) \tag{19.22}$$

The new Hamiltonian will be expressible as

$$F^* = F^*(L', G', H', -, g', -) \tag{19.23}$$

if we choose an S so as to eliminate ℓ'. Then

$$F(L, G, H, \ell, g, -) = F^*(L', G', H, -, g', -) \tag{19.24}$$

where H' has been replaced by H, according to Eq. (19.18).

With ℓ' eliminated from F^*, we have from Eq. (19.15) that $\dot{L}' = 0$ or

$$L' = \text{const} \tag{19.25}$$

Next, insert Eqs. (19.18) and (19.19) into Eq. (19.24), making use of Eqs. (19.20) and (19.21). Then

$$F_0\left(L' + \frac{\partial S_1}{\partial \ell} + \frac{\partial S_2}{\partial \ell}\right) + F_1\left(L' + \frac{\partial S_1}{\partial \ell} + \frac{\partial S_2}{\partial \ell}, G' + \frac{\partial S_1}{\partial g} + \frac{\partial S_2}{\partial g}, H, \ell, g, -\right)$$
$$= F_0^* + F_1^*\left(L', G', H, -, g + \frac{\partial S_1}{\partial G'} + \frac{\partial S_2}{\partial G'}, -\right)$$
$$+ F_2^*\left(L', G', H, -, g + \frac{\partial S_1}{\partial G'} + \frac{\partial S_2}{\partial G'}, -\right) \tag{19.26}$$

The next step is the crucial one, a Taylor expansion of a function f about f_0. Let

$$\boldsymbol{x}_0 = \boldsymbol{x}(x_1, x_2, \ldots, x_i, \ldots) \qquad \boldsymbol{x} = \boldsymbol{x}(x_1 + h_1, x_2 + h_2, \ldots, x_i + h_i, \ldots)$$
$$f(\boldsymbol{x}_0) = f(x_1, x_2, \ldots, x_i, \ldots) \qquad f(\boldsymbol{x}) = f(x_1 + h_1, x_2 + h_2, \ldots, x_i + h_i, \ldots)$$

Then

$$f(\boldsymbol{x}) = f(\boldsymbol{x}_0) + \left[\left(h_i \frac{\partial}{\partial x_i}\right) f + \frac{1}{2}\left(h_i^2 \frac{\partial^2}{\partial x_i^2}\right) f + \cdots\right]\Bigg|_{\boldsymbol{x}_0} \tag{19.27}$$

where h_i is the small increment of the element x_i about $\boldsymbol{x}_0$.

Apply Eq. (19.27) to Eq. (19.26), retaining terms only through order J_2^2. The result is

$$F_0(L') + \frac{\partial S_1}{\partial \ell}\frac{\mathrm{d}F_0}{\mathrm{d}L'} + \frac{\partial S_2}{\partial \ell}\frac{\mathrm{d}F_0}{\mathrm{d}L'} + \frac{1}{2}\left(\frac{\partial S_1}{\partial \ell}\right)^2 \frac{\mathrm{d}^2 F_0}{\mathrm{d}L'^2}$$
$$+ F_1(L', G', H, \ell, g) + \frac{\partial S_1}{\partial \ell}\frac{\partial F_1}{\partial L'} + \frac{\partial S_1}{\partial g}\frac{\partial F_1}{\partial G'}$$
$$= F_0^* + F_1^*(L', G', H, g) + \frac{\partial S_1}{\partial G'}\frac{\partial F_1^*}{\partial g} + F_2^*(L', G', H, g) \tag{19.28}$$

This is an expansion in the "mixed" variables L'_k and ℓ_k, in the neighborhood of L', G', $H' = H$, ℓ, and g.

The next step is to resolve Eq. (19.28) into separate equations for the orders of $J_2(0, 1, 2)$. All this is to find S_1 and S_2 so as to eliminate ℓ' from F^*.

Zero order:

$$F_0^* = F_0(L') = \frac{\mu^2}{2L'^2} \tag{19.29}$$

First order:

$$\frac{\partial S_1}{\partial \ell}\frac{\mathrm{d}F_0}{\mathrm{d}L'} + F_1(L', G', H, \ell, g) = F_1^* \tag{19.30}$$

Second order:

$$\frac{\partial S_2}{\partial \ell}\frac{\mathrm{d}F_0}{\mathrm{d}L'} + \frac{1}{2}\left(\frac{\partial S_1}{\partial \ell}\right)^2 \frac{\mathrm{d}^2 F_0}{\mathrm{d}L'^2} + \frac{\partial S_1}{\partial \ell}\frac{\partial F_1}{\partial L'} + \frac{\partial S_1}{\partial g}\frac{\partial F_1}{\partial G'} = \frac{\partial S_1}{\partial G'}\frac{\partial F_1^*}{\partial g} + F_2^*(L', G', H, g) \tag{19.31}$$

To handle the first order, we use $F_1 = \bar{F}_1 + F_{1\ell}$, where $\bar{F}_1$ is given by Eq. (19.9) and $F_{1\ell}$ by Eq. (19.11). In doing so, however, we must replace L and G by L' and G', according to Eq. (19.30). By Eq. (19.9)

$$\bar{F}_1(L', G', H, -, g) = \frac{\mu r_e^2 J_2}{2a'^3}\left(\frac{L'}{G'}\right)^3\left(-\frac{1}{2} + \frac{3}{2}\frac{H^2}{G'^2}\right) \tag{19.32}$$

where

$$a' = L'^2/\mu \tag{19.33}$$

By Eq. (19.11)

$$F_{1\ell}(L', G', H, \ell, g) = \frac{\mu r_e^2 J_2}{2a'^3}\left\{\left[-\frac{1}{2} + \frac{3}{2}\frac{H^2}{G'^2}\right]\left[\frac{a'^3}{r'^3} - \frac{L'^3}{G'^3}\right]\right.$$
$$\left. + \frac{3}{2}\left[1 - \frac{H^2}{G'^2}\right]\left(\frac{a'}{r'}\right)^3 \cos(2g + 2f')\right\} \tag{19.34}$$

To find r', use

$$e'^2 = 1 - (G'^2/L'^2) \tag{19.35}$$

to solve for E' in

$$E' - e' \sin E' = \ell \tag{19.36}$$

where ℓ is unprimed because we are working in the neighborhood of L', G', H, ℓ, g. Then

$$r' = a'(1 - e' \cos E') \tag{19.37}$$

To find f', use

$$\cos f' = \frac{\cos E' - e'}{1 - e' \cos E'} \tag{19.38a}$$

$$\sin f' = \frac{\sqrt{1 - e'^2}\sin E'}{1 - e' \cos E'} \tag{19.38b}$$

Now return to Eq. (19.30). We now have

$$\frac{dF_0}{dL'}\frac{\partial S_1}{\partial \ell} + \bar{F}_1 + F_{1\ell} = F_1^* \tag{19.39}$$

The best way to find both F_1^* and $\partial S_1/\partial \ell$ is to choose S_1, so that

$$\frac{dF_0}{dL'}\frac{\partial S_1}{\partial \ell} + F_{1\ell} = 0 \tag{19.40}$$

$$F_1^* = \bar{F}_1 \tag{19.41}$$

From Eqs. (19.32) and (19.41)

$$F_1^* = \frac{\mu^4 r_e^2 J_2}{2L'^3 G'^3}\left(-\frac{1}{2} + \frac{3}{2}\frac{H^2}{G'^2}\right) \tag{19.42}$$

This gives F_1^* explicitly independent of time, as desired, and also independent of g, so that the term $(\partial F_1^*/\partial g)(\partial S_1/\partial G')$ drops out of Eq. (19.31), the second-order equation.

To find S_1, use $F_0(L') = \mu^2/2L'^2$, so that

$$\frac{dF_0}{dL'} = -\frac{\mu^2}{L'^3} \tag{19.43}$$

By Eqs. (19.40) and (19.43)

$$\frac{\partial S_1}{\partial \ell} = \frac{L'^3}{\mu^2} F_{1\ell} \tag{19.44}$$

By Eqs. (19.44), (19.33), and (19.34)

$$\frac{\partial S_1}{\partial \ell} = \frac{\mu r_e^2 J_2}{2L'^3}\{A'\sigma_1 + B'\sigma_2\} \tag{19.45}$$

where

$$A' = -\frac{1}{2} + \frac{3}{2}\frac{H^2}{G'^2} \tag{19.46}$$

$$B' = \frac{3}{2}\left[1 - \frac{H^2}{G'^2}\right] \tag{19.47}$$

$$\sigma_1 = \frac{a'^3}{r'^3} - \frac{L'^3}{G'^3} \tag{19.48}$$

$$\sigma_2 = (a'/r')^3 \cos(2g + 2f') \tag{19.49}$$

Integration of Eq. (19.45) yields

$$S_1 = \frac{\mu^2 r_e^2 J_2}{2L'^3}\int [A'\sigma_1 + B'\sigma_2]\, d\ell + \Phi(L', G', g) \tag{19.50}$$

The formulas connecting a', r', e', f', and ℓ (unprimed) are those of elliptic motion,

so that

$$d\ell = (r'/a')^2(1-e'^2)^{-\frac{1}{2}}\,df' \tag{19.51}$$

$$\int \sigma_1\,d\ell = (1-e'^2)^{-\frac{1}{2}}\int\left(\frac{a'}{r'}\right)df' - \frac{L'^3}{G'^3}\ell \tag{19.52}$$

$$\int\left(\frac{a'}{r'}\right)df' = (1-e'^2)^{-1}\int(1+e'\cos f')\,df' = \frac{f'+e'\sin f'}{1-e'^2} \tag{19.53}$$

$$\int \sigma_1\,d\ell = \frac{f'+e'\sin f'}{(1-e'^2)^{\frac{3}{2}}} - \frac{L'^3}{G'^3}\ell = \frac{L'^3}{G'^3}(f'-\ell+e'\sin f') \tag{19.54}$$

since $L'/G' = (1-e'^2)^{-1/2}$.

Now

$$\begin{aligned}\int \sigma_2\,d\ell &= (1-e'^2)^{-\frac{1}{2}}\int(a'/r')\cos(2g+2f')\,df'\\ &= (1-e'^2)^{-\frac{3}{2}}\int(1+e'\cos\ f')\cos(2g+2f')\,df'\\ &= (1-e'^2)^{-\frac{3}{2}}\int\left[\cos(2g+2f')+\frac{e'}{2}\cos(2g+f')\right.\\ &\quad\left.+\frac{e'}{2}\cos(2g+3f')\right]df'\\ &= (1-e'^2)^{-\frac{3}{2}}\left[\frac{1}{2}\sin(2g+2f')+\frac{e'}{2}\sin(2g+f')+\frac{e'}{6}\sin(2g+3f')\right]\end{aligned} \tag{19.55}$$

Thus

$$\int \sigma_2\,d\ell = \frac{1}{2}\frac{L'^3}{G'^3}\left[\sin(2g+f')+e'\sin(2g+f')+\frac{e'}{3}\sin(2g+3f')\right] \tag{19.56}$$

Substituting Eqs. (19.46), (19.47), (19.54), and (19.56) into Eq. (19.50)

$$\begin{aligned}S_1 &= \frac{\mu^2 r_e^2 J_2}{2G'^3}\left[A'\{f'-\ell+e'\sin\ f'\}\right.\\ &\quad\left.+\frac{B'}{2}\left\{\sin(2g+2f')+e'\sin(2g+f')+\frac{e'}{3}\sin(2g+3f')\right\}\right]+\Phi(g)\\ &= \frac{\mu^2 r_e^2 J_2}{2G'^3}\left[\left\{-\frac{1}{2}+\frac{3}{2}\frac{H^2}{G'^2}\right\}\{f'-\ell+e'\sin\ f'\}+\frac{1}{2}\left\{\frac{3}{2}-\frac{3}{2}\frac{H^2}{G'^2}\right\}\right.\\ &\quad\left.\times\left\{\sin(2g+2f')+e'\sin(2g+f')+\frac{e'}{3}\sin(2g+3f')\right\}\right]+\Phi(g)\end{aligned} \tag{19.57}$$

Although $\partial S_1/\partial \ell$ is purely short periodic, being proportional to $F_{1\ell}$, it happens that S_1 is not unless one chooses $\Phi(g) = -\bar{S}_1$. Brouwer did not do this, so that there are some long periodic impurities in some of his short periodic terms.[1] There is no overall error, however, because the later developed long periodic terms are automatically adjusted to take this fact into account. We shall follow the same procedure to avoid any extra labor.

IV. Short Periodic Terms of Order J_2

From Eqs. (19.18) and (19.19) with order J_2

$$L - L' = \frac{\partial S_1}{\partial \ell}$$

$$G - G' = \frac{\partial S_1}{\partial g} \tag{19.58}$$

$$\ell' - \ell = \frac{\partial S_1}{\partial L'}$$

$$g' - g = \frac{\partial S_1}{\partial G'} \tag{19.59}$$

$$h' - h = \frac{\partial S_1}{\partial H'}$$

From Eqs. (19.34), (19.44), and (19.58)

$$L - L' = \frac{\mu^2 r_e^2 J_2}{2L'^3}\left[\left(-\frac{1}{2} + \frac{3}{2}\frac{H^2}{G'^2}\right)\left(\frac{a'^3}{r'^3} - \frac{L'^3}{G'^3}\right) + \frac{3}{2}\left(1 - \frac{H^2}{G'^2}\right)\left(\frac{a'}{r'}\right)^3 \cos(2g + 2f')\right] \tag{19.60}$$

From Eqs. (19.57) and (19.58)

$$G - G' = \frac{\mu^2 r_e^2 J_2}{2G'^3}\left[\left\{\frac{3}{2} - \frac{3}{2}\frac{H^2}{G'^2}\right\} \times \left\{\cos(2g + 2f') + e'\cos(2g + f') + \frac{e'}{3}\cos(2g + 3f')\right\}\right] \tag{19.61}$$

However,

$$\frac{1}{G'^4} = \frac{L'^4}{G'^4}\frac{1}{L'^4} = \frac{(1 - e'^2)^{-\frac{1}{2}}}{\mu^2 a'^2}$$

so that

$$G = G'\left[1 + \frac{r_e^2 J_2}{2a'^2}(1 - e'^2)^{-\frac{1}{2}}\left\{\frac{3}{2} - \frac{3}{2}\frac{H^2}{G'^2}\right\} \times \left\{\cos(2g + 2f') + e'\cos(2g + f') + \frac{e'}{3}\cos(2g + 3f')\right\}\right] \tag{19.62}$$

From Eqs. (19.57) and (19.59)

$$h' - h = \frac{\mu^2 r_e^2 J_2}{2G'^3}\left[\frac{3H}{G'^2}\{f' - \ell + e'\sin f'\} - \frac{3H}{2G'^2}\left\{\sin(2g + 2f')\right.\right.$$

$$\left.\left.+ e'\sin(2g + f') + \frac{e'}{3}\sin(2g + 3f')\right\}\right] \tag{19.63}$$

To find $\ell' - \ell = \partial S_1/\partial L'$, we note that L' occurs in S_1 in Eq. (19.57) only through e' and f'. In turn, f' depends only on e', of the primed Keplerian variables. This statement follows from $f' = f'(e', E')$ and $E' = E'(e', \ell)$. Thus

$$\frac{\partial S_1}{\partial L'} = \frac{\partial S_1}{\partial e'}\frac{\partial e'}{\partial L'} \tag{19.64}$$

From $1 - e'^2 = G'^2/L'^2$, we have

$$\frac{\partial e'}{\partial L'} = \frac{G'^2}{e'L'^3} \tag{19.65}$$

From Eq. (19.57),

$$\frac{\partial S_1}{\partial e'} = \frac{\mu^2 r_e^2 J_2}{2G'^3}\left[\left(-\frac{1}{2} + \frac{3}{2}\frac{H^2}{G'^2}\right)\left((1 + e'\cos f')\frac{\partial f'}{\partial e'} + \sin f'\right)\right.$$

$$+ \left(\frac{3}{2} - \frac{3}{2}\frac{H^2}{G'^2}\right)\left\{\left[\cos(2g + 2f') + \frac{e'}{2}\cos(2g + f') + \frac{e'}{2}\cos(2g + 3f')\right]\right.$$

$$\left.\left.\times \frac{\partial f'}{\partial e'}\frac{1}{2}\sin(2g + f') + \frac{1}{6}\sin(2g + 3f')\right\}\right] \tag{19.66}$$

Introduce the simplification

$$\cos(2g + 2f') + \frac{e'}{2}\cos(2g + f') + \frac{e'}{2}\cos(2g + 3f')$$

$$= (1 + e'\cos f')\cos(2g + 2f') \tag{19.67}$$

Then

$$\frac{\partial S_1}{\partial e'} = \frac{\mu^2 r_e^2 J_2}{2G'^3}\left[\left(-\frac{1}{2} + \frac{3}{2}\frac{H^2}{G'^2}\right)\left((1 + e'\cos f')\frac{\partial f'}{\partial e'} + \sin f'\right)\right.$$

$$+ \left(\frac{3}{2} - \frac{3}{2}\frac{H^2}{G'^2}\right)\left\{(1 + e'\cos f')\cos(2g + 2f')\frac{\partial f'}{\partial e'}\right.$$

$$\left.\left.+ \frac{1}{2}\sin(2g + f') + \frac{1}{6}\sin(2g + 3f')\right\}\right] \tag{19.68}$$

From Eq. (17.15), applied to primed variables:

$$\frac{\partial f'}{\partial e'} = \left(\frac{a'}{r'} + \frac{1}{1 - e'^2}\right)\sin f' \tag{19.69}$$

Then

$$(1 + e' \cos f')\frac{\partial f'}{\partial e'} = \frac{a'(1 - e'^2)}{r'}\left(\frac{a'}{r'} + \frac{1}{1 - e'^2}\right)\sin f'$$

$$= \left(\frac{a'^2(1 - e'^2)}{r'^2} + \frac{a'}{r'}\right)\sin f' \tag{19.70}$$

Insert Eq. (19.70) into Eq. (19.68) to obtain

$$\frac{\partial S_1}{\partial e'} = \frac{\mu^2 r_e^2 J_2}{2G'^3}\Bigg[\left(-\frac{1}{2} + \frac{3}{2}\frac{H^2}{G'^2}\right)\left(\frac{a'^2(1 - e'^2)}{r'^2} + \frac{a'}{r'} + 1\right)\sin f'$$
$$+ \left(\frac{3}{2} - \frac{3}{2}\frac{H^2}{G'^2}\right)\Bigg\{\left(\frac{a'^2(1 - e'^2)}{r'^2} + \frac{a'}{r'}\right)\sin f' \cos(2g + 2f')$$
$$+ \frac{1}{2}\sin(2g + f') + \frac{1}{6}\sin(2g + 3f')\Bigg\}\Bigg] \tag{19.71}$$

Next use

$$\sin f' \cos(2g + 2f') = \tfrac{1}{2}\sin(2g + 3f') - \tfrac{1}{2}\sin(2g + f') \tag{19.72}$$

in Eq. (19.71) to get

$$\frac{\partial S_1}{\partial e'} = \frac{\mu^2 r_e^2 J_2}{2G'^3}\Bigg[\left(-\frac{1}{2} + \frac{3}{2}\frac{H^2}{G'^2}\right)\left(\frac{a'^2(1 - e'^2)}{r'^2} + \frac{a'}{r'} + 1\right)\sin f'$$
$$+ \left(\frac{3}{2} - \frac{3}{2}\frac{H^2}{G'^2}\right)\Bigg\{\left(-\frac{a'^2(1 - e'^2)}{2r'^2} - \frac{a'}{2r'} + \frac{1}{2}\right)\sin(2g + f')$$
$$+ \left(\frac{a'^2(1 - e'^2)}{2r'^2} + \frac{a'}{2r'} + \frac{1}{6}\right)\sin(2g + 3f')\Bigg\}\Bigg] \tag{19.73}$$

Then by Eqs. (19.64) and (19.65)

$$\frac{\partial S_1}{\partial L'} = \frac{\partial S_1}{\partial e'}\frac{\partial e'}{\partial L'} = \frac{G'^2}{e'L'^3}\frac{\partial S_1}{\partial e'}$$

and

$$\ell' - \ell = \frac{\partial S_1}{\partial L'} = \frac{G'^2}{e'L'^3}\frac{\partial S_1}{\partial e'}$$

$$\ell' - \ell = \frac{\mu^2 r_e^2 J_2}{2e'G'L'^3}\Bigg[\left(-\frac{1}{2} + \frac{3}{2}\frac{H^2}{G'^2}\right)\left(\frac{a'^2(1 - e'^2)}{r'^2} + \frac{a'}{r'} + 1\right)\sin f'$$
$$+ \left(\frac{3}{2} - \frac{3}{2}\frac{H^2}{G'^2}\right)\Bigg\{\left(-\frac{a'^2(1 - e'^2)}{2r'^2} - \frac{a'}{2r'} + \frac{1}{2}\right)\sin(2g + f')$$
$$+ \left(\frac{a'^2(1 - e'^2)}{2r'^2} + \frac{a'}{2r'} + \frac{1}{6}\right)\sin(2g + 3f')\Bigg\}\Bigg] \tag{19.73a}$$

We have next

$$g' - g = \frac{\partial S_1}{\partial G'} \tag{19.74}$$

where

$$S_1 = \frac{\mu^2 r_e^2 J_2}{2G'^3}\left[\left\{-\frac{1}{2}+\frac{3}{2}\frac{H^2}{G'^2}\right\}\{f' - \ell + e' \sin\, f'\} + \frac{1}{2}\left\{\frac{3}{2}-\frac{3}{2}\frac{H^2}{G'^2}\right\}\right.$$
$$\left.\times\{\sin(2g+2f') + e'\sin(2g+f') + \frac{e'}{3}\sin(2g+3f')\}\right]_1 \tag{19.57}$$

Here G' occurs explicitly and also implicitly through e'. Thus, if the explicit derivative is $[\partial S_1/\partial G']$, we have

$$\frac{\partial S_1}{\partial G'} = \left[\frac{\partial S_1}{\partial G'}\right] + \frac{\partial S_1}{\partial e'}\frac{\partial e'}{\partial G'} \tag{19.75}$$

Since $1 - e'^2 = G'^2/L'^2$

$$\frac{\partial e'}{\partial G'} = -\frac{G'}{e'L'^2} \tag{19.76}$$

and

$$\frac{\partial S_1}{\partial G'} = \left[\frac{\partial S_1}{\partial G'}\right] - \frac{G'}{e'L'^2}\frac{\partial S_1}{\partial e'} \tag{19.77}$$

By using Eq. (19.57)

$$\left[\frac{\partial S_1}{\partial G'}\right] = -\frac{3\mu^2 r_e^2 J_2}{2G'^4}[\,]_1 + \frac{\mu^2 r_e^2 J_2}{2G'^3}\left[-\frac{3H^2}{G'^3}\{f' - \ell + e'\sin f'\}\right.$$
$$\left. + \frac{3H^2}{2G'^3}\left\{\sin(2g+2f') + e'\sin(2g+f') + \frac{e'}{3}\sin(2g+3f')\right\}\right]_2 \tag{19.78}$$

In Eq. (19.78) the coefficients of $f' - \ell + e'\sin f'$

$$= -\frac{3\mu^2 r_e^2 J_2}{2G'^4}\left[-\frac{1}{2}+\frac{3}{2}\cos^2 I' + \cos^2 I'\right]$$
$$= -\frac{3\mu^2 r_e^2 J_2}{2G'^4}\left[-\frac{1}{2}+\frac{5}{2}\cos^2 I'\right] \tag{19.79}$$

The rest of $[\partial S_1/\partial G']$

$$= -\frac{3\mu^2 r_e^2 J_2}{2G'^4}\left[\frac{1}{2}\left\{\frac{3}{2}-\frac{3}{2}\cos^2 I'\right\}\right.$$
$$\times\left\{\sin(2g+2f') + e'\sin(2g+f') + \frac{e'}{3}\sin(2g+3f')\right\}$$
$$\left.-\frac{1}{2}\cos^2 I'\left\{\sin(2g+2f') + e'\sin(2g+f') + \frac{e'}{3}\sin(2g+3f')\right\}\right] \tag{19.80}$$

Thus

$$\left[\frac{\partial S_1}{\partial G'}\right] = -\frac{3\mu^2 r_e^2 J_2}{2G'^4}\left[\left\{-\frac{1}{2}+\frac{5}{2}\cos^2 I'\right\}\{f'-\ell+e'\sin f'\}\right.$$
$$+\frac{1}{2}\left\{\frac{3}{2}-\frac{5}{2}\cos^2 I'\right\}\left\{\sin(2g+2f')+e'\sin(2g+f')\right.$$
$$\left.\left.+\frac{e'}{3}\sin(2g+3f')\right\}\right] \tag{19.81}$$

From Eqs. (19.73) and (19.76),

$$\frac{\partial S_1}{\partial e'}\frac{\partial e'}{\partial G'} = -\frac{\mu^2 r_e^2 J_2}{2e'G'^2L'^2}\left[\left(-\frac{1}{2}+\frac{3}{2}\cos^2 I'\right)\left(\frac{a'^2(1-e'^2)}{r'^2}+\frac{a'}{r'}+1\right)\sin f'\right.$$
$$+\left(\frac{3}{2}-\frac{3}{2}\cos^2 I'\right)\left\{\left(-\frac{a'^2(1-e'^2)}{2r'^2}-\frac{a'}{2r'}+\frac{1}{2}\right)\sin(2g+f')\right.$$
$$\left.\left.+\left(\frac{a'^2(1-e'^2)}{2r'^2}+\frac{a'}{2r'}+\frac{1}{6}\right)\sin(2g+3f')\right\}\right] \tag{19.82}$$

By Eq. (19.77)

$$g'-g = \frac{\partial S_1}{\partial G'} = \left[\frac{\partial S_1}{\partial G'}\right] + \frac{\partial S_1}{\partial e'}\frac{\partial e'}{\partial G'} \tag{19.83}$$

This completes the evaluation of the first-order periodic terms.

V. Second-Order Terms, General

We now go back to Eq. (19.31). By Eq. (19.42), F_1^* does not depend on g. Thus, Eq. (19.31) becomes

$$\frac{\mathrm{d}F_0}{\mathrm{d}L'}\frac{\partial S_2}{\partial \ell} + \frac{1}{2}\frac{\mathrm{d}^2F_0}{\mathrm{d}L'^2}\left(\frac{\partial S_1}{\partial \ell}\right)^2 + \frac{\partial F_1}{\partial L'}\frac{\partial S_1}{\partial \ell} + \frac{\partial F_1}{\partial G'}\frac{\partial S_1}{\partial g} = F_2^*(L', G', H, g') \tag{19.84}$$

where we have replaced g by g' on the right side of Eq. (19.84). Because $g'-g = O(J_2)$ and F_2^* has a factor J_2^2, the error from this substitution is of order J_2^3.

Next, resolve

$$\frac{1}{2}\frac{\mathrm{d}^2F_0}{\mathrm{d}L'^2}\left(\frac{\partial S_1}{\partial \ell}\right)^2 + \frac{\partial F_1}{\partial L'}\frac{\partial S_1}{\partial \ell} + \frac{\partial F_1}{\partial G'}\frac{\partial S_1}{\partial g} \equiv N \tag{19.85}$$

into two parts: 1)

$$\bar{N} \equiv \frac{1}{2\pi}\int_0^{2\pi} N\,\mathrm{d}\ell \tag{19.86}$$

and 2)

$$N_p = \text{the short periodic part of } N$$
$$= N - \bar{N} \tag{19.87}$$

Then

$$\frac{dF_0}{dL'}\frac{\partial S_2}{\partial \ell} + N_p + \bar{N} = F_2^* \tag{19.88}$$

Brouwer's step here is to resolve Eq. (19.88) as follows.[1]

$$\frac{dF_0}{dL'}\frac{\partial S_2}{\partial \ell} + N_p = 0 \tag{19.88a}$$

$$F_2^* = \bar{N} \tag{19.89}$$

This is a reasonable resolution, since the new Hamiltonian $F_0^* + F_1^* + F_2^*$ will not depend explicitly on the time.

Brouwer does not attempt to solve Eq. (19.88a) for S_2, which would yield short periodic terms of the second order. He evaluates Eq. (19.89) in a long derivation (see Ref. 1), which permits evaluation of secular terms through the second order and long periodic terms of the first order. The result is

$$\begin{aligned} F_2^* = {} & \frac{\mu^6 r_e^4 J_2^2}{4(L')^{10}}\left[\frac{3}{32}\left(\frac{L'}{G'}\right)^5\left(5 - \frac{18H^2}{G'^2} + \frac{5H^4}{G'^4}\right) + \frac{3}{8}\left(\frac{L'}{G'}\right)^6\right. \\ & \left. \times\left(1 - \frac{6H^2}{G'^2} + \frac{9H^4}{G'^4}\right) - \frac{15}{32}\left(\frac{L'}{G'}\right)^7\left(1 - \frac{2H^2}{G'^2} - \frac{7H^4}{G'^4}\right)\right] \\ & + \frac{\mu^6 r_e^4 J_2^2}{4(L')^{10}}\left[\frac{3}{16}\left(\frac{L'}{G'}\right)^5\left(\frac{L'^2}{G'^2} - 1\right)\left(1 - \frac{16H^2}{G'^2} + \frac{15H^4}{G'^4}\right)\cos 2g'\right] \end{aligned} \tag{19.90}$$

The calculation actually gives $\cos 2g$, but we can replace g by g' with an error of $O(J_2^3)$. Here, the first group of terms, F_{2s}^*, is the secular term, and the second, F_{2p}^*, is a long periodic term.

Summary: By transforming from L, G, H, ℓ, g, h to $L', G', H', \ell', g', h'$, we have eliminated short periodic terms and have gone from the Hamiltonian

$$F = F_0 + F_1 \qquad \text{to} \qquad F^* = F_0^* + F_1^* + F_2^*$$

where

$$F_0^* = \frac{\mu^2}{2L'^2} \qquad F_1^* = \frac{\mu^4 r_e^2 J_2}{2L'^3 G'^3}\left(-\frac{1}{2} + \frac{3}{2}\frac{H^2}{G'^2}\right)$$

and F_2^* is given by Eq. (19.90). This was a canonical transformation that changed the Hamiltonian

$$F = F(L, G, H, \ell, g, -) \qquad \text{to} \qquad F^* = F^*(L'G', H', -, g', -)$$

VI. A Second Canonical Transformation

We now make another canonical transformation from $L', G', H', \ell', g', h'$ to $L'', G'', H'', \ell'', g'', h''$ of such a kind that

$$F^*(L', G', H', -, g', -) = F^{**}(L'', G'', H'', -, -, -)$$

To do so, let us introduce the new generating function

$$\begin{aligned} S^* &= S_0^* + S_1^* = \Sigma_k L_k'' \ell_k' + S_1^* \\ &= L''\ell' + G''g' + H''h' + S_1^*(L'', G'', H'', g', -, -) \end{aligned} \tag{19.91}$$

where ℓ', g', h', are the old p's and L'', G'', H'' are the new Q's. Thus, from

$$q_k = \frac{\partial S^*}{\partial p_k} \qquad P_k = \frac{\partial S^*}{\partial Q_k}$$

we have

$$L' = \frac{\partial S^*}{\partial \ell'} = L''$$

$$G' = \frac{\partial S^*}{\partial g'} = G'' + \frac{\partial S_1^*}{\partial g'} \tag{19.92a}$$

$$H' = \frac{\partial S^*}{\partial h'} = H''(= H)$$

$$\ell'' = \frac{\partial S^*}{\partial L''} = \ell' + \frac{\partial S_1^*}{\partial L''}$$

$$g'' = \frac{\partial S^*}{\partial G''} = g' + \frac{\partial S_1^*}{\partial G''} \tag{19.92b}$$

$$h'' = \frac{\partial S^*}{\partial H''} = h' + \frac{\partial S_1^*}{\partial H''} = h' + \frac{\partial S_1^*}{\partial H}$$

Also

$$\frac{\mathrm{d}L''}{\mathrm{d}t} = \frac{\partial F^{**}}{\partial \ell''} = 0$$

$$\frac{\mathrm{d}G''}{\mathrm{d}t} = \frac{\partial F^{**}}{\partial g''} = 0 \tag{19.93a}$$

$$\frac{\mathrm{d}H''}{\mathrm{d}t} = \frac{\partial F^{**}}{\partial h''} = 0$$

$$\frac{\mathrm{d}\ell''}{\mathrm{d}t} = -\frac{\partial F^{**}}{\partial L''} = -\frac{\partial F^{**}}{\partial L'}$$

$$\frac{\mathrm{d}g''}{\mathrm{d}t} = -\frac{\partial F^{**}}{\partial G''} \tag{19.93b}$$

$$\frac{\mathrm{d}h''}{\mathrm{d}t} = -\frac{\partial F^{**}}{\partial H''} = -\frac{\partial F^{**}}{\partial H}$$

so that

$$L'' = L' = \text{const}$$
$$G'' = \text{const} \tag{19.94a}$$
$$H'' = H' = H = \text{const}$$

$$\ell'' = \ell''_0 - \frac{\partial F^{**}}{\partial L'} t$$
$$g'' = g''_0 - \frac{\partial F^{**}}{\partial G'} t \tag{19.94b}$$
$$h'' = h''_0 - \frac{\partial F^{**}}{\partial H} t$$

where Eqs. (19.94b) yield the secular terms. Then, L', G', H', ℓ''_0, g''_0, h''_0 are the constants of the motion to be determined by comparison with observations. The partial derivatives of $F^{**} = F^{**}(L'', G'', H'')$ are constant because L'', G'', H are constant.

To find the new canonical transformation, write

$$F^*(L', G', H', -, g', -) = F^{**}(L'', G'', H, -, -, -)$$

as

$$F_0^*(L') + F_1^*\left(L', G'' + \frac{\partial S_1^*}{\partial g'}, H\right) + F_{2s}^*\left(L', G'' + \frac{\partial S_1^*}{\partial g'}, H\right)$$
$$+ F_{2p}^*\left(L', G'' + \frac{\partial S_1^*}{\partial g'}, H, g'\right) = F_0^{**} + F_1^{**} + F_2^{**} \tag{19.95}$$

Expand this in a Taylor's series in the neighborhood of L', G'', H, g', rejecting all terms of order higher than J_2^2. We find

$$F_0^*(L') + F_1^*(L', G'', H) + \frac{\partial F_1^*}{\partial G''}\frac{\partial S_1^*}{\partial g'} + F_{2s}^*(L', G'', H)$$
$$+ F_{2p}^*(L', G'', H, g') = F_0^{**} + F_1^{**} + F_2^{**} \tag{19.96}$$

The resolution by orders of J_2 is
Zero order:

$$F_0^{**} = F_0^*(L') = \mu^2/2L'^2 \tag{19.97}$$

First order:

$$F_1^{**} = F_1^*(L', G'', H) = \frac{\mu^4 r_e^2 J_2}{2L'^3 G''^3}\left(-\frac{1}{2} + \frac{3}{2}\frac{H^2}{G''^2}\right) \tag{19.98}$$

Second order:

$$F_2^{**} = \frac{\partial F_1^*}{\partial G''}\frac{\partial S_1^*}{\partial g'} + F_{2s}^*(L', G'', H) + F_{2p}^*(L', G'', H, g') \tag{19.99}$$

The resolution of the second-order equation into secular and long periodic terms is

$$F_2^{**} = F_{2s}^*(L', G'', H)$$

$$F_2^{**} = \frac{\mu^6 r_e^4 J_2^2}{4(L')^{10}}\left[\frac{3}{32}\left(\frac{L'}{G''}\right)^5\left(5 - \frac{18H^2}{G''^2} + \frac{5H^4}{G''^4}\right) + \frac{3}{8}\left(\frac{L'}{G''}\right)^6 \times\left(1 - \frac{6H^2}{G''^2} + \frac{9H^4}{G''^4}\right) - \frac{15}{32}\left(\frac{L'}{G''}\right)^7\left(1 - \frac{2H^2}{G''^2} - \frac{7H^4}{G''^4}\right)\right] \tag{19.100}$$

$$\frac{\partial F_1^*}{\partial G''}\frac{\partial S_1^*}{\partial g'} + F_{2p}^*(L', G'', H, g') = 0 \tag{19.101}$$

By Eqs. (19.97), (19.98), and (19.100)

$$F^{**} = \frac{\mu^2}{2L'^2} + \frac{\mu^4 r_e^2 J^4}{2L'^3 G''^3}\left(-\frac{1}{2} + \frac{3}{2}\frac{H^2}{G''^2}\right) + \frac{\mu^6 r_e^4 J_2^2}{4(L')^{10}}\left[\frac{3}{32}\left(\frac{L'}{G''}\right)^5\left(5 - \frac{18H^2}{G''^2} + \frac{5H^4}{G''^4}\right) + \frac{3}{8}\left(\frac{L'}{G''}\right)^6 \times\left(1 - \frac{6H^2}{G''^2} + \frac{9H^4}{G''^4}\right) - \frac{15}{32}\left(\frac{L'}{G''}\right)^7\left(1 - \frac{2H^2}{G''^2} - \frac{7H^4}{G''^4}\right)\right] \tag{19.102}$$

This is the new Hamiltonian. The new generating function is given by Eq. (19.101). From Eq. (19.98)

$$\frac{\partial F_1^*}{\partial G''} = \frac{\mu^4 r_e^2 J_2}{2L'^3}\left(\frac{3}{2G''^4} - \frac{15}{2}\frac{H^2}{G''^6}\right) \tag{19.103}$$

or

$$\frac{\partial F_1^*}{\partial G''} = \frac{3\mu^4 r_e^2 J_2}{4L'^3 G''^4}\left(1 - \frac{5H^2}{G''^2}\right) \tag{19.104}$$

Insert Eq. (19.104) into Eq. (19.101) and use Eq. (19.90) to obtain $F_{2p}^*(L', G'', H, g')$. The result is

$$\frac{3\mu^4 r_e^2 J_2}{4L'^3 G''^4}\left(1 - \frac{5H^2}{G''^2}\right)\frac{\partial S_1^*}{\partial g'} = \frac{3\mu^6 r_e^4 J_2^2}{64(L')^{10}}\left(\frac{L'^5}{G''^5} - \frac{L'^7}{G''^7}\right) \times\left(1 - \frac{16H^2}{G''^2} + \frac{15H^4}{G''^4}\right)\cos 2g' \tag{19.105}$$

or

$$\frac{\partial S_1^*}{\partial g'} = \frac{\mu^2 r_e^2 J_2 G''}{16(L')^4}\left(\frac{L'^2}{G''^2} - \frac{L'^4}{G''^4}\right)\left(1 - \frac{16H^2}{G''^2} + \frac{15H^4}{G''^4}\right)\left(1 - \frac{5H^2}{G''^2}\right)^{-1}\cos 2g' \tag{19.106}$$

To integrate this partial differential equation, we simply change $\cos 2g'$ to $(\sin 2g'/2)$ and replace g' by g'', the resulting error being of order in J_2 higher than what we are keeping. There is, of course, a constant of integration, viz., $\psi(L', G'', H)$. By Eq. (19.92), however, this would give terms for ℓ'', g'', and h'' that can be absorbed into the ℓ_0'', g_0'', and h_0''; these will appear when we find the secular terms. Thus

$$S_1^* = \frac{\mu^2 r_e^2 J_2 G''}{32(L')^4}\left(\frac{L'^2}{G''^2} - \frac{L'^4}{G''^4}\right)\left(1 - \frac{16H^2}{G''^2} + \frac{15H^4}{G''^4}\right)\left(1 - \frac{5H^2}{G''^2}\right)^{-1} \sin 2g'' \tag{19.107}$$

From Eqs. (19.107) and (19.92), we can find the long periodic terms $\ell' - \ell''$, $g' - g''$, $h' - h''$, and $G' - G''$. Note that they are of the first order in J_2, even though we had to go to a second-order calculation to find them. Also note that there is a "resonance denominator" $1 - 5H^2/G''^2 = 1 - \cos^2 I''$. The value of I for which this resonance denominator vanishes, 63.4° or its supplement, is called the "critical inclination." The solution is not valid in the immediate neighborhood of $I = 63.4°$.

VII. Results to This Point

Let us collect the results. We have

$$
\begin{aligned}
L &= L' + \frac{\partial S_1}{\partial \ell} \\
G &= G' + \frac{\partial S_1}{\partial g} = G'' + \frac{\partial S_1}{\partial g} + \frac{\partial S_1^*}{\partial g'} \\
H &= H'' \\
\ell &= \ell' - \frac{\partial S_1}{\partial L'} = \ell'' - \frac{\partial S_1}{\partial L'} - \frac{\partial S_1^*}{\partial L'} = \ell_0'' + c_1(L', G'', H)t - \frac{\partial S_1}{\partial L'} - \frac{\partial S_1^*}{\partial L'} \\
g &= g' - \frac{\partial S_1}{\partial G'} = g'' - \frac{\partial S_1}{\partial G'} - \frac{\partial S_1^*}{\partial G''} = g_0'' + c_2(L', G'', H)t - \frac{\partial S_1}{\partial G'} - \frac{\partial S_1^*}{\partial G''} \\
h &= h' - \frac{\partial S_1}{\partial H} = h'' - \frac{\partial S_1}{\partial H} - \frac{\partial S_1^*}{\partial H} = h_0'' + c_3(L', G'', H)t - \frac{\partial S_1}{\partial H} - \frac{\partial S_1^*}{\partial H}
\end{aligned}
\tag{19.108}
$$

Here, S_1 is given by Eq. (19.57) and S_1^* by Eq. (19.107). Also

$$
\begin{aligned}
c_1(L', G'', H) &= \frac{\mathrm{d}\ell''}{\mathrm{d}t} = -\frac{\partial F^{**}}{\partial L'} \\
c_2(L', G'', H) &= \frac{\mathrm{d}g''}{\mathrm{d}t} = -\frac{\partial F^{**}}{\partial G''} \\
c_3(L', G'', H) &= \frac{\mathrm{d}h''}{\mathrm{d}t} = -\frac{\partial F^{**}}{\partial H}
\end{aligned}
\tag{19.109}
$$

Given L' $(=L'')$, G'', H $(=H'')$, ℓ_0'', g_0'', h_0'', we have here the complete schedule for calculating L, G, H, ℓ, g, h as functions of t, so that we can find x, y, z, $\dot{x}$, $\dot{y}$, $\dot{z}$ at any time t.

VIII. Secular Terms

Let us calculate the secular terms only through the first order in J_2. Then

$$F_0^{**} = \frac{\mu^2}{2L'^2} \rightarrow \frac{\partial F_0^{**}}{\partial L'} = -\frac{\mu^2}{L'^3} = -n'$$

$$F_1^{**} = \frac{\mu^4 r_e^2 J_2}{2L'^3 G''^3}\left(-\frac{1}{2} + \frac{3}{2}\frac{H^2}{G''^2}\right) \rightarrow \frac{\partial F_1^{**}}{\partial L'} = -\frac{3\mu^4 r_e^2 J_2}{2L'^4 G''^3}\left(-\frac{1}{2} + \frac{3}{2}\frac{H^2}{G''^2}\right)$$

$$= -\frac{3n' r_e^2 J_2}{2a'^2 G''^3}\left(-\frac{1}{2} + \frac{3}{2}\frac{H^2}{G''^2}\right)$$

using $L' = (\mu a')^{1/2} = n'a'^2$ and $\mu = n'^2 a'^3$. Thus

$$c_1(L', G'', H) = n'\left[1 + \frac{3}{2} J_2 \frac{r_e^2}{a'^2}\frac{L'^3}{G''^3}\left(-\frac{1}{2} + \frac{3}{2}\frac{H^2}{G''^2}\right)\right] \tag{19.110}$$

With use of

$$L'^2/G''^2 = (1 - e''^2)^{-1} \qquad p'' = a'(1 - e''^2)$$

we have

$$c_1(L', G'', H) = n'\left[1 + \frac{3}{2} J_2 \frac{r_e^2}{p''^2}(1 - e''^2)^{\frac{1}{2}}\left(-\frac{1}{2} + \frac{3}{2}\frac{H^2}{G''^2}\right)\right] + O\left(J_2^2\right) \tag{19.111}$$

To show that this agrees with the secular part of $\dot{\ell}$ found in Eq. (17.81), we proceed as follows. To find c_1 in terms of initial values, we use

$$n_0 = \mu^{\frac{1}{2}} a_0^{-\frac{3}{2}} \qquad a_0 = L_0^2/\mu$$

However,

$$\frac{c_1}{n_0} = \frac{c_1}{n'}\frac{n'}{n_0} \tag{19.112}$$

$$n' = \mu^{\frac{1}{2}} a'^{-\frac{3}{2}} = \mu^{\frac{1}{2}}(L'^2\mu)^{-\frac{3}{2}} = \mu^2 L'^{-3}$$

Then

$$\frac{n'}{n_0} = \left(\frac{L_0}{L'}\right)^3 \tag{19.113}$$

To find L_0/L', we need $L - L'$. By Eq. (19.60)

$$L - L' = \frac{\mu^2 r_e^2 J_2}{2L'^3}\left[\left(-\frac{1}{2} + \frac{3}{2}\frac{H^2}{G'^2}\right)\left(\frac{a'^3}{r'^3} - \frac{L'^3}{G'^3}\right)\right.$$

$$\left. + \frac{3}{2}\left(1 - \frac{H^2}{G'^2}\right)\left(\frac{a'}{r'}\right)^3 \cos(2g + 2f')\right] + O\left(J_2^2\right) \tag{19.114}$$

Replace $\cos(2g + 2f')$ by $1 - 2\sin^2(g + f')$. Inserting this in Eq. (19.114), we may drop the primes on the right side and still keep the error down to $O(J_2^2)$. The result is

$$L - L' = \frac{\mu^2 r_e^2 J_2}{2L^3}\left[\left(\frac{1}{2} - \frac{3}{2}\frac{H^2}{G'^2}\right)\frac{L^3}{G^3} + \frac{a^3}{r^3}\left\{1 - 3\left(1 - \frac{H^2}{G'^2}\right)\sin^2(g + f)\right\}\right] + O(J_2^2) \tag{19.115}$$

Put $H/G = \cos I$, $L^2/G^2 = (1 - e^2)^{-1}$, and $\sin\theta = \sin I \sin(g + f)$. Equation (19.115) becomes

$$L - L' = \frac{\mu^2 r_e^2 J_2}{2L^3}\left[\frac{1}{2}(1 - 3\cos^2 I)(1 - e^2)^{-\frac{3}{2}} + \frac{a^3}{r^3}(1 - 3\sin^2\theta)\right] + O(J_2^2) \tag{19.116}$$

and

$$\frac{L_0}{L'} = 1 + \frac{\mu^2 r_e^2 J_2}{2L_0^4}\left[\frac{1}{2}(1 - 3\cos^2 I_0)\left(1 - e_0^2\right)^{-\frac{3}{2}} + \frac{a_0^3}{r_0^3}\left(1 - 3\sin^2\theta_0\right)\right] + O(J_2^2) \tag{19.117}$$

To compare with Eq. (17.81), we need c_1/n_0, and by Eq. (19.117) we need c_1/n' and n'/n_0. By Eqs. (19.113) and (19.117)

$$\frac{n'}{n_0} = 1 + \frac{3\mu^2 r_e^2 J_2}{2L_0^4}\left[\frac{1}{2}(1 - 3\cos^2 I_0)\left(1 - e_0^2\right)^{-\frac{3}{2}} + \frac{a_0^3}{r_0^3}\left(1 - 3\sin^2\theta_0\right)\right] + O(J_2^2) \tag{19.118}$$

or

$$\frac{n'}{n_0} = 1 + \frac{3r_e^2 J_2}{4p_0^2}\left(1 - 3\cos^2 I_0\right)\left(1 - e_0^2\right)^{\frac{1}{2}} + \frac{3r_e^2 J_2}{2a_0^2}\frac{a_0^3}{r_0^3}\left(1 - 3\sin^2\theta_0\right) + O(J_2^2) \tag{19.119}$$

By Eq. (19.110),

$$\frac{c_1}{n'} = 1 + \frac{3}{2}\frac{r_e^2 J_2}{p_0^2}\left(1 - e_0^2\right)^{\frac{1}{2}}\left(-\frac{1}{2} + \frac{3}{2}\cos^2 I\right) + O(J_2^2) \tag{19.120}$$

On multiplication of Eqs. (19.119) and (19.120), we find that the second terms on each right-hand side cancel, so that

$$\frac{c_1}{n_0} = 1 + \frac{3a_0 r_e^2 J_2}{2r_0^3}\left(1 - 3\sin^2\theta_0\right) + O(J_2^2) \tag{19.121}$$

This agrees with Eq. (17.81) because c_1 in Eq. (19.121) is the same as n' in Eq. (17.81).

Next, we need

$$c_2(L', G'', H) = \frac{\mathrm{d}g''}{\mathrm{d}t} = -\frac{\partial F^{**}}{\partial G} \tag{19.122}$$

Here

$$F_0^{**} = \frac{\mu^2}{2L'^2} \tag{19.97}$$

$$F_1^{**} = \frac{\mu^4 r_e^2 J_2}{2L'^3}\left(-\frac{1}{2G''^3} + \frac{3}{2}\frac{H^2}{G''^5}\right) \tag{19.98}$$

We find

$$c_2 = -\frac{3\mu^4 r_e^2 J_2}{4L'^3 G''^4}\left(1 - \frac{5H^2}{G''^2}\right) \tag{19.123}$$

$$c_2 = \frac{3\mu^4 r_e^2 J_2}{4L'^3 G''^4}(5\cos^2 I'' - 1) \tag{19.124}$$

so that

$$g'' = g_0'' + \frac{3\mu^4 r_e^2 J_2}{4L'^3 G''^4}(5\cos^2 I'' - 1)t \tag{19.125}$$

This agrees with the result (17.48) for the secular change of $g = \omega$. If P_L is the long period, we find for a close orbit that

$$\frac{2\pi}{P_L} \approx \frac{2\pi}{P}\frac{3J_2}{4}(5\cos^2 I - 1) \tag{19.125a}$$

$$\frac{P_L}{P} \approx \frac{4}{3J_2(5\cos^2 I - 1)} \tag{19.125b}$$

where P is the short period. (Short periods are on the order of time of one satellite passage around the Earth. Long periods are on the order of time of one complete perigee passage around the Earth.)

For a close satellite of the Earth, with $P \approx 1.5$ h, this gives $P_L \approx 450$ h for an equatorial orbit, 1800 h for a polar orbit, and infinity at the critical inclination.

Finally

$$c_3(L', G'', H) = \frac{\mathrm{d}h}{\mathrm{d}t} = -\frac{\partial F^{**}}{\partial H}$$

By Eq. (19.98), it follows that

$$c_3 = -\frac{3\mu^4 r_e^2 J_2}{2L'^3 G''^4}\frac{H}{G''} \tag{19.126}$$

$$c_3 = -\frac{3n' r_e^2 J_2}{2p''^2}\cos I'' \tag{19.127}$$

so that

$$h'' = h_0'' - \frac{3n' r_e^2 J_2}{2p''^2}(\cos I'')t \tag{19.128}$$

in agreement with Eq. (17.37). Of course, the present treatment has the advantage that it permits the evaluation of the second-order secular terms.

IX. Algorithm

Given μ, r_e, J_2, and the six *mean* orbital elements, viz., $L', G'', H, \ell''_0, g''_0, h''_0$, calculate the position and velocity vectors at time t.

1) Calculate $a' = L'^2/\mu, n' = \mu^{1/2}(a')^{-3/2}$, c_1, c_2, c_3, and t as in Sec. VIII. Then

$$\ell'' = \ell''_0 + c_1 t \qquad g'' = g''_0 + c_2 t \qquad h'' = h''_0 + c_3 t$$

2) Calculate

$$\frac{\partial S_1^*}{\partial g'} \qquad \frac{\partial S_1^*}{\partial L'} \qquad \frac{\partial S_1^*}{\partial G''} \qquad \frac{\partial S_1^*}{\partial H}$$

The long periodic terms $G' - G''$, $\ell' - \ell''$, $g' - g''$, $h' - h''$ are given by

$$G' = G'' + \frac{\partial S_1^*}{\partial g'} \qquad \ell' = \ell'' - \frac{\partial S_1^*}{\partial L'} \qquad g' = g'' - \frac{\partial S_1^*}{\partial G''} \qquad h' = h'' - \frac{\partial S_1^*}{\partial H}$$

Here, note that one puts $g' = g''$ in the expressions for the derivatives of S_1^*. If one did not do so, one would have to solve a transcendental equation for g', viz.,

$$g' = g'' - \psi(g')$$

where $\psi(g')$ is $\partial S_1^*/\partial G''$ expressed in terms of g'. To the accuracy at which we are working, however, this is not necessary because substitution of g'' for g' yields an error of $O(J_2^2)$. We are calculating long periodic terms only through order J_2.

3) Calculate $e' = (1 - G'^2/L'^2)^{1/2}$. We then have $L',\ G',\ H', \ell', g', h'$, and e'. Then calculate

$$\frac{\partial S_1}{\partial \ell} \qquad \frac{\partial S_1}{\partial g} \qquad \frac{\partial S_1}{\partial h} \qquad \frac{\partial S_1}{\partial L'} \qquad \frac{\partial S_1}{\partial G'} \qquad \frac{\partial S_1}{\partial H}$$

The short periodic terms $L - L'$, $G - G'$, $\ell - \ell'$, $g - g'$, and $h - h'$ are given by

$$L - L' = \frac{\partial S_1}{\partial \ell} \qquad \ell - \ell' = -\frac{\partial S_1}{\partial L'}$$

$$G - G' = \frac{\partial S_1}{\partial g} \qquad g - g' = -\frac{\partial S_1}{\partial G'}$$

$$h - h' = -\frac{\partial S_1}{\partial H'}$$

Note that we replace ℓ and g on the right sides of these equations by ℓ' and g'. Otherwise, we should have to solve the pair of equations

$$\ell - \ell' = -\frac{\partial S_1}{\partial L'} \qquad \ell - \ell' = -\frac{\partial S_1}{\partial L'}$$

simultaneously for ℓ and g. The error introduced by this substitution is of $O(J_2^2)$. This is acceptable because we are calculating short periodic terms only through order J_2.

4) We now have the full set of Delaunay variables at time t, viz., L, G, H, ℓ, g, h. The next procedure is to calculate

$$
\begin{array}{llll}
a & = L^2/\mu & & e = (1 - G^2/L^2)^{\frac{1}{2}} \\
b & = (1 - e^2)^{\frac{1}{2}} & & n = \mu^{\frac{1}{2}} a^{-\frac{3}{2}} \\
E & \text{from} & E - e \sin E = \ell & \\
r & \text{from} & r = a(1 - e \cos E) & \\
I & \text{from} & I = \cos^{-1}(H/G) &
\end{array}
$$

Then

$$\boldsymbol{r} = \boldsymbol{A}(\cos E - e) + \boldsymbol{B} \sin E$$

$$\dot{\boldsymbol{r}} = na/r(-\boldsymbol{A} \sin E + \boldsymbol{B} \cos E)$$

where

$$\frac{\boldsymbol{A}}{a} = \begin{bmatrix} \cos g \cos h - \sin g \sin h \cos I \\ \cos g \sin h + \sin g \cos h \cos I \\ \sin g \sin I \end{bmatrix}$$

$$\frac{\boldsymbol{B}}{b} = \begin{bmatrix} -\sin g \cos h - \cos g \sin h \cos I \\ \sin g \sin h + \cos g \cos h \cos I \\ \cos g \sin I \end{bmatrix}$$

The advantage of Brouwer's method over that of Chapter 17 is that it yields the long periodic terms through order J_2. It also yields secular terms through $O(J_2^2)$, although we have only indicated how to find them and not actually written them down.

References

[1]Brouwer, D., "Solution of Problem of Artificial Satellite Theory Without Drag," *Astronomical Journal,* Vol. 64, No. 9, 1959, pp. 378–397.

[2]Brouwer, D., and Clemence, G., *Methods of Celestial Mechanics,* Academic Press, New York, 1961, p. 562.

Chapter 20

The Brouwer–von Zeipel Method II

I. Introduction

THIS chapter will show how to incorporate the third and fourth zonal harmonics into the Brouwer solution. Because J_3 and J_4 are both of order J_2^2, they will not affect the function S_1, which we carry (following Ref. 1) only through order J_2.

If one traces through the previous derivations, one sees that the Hamiltonian contributions F_3 and F_4, which we can write as $\Delta_3 F$ and $\Delta_4 F$, affect only F_{2s}^*, F_{2p}^*, and S_1^*.

II. The Effects of J_3

We have

$$\Delta_3 F = -\Delta_3 V = -\frac{\mu}{r}\left(\frac{r_e}{r}\right)^3 J_3 P_3(\sin\theta) \tag{20.1}$$

We first split this into $\overline{\Delta_3 F}$ and $(\Delta_3 F)_\ell$, where $\overline{\Delta_3 F}$ is the average of $\Delta_3 F$ over the osculating orbit and $(\Delta_3 F)_\ell$ is the short periodic part. This short periodic part is of order J_2^2; we shall not have any use for it since the von Zeipel method is not suitable for the calculation of second-order short periodic terms. It turns out that $\overline{\Delta_3 F}$ has no secular part, only a long periodic part proportional to sin g.

Because

$$P_3(\sin\theta) = \tfrac{5}{2}\sin^3\theta - \tfrac{3}{2}\sin\theta \tag{20.2}$$

$$\sin\theta = \sin I \sin(g+f) \tag{20.3}$$

we can calculate P_3 as a function of the true anomaly f. One shows readily that

$$\sin 3x = 3\sin x - 4\sin^3 x$$

so that

$$\sin^3 x = \tfrac{3}{4}\sin x - \tfrac{1}{4}\sin 3x$$

Then

$$P_3(\sin\theta) = \frac{5}{2}\sin^3 I\left[\frac{3}{4}\sin(g+f) - \frac{1}{4}\sin(3g+3f)\right] - \frac{3}{2}\sin I\sin(g+f)$$

$$= \left(\frac{15}{8}\sin^3 I - \frac{3}{2}\sin I\right)\sin(g+f) - \frac{5}{8}\sin^3 I\sin(3g+3f) \tag{20.4}$$

and

$$\Delta_3 F = -\frac{\mu r_e^3 J_3}{a^4}\left[\left(\frac{15}{8}\sin^3 I - \frac{3}{2}\sin I\right)\left(\frac{a}{r}\right)^4 \sin(g+f) - \frac{5}{8}\sin^3 I\left(\frac{a}{r}\right)^4 \sin(3g+3f)\right] \tag{20.5}$$

and

$$\overline{\Delta_3 F} = \frac{1}{2\pi}\int_0^{2\pi} \Delta_3 F \mathrm{d}\ell \tag{20.6}$$

We do not need $(\Delta_3 F)_\ell = \Delta_3 F - \overline{\Delta_3 F}$. For the osculating orbit

$$\mathrm{d}\ell = \left(\frac{r}{a}\right)^2 (1-e^2)^{-\frac{1}{2}} \mathrm{d}f$$

so that

$$\overline{\Delta_3 F} = \frac{(1-e^2)^{-\frac{1}{2}}}{2\pi}\int_0^{2\pi}\left(\frac{r}{a}\right)^2 \Delta_3 F \mathrm{d}\ell \tag{20.6a}$$

or

$$\overline{\Delta_3 F} = -\frac{\mu r_e^3 J_3}{2\pi a^4}(1-e^2)^{-\frac{1}{2}}\left[\left(\frac{15}{8}\sin^3 I - \frac{3}{2}\sin I\right)\int_0^{2\pi}\left(\frac{a}{r}\right)^2 \sin(g+f)\,\mathrm{d}f - \frac{5}{8}\sin^3 I\int_0^{2\pi}\left(\frac{a}{r}\right)^2 \sin(3g+3f)\,\mathrm{d}f\right] \tag{20.7}$$

Now

$$(a/r)^2 = (1-e^2)^{-2}(1+e\cos f)^2 \tag{20.8}$$

which gives a constant plus terms in $\cos f$ and $\cos 2f$. When these are multiplied by $\sin(3g+3f)$, sines of $3g+f$, $3g+2f$, $3g+3f$, $3g+4f$, and $3g+5f$ are the results. The term in $\sin(3g+3f)$ does not contribute to the integral.

We now need

$$\int_0^{2\pi}\left(\frac{a}{r}\right)^2 \sin(g+f)\,\mathrm{d}f = (1-e^2)^{-2}\int_0^{2\pi}(1+e\cos f)^2\sin(g+f)\,\mathrm{d}f$$
$$= (1-e^2)^{-2}\int_0^{2\pi}\left(1+\frac{e^2}{2}+2e\cos f+\frac{e^2}{2}\cos 2f\right)\sin(g+f)\,\mathrm{d}f \tag{20.9}$$

Here, only the term in $2e\cos f$ contributes to the integral. We have

$$\cos f \sin(g+f) = \tfrac{1}{2}\sin g + \tfrac{1}{2}\sin(g+2f)$$

so that

$$\int_0^{2\pi} \left(\frac{a}{r}\right)^2 \sin(g+f)\,\mathrm{d}f = (1-e^2)^{-2} 2\pi e \sin g \tag{20.10}$$

Thus

$$\overline{\Delta_3 F} = -\frac{\mu r_e^3 J_3}{a^4}(1-e^2)^{-\frac{5}{2}}\left(\frac{15}{8}\sin^3 I - \frac{3}{2}\sin I\right) e \sin g \tag{20.11}$$

This is purely long periodic, of order J_2^2. It is to be added to F_{2p}^* in Eq. (19.102). Here, one must put L' in place of L, G'' in place of G, and g' in place of g. From Eq. (19.101), we obtain

$$\frac{\partial F_1^*}{\partial G''}\frac{\partial S_1^*}{\partial g'} + F_{2p}^*(L', G'', H, g') + \overline{\Delta_3 F}(L', G'', H, g') = 0 \tag{20.12}$$

The change $\Delta_3 S_1^*$, produced by J_3, satisfies

$$\frac{\partial F_1^*}{\partial G''}\frac{\partial \Delta_3 S_1^*}{\partial g'} = -\overline{\Delta_3 F}(L', G'', H, g') \tag{20.13}$$

where

$$\frac{\partial F_1^*}{\partial G''} = \frac{3\mu^4 r_e^2 J_2}{4L'^3 G''^4}\left(1 - \frac{5H^2}{G''^2}\right) \tag{20.14}$$

from Eq. (19.104). Making appropriate changes in Eq. (20.11), we also write

$$\overline{\Delta_3 F} = -\frac{\mu r_e^2 J_3}{a'^4}(1-e''^2)^{-\frac{5}{2}} e''\left(\frac{15}{8}\sin^3 I'' - \frac{3}{2}\sin I''\right)\sin g' \tag{20.15}$$

With use of $a' = L'^2/\mu$, this becomes

$$\overline{\Delta_3 F} = -\frac{3\mu^5 r_e^3 J_3}{8L'^3 G''^5} e'' \sin I''(1 - 5\cos^2 I'')\sin g' \tag{20.16}$$

From Eqs. (20.13), (20.14), and (20.16), we find

$$\frac{3\mu^4 r_e^2 J_2}{4L'^3 G''^4}(1-5\cos^2 I'')\frac{\partial \Delta_3 S_1^*}{\partial g'} = \frac{3\mu^5 r_e^3 J_3}{8L'^3 G''^5} e'' \sin I''(1-5\cos^2 I'')\sin g' \tag{20.17}$$

The "resonance denominator" $1 - 5\cos^2 I''$ cancels out. It is remarkable that such a cancellation occurs only for the third zonal harmonic. Equation (20.17) becomes

$$\frac{\partial \Delta_3 S_1^*}{\partial g'} = \frac{e''}{2}\frac{J_3}{J_2}\frac{\mu r_e}{G''}\sin I'' \sin g' \tag{20.18}$$

Then

$$\Delta_3 S_1^* = -\frac{e''}{2}\frac{J_3}{J_2}\frac{\mu r_e}{G''}\sin I'' \cos g' \tag{20.19}$$

Since J_3 is of order J_2^2, it follows that $\Delta_3 S_1^*$ is of order J_2. Since

$$L' - L'' = \frac{\partial S_1^*}{\partial \ell'}$$

$$G' - G'' = \frac{\partial S_1^*}{\partial g'} \tag{20.19a}$$

$$H' - H'' = \frac{\partial S_1^*}{\partial h'}$$

$$\ell' - \ell'' = -\frac{\partial S_1^*}{\partial L'}$$

$$g' - g'' = -\frac{\partial S_1^*}{\partial G'} \tag{20.19b}$$

$$h' - h'' = -\frac{\partial S_1^*}{\partial H'}$$

it follows that

$$\delta_3 L = \frac{\partial \Delta_3 S_1^*}{\partial \ell'} = 0$$

$$\delta_3 G = \frac{\partial \Delta_3 S_1^*}{\partial g'} \tag{20.20a}$$

$$\delta_3 H = \frac{\partial \Delta_3 S_1^*}{\partial h'} = 0$$

$$\delta_3 \ell = -\frac{\partial \Delta_3 S_1^*}{\partial L'}$$

$$\delta_3 g = -\frac{\partial \Delta_3 S_1^*}{\partial G''} \tag{20.20b}$$

$$\delta_3 h = -\frac{\partial \Delta_3 S_1^*}{\partial H}$$

In Eq. (20.19), we may change g' to g'' without affecting the order of the accuracy. From Eqs. (20.18) and (20.20a)

$$\delta_3 G = \frac{e''}{2} \frac{J_3}{J_2} \frac{\mu r_e}{G''} \sin I'' \sin g'' \tag{20.21}$$

Next, from Eq. (20.20b)

$$\delta_3 \ell = -\frac{\partial \Delta_3 S_1^*}{\partial L'} \tag{20.22}$$

In Eq. (20.19), the only quantity that depends on L' is e''. From

$$e''^2 = 1 - (G''^2/L'^2)$$

we find

$$\frac{\partial e''}{\partial L'} = \frac{1 - e''^2}{e'' L'} \tag{20.23}$$

From Eqs. (20.19) and (20.23)

$$\delta_3 \ell = -\frac{\partial \Delta_3 S_1^*}{\partial L'} = \frac{1}{2} \frac{J_3}{J_2} \frac{\mu r_e}{G''} \frac{(1 - e''^2)}{e'' L'} \sin I'' \cos g'' \tag{20.24}$$

Next, from Eqs. (20.19) and (20.20b)

$$\delta_3 g = -\frac{\partial \Delta_3 S_1^*}{\partial G'} = \frac{\mu r_e}{2} \frac{J_3}{J_2} \cos g'' \frac{\partial}{\partial G''} \left(\frac{e'' \sin I''}{G''} \right) \tag{20.25}$$

From

$$1 - e''^2 = G''^2 / L'^2 \qquad \cos^2 I'' = H^2 / G''^2$$

we find

$$\frac{\partial}{\partial G''} \left(\frac{e'' \sin I''}{G''} \right) = (\mu p'')^{-1} \left(\frac{e'' \cos^2 I''}{\sin I''} - \frac{\sin I''}{e''} \right) \tag{20.26}$$

where

$$p'' = G''^2 / \mu \tag{20.26a}$$

Thus

$$\delta_3 g = \frac{r_e}{2p''} \frac{J_3}{J_2} \cos g'' \left(\frac{e'' \cos^2 I''}{\sin I''} - \frac{\sin I''}{e''} \right) \tag{20.27}$$

Finally

$$\delta_3 h = -\frac{\partial \Delta_3 S_1^*}{\partial H} = \frac{e''}{2} \frac{J_3}{J_2} \frac{\mu r_e}{G''} \cos g'' \frac{\partial}{\partial H} (\sin I'') \tag{20.28}$$

From

$$\sin^2 I'' = 1 - (H^2 / G''^2)$$

we have

$$\frac{\partial}{\partial H} (\sin I'') = -\frac{1}{G''} \cot I'' \tag{20.29}$$

Thus

$$\begin{aligned} \delta_3 h &= -\frac{e''}{2} \frac{J_3}{J_2} \frac{\mu r_e}{G''^2} \cos g'' \cot I'' \\ &= -\frac{e'' r_e}{2p''} \frac{J_3}{J_2} \cot I'' \cos g'' \end{aligned} \tag{20.30}$$

Note that, in the algorithm for the orbit, we must add $\delta_3 G$ to $G' - G''$, $\delta_3 \ell$ to $\ell' - \ell''$, $\delta_3 g$ to $g' - g''$, and $\delta_3 h$ to $h' - h''$.

III. The Effects of J_4

The fourth zonal harmonic differs from the third by giving rise to secular terms of order J_2^2 and to long periodic terms of order $J_4/J_2 = O(J_2)$, which have a resonance denominator $1 - 5\cos^2 I''$.

Here

$$\Delta_4 F = -\Delta_4 V = -\frac{\mu}{r}\left(\frac{r_e}{r}\right)^4 J_4 P_4(\sin\theta) \tag{20.31}$$

We set up the problem as we did for J_3 and find

$$F_2^* = \text{old } F_2^* + \overline{\Delta_4 F} \tag{20.32}$$

so that

$$\Delta_4 F_2^* = \overline{\Delta_4 F} = \Delta_4 F_{2s}^* + \Delta_4 F_{2p}^* \tag{20.33}$$

In this case, $\Delta_4 F_{2s}^*$ is a secular correction to the Hamiltonian, and $\Delta_4 F_{2p}^*$ is a long periodic term.

The correction to the Hamiltonian will give the secular terms

$$\begin{aligned} \Delta_4 \dot{\ell}'' &= \frac{\partial}{\partial L'}\Delta_4 F_{2s}^* = \Delta_4 c_1 \\ \Delta_4 \dot{g}'' &= \frac{\partial}{\partial G''}\Delta_4 F_{2s}^* = \Delta_4 c_2 \\ \Delta_4 \dot{h}'' &= \frac{\partial}{\partial H}\Delta_4 F_{2s}^* = \Delta_4 c_3 \end{aligned} \tag{20.34}$$

We also obtain as before

$$\frac{\partial F_1^*}{\partial G''}\frac{\partial \Delta_4 S_1^*}{\partial g'} + \Delta_4 F_{2p}^*(L', G'', H, g') = 0 \tag{20.35}$$

where

$$F_1^* = \frac{\mu^4 r_e^2 J_2}{2L'^3 G''^3}\left(-\frac{1}{2} + \frac{3}{2}\frac{H^2}{G''^2}\right) \tag{20.36}$$

Then

$$\begin{aligned} \Delta_4(G' - G'') &= \frac{\partial \Delta_4 S_1^*}{\partial g'} \\ \Delta_4(\ell' - \ell'') &= -\frac{\partial \Delta_4 S_1^*}{\partial L'} \\ \Delta_4(g' - g'') &= -\frac{\partial \Delta_4 S_1^*}{\partial G''} \\ \Delta_4(h' - h'') &= -\frac{\partial \Delta_4 S_1^*}{\partial H} \end{aligned} \tag{20.37}$$

give the long periodic terms.

IV. The Average $\overline{\Delta_4 F}$

We have

$$P_4(\sin\theta) = \frac{3}{8}\left(1 - 10\sin^2\theta + \frac{35}{3}\sin^4\theta\right) \tag{20.38}$$

where

$$\sin\theta = \sin I \sin(g + f)$$

Then

$$\sin^2\theta = \tfrac{1}{2}\sin^2 I[1 - \cos(2g + 2f)]$$

$$\sin^4\theta = \tfrac{1}{4}\sin^4 I\left[\tfrac{3}{2} - 2\cos(2g + 2f) + \tfrac{1}{2}\cos(4g + 4f)\right]$$

Thus

$$P_4(\sin\theta) = \frac{3}{8} - \frac{15}{8}\sin^2 I + \frac{105}{64}\sin^4 I + \left[\frac{15}{8}\sin^2 I - \frac{35}{16}\sin^4 I\right]$$
$$\times \cos(2g + 2f) + \frac{35}{64}\sin^4 I\cos(4g + 4f) \tag{20.38a}$$

$$P_4(\sin\theta) = \frac{9}{64} - \frac{45}{32}\cos^2 I + \frac{105}{64}\cos^4 I$$
$$+ \left[-\frac{5}{16} + \frac{5}{2}\cos^2 I - \frac{35}{16}\cos^4 I\right]\cos(2g + 2f)$$
$$+ \frac{35}{64}\left[1 - 2\cos^2 I + \cos^4 I\right]\cos(4g + 4f) \tag{20.38b}$$

Now, by Eq. (20.31)

$$\Delta_4 F = -\frac{\mu r_e^4}{a^5}\left(\frac{a}{r}\right)^5 J_4 P_4(\sin\theta)$$

or, since $a = L^2/\mu$,

$$\Delta_4 F = -\frac{\mu r_e^4}{L^{10}} J_4\left(\frac{a}{r}\right)^5 P_4(\sin\theta) \tag{20.38c}$$

From Eqs. (20.38c) and (20.38b)

$$\Delta_4 F = -\frac{\mu r_e^4}{L^{10}} J_4\left(\frac{a}{r}\right)^5\left[\frac{9}{64} - \frac{45}{32}\cos^2 I + \frac{105}{64}\cos^4 I\right.$$
$$+ \left[-\frac{5}{16} + \frac{5}{2}\cos^2 I - \frac{35}{16}\cos^4 I\right]\cos(2g + 2f)$$
$$\left.+ \frac{35}{64}[1 - 2\cos^2 I + \cos^4 I]\cos(4g + 4f)\right] \tag{20.39}$$

Now use

$$\overline{\Delta_4 F} = \frac{1}{2\pi}\int_0^{2\pi} \Delta_4 F \mathrm{d}\ell$$

with

$$\mathrm{d}\ell = \left(\frac{r}{a}\right)^2 (1-e^2)^{-\frac{1}{2}} \mathrm{d}f = \left(\frac{r}{a}\right)^2 \frac{L}{G} \mathrm{d}f$$

Then

$$\overline{\Delta_4 F} = \frac{L}{2\pi G}\int_0^{2\pi} \left(\frac{r}{a}\right)^2 \Delta_4 F \,\mathrm{d}f \tag{20.40}$$

From Eqs. (20.39) and (20.40)

$$\begin{aligned}\overline{\Delta_4 F} = -\frac{3\mu^6 r_e^4 J_4}{16\pi L^9 G}\Bigg[&\left\{\frac{3}{8} - \frac{15}{4}\cos^2 I + \frac{35}{8}\cos^4 I\right\}\int_0^{2\pi}\left(\frac{a}{r}\right)^3 \mathrm{d}f \\ &+ \left\{-\frac{5}{6} + \frac{20}{3}\cos^2 I - \frac{35}{6}\cos^4 I\right\}\int_0^{2\pi}\left(\frac{a}{r}\right)^3 \cos(2g+2f)\,\mathrm{d}f \\ &+ \left\{\frac{35}{24} - \frac{35}{12}\cos^2 I - \frac{35}{24}\cos^4 I\right\}\int_0^{2\pi}\left(\frac{a}{r}\right)^3 \cos(4g+4f)\,\mathrm{d}f\Bigg]\end{aligned} \tag{20.41}$$

Now

$$\frac{a}{r} = (1-e^2)^{-1}(1+e\cos f) = \frac{L^2}{G^2}(1+e\cos f)$$

$$\left(\frac{a}{r}\right)^3 = \frac{L^6}{G^6}\left\{1 + \frac{3e^2}{2} + \left(3e + \frac{3e^3}{4}\right)\cos f + \frac{3e^2}{2}\cos 2f + \frac{e^3}{4}\cos 3f\right\} \tag{20.42}$$

Multiplication of $\cos(4g+4f)$ by $(a/r)^3$ gives terms in $\cos(2g+kf)$, where $k = 1, 2, 3, 4, 5, 6, 7$. Thus, the integral involving $\cos(2g+4f)$ gives no contribution to Eq. (20.41).

We also have

$$\int_0^{2\pi}\left(\frac{a}{r}\right)^3 \mathrm{d}f = \frac{2\pi L^6}{G^6}\left[1 + \frac{3e^2}{2}\right] = \frac{2\pi L^6}{G^6}\left[\frac{5}{2} - \frac{3}{2}\frac{G^2}{L^2}\right] \tag{20.43}$$

$$\begin{aligned}\int_0^{2\pi}\left(\frac{a}{r}\right)^3 \cos(2g+2f)\,\mathrm{d}f &= \frac{L^6}{G^6}\frac{3e^2}{2}\int_0^{2\pi}\cos 2f\cos(2g+2f)\,\mathrm{d}f \\ &= \frac{L^6}{G^6}\frac{3e^2}{2}\pi\cos 2g \\ &= \frac{\pi L^6}{G^6}\left[\frac{3}{2} - \frac{3}{2}\frac{G^2}{L^2}\right]\cos 2g\end{aligned} \tag{20.44}$$

so that

$$\overline{\Delta_4 F} = -\frac{3\mu^6 r_e^4 J_4}{8L^3 G^7}\left[\left\{\frac{3}{8} - \frac{15}{4}\frac{H^2}{G^2} + \frac{35}{8}\frac{H^4}{G^4}\right\}\left[\frac{5}{2} - \frac{3}{2}\frac{G^2}{L^2}\right]\right.$$
$$\left. - \frac{5}{6}\left\{-1 + \frac{8H^2}{G^2} - \frac{7H^4}{G^4}\right\}\left[\frac{3}{4} - \frac{3}{4}\frac{G^2}{L^2}\right]\cos 2g\right] \tag{20.45}$$

When we split $\overline{\Delta_4 F}$ into secular and long periodic terms, we must put a prime on the L and the g and a double prime on the G. Then

$$\Delta_4 F_{2s}^* = -\frac{3\mu^6 r_e^4 J_4}{8L'^{10}}\left\{\frac{3}{8} - \frac{15}{4}\frac{H^2}{G''^2} + \frac{35}{8}\frac{H^4}{G''^4}\right\}\left[\frac{5L'^7}{2G''^7} - \frac{3}{2}\frac{L'^5}{G''^5}\right] \tag{20.46}$$

$$\Delta_4 F_{2p}^* = \frac{15\mu^6 r_e^4 J_4}{64L'^{10}}\left\{1 - \frac{8H^2}{G^2} + \frac{7H^4}{G^4}\right\}\left[\frac{L'^7}{G''^7} - \frac{L'^5}{G''^5}\right]\cos 2g' \tag{20.47}$$

We now use Eq. (20.35)

$$\frac{\partial F_1^*}{\partial G''}\frac{\partial \Delta_4 S_1^*}{\partial g'} = \Delta_4 F_{2p}^* \tag{20.35}$$

where

$$F_1^* = \frac{\mu^4 r_e^2 J_2}{2L'^3 G''^3}\left(-\frac{1}{2} + \frac{3}{2}\frac{H^2}{G''^2}\right) \tag{20.36}$$

Then

$$\frac{\partial F_1^*}{\partial G''} = \frac{3\mu^4 r_e^2 J_2}{4L'^3 G''^4}\left(1 - 5\frac{H^2}{G''^2}\right) \tag{20.14}$$

From Eqs. (20.35), (20.14), and (20.47)

$$\frac{3\mu^4 r_e^2 J_2}{4L'^3 G''^4}\left(1 - \frac{5H^2}{G''^2}\right)\frac{\partial \Delta_4 S_1^*}{\partial g'}$$
$$= -\frac{15\mu^6 r_e^4 J_4}{64L'^{10}}\left\{1 - \frac{8H^2}{G^2} + \frac{7H^4}{G^4}\right\}\left[\frac{L'^7}{G''^7} - \frac{L'^5}{G''^5}\right]\cos 2g' \tag{20.48}$$

Then

$$\frac{\partial \Delta_4 S_1^*}{\partial g'} = -\frac{5\mu^2 r_e^2 J_4}{16L'^4 J_2}G''\left\{1 - \frac{8H^2}{G''^2} + \frac{7H^4}{G''^4}\right\}\left[\frac{L'^4}{G''^4} - \frac{L'^2}{G''^2}\right]\left(1 - \frac{5H^2}{G''^2}\right)^{-1}\cos 2g' \tag{20.49}$$

so that

$$\Delta_4 S_1^* = -\frac{5\mu^2 r_e^2 J_4}{32L'^4 J_2}G''\left\{1 - \frac{8H^2}{G''^2} + \frac{7H^4}{G''^4}\right\}\left[\frac{L'^4}{G''^4} - \frac{L'^2}{G''^2}\right]\left(1 - \frac{5H^2}{G''^2}\right)^{-1}\sin 2g' \tag{20.50}$$

Then the secular terms from J_4 are

$$\Delta_4 c_1 = -\frac{\partial}{\partial L'}\Delta_4 F^*_{2s}$$
$$\Delta_4 c_2 = -\frac{\partial}{\partial G''}\Delta_4 F^*_{2s} \tag{20.51}$$
$$\Delta_4 c_3 = -\frac{\partial}{\partial H}\Delta_4 F^*_{2s}$$

where

$$\ell'' = \ell''_0 + c_1 t$$
$$g'' = g''_0 + c_2 t \tag{20.52}$$
$$h'' = h''_0 + c_3 t$$

The long periodic terms are given by

$$\Delta_4(G' - G'') = \frac{\partial \Delta_4 S^*_1}{\partial g'}$$
$$\Delta_4(\ell' - \ell'') = -\frac{\partial \Delta_4 S^*_1}{\partial L'}$$
$$\Delta_4(g' - g'') = -\frac{\partial \Delta_4 S^*_1}{\partial G''} \tag{20.53}$$
$$\Delta_4(h' - h'') = -\frac{\partial \Delta_4 S^*_1}{\partial H}$$

In the preceding formulas, g' is to be replaced by g''. Also it is instructive to add $\Delta_4 S^*_1$ to S^*_1 for the main problem. For the main problem we had, from Eq. (19.107),

$$S^*_1 = \frac{\mu^2 r_e^2 J_2 G''}{32(L')^4}\left(\frac{L'^2}{G''^2} - \frac{L'^4}{G''^4}\right)\left(1 - \frac{16H^2}{G''^2} + \frac{15H^4}{G''^4}\right)\left(1 - \frac{5H^2}{G''^2}\right)^{-1}\sin 2g'' \tag{20.54}$$

From Eq. (20.50)

$$\Delta_4 S^*_1 = \frac{5\mu^2 r_e^2 J_4}{32 L'^4 J_2} G''\left\{1 - \frac{8H^2}{G''^2} + \frac{7H^4}{G''^4}\right\}\left[\frac{L'^2}{G''^2} - \frac{L'^4}{G''^4}\right]\left(1 - \frac{5H^2}{G''^2}\right)^{-1}\sin 2g'' \tag{20.55}$$

Addition gives

$$S^*_1 + \Delta_4 S^*_1 = \frac{\mu^2 r_e^2 G''}{32(L')^4}\left[\frac{L'^2}{G''^2} - \frac{L'^4}{G''^4}\right]\left(1 - \frac{5H^2}{G''^2}\right)^{-1}\sin 2g'' Q \tag{20.56}$$

where

$$Q = \left(1 - \frac{16H^2}{G''^2} + \frac{15H^4}{G''^4}\right)J_2 + 5\left(1 - \frac{8H^2}{G''^2} + \frac{7H^4}{G''^4}\right)\frac{J_4}{J_2} \tag{20.57}$$

Put

$$H^2/G''^2 = c^2$$

then

$$Q = (1 - 16c^2 + 15c^4)J_2 + 5(1 - 8c^2 + 7c^4)\frac{J_4}{J_2} \tag{20.58}$$

$$Q = (1 - c^2)(1 - 15c^2)J_2 + 5(1 - c^2)(1 - 7c^2)\frac{J_4}{J_2} \tag{20.59}$$

$$Q = (1 - c^2)\left[(1 - 15c^2)J_2 + 5(1 - 7c^2)\frac{J_4}{J_2}\right] \tag{20.60}$$

Now

$$1 - 15c^2 = 1 - 5c^2 - 10c^2$$
$$1 - 7c^2 = 1 - 5c^2 - 2c^2$$

Thus

$$Q = (1 - c^2)\left[(1 - 5c^2)J_2 - 10c^2 J_2 + 5(1 - 5c^2)\frac{J_4}{J_2} - 10c^2\frac{J_4}{J_2}\right]$$

$$Q = (1 - c^2)\left[(1 - 5c^2)\left(J_2 + 5\frac{J_4}{J_2}\right) - 10c^2\left(\frac{J_2^2 + J_4}{1 - 5c^2}\right)\right] \tag{20.61}$$

Take

$$\frac{Q}{1 - 5c^2} = \frac{1 - c^2}{J_2}\left[J_2^2 + 5J_4 - 10\frac{H^2}{G''^2}\left(\frac{J_2^2 + J_4}{1 - 5c^2}\right)\right] \tag{20.62}$$

and insert this into Eq. (20.56). Then

$$S_1^* + \Delta_4 S_1^* = \frac{\mu^2 r_e^2 G''}{32L'^4 J_2}\left(\frac{L'^2}{G''^2} - \frac{L'^4}{G''^4}\right)\left(1 - \frac{H^2}{G''^2}\right)$$
$$\times\left[J_2^2 + 5J_4 - \frac{10H^2(J_2^2 + J_4)}{G''^2(1 - 5\cos^2 I'')}\right]\sin 2g'' \tag{20.63}$$

because $H^2/G''^2 = \cos^2 I''$.

Thus, the resonance denominator $1 - 5\cos^2 I''$ has a numerator $J_2^2 + J_4$. This statement is true for all the long periodic terms, which are obtained by differentiation of $S_1^* + \Delta_4 S_1^*$. A potential for which $J_4 = -J_2^2$ would not give rise to a critical inclination.

Reference

[1]Brouwer, D., "Solution of Problem of Artificial Satellite Theory Without Drag," *Astronomical Journal,* Vol. 64, No. 9, 1959, pp. 378–397.

Chapter 21

Lagrange and Poisson Brackets

I. Introduction

AN IMPORTANT method for doing perturbation theory for canonical systems is the method of Lie series. To develop it, we need to know more about canonical transformations. In Chapter 5, we saw that

$$(\Sigma_k p_k\, \mathrm{d}q_k - H\, \mathrm{d}t) - (\Sigma_k P_k\, \mathrm{d}Q_k - K \mathrm{d}t) = \mathrm{d}F \tag{21.1}$$

is a sufficient condition for the transformation $(q, p) \to (Q, P)$ to be canonical. It is not a necessary condition. If we insist, however, that we shall deal only with canonical transformations that satisfy Eq. (21.1), we shall need a special name for such a subspecies. We shall call it a "contact transformation." (See Refs. 1 and 2.)

If the transformation

$$\begin{aligned} q_k &= q_k(Q_1 \ldots Q_n, P_1 \ldots P_n, t) \\ p_k &= p_k(Q_1 \ldots Q_n, P_1 \ldots P_n, t) \end{aligned} \tag{21.2}$$

has a Jacobian that does not vanish anywhere in the domain of the Q's and P's that we are considering, we can solve Eq. (21.2) freely, back and forth between the q's and p's and Q's and P's. In that case, no matter what functional dependence may be indicated for F in Eq. (21.1), we can express it as

$$F = F(Q, P, t) \tag{21.3}$$

With use of the summation convention, we find from Eq. (21.2)

$$\mathrm{d}q_i = \frac{\partial q_i}{\partial Q_j}\mathrm{d}Q_j + \frac{\partial q_i}{\partial P_j}\mathrm{d}P_j + \frac{\partial q_i}{\partial t}\mathrm{d}t \tag{21.4}$$

so that

$$p_i\, \mathrm{d}q_i = p_i\frac{\partial q_i}{\partial Q_j}\mathrm{d}Q_j + p_i\frac{\partial q_i}{\partial P_j}\mathrm{d}P_j + p_i\frac{\partial q_i}{\partial t}\mathrm{d}t \tag{21.5}$$

The condition (21.1) becomes

$$\left(p_i\frac{\partial q_i}{\partial Q_j} - P_j\right)\mathrm{d}Q_j + p_i\frac{\partial q_i}{\partial P_j}\mathrm{d}P_j + \left(p_i\frac{\partial q_i}{\partial t} + K - H\right)\mathrm{d}t = \mathrm{d}F \tag{21.6a}$$

$$= \frac{\partial F}{\partial Q_j}\mathrm{d}Q_j + \frac{\partial F}{\partial P_j}\mathrm{d}P_j + \frac{\partial F}{\partial t}\mathrm{d}t \tag{21.6b}$$

Equate coefficients of $\mathrm{d}Q_j$, $\mathrm{d}P_j$, and $\mathrm{d}t$ on both sides of Eq. (21.6b). Then for a contact transformation

$$\frac{\partial F}{\partial Q_j} = p_i \frac{\partial q_i}{\partial Q_j} - P_j \tag{21.7}$$

$$\frac{\partial F}{\partial P_j} = p_i \frac{\partial q_i}{\partial P_j} \tag{21.8}$$

$$\frac{\partial F}{\partial t} = p_i \frac{\partial q_i}{\partial t} + K - H \tag{21.9}$$

Conversely, if Eqs. (21.7) and (21.8) hold for the mapping (21.2), then Eq. (21.6a) holds, and Eq. (21.1) is true, provided that the new Hamiltonian is given by Eq. (21.9). Thus, Eqs. (21.7) and (21.8) are necessary and sufficient that the mapping (21.2) be a contact transformation, with H the Hamiltonian $H(q, p, t)$,

$$K(Q, P, t) = H + \frac{\partial F(Q, P, t)}{\partial t} - p_i \frac{\partial q_i}{\partial t} \tag{21.9a}$$

and $\mathrm{d}F(Q, P, t)$ the perfect differential of the contact transformation.

Now, from Eqs. (21.7) and (21.8), we can express each of the second derivatives of F in two ways. First, $\partial^2 F/\partial Q_s \partial Q_r$ is given by either of

$$\frac{\partial}{\partial Q_s}\left(p_i \frac{\partial q_i}{\partial Q_r} - P_r\right) = \frac{\partial}{\partial Q_r}\left(p_i \frac{\partial q_i}{\partial Q_s} - P_s\right) \tag{21.10}$$

Next, $\partial^2 F/\partial P_s \partial P_r$ is given by either of

$$\frac{\partial}{\partial P_s}\left(p_i \frac{\partial q_i}{\partial P_r}\right) = \frac{\partial}{\partial P_r}\left(p_i \frac{\partial q_i}{\partial P_s}\right) \tag{21.11}$$

Finally, $\partial^2 F/\partial P_s \partial Q_r$ is given by either of

$$\frac{\partial}{\partial P_s}\left(p_i \frac{\partial q_i}{\partial Q_r} - P_r\right) = \frac{\partial}{\partial Q_r}\left(p_i \frac{\partial q_i}{\partial P_s}\right) \tag{21.12}$$

Equations (21.10)–(21.12) are necessary and sufficient for the validity of Eqs. (21.7) and (21.8) and, thus, for the validity of Eqs. (21.6a) and (21.1). However, Eq. (21.1) defines a contact transformation. Thus, Eqs. (21.10)–(21.12) are necessary and sufficient that the mapping (21.2) be a contact transformation.

II. Lagrange Brackets

Consider a set of $q_i, p_i, i = 1, \ldots, n$, and let u and v be any two parameters on which they may depend. Define the Lagrange bracket $[u, v]$ of u and v as

$$[u, v] \equiv \sum_{i=1}^{n} \left[\frac{\partial q_i}{\partial u}\frac{\partial p_i}{\partial v} - \frac{\partial q_i}{\partial v}\frac{\partial p_i}{\partial u}\right] \tag{21.13}$$

At once

$$\begin{aligned} [q_j, q_k] &= 0 \\ [p_j, p_k] &= 0 \\ [q_j, p_k] &= \delta_{jk} \end{aligned} \tag{21.14}$$

the Kronecker delta.

Now, consider Eq. (21.10), which is equivalent to

$$\sum_{i=1}^{n}\left[\frac{\partial q_i}{\partial Q_r}\frac{\partial p_i}{\partial Q_s}-\frac{\partial q_i}{\partial Q_s}\frac{\partial p_i}{\partial Q_r}\right]=0$$

or

$$[Q_r, Q_s] = 0 \tag{21.15a}$$

Consider Eq. (21.11), which is equivalent to

$$\sum_{i=1}^{n}\left[\frac{\partial q_i}{\partial P_r}\frac{\partial p_i}{\partial P_s}-\frac{\partial q_i}{\partial P_s}\frac{\partial p_i}{\partial P_r}\right]=0$$

or

$$[P_r, P_s] = 0 \tag{21.15b}$$

Finally, consider Eq. (21.12), which is equivalent to

$$\sum_{i=1}^{n}\left[\frac{\partial q_i}{\partial Q_r}\frac{\partial p_i}{\partial P_s}-\frac{\partial q_i}{\partial P_s}\frac{\partial p_i}{\partial Q_r}\right]=\delta_{rs}$$

or

$$[Q_r, P_s] = \delta_{rs} \tag{21.15c}$$

The Lagrange bracket relations (21.15) are necessary and sufficient for the validity of Eqs. (21.10)–(21.12) and, thus, for the validity of Eq. (21.1). The mapping (21.2) is a contact transformation if and only if the Lagrange brackets of the Q's and P's, relative to the q's and p's, satisfy Eqs. (21.15).

III. The Jacobi Relations

If $q_k = q_k(Q, P, t)$, $p_k = p_k(Q, P, t)$ is a contact transformation, the Jacobi relations are

$$\frac{\partial Q_r}{\partial q_s} = \frac{\partial p_s}{\partial P_r} \tag{21.16a}$$

$$\frac{\partial Q_r}{\partial p_s} = -\frac{\partial q_s}{\partial P_r} \tag{21.16b}$$

$$\frac{\partial P_r}{\partial q_s} = -\frac{\partial p_s}{\partial Q_r} \tag{21.16c}$$

$$\frac{\partial P_r}{\partial p_s} = \frac{\partial q_s}{\partial Q_r} \tag{21.16d}$$

To prove these, first write with the summation convention,

$$\mathrm{d}q_i = \frac{\partial q_i}{\partial Q_s}\,\mathrm{d}Q_s + \frac{\partial q_i}{\partial P_s}\,\mathrm{d}P_s + \frac{\partial q_i}{\partial t}\,\mathrm{d}t \tag{21.17a}$$

$$\mathrm{d}p_i = \frac{\partial p_i}{\partial Q_s}\,\mathrm{d}Q_s + \frac{\partial p_i}{\partial P_s}\,\mathrm{d}P_s + \frac{\partial p_i}{\partial t}\,\mathrm{d}t \tag{21.17b}$$

Then

$$\frac{\partial p_i}{\partial P_r}\,\mathrm{d}q_i - \frac{\partial q_i}{\partial P_r}\,\mathrm{d}p_i = \frac{\partial p_i}{\partial P_r}\left(\frac{\partial q_i}{\partial Q_s}\,\mathrm{d}Q_s + \frac{\partial q_i}{\partial P_s}\,\mathrm{d}P_s + \frac{\partial q_i}{\partial t}\,\mathrm{d}t\right)$$

$$- \frac{\partial q_i}{\partial P_r}\left(\frac{\partial p_i}{\partial Q_s}\,\mathrm{d}Q_s + \frac{\partial p_i}{\partial P_s}\,\mathrm{d}P_s + \frac{\partial p_i}{\partial t}\,\mathrm{d}t\right) \tag{21.18}$$

$$= [Q_s, P_r]\,\mathrm{d}Q_s + [P_s, P_r]\,\mathrm{d}P_s + G_1\,\mathrm{d}t$$

$$= \mathrm{d}Q_r + G_1\,\mathrm{d}t$$

by use of the definitions (21.13) and (21.14). From the Lagrange bracket conditions, this becomes

$$\mathrm{d}Q_r = \frac{\partial p_i}{\partial P_r}\,\mathrm{d}q_i - \frac{\partial q_i}{\partial P_r}\,\mathrm{d}p_i - G_1\,\mathrm{d}t \tag{21.18a}$$

However,

$$\mathrm{d}Q_r = \frac{\partial Q_r}{\partial q_s}\,\mathrm{d}q_s + \frac{\partial Q_r}{\partial p_s}\,\mathrm{d}p_s + \frac{\partial Q_r}{\partial t}\,\mathrm{d}t \tag{21.18b}$$

Comparison of Eqs. (21.18a) and (21.18b) shows that

$$\frac{\partial Q_r}{\partial q_s} = \frac{\partial p_s}{\partial P_r} \tag{21.16a}$$

$$\frac{\partial Q_r}{\partial p_s} = -\frac{\partial q_s}{\partial P_r} \tag{21.16b}$$

This completes the proof of the first two Jacobi relations.

To prove the other two from Eqs. (21.17), form

$$\frac{\partial p_i}{\partial Q_r}\,\mathrm{d}q_i - \frac{\partial q_i}{\partial Q_r}\,\mathrm{d}p_i = \frac{\partial p_i}{\partial Q_r}\left(\frac{\partial q_i}{\partial Q_s}\,\mathrm{d}Q_s + \frac{\partial q_i}{\partial P_s}\,\mathrm{d}P_s + \frac{\partial q_i}{\partial t}\,\mathrm{d}t\right)$$

$$- \frac{\partial q_i}{\partial Q_r}\left(\frac{\partial p_i}{\partial Q_s}\,\mathrm{d}Q_s + \frac{\partial p_i}{\partial P_s}\,\mathrm{d}P_s + \frac{\partial p_i}{\partial t}\,\mathrm{d}t\right) \tag{21.18c}$$

$$= [Q_s, Q_r]\,\mathrm{d}Q_s + [P_s, Q_r]\,\mathrm{d}P_s + G_2\,\mathrm{d}t$$

$$= -\mathrm{d}P_r + G_2\,\mathrm{d}t$$

Thus

$$\mathrm{d}P_r = -\frac{\partial p_i}{\partial Q_r}\,\mathrm{d}q_i + \frac{\partial q_i}{\partial Q_r}\,\mathrm{d}p_i + G_2\,\mathrm{d}t \tag{21.18d}$$

However,

$$\mathrm{d}P_r = \frac{\partial P_r}{\partial q_s}\,\mathrm{d}q_s + \frac{\partial P_r}{\partial p_s}\,\mathrm{d}p_s + \frac{\partial P_r}{\partial t}\,\mathrm{d}t \tag{21.18e}$$

Comparison of Eqs. (21.18d) and (21.18e) yields

$$\frac{\partial P_r}{\partial q_s} = -\frac{\partial p_s}{\partial Q_r} \tag{21.16c}$$

$$\frac{\partial P_r}{\partial p_s} = \frac{\partial q_s}{\partial Q_r} \tag{21.16d}$$

which are the other Jacobi relations. Note that the Jacobi relations of Chapter 12 are special cases of these, with $Q_k = \beta_k$ and $P_k = \alpha_k$.

IV. Poisson Brackets

Let u and v be functions of $q_i, p_i, i = 1, \ldots, n$, and t. The Poisson bracket (u, v) is defined by

$$(u, v) = \sum_{i=1}^{n} \left(\frac{\partial u}{\partial q_i}\frac{\partial v}{\partial p_i} - \frac{\partial u}{\partial p_i}\frac{\partial v}{\partial q_i} \right) \tag{21.19}$$

We first use the Jacobi relations to derive some relations pertinent to Lagrange and Poisson brackets in connection with contact transformations.

Theorem: For a contact transformation

$$[Q_r, Q_s] = (P_r, P_s) \tag{21.20}$$

Proof: With use of the summation convention,

$$[Q_r, Q_s] = \frac{\partial q_i}{\partial Q_r}\frac{\partial p_i}{\partial Q_s} - \frac{\partial q_i}{\partial Q_s}\frac{\partial p_i}{\partial Q_r} \tag{21.21}$$

from the definition of a Lagrange bracket. However, for a contact transformation, we have

$$\frac{\partial q_i}{\partial Q_r} = \frac{\partial P_r}{\partial p_i} \tag{21.16d}$$

and

$$\frac{\partial p_i}{\partial Q_s} = -\frac{\partial P_s}{\partial q_i} \tag{21.16c}$$

Insertion of Eqs. (21.16d) and (21.16c) into Eq. (21.21) yields

$$[Q_r, Q_s] = -\frac{\partial P_r}{\partial p_i}\frac{\partial P_s}{\partial q_i} + \frac{\partial P_s}{\partial p_i}\frac{\partial P_r}{\partial q_i} = (P_r, P_s) \tag{21.22a}$$

Similarly

$$[P_r, P_s] = (Q_r, Q_s) \tag{21.22b}$$

$$[Q_r, P_s] = (Q_s, P_r) = \delta_{rs} \tag{21.22c}$$

The Lagrange brackets conditions immediately become the Poisson brackets conditions $(Q_r, Q_s) = 0, (P_r, P_s) = 0, (Q_s, P_r) = \delta_{rs}$. These Poisson brackets conditions are necessary and sufficient for the mapping (21.2) to be contact transformation.

V. Invariance of a Poisson Bracket to a Contact Transformation

Suppose we have two functions $u(q_i, p_i, t)$, $v(q_i, p_i, t)$, $i = 1, \ldots, n$, and transform them by means of a contact transformation. They will appear as some other functions U and V of Q_i, P_i, $i = 1, \ldots, n$. That is

$$u(q, p, t) = U(Q, P, t) \tag{21.23a}$$

$$v(q, p, t) = V(Q, P, t) \tag{21.23b}$$

The invariance theorem states that

$$(u, v) = (U, V) \tag{21.24}$$

which can be written as

$$\frac{\partial u}{\partial q_i}\frac{\partial v}{\partial p_i} - \frac{\partial u}{\partial p_i}\frac{\partial v}{\partial q_i} = \frac{\partial U}{\partial Q_i}\frac{\partial V}{\partial P_i} - \frac{\partial U}{\partial P_i}\frac{\partial V}{\partial Q_i} \tag{21.25}$$

with use of the summation convention. At any given time t, we have from Eqs. (21.23)

$$\mathrm{d}u = \frac{\partial U}{\partial Q_r}\mathrm{d}Q_r + \frac{\partial U}{\partial P_r}\mathrm{d}P_r \tag{21.26a}$$

$$\mathrm{d}v = \frac{\partial V}{\partial Q_r}\mathrm{d}Q_r + \frac{\partial V}{\partial P_r}\mathrm{d}P_r \tag{21.26b}$$

so that

$$\begin{aligned} \frac{\partial u}{\partial q_s} &= \frac{\partial U}{\partial Q_r}\frac{\partial Q_r}{\partial q_s} + \frac{\partial U}{\partial P_r}\frac{\partial P_r}{\partial q_s} & \frac{\partial v}{\partial q_s} &= \frac{\partial V}{\partial Q_j}\frac{\partial Q_j}{\partial q_s} + \frac{\partial V}{\partial P_j}\frac{\partial P_j}{\partial q_s} \\ \frac{\partial u}{\partial p_s} &= \frac{\partial U}{\partial Q_r}\frac{\partial Q_r}{\partial p_s} + \frac{\partial U}{\partial P_r}\frac{\partial P_r}{\partial p_s} & \frac{\partial v}{\partial p_s} &= \frac{\partial V}{\partial Q_j}\frac{\partial Q_j}{\partial p_s} + \frac{\partial V}{\partial P_j}\frac{\partial P_j}{\partial p_s} \end{aligned} \tag{21.27}$$

Here, the derivatives of U and V are obtained from the functions indicated in Eqs. (21.23). The derivatives of the Q's and P's come from the canonical mapping (21.2), which can be inverted when its Jacobian does not vanish.

Now

$$(u, v) = \frac{\partial u}{\partial q_s}\frac{\partial v}{\partial p_s} - \frac{\partial u}{\partial p_s}\frac{\partial v}{\partial q_s} \tag{21.28}$$

Insert Eqs. (21.27) into Eq. (21.28). Then

$$\begin{aligned} (u, v) = &\left(\frac{\partial U}{\partial Q_r}\frac{\partial Q_r}{\partial q_s} + \frac{\partial U}{\partial P_r}\frac{\partial P_r}{\partial q_s}\right)\left(\frac{\partial V}{\partial Q_j}\frac{\partial Q_j}{\partial p_s} + \frac{\partial V}{\partial P_j}\frac{\partial P_j}{\partial p_s}\right) \\ &- \left(\frac{\partial U}{\partial Q_r}\frac{\partial Q_r}{\partial p_s} + \frac{\partial U}{\partial P_r}\frac{\partial P_r}{\partial p_s}\right)\left(\frac{\partial V}{\partial Q_j}\frac{\partial Q_j}{\partial q_s} + \frac{\partial V}{\partial P_j}\frac{\partial P_j}{\partial q_s}\right) \end{aligned} \tag{21.29}$$

Regroup terms to obtain

$$(u, v) = \frac{\partial U}{\partial Q_r}\frac{\partial V}{\partial Q_j}\left(\frac{\partial Q_r}{\partial q_s}\frac{\partial Q_j}{\partial p_s} - \frac{\partial Q_r}{\partial p_s}\frac{\partial Q_j}{\partial q_s}\right)_1$$
$$+ \frac{\partial U}{\partial Q_r}\frac{\partial V}{\partial P_j}\left(\frac{\partial Q_r}{\partial q_s}\frac{\partial P_j}{\partial p_s} - \frac{\partial Q_r}{\partial p_s}\frac{\partial P_j}{\partial q_s}\right)_2$$
$$+ \frac{\partial U}{\partial P_r}\frac{\partial V}{\partial Q_j}\left(\frac{\partial P_r}{\partial q_s}\frac{\partial Q_j}{\partial p_s} - \frac{\partial P_r}{\partial p_s}\frac{\partial Q_j}{\partial q_s}\right)_3$$
$$+ \frac{\partial U}{\partial P_r}\frac{\partial V}{\partial P_j}\left(\frac{\partial P_r}{\partial q_s}\frac{\partial P_j}{\partial p_s} - \frac{\partial P_r}{\partial p_s}\frac{\partial P_j}{\partial q_s}\right)_4$$

However,

$$[\ ]_1 = (Q_r, Q_s) = 0 \qquad [\ ]_2 = (Q_r, P_j) = \delta_{rj}$$
$$[\ ]_3 = (P_r, Q_j) = -\delta_{rj} \qquad [\ ]_4 = (P_r, P_s) = 0$$

because the transformation is of the contact type. Thus

$$(u, v) = \frac{\partial U}{\partial Q_i}\frac{\partial V}{\partial P_i} - \frac{\partial U}{\partial P_i}\frac{\partial V}{\partial Q_i} = (U, V) \tag{21.24}$$

as was to be proved.

VI. Other Relations for Poisson Brackets

We have, very easily,

$$(u, u) = 0 \tag{21.30}$$

$$(u, c) = 0 \tag{21.31}$$

$$(u, v) = -(v, u) \tag{21.32}$$

where c is a constant. The reader should also verify that

$$(uv, w) = u(v, w) + v(u, w) \tag{21.33}$$

For a Hamiltonian system with Hamiltonian $H(q, p, t)$

$$(q_i, H) = \frac{\partial q_i}{\partial q_j}\frac{\partial H}{\partial p_j} - \frac{\partial q_i}{\partial p_j}\frac{\partial H}{\partial q_j} = \frac{\partial H}{\partial p_i} = \dot{q}_i \tag{21.34}$$

$$(p_i, H) = \frac{\partial p_i}{\partial q_j}\frac{\partial H}{\partial p_j} - \frac{\partial p_i}{\partial p_j}\frac{\partial H}{\partial q_j} = -\frac{\partial H}{\partial q_i} = -\dot{p}_i \tag{21.35}$$

Any function $u(q, p, t)$ of such canonical variables satisfies

$$\begin{aligned} u &= \frac{\partial u}{\partial q_i}\dot{q}_i + \frac{\partial u}{\partial p_i}\dot{p}_i + \frac{\partial u}{\partial t} = \frac{\partial u}{\partial q_i}\frac{\partial H}{\partial p_i} - \frac{\partial u}{\partial p_i}\frac{\partial H}{\partial q_i} + \frac{\partial u}{\partial t} \\ \dot{u} &= (u, H) + \frac{\partial u}{\partial t} \end{aligned} \tag{21.36}$$

Also

$$\frac{\partial}{\partial q_i}(u, v) = \left(u, \frac{\partial v}{\partial q_i}\right) + \left(\frac{\partial u}{\partial q_i}, v\right) \tag{21.37}$$

$$\frac{\partial}{\partial p_i}(u, v) = \left(u, \frac{\partial v}{\partial p_i}\right) + \left(\frac{\partial u}{\partial p_i}, v\right) \tag{21.38}$$

Finally, we need Poisson's identity

$$(u, (v, w)) + (v, (w, u)) + (w, (u, v)) = 0 \tag{21.39}$$

This is reminiscent of a similar cyclic rule for vectors,

$$\boldsymbol{A} \times (\boldsymbol{B} \times \boldsymbol{C}) + \boldsymbol{B} \times (\boldsymbol{C} \times \boldsymbol{A}) + \boldsymbol{C} \times (\boldsymbol{A} \times \boldsymbol{B}) = \boldsymbol{0}$$

the "snake-biting-its-tail" relation.

To prove Poisson's identity, we first note that

$$(\phi, \psi) = -(\psi, \phi)$$

so that

$$(u, (v, w)) + (v, (w, u)) = (u, (v, w)) - (v, (u, w)) \tag{21.40}$$

Now

$$(v, w) = \sum_{i=1}^{n} \left(\frac{\partial v}{\partial q_i}\frac{\partial w}{\partial p_i} - \frac{\partial v}{\partial p_i}\frac{\partial w}{\partial q_i}\right) = D_v w \tag{21.41a}$$

and

$$(u, w) = \sum_{i=1}^{n} \left(\frac{\partial u}{\partial q_i}\frac{\partial w}{\partial p_i} - \frac{\partial u}{\partial p_i}\frac{\partial w}{\partial q_i}\right) = D_u w \tag{21.41b}$$

where

$$D_v = \sum_{i=1}^{n} \left(\frac{\partial v}{\partial q_i}\frac{\partial}{\partial p_i} - \frac{\partial v}{\partial p_i}\frac{\partial}{\partial q_i}\right) \tag{21.42a}$$

$$D_u = \sum_{i=1}^{n} \left(\frac{\partial u}{\partial q_i}\frac{\partial}{\partial p_i} - \frac{\partial u}{\partial p_i}\frac{\partial}{\partial q_i}\right) \tag{21.42b}$$

The operators D_v and D_u can be expressed as

$$D_v = \sum_{i=1}^{2n} \alpha_i \frac{\partial}{\partial x_i} \tag{21.43a}$$

$$D_u = \sum_{i=1}^{2n} \beta_i \frac{\partial}{\partial x_i} \tag{21.43b}$$

where we denote

$$q_1, q_2, \ldots, q_n, p_1, p_2, \ldots, p_n$$

by

$$x_1, x_2, \ldots, x_n, x_{n+1}, x_{n+2}, \ldots, x_{2n}$$

and where the α's and β's are as follows.

α_1	α_2	$\ldots$	α_n	α_{n+1}	$\ldots$	α_{2n}
$-\dfrac{\partial v}{\partial p_1}$	$-\dfrac{\partial v}{\partial p_2}$		$-\dfrac{\partial v}{\partial p_n}$	$\dfrac{\partial v}{\partial q_1}$		$\dfrac{\partial v}{\partial q_n}$
β_1	β_2	$\ldots$	β_n	β_{n+1}	$\ldots$	β_{2n}
$-\dfrac{\partial u}{\partial p_1}$	$-\dfrac{\partial u}{\partial p_2}$		$-\dfrac{\partial u}{\partial p_n}$	$\dfrac{\partial u}{\partial q_1}$		$\dfrac{\partial u}{\partial q_n}$

Now by Eqs. (21.41)

$$(v, w) = D_v w \qquad (u, w) = D_u w$$

Then

$$(u, (v, w)) = D_u D_v w \tag{21.44a}$$

$$(v, (u, w)) = D_v D_u w \tag{21.44b}$$

Thus

$$(u, (v, w)) - (v, (u, w)) = \{D_u D_v - D_v D_u\} w \tag{21.45}$$

Apply Eq. (21.43) to Eq. (21.45). Then

$$(u, (v, w)) - (v, (u, w)) = \sum_{j=1}^{2n} \beta_j \frac{\partial}{\partial x_j} \sum_{i=1}^{2n} \alpha_i \frac{\partial w}{\partial x_i} - \sum_{i=1}^{2n} \alpha_i \frac{\partial}{\partial x_i} \sum_{j=1}^{2n} \beta_j \frac{\partial w}{\partial x_j} \tag{21.46}$$

The second derivative terms vanish immediately, and we are left with

$$(u, (v, w)) - (v, (u, w)) = \sum_{i=1}^{2n} \sum_{j=1}^{2n} \left(\beta_j \frac{\partial \alpha_i}{\partial x_j} \frac{\partial w}{\partial x_i} - \alpha_i \frac{\partial \beta_j}{\partial x_i} \frac{\partial w}{\partial x_j} \right) \tag{21.47}$$

This is simply a sum with coefficients of all the $\partial w / q_k$ and all the $\partial w / p_k$, so that

$$(u, (v, w)) - (v, (u, w)) = \sum_{k=1}^{n} \left(A_k \frac{\partial w}{\partial q_k} + B_k \frac{\partial w}{\partial p_k} \right) \tag{21.48}$$

where the A's and B's do not depend on w. We may, therefore, determine the A's and B's by giving special values to w.

To determine the B's, let $w = p_i$. Then

$$(v, w) = (v, p_i) = \frac{\partial v}{\partial q_k} \frac{\partial p_i}{\partial p_k} - \frac{\partial v}{\partial p_k} \frac{\partial p_i}{\partial q_k} = \frac{\partial v}{\partial q_i} \tag{21.49}$$

Similarly

$$(u, w) = (u, p_i) = \frac{\partial u}{\partial q_i} \tag{21.50}$$

Now, insert $w = p_i$ in the right side of Eq. (21.48) and Eqs. (21.49) and (21.50) on the left side. We find

$$B_i = \left(u, \frac{\partial v}{\partial q_i}\right) - \left(v, \frac{\partial u}{\partial q_i}\right) \tag{21.51}$$

However, from Eq. (21.32)

$$\left(v, \frac{\partial u}{\partial q_i}\right) = -\left(\frac{\partial u}{\partial q_i}, v\right)$$

so that

$$B_i = \left(u, \frac{\partial v}{\partial q_i}\right) + \left(\frac{\partial u}{\partial q_i}, v\right) = \frac{\partial}{\partial q_i}(u, v) \tag{21.52}$$

by Eq. (21.37).

To determine the A's, let $w = q_i$. Equation (21.48) becomes

$$A_i = (u, (v, q_i)) - (v, (u, q_i))$$

but

$$(v, q_i) = \frac{\partial v}{\partial q_j}\frac{\partial q_i}{\partial p_j} - \frac{\partial v}{\partial p_j}\frac{\partial q_i}{\partial q_j} = -\frac{\partial v}{\partial p_i}$$

$$(u, q_i) = -\frac{\partial u}{\partial p_i}$$

Thus

$$A_i = -\left(u, \frac{\partial v}{\partial p_i}\right) - \left(\frac{\partial u}{\partial p_i}, v\right) = -\frac{\partial}{\partial p_i}(u, v) \tag{21.53}$$

Now, insert Eq. (21.52) for B_i and Eq. (21.53) for A_i into Eq. (21.48). The result is

$$(u, (v, w)) - (v, (u, w)) = \sum_{k=1}^{n}\left(\frac{\partial w}{\partial p_k}\frac{\partial}{\partial q_k}(u, v) - \frac{\partial w}{\partial q_k}\frac{\partial}{\partial p_k}(u, v)\right) \tag{21.54}$$

$$(u, (v, w)) - (v, (u, w)) = -(w, (u, v))$$

Thus

$$(u, (v, w)) - (v, (u, w)) + (w, (u, v)) = 0$$

or

$$(u, (v, w)) + (v, (w, u)) + (w, (u, v)) = 0 \tag{21.39}$$

which is Poisson's identity.

References

[1]Goldstein, H., *Classical Mechanics,* 2nd ed., Addison–Wesley, Reading, MA, 1980, Chap. 9.

[2]Pars, L. A., *A Treatise on Analytical Dynamics,* Wiley, New York, 1965, pp. 493–497.

Chapter 22

Lie Series

I. Introduction

THERE are series of nested Poisson brackets that can be used to form contact transformations. Such series have been used by Hori[1] and many others to formulate methods of doing perturbation theory. In this chapter, we shall use other methods to carry out the first section of Hori's paper[1] because his first section is hard to understand and because of his use of a pseudo-time τ.

In Chapter 23, we shall follow Hori's methods[1] but avoid his use of certain artificial times t^* and t. This avoidance of artificial times is facilitated by using some of Brouwer's methods[2] in doing perturbation theory for artificial satellites.

This method of Lie series has a decided advantage over the Brouwer–von Zeipel method in that it does not use "mixed variables" to build up contact transformations. It, therefore, proceeds in a purely recursive fashion, well adapted to the use of machine algebra for the higher approximations.

We shall formulate only that much of perturbation theory that can be done by the Lie series of Hori. For a comparison of the theories of Hori and others, see Ref. 3.

II. Hori's Section 1

Let ξ_j, η_j be a set of $2N$ variables, and let $f(\xi, \eta)$ and $S(\xi, \eta)$ be arbitrary functions of them. Let (f, S) be the Poisson bracket of f and S. We define the operator D_s as follows:

$$D_s^0 f = f \qquad D_s f = (f, S) \qquad D_s^n f = D_s^{n-1}(D_s f) \tag{22.1}$$

Define $2N$ variables x_j, y_j by

$$f(x, y) = \sum_{n=0}^{\infty} \frac{\varepsilon^n}{n!} D_s^n f(\xi, \eta) \tag{22.2}$$

where ε is a small parameter arising from the physics of the problem. In artificial satellite theory, it might be J_2. At this point, Hori uses a pseudo-time τ to show that, if ξ_j, η_j are canonical with respect to some Hamiltonian $F(\xi, \eta, \varepsilon)$, then x_j, y_j will also be canonical with respect to $F(\xi, \eta, \varepsilon)$.[1] We shall also prove this, but it will take many steps to avoid the τ.

III. Theorems

With the definition (22.1), if $S = S(\xi, \eta)$, $S^* = S^*(\xi, \eta)$, $f = f(\xi, \eta)$, and $g = g(\xi, \eta)$ and if α and β are constants, we have some theorems.

Theorem 1:

$$D_s(\alpha f + \beta g) = \alpha D_s f + \beta D_s g$$

Proof:

$$(\alpha f + \beta g, S) = (\alpha f, S) + (\beta g, S)$$
$$= \alpha(f, S) + \beta(g, S)$$

Thus

$$D_s(\alpha f + \beta g) = \alpha D_s f + \beta D_s g$$

Theorem 2:

$$D_s(fg) = f D_s g + g D_s f$$

Proof:

$$D_s(fg) = (fg, S) = \sum_{i=1}^{N} \left(\frac{\partial(fg)}{\partial \xi_i} \frac{\partial S}{\partial \eta_i} - \frac{\partial(fg)}{\partial \eta_i} \frac{\partial S}{\partial \xi_i} \right)$$

Since

$$(f, S) = \sum_{i=1}^{N} \left(\frac{\partial f}{\partial \xi_i} \frac{\partial S}{\partial \eta_i} - \frac{\partial f}{\partial \eta_i} \frac{\partial S}{\partial \xi_i} \right)$$

this becomes

$$D_s(fg) = f(g, S) + g(f, S) = f D_s g + g D_s f$$

Theorem 3:

$$D_s(f, g) = (f, D_s g) + (D_s f, g) \tag{22.3}$$

Proof: By Poisson's identity

$$(f, (g, S)) + (g, (S, f)) + (S, (f, g)) = 0$$
$$(f, D_s g) - (g, D_s f) - D_s(f, g) = 0$$

This proves the theorem.

Theorem 4:

$$D_{s^*} D_s - D_s D_{s^*} = D_{(s,s^*)} \tag{22.4}$$

Proof: By Poisson's identity

$$(f, (S, S^*)) + (S, (S^*, f)) + (S^*, (f, S)) = 0$$
$$(f, (S, S^*)) + ((f, S^*), S) - ((f, S), S^*) = 0$$
$$D_{(s,s^*)} f + D_s D_{s^*} f - D_{s^*} D_s = 0$$

This proves the theorem.

Theorem 5:

$$D_s^n(fg) = \sum_{m=0}^{n} \binom{n}{m} D_s^m f D_s^{n-m} g \tag{22.5}$$

where

$$\binom{n}{m} = \frac{n!}{(n-m)!m!}$$

the binomial coefficient.

Proof:

$$\begin{aligned} D_s(fg) &= f D_s g + g D_s f \\ D_s^2(fg) &= f D_s^2 g + 2 D_s f D_s g + g D_s^2 f \end{aligned} \tag{22.6}$$

The theorem holds for $n = 2$. We now use mathematical induction. If it holds for n, then application of D_s to Eq. (22.5) gives

$$D_s^{n+1}(fg) = \sum_{m=0}^{n} \binom{n}{m} \left[D_s^m f D_s^{n-m+1} g + D_s^{m+1} f D_s^{n-m} g \right] \tag{22.7}$$

Now, break up Eq. (22.7) into two parts. In the first, let m run from 0 to n. In the second, put $m = m' - 1$, and let m' run from 1 to $n + 1$. Then split off the term $m = 0$ from the first sum and $m = n + 1$ from the second sum. We obtain

$$\begin{aligned} D_s^{n+1}(fg) = f D_s^{n+1} g + g D_s^{n+1} f + \sum_{m=1}^{n} D_s^n f D_s^{n-m+1} g \\ \times \left[\frac{n!}{(n-m)!m!} + \frac{n!}{(n+1-m)!(m-1)!} \right] \end{aligned} \tag{22.8}$$

where we have switched m' back to m in this second sum. However, Eq. (22.8),

$$\begin{aligned} &\left[\frac{n!}{(n-m)!m!} + \frac{n!}{(n+1-m)!(m-1)!} \right] \\ &= \frac{n!}{(n-m)!(m-1)!} \left[\frac{1}{m} + \frac{1}{(n+1-m)} \right] = \frac{(n+1)!}{(n+1-m)!m!} \end{aligned}$$

Thus

$$D_s^{n+1}(fg) = \sum_{m=0}^{n+1} \frac{(n+1)!}{(n+1-m)!m!} D_s^n f D_s^{n-m+1} g \tag{22.9}$$

If Theorem 5 holds for n, it holds for $n + 1$. However, it holds for $n = 2$, so that it holds for all n.

Theorem 6:

$$D_s^n(f, g) = \sum_{m=0}^{n} \binom{n}{m} \left(D_s^m f, D_s^{n-m} g \right) \tag{22.10}$$

Here, the comma denotes a Poisson bracket.

Proof: By Theorem 3

$$D_s(f, g) = (f, D_s g) + (D_s f, g) \tag{22.11}$$

Another application of D_s with use of Theorem 3 gives

$$D_s^2(f, g) = \left(f, D_s^2 g\right) + 2(D_s f, D_s g) + \left(D_s^2 f, g\right) \tag{22.12}$$

Thus, Eq. (22.10) holds for $n = 2$. The proof of Eq. (22.10) by mathematical induction proceeds just like the proof of Theorem 5; therefore Theorem 6 holds.

We now define the operator $\exp \varepsilon D_s$ by

$$\exp \varepsilon D_s f = \sum_{n=0}^{\infty} \frac{\varepsilon^n}{n!} D_s^n f \tag{22.13}$$

Theorem 7:

$$\exp \varepsilon D_s(fg) = (\exp \varepsilon D_s f)(\exp \varepsilon D_s g) \tag{22.14}$$

where

$$\exp \varepsilon D_s(fg) = \sum_{n=0}^{\infty} \frac{\varepsilon^n}{n!} D_s^n(fg) \tag{22.15}$$

Proof: Apply Theorem 5 to Eq. (22.15). Then

$$\exp \varepsilon D_s(fg) = \sum_{n=0}^{\infty} \frac{\varepsilon^n}{n!} \sum_{m=0}^{n} \binom{n}{m} D_s^m f D_s^{n-m} g \tag{22.16}$$

Also

$$(\exp \varepsilon D_s f)(\exp \varepsilon D_s g) = \sum_{k=0}^{\infty} \frac{\varepsilon^k}{k!} D_s^k f \sum_{j=0}^{\infty} \frac{\varepsilon^j}{j!} D_s^j g \tag{22.17}$$

Here, Eq. (22.17) is a sum over the first lattice quadrant. To perform the summation, draw all the lattice lines perpendicular to the 45°, sum over each of these lines ($k = 0$ to n), and then sum over all the lines in the quadrant ($n = 0$ to ∞). With $k + j = n$, we obtain

$$\begin{aligned} \Sigma_k \Sigma_j \frac{\varepsilon^n}{k!j!} D_s^k f D_s^{n-k} g &= \sum_{n=0}^{\infty} \sum_{k=0}^{n} \frac{\varepsilon^n}{(n-k)!k!} D_s^k f D_s^{n-k} g \\ &= \sum_{n=0}^{\infty} \frac{\varepsilon^n}{n!} \sum_{k=0}^{n} \frac{n!}{(n-k)!k!} D_s^k f D_s^{n-k} g \end{aligned} \tag{22.18}$$

which is the same as Eq. (22.16). Thus, Eq. (22.16) equals Eq. (22.17), so that Theorem 7 is proved.

Theorem 8: With (,) denoting a Poisson bracket,

$$\exp \varepsilon D_s(f, g) = (\exp \varepsilon D_s f, \exp \varepsilon D_s g) \tag{22.19}$$

Here

$$\exp \varepsilon D_s(f, g) = \sum_{n=0}^{\infty} \frac{\varepsilon^n}{n!} D_s^n(f, g) \tag{22.20}$$

Apply Theorem 6 to Eq. (22.20). Then

$$\exp \varepsilon D_s(f, g) = \sum_{n=0}^{\infty} \frac{\varepsilon^n}{n!} \sum_{m=0}^{n} \binom{n}{m} (D_s^m f, D_s^{n-m} g) \tag{22.21}$$

but

$$(\exp \varepsilon D_s f, \exp \varepsilon D_s g) = \left(\sum_{k=0}^{\infty} \frac{\varepsilon^k}{k!} D_s^k f, \sum_{j=0}^{\infty} \frac{\varepsilon^j}{j!} D_s^j g \right) \tag{22.22}$$

The proof proceeds just like the proof of Theorem 7 with the appropriate insertion of commas; so we shall regard Theorem 8 as proved.

Applications to Canonical Transformation

Let the variables $\xi_j, \eta_j,\ j = 1, \ldots, N$, be canonical with respect to some Hamiltonian $F(\xi, \eta)$. Relative to the ξ's and η's, their Poisson brackets satisfy

$$(\xi_j, \xi_k) = 0 \qquad (\eta_j, \eta_k) = 0 \qquad (\xi_j, \eta_k) = \delta_{jk} \tag{22.23}$$

These relations Eq. (22.23), of course, follow at once from the definition of a Poisson bracket and do not depend on the canonicity of the ξ's and η's, with respect to $F(\xi, \eta)$.

Now, suppose we introduce new variables $x_j, y_j,\ j = 1, \ldots, N$, by

$$x_j = \sum_{n=0}^{\infty} \frac{\varepsilon^n}{n!} D_s^n \xi_j \qquad y_j = \sum_{n=0}^{\infty} \frac{\varepsilon^n}{n!} D_s^n \eta_j \tag{22.24}$$

where $S = S(\xi, \eta)$, independent of t. We shall show that these x's and y's will also be canonical with respect to some Hamiltonian K, where $K = F$ if F is explicitly independent of t. The proof goes as follows.

From Eqs. (22.24)

$$x_j = \exp \varepsilon D_s \xi_j \qquad y_j = \exp \varepsilon D_s \eta_j \tag{22.25}$$

The Poisson brackets of the x's and y's are given by

$$\begin{aligned}(x_j, x_k) &= (\exp \varepsilon D_s \xi_j, \exp \varepsilon D_s \xi_k) \\ &= \exp \varepsilon D_s (\xi_j, \xi_k)\end{aligned} \tag{22.26}$$

by Theorem 8. Since $(\xi_j, \xi_k) = 0$,

$$(x_j, x_k) = 0 \tag{22.27a}$$

Similarly

$$(y_j, y_k) = 0 \tag{22.27b}$$

and

$$(x_j, y_k) = \delta_{jk} \tag{22.27c}$$

However, Eqs. (22.27) are the necessary and sufficient conditions that Eqs. (22.24) should be a contact transformation. Because the ξ's and η's are canonical, the result is that the x's and y's are also canonical.

We may write Eqs. (22.24) in the form of Hori's equations (5a) and (5b).[1] To do so, note that $D_s^0\xi_j = \xi_j$, $D_s^0\eta_j = \eta_j$ and that

$$
\begin{aligned}
D_s\xi_j &= (\xi_j, S) = \sum_{i=1}^{N}\left(\frac{\partial \xi_j}{\partial \xi_i}\frac{\partial S}{\partial \eta_i} - \frac{\partial \xi_j}{\partial \eta_i}\frac{\partial S}{\partial \xi_i}\right) = \frac{\partial S}{\partial \eta_j} \\
D_s\eta_j &= (\eta_j, S) = \sum_{i=1}^{N}\left(\frac{\partial \eta_j}{\partial \xi_i}\frac{\partial S}{\partial \eta_i} - \frac{\partial \eta_j}{\partial \eta_i}\frac{\partial S}{\partial \xi_i}\right) = -\frac{\partial S}{\partial \xi_j}
\end{aligned}
\tag{22.28}
$$

By Eqs. (22.24) and (22.28),

$$x_j = \xi_j + \sum_{n=1}^{\infty}\frac{\varepsilon^n}{n!}D_s^{n-1}\frac{\partial S}{\partial \eta_j} \tag{22.29a}$$

$$y_j = \eta_j - \sum_{n=1}^{\infty}\frac{\varepsilon^n}{n!}D_s^{n-1}\frac{\partial S}{\partial \xi_j} \tag{22.29b}$$

These are the same as Hori's equations (5a) and (5b). They are equivalent to the series (22.24), which are the series of nested Poisson brackets previously mentioned. Thus

$$D_s\xi = (\xi, S), \quad D_s^2\xi = ((\xi, S), S), \quad D_s^3\xi = (((\xi, S), S), S), \quad \ldots$$

Suppose now that we have a function of the x's and y's that does not depend explicitly on ε. Call it $f(x, y)$, where the comma does not indicate a Poisson bracket.

$$x_j = \exp \varepsilon D_s \xi_j \qquad y_j = \exp \varepsilon D_s \eta_j \tag{22.30}$$

Theorem 9:

$$f(x, y) = \exp \varepsilon D_s f(\xi, \eta) \tag{22.31}$$

where $f(\xi, \eta)$ is the same function of the ξ's and η's that $f(x, y)$ is of the x's and y's.

Proof: From Eqs. (22.30) and (22.31)

$$f(x, y) = g(\xi, \eta, \varepsilon) \tag{22.32}$$

where $x_j = \xi_j$ and $y_j = \eta_j$ when $\varepsilon = 0$. From Eq. (22.32)

$$\frac{\partial g}{\partial \xi_i} = \sum_{i=1}^{N}\left(\frac{\partial f}{\partial x_k}\frac{\partial x_k}{\partial \xi_i} + \frac{\partial f}{\partial y_k}\frac{\partial y_k}{\partial \xi_i}\right) \tag{22.33a}$$

$$\frac{\partial g}{\partial \eta_i} = \sum_{i=1}^{N}\left(\frac{\partial f}{\partial x_k}\frac{\partial x_k}{\partial \eta_i} + \frac{\partial f}{\partial y_k}\frac{\partial y_k}{\partial \eta_i}\right) \tag{22.33b}$$

We then insert Eq. (22.33) into the expression for the Poisson bracket of g and S:

$$(g, S) = \sum_{i=1}^{N} \left(\frac{\partial g}{\partial \xi_i} \frac{\partial S}{\partial \eta_i} - \frac{\partial g}{\partial \eta_i} \frac{\partial S}{\partial \xi_i} \right) \tag{22.34}$$

with the result

$$(g, S) = \sum_{k=1}^{N} \sum_{i=1}^{N} \left(\frac{\partial f}{\partial x_k} \frac{\partial x_k}{\partial \xi_i} + \frac{\partial f}{\partial y_k} \frac{\partial y_k}{\partial \xi_i} \right) \frac{\partial S}{\partial \eta_i} - \sum_{k=1}^{N} \sum_{i=1}^{N} \left(\frac{\partial f}{\partial x_k} \frac{\partial x_k}{\partial \eta_i} + \frac{\partial f}{\partial y_k} \frac{\partial y_k}{\partial \eta_i} \right) \frac{\partial S}{\partial \xi_i}$$

Regrouping terms, we obtain

$$(g, S) = \sum_{k=1}^{N} \left((x_k, S) \frac{\partial f}{\partial x_k} + (y_k, S) \frac{\partial f}{\partial y_k} \right) \tag{22.35}$$

$$= \sum_{k=1}^{N} \left(\frac{\partial f}{\partial x_k} D_s x_k + \frac{\partial f}{\partial y_k} D_s y_k \right) \tag{22.36}$$

By Eqs. (22.30),

$$x_k = \xi_k + \sum_{n=1}^{\infty} \frac{\varepsilon^n}{n!} D_s^n \xi_k$$

so that

$$\frac{\partial x_k}{\partial \varepsilon} = \sum_{n=1}^{\infty} \frac{n \varepsilon^{n-1}}{n!} D_s^n \xi_k = D_s \sum_{n=1}^{\infty} \frac{\varepsilon^{n-1}}{(n-1)!} D_s^{n-1} \xi_k = D_s \sum_{n=0}^{\infty} \frac{\varepsilon^n}{n!} D_s^n \xi_k = D_s x_k$$

Thus

$$\frac{\partial x_k}{\partial \varepsilon} = D_s x_k \tag{22.37a}$$

Similarly

$$\frac{\partial y_k}{\partial \varepsilon} = D_s y_k \tag{22.37b}$$

and from Eqs. (22.36) and (22.37)

$$(g, S) = \sum_{k=1}^{N} \left(\frac{\partial f}{\partial x_k} \frac{\partial x_k}{\partial \varepsilon} + \frac{\partial f}{\partial y_k} \frac{\partial y_k}{\partial \varepsilon} \right) = \frac{\partial g}{\partial \varepsilon} \tag{22.38}$$

The last step follows from Eqs. (22.32). Thus

$$\frac{\partial g}{\partial \varepsilon} = D_s g, \quad \frac{\partial^2 g}{\partial \varepsilon^2} = D_s^2 g, \quad \ldots, \quad \frac{\partial^n g}{\partial \varepsilon^n} = D_s^n g \tag{22.39}$$

and

$$\left(\frac{\partial^n g}{\partial \varepsilon^n} \right)_{\varepsilon=0} = \left(D_s^n g \right)_{\varepsilon=0} = D_s^n f \tag{22.40}$$

because $g = f$ for $\varepsilon = 0$. However, by a McLaurin expansion

$$g(\xi, \eta, \varepsilon) = \sum_{n=0}^{\infty} \frac{\varepsilon^n}{n!} \left(\frac{\partial^n g}{\partial \varepsilon^n} \right)_{\varepsilon=0} \tag{22.41}$$

By Eqs. (22.40) and (22.41)

$$g(\xi, \eta, \varepsilon) = \sum_{n=0}^{\infty} \frac{\varepsilon^n}{n!} D_s^n f(\xi, \eta) \tag{22.42}$$

but $g(\xi, \eta, \varepsilon) = f(x, y)$, so that

$$f(x, y) = \sum_{n=0}^{\infty} \frac{\varepsilon^n}{n!} D_s^n f(\xi, \eta)$$

or

$$f(x, y) = \exp \varepsilon D_s f(\xi, \eta) \tag{22.43}$$

Theorem 9 may also be expressed as

$$f(\exp \varepsilon D_s \xi, \exp \varepsilon D_s \eta) = \exp \varepsilon D_s f(\xi, \eta) \tag{22.43a}$$

If $f(\xi, \eta) = S(\xi, \eta)$, this becomes

$$f(x, y) = S(\xi, \eta) \tag{22.44}$$

which means that the generator is conserved under the mapping (22.29).

Compounding Transformations

Suppose we go from (x, y) to (ξ, η) by means of the transformation function $S(\xi, \eta)$, i.e., by

$$x_k = \exp \varepsilon D_s \xi_k \qquad y_k = \exp \varepsilon D_s \eta_k \tag{22.45}$$

and then from (ξ, η) to (p, q) by means of the transformation function $S^*(p, q)$, i.e., by

$$\xi_k = \exp \varepsilon D_{s^*} q_k \qquad \eta_k = \exp \varepsilon D_{s^*} p_k \tag{22.46}$$

How then can we express the x's and y's directly in terms of the q's and p's?

Equations (22.45) imply Eqs. (22.43), so that

$$f(x, y) = \sum_{k=0}^{\infty} \frac{\varepsilon^k}{k!} D_{s(\xi,\eta)}^k f(\xi, \eta) \tag{22.47}$$

Similarly Eqs. (22.46) imply

$$g(\xi, \eta) = \sum_{m=0}^{\infty} \frac{\varepsilon^m}{m!} D_{s^*(q,p)}^m g(q, p) \tag{22.48}$$

In Eq. (22.47) put

$$D_{s(\xi,\eta)}^k f(\xi, \eta) = g(\xi, \eta) \tag{22.49}$$

Then by Eqs. (22.48) and (22.49)

$$D^k_{s(\xi,\eta)} f(\xi, \eta) = \sum_{m=0}^{\infty} \frac{\varepsilon^m}{m!} D^m_{s^*(q,p)} D^k_{s(q,p)} f(q, p) \tag{22.50}$$

Now, insert Eqs. (22.50) into Eq. (22.47). We obtain

$$f(x, y) = \sum_{k=0}^{\infty} \sum_{m=0}^{\infty} \frac{\varepsilon^{k+m}}{k!m!} D^m_{s^*(q,p)} D^k_{s(q,p)} f(q, p) \tag{22.51}$$

The sum is over the first lattice quadrant. As in proving Theorem 7, put $m + k = n$, sum over m from 0 to n and then sum over n from 0 to ∞. We obtain

$$f(x, y) = \sum_{n=0}^{\infty} \frac{\varepsilon^n}{n!} \sum_{m=0}^{n} \frac{n!}{(n-m)!m!} D^m_{s^*(q,p)} D^{n-m}_{s(q,p)} f(q, p) \tag{22.52}$$

This is the desired compound transformation, the same as Hori's equation (7).[1]

For the special cases $f = x_j$ or $f = y_j$, and omitting (q, p) in S^* and S for clarity, we obtain

$$\begin{aligned} x_j &= \sum_{n=0}^{\infty} \frac{\varepsilon^n}{n!} \sum_{m=0}^{n} \binom{n}{m} D^m_{s^*} D^{n-m}_{s} q_j \\ y_j &= \sum_{n=0}^{\infty} \frac{\varepsilon^n}{n!} \sum_{m=0}^{n} \binom{n}{m} D^m_{s^*} D^{n-m}_{s} p_j \end{aligned} \tag{22.53}$$

or

$$\begin{aligned} x_j &= q_j + \sum_{n=1}^{\infty} \frac{\varepsilon^n}{n!} \sum_{m=0}^{n} \binom{n}{m} D^m_{s^*} D^{n-m}_{s} q_j \\ y_j &= p_j + \sum_{n=1}^{\infty} \frac{\varepsilon^n}{n!} \sum_{m=0}^{n} \binom{n}{m} D^m_{s^*} D^{n-m}_{s} p_j \end{aligned} \tag{22.54}$$

From these equations, we can show that $-S(\xi, \eta)$ produces the inverse of the transformation produced by $S(\xi, \eta)$. To show this, put $S^*(q, p) = -S(q, p)$ in Eq. (22.54). By Theorem 4

$$D_{s^*} D_s - D_s D_{s^*} = D_{(s,s^*)}$$

Because $(S, -S) = 0$, we have $D_{(s,s^*)} = 0$, so that

$$D_{-s} D_s = D_s D_{-s}$$

In Eq. (22.54)

$$\sum_{m=0}^{n} \binom{n}{m} D^m_{s^*} D^{n-m}_{s} = (D_{-s} + D_s)^n$$

by the binomial theorem, since D_s and D_{-s} commute. However, $D_{-s} + D_s = 0$, so that the sums from $m = 0$ to n in Eqs. (22.54) vanish. Thus, $S^* = -S$ yields

$x_j = q_j$ and $y_j = p_j$. Changing the sign of S reverses the transformation, as was to be proved.

We may now put the compound transformation $x, y \to q, p$ of Eq. (22.52) into another form.

Theorem 10: For Lie series mapping, if $x, y \to \xi, \eta$ and $\xi, \eta \to q, p$, then

$$f(x,y) = f(q,p) + \varepsilon(f, S+S^*) + \frac{\varepsilon^2}{2}((f, S+S^*), S+S^*) + \frac{\varepsilon^2}{2}(f,(S,S^*))$$
$$+ \frac{\varepsilon^3}{6}((f,(S,S^*)), S+S^*), S+S^*) + \frac{\varepsilon^3}{6}((f, 2S+S^*), (S,S^*))$$
$$+ \frac{\varepsilon^3}{6}((f,(S,S^*)), S+2S^*) + \cdots \tag{22.55}$$

Here $S = S(\xi, \eta)$ and $S^* = S^*(q, p)$, but both are to be expressed in Eq. (22.55) as functions of q and p, according to Eqs. (22.54).

Proof: From Eq. (22.52), we have for the terms
$n = 0$:

$$f(q,p)$$

$n = 1$:

$$\varepsilon\left(D_{s^*}^0 D_s + D_{s^*} D_s^0\right) f(q,p) = \varepsilon(D_s + D_{s^*}) f = \varepsilon(f, S+S^*)$$

$n = 2$:

$$\frac{\varepsilon^2}{2}\left(D_s^2 + 2D_{s^*} D_s + D_{s^*}^2\right) f(q,p) = \frac{\varepsilon^2}{2} Q_2 f$$

Here, we have to do some noncommutative algebra. Put

$$D_s = \alpha \qquad D_{s^*} = \beta$$

Then

$$Q_2 = \alpha^2 + 2\beta\alpha + \beta^2$$

Now

$$(\alpha+\beta)^2 = \alpha(\alpha+\beta) + \beta(\alpha+\beta) = \alpha^2 + \alpha\beta + \beta\alpha + \beta^2$$

Then

$$\begin{aligned} Q_2 - (\alpha+\beta)^2 &= \alpha^2 + 2\beta\alpha + \beta^2 - \alpha^2 - \alpha\beta - \beta\alpha - \beta^2 \\ &= \beta\alpha - \alpha\beta \\ &= D_{s^*} D_s - D_s D_{s^*} \\ &= D_{(s,s^*)} \end{aligned}$$

by Theorem 4. Thus

$$Q_2 = (D_s + D_{s^*})^2 + D_{(s,s^*)}$$

The $n = 2$ term becomes

$$\frac{\varepsilon^2}{2}((f, S + S^*), S + S^*) + \frac{\varepsilon^2}{2}(f, (S, S^*))$$

For $n = 3$, the binomial coefficients are 1, 3, 3, 1, so that this term is

$$\frac{\varepsilon^3}{6}\left(D_s^3 + 3D_{s^*}D_s^2 + 3D_{s^*}^2 D_s + D_{s^*}^3\right) f = \frac{\varepsilon^3}{6}(\alpha^3 + 3\beta\alpha^2 + 3\beta^2\alpha + \beta^3) f = \frac{\varepsilon^3}{6} Q_3 f$$

Now

$$\begin{aligned}(\alpha + \beta)^3 &= \alpha(\alpha + \beta)^2 + \beta(\alpha + \beta)^2 \\ &= \alpha(\alpha^2 + \alpha\beta + \beta\alpha + \beta^2) + \beta(\alpha^2 + \alpha\beta + \beta\alpha + \beta^2) \\ &= \alpha^3 + \alpha^2\beta + \alpha\beta\alpha + \alpha\beta^2 + \beta\alpha^2 + \beta\alpha\beta + \beta^2\alpha + \beta^3\end{aligned}$$

Then

$$\begin{aligned}Q_3 - (\alpha + \beta)^3 &= 2\beta\alpha^2 + 2\beta^2\alpha - \alpha^2\beta - \alpha\beta^2 - \alpha\beta\alpha - \beta\alpha\beta \\ &= (\beta\alpha - \alpha\beta)(2\alpha + \beta) + (\alpha + 2\beta)(\beta\alpha - \alpha\beta)\end{aligned}$$

Thus

$$Q_3 = (D_s + D_{s^*})^3 + D_{(s,s^*)}(2D_s + D_{s^*}) + (D_s + 2D_{s^*})D_{(s,s^*)}$$

This gives for the $n = 3$ terms

$$\frac{\varepsilon^3}{6}((f, (S, S^*), S + S^*), S + S^*) + \frac{\varepsilon^3}{6}((f, 2S + S^*), (S, S^*))$$
$$+ \frac{\varepsilon^3}{6}((f, (S, S^*)), S + 2S^*)$$

This concludes the proof of Theorem 10.

References

[1]Hori, G., *Publications of the Astronomical Society of Japan,* Vol. 18, 1966, pp. 287–296.

[2]Brouwer, D., "Solution of Problem of Artificial Satellite Theory Without Drag," *Astronomical Journal,* Vol. 64, No. 9, 1959, p. 380, Eq. (10) and p. 385, Eqs. (33) and (34).

[3]Campbell, J. A., and Jefferys, W. H., *Celestial Mechanics,* Vol. 2, 1970, pp. 467–473.

Chapter 23

Perturbations by Lie Series

I. Introduction

IN THIS chapter, we shall use the method of Lie series to solve a perturbations problem defined by a time-independent Hamiltonian

$$F = F_0(x, y) + \sum_{k=1} F_k(x, y) \tag{23.1}$$

Here, F_k has a factor ε^k, ε being a small parameter, and the x's and y's form a canonical system

$$\frac{\mathrm{d}x_j}{\mathrm{d}t} = \frac{\partial F}{\partial y_j} \qquad \frac{\mathrm{d}y_j}{\mathrm{d}t} = -\frac{\partial F}{\partial x_j} \qquad j = 1, \ldots, N \tag{23.2}$$

We shall follow Hori[1] up to the point where he introduces artificial times. After that, we shall use the methods of Brouwer[2] to indicate the solution of the problem of an artificial satellite, when only zonal harmonics are considered. The results will go beyond that of Brouwer, but we shall show how they include Brouwer's results.

II. Lie Transformations

Since $F(x, y)$ is time independent, it is constant. Suppose we transform to new variables ξ and η by means of a Lie series with a generating function $S(\xi, \eta, \varepsilon)$. We obtain a new Hamiltonian $F^*(\xi, \eta)$ with (ξ, η) canonical with respect to it, so that

$$\frac{\mathrm{d}\xi_j}{\mathrm{d}t} = \frac{\partial F^*(\xi, \eta)}{\partial \eta_j} \qquad \frac{\mathrm{d}\eta_j}{\mathrm{d}t} = -\frac{\partial F^*(\xi, \eta)}{\partial \xi_j} \qquad j = 1, \ldots, N \tag{23.3}$$

Then

$$F(x, y) = F^*(\xi, \eta) = \text{const} \tag{23.4}$$

is an integral of the motion. We can write this as

$$\sum_{k=0} F_k(x, y) = \sum_{k=0} F_k^*(\xi, \eta) \tag{23.5}$$

where the subscript k means that the term contains ε^k as a factor.

A Lie series with $S(\xi, \eta, \varepsilon)$ as a generating function has the form

$$f(x, y) = \sum_{n=0}^{\infty} \frac{\varepsilon^n}{n!} D_s^n f(\xi, \eta) \tag{23.6}$$

Here

$$D_s^0 f = f \qquad D_s^1 f = (f, S) \qquad D_s^n f = D_s^{n-1}(f, S) \tag{23.7}$$

where the Poisson bracket

$$(f, S) = \sum_{i=1}^{N} \left(\frac{\partial f}{\partial \xi_i} \frac{\partial S}{\partial \eta_i} - \frac{\partial f}{\partial \eta_i} \frac{\partial S}{\partial \xi_i} \right) \tag{23.8}$$

It is convenient to define functions $S_k(\xi, \eta, \varepsilon)$ by means of

$$\varepsilon S = \Sigma_k S_k = S_1 + S_2 + S_3 + \cdots \tag{23.8a}$$

where each S_k has ε^k as a factor.

We next apply Eq. (23.6) to the left side of Eq. (23.5) and equate terms with equal powers of ε on either side. What happens to $\varepsilon^n D_s^n F_k(\xi, \eta)$? If $n = 0$, we obtain F_k. If $n = 1$, we obtain

$$\varepsilon D_s F_k = \varepsilon(F_k, S) = (F_k, \varepsilon S) = (F_k, S_1 + S_2 + S_3 + \cdots) \tag{23.9}$$

For general n

$$\begin{aligned} \varepsilon^n D_s^n F_k &= \varepsilon^n(\cdots(((F_k, S), S), S)\cdots) \\ &= (\cdots(((F_k, \varepsilon S), \varepsilon S), \varepsilon S)\cdots) \end{aligned} \tag{23.10}$$

an n-fold nested Poisson bracket. However, this is

$$\varepsilon^n D_s^n F_k = (\cdots(((F_k, S_1+S_2+S_3+\cdots), S_1+S_2+S_3+\cdots), S_1+S_2+S_3+\cdots)\cdots) \tag{23.10a}$$

also n fold. The left side of Eq. (23.5) becomes

$$\begin{aligned} &\Sigma_k F_k(\xi, \eta) + \Sigma_k(F_k, S_1 + S_2 + \cdots) \\ &\quad + \tfrac{1}{2}\Sigma_k((F_k, S_1 + S_2 + \cdots), S_1 + S_2 + \cdots) \\ &\quad + \tfrac{1}{6}\Sigma_k(((F_k, S_1 + S_2 + \cdots), S_1 + S_2 + \cdots), S_1 + S_2 + \cdots) \\ &\quad + \tfrac{1}{24}\Sigma_k(\text{quadruple nest}) + \cdots = \Sigma_m F_m^*(\xi, \eta) \end{aligned} \tag{23.11}$$

Thus, F_m^* is equal to the sum of all those terms on the left side of Eq. (23.11) for which the sum of k and the subscripts of the S's is equal to m. We obtain

$$F_0 = F_0^* \tag{23.12a}$$

$$F_1 + (F_0, S_1) = F_1^* \tag{23.12b}$$

$$F_2 + (F_0, S_2) + (F_1, S_1) + \tfrac{1}{2}((F_0, S_1), S_1) = F_2^* \tag{23.12c}$$

$$\begin{aligned} &F_3 + (F_0, S_3) + (F_1, S_2) + (F_2, S_1) + \tfrac{1}{2}((F_0, S_1), S_2) + \tfrac{1}{2}((F_0, S_2), S_1) \\ &\quad + \tfrac{1}{2}((F_1, S_1), S_1) + \tfrac{1}{6}(((F_0, S_1), S_1)S_1) = F_3^* \end{aligned} \tag{23.12d}$$

If we insert Eq. (23.12b) into Eq. (23.12c) and Eqs. (23.12b) and (23.12c) into

Eq. (23.12d), we can express these equations in sequential form

$$\varepsilon^0 \quad F_0 = F_0^* \tag{23.13a}$$

$$\varepsilon^1 \quad F_1 + (F_0, S_1) = F_1^* \tag{23.13b}$$

$$\varepsilon^2 \quad F_2 + (F_0, S_2) + \tfrac{1}{2}(F_1 + F_1^*, S_1) = F_2^* \tag{23.13c}$$

$$\begin{aligned} \varepsilon^3 \quad & F_3 + (F_0, S_3) + \tfrac{1}{2}(F_1 + F_1^*, S_2) \\ & + \tfrac{1}{2}(F_2 + F_2^*, S_1) + \tfrac{1}{12}((F_1 - F_1^*, S_1), S_1) = F_3^* \end{aligned} \tag{23.13d}$$

These equations are canonically invariant, so that any particular set of canonical variables can be used in them.

III. Application to Satellite Orbits

To apply the preceding to artificial satellites, we use Delaunay variables L, G, H, ℓ, g, h. According to Delaunay's choice, the Hamiltonian F is minus the energy; L, G, H are the x's; and ℓ, g, h are the y's in

$$\frac{\mathrm{d}x_k}{\mathrm{d}t} = \frac{\partial F}{\partial y_k} \qquad \frac{\mathrm{d}y_k}{\mathrm{d}t} = -\frac{\partial F}{\partial x_k} \qquad k = 1, \ldots, N \tag{23.14}$$

We shall treat only zonal harmonics as perturbations. Then $h = \Omega$ is absent from F. If

$$F_0 = \mu^2/2L^2 \tag{23.15a}$$

$$F_1 = -\frac{\mu r_e^2}{r^3} J_2 P_2(\sin\theta) \tag{23.15b}$$

we have

$$F = F_0 + F_1 + \text{higher zonals} \tag{23.15c}$$

Here, F_2 would appear only if we were to include higher zonal harmonics. Also, $\varepsilon = J_2$. Since the higher zonals are of order J_2^2 up to rather high zonals, we should have

$$F_2 = k_3\varepsilon^2 f_3 + k_4\varepsilon^2 f_4 + \cdots \tag{23.16}$$

Here, k_3 and k_4 are of order unity, and f_3 and f_4 come from expressions for the third and fourth zonal harmonics in the potential. With the notation we have used, the ξ's are then L', G', H', and the η's are ℓ', g', h'. Equations (23.14) are

$$\begin{aligned} \frac{\mathrm{d}L}{\mathrm{d}t} &= \frac{\partial F}{\partial \ell} & \frac{\mathrm{d}\ell}{\mathrm{d}t} &= -\frac{\partial F}{\partial L} \\ \frac{\mathrm{d}G}{\mathrm{d}t} &= \frac{\partial F}{\partial g} & \frac{\mathrm{d}g}{\mathrm{d}t} &= -\frac{\partial F}{\partial G} \\ \frac{\mathrm{d}H}{\mathrm{d}t} &= \frac{\partial F}{\partial h} & \frac{\mathrm{d}h}{\mathrm{d}t} &= -\frac{\partial F}{\partial H} \end{aligned} \tag{23.17}$$

Since h does not appear when zonals only appear in F, we have

$$H = \text{const} \tag{23.18}$$

After the Lie transformation $(x, y) \to (\xi, \eta)$, Eqs. (23.17) become

$$\begin{aligned} \frac{\mathrm{d}L'}{\mathrm{d}t} &= \frac{\partial F^*}{\partial \ell'} & \frac{\mathrm{d}\ell'}{\mathrm{d}t} &= -\frac{\partial F^*}{\partial L'} \\ \frac{\mathrm{d}G'}{\mathrm{d}t} &= \frac{\partial F^*}{\partial g'} & \frac{\mathrm{d}g'}{\mathrm{d}t} &= -\frac{\partial F^*}{\partial G'} \\ \frac{\mathrm{d}H'}{\mathrm{d}t} &= \frac{\partial F^*}{\partial h'} & \frac{\mathrm{d}h'}{\mathrm{d}t} &= -\frac{\partial F^*}{\partial H'} \end{aligned} \tag{23.19}$$

We anticipate the fact that the Lie transformation will make the F_k^* and the S_k independent of h', so that we shall have $H = H'$.

IV. Elimination of the Mean Anomaly

Consider

$$F_1 + (F_0, S_1) = F_1^* \tag{23.13b}$$

Split F_1 into two parts:

$$F_1 = \bar{F}_1 + F_{1p} \tag{23.20}$$

where

$$\bar{F}_1 = \frac{1}{2\pi}\int_0^{2\pi} F_1(L', G', H', \ell', g')\,\mathrm{d}\ell' \tag{23.21}$$

and

$$F_{1p} = F_1 - \bar{F}_1 \tag{23.22}$$

By Eqs. (23.13b), (23.20), and (23.22)

$$F_1 + (F_0, S_1) = F_1^* \tag{23.23}$$

To eliminate ℓ' from the new Hamiltonian, use Brouwer's procedure[2]:

$$F_1^* = \bar{F}_1 \tag{23.24}$$

$$(F_0, S_1) = -F_{1p} \tag{23.25}$$

Here, $\bar{F}_1$ and F_{1p} are given in Chapter 19 on the Brouwer theory. In the present case, they become

$$F_1^* = \bar{F}_1 = \frac{\mu r_e^2 J_2}{2a'^3}\frac{L'^3}{G'^3}\left(-\frac{1}{2} + \frac{3}{2}\frac{H'^2}{G'^2}\right) \tag{23.26}$$

$$F_{1p} = \frac{\mu r_e^2 J_2}{2a'^3}\left\{\left[-\frac{1}{2} + \frac{3}{2}\frac{H'^2}{G'^2}\right]\left[\frac{a'^3}{r'^3} - \frac{L'^3}{G'^3}\right] + \frac{3}{2}\left[1 - \frac{H'^2}{G'^2}\right]\frac{a'^3}{r'^3}\cos(2g' + 2f')\right\} \tag{23.27}$$

In contradistinction to Chapter 19, all quantities in Eq. (23.27) are primed.

Also, in Eq. (23.25)

$$(F_0, S_1) = \frac{\mathrm{d}F_0}{\mathrm{d}L'}\frac{\partial S_1}{\partial \ell'} = -F_{1p} \tag{23.28}$$

Then

$$S_1 = S_1(L', G', H', \ell', g') = -\left(\frac{\mathrm{d}F_0}{\mathrm{d}L'}\right)^{-1} \int F_{1p}\,\mathrm{d}\ell' \tag{23.29}$$

We have not had to introduce an artificial time as Hori did.[1]

To find F_2^* and S_2, we use Eq. (23.13c), omitting the F_2 if we choose not to include effects of zonal harmonics higher than the second. On resolving F_1 as before, we obtain

$$F_2^* = (F_0, S_2) + \tfrac{1}{2}(F_1 + F_1^*, S_1)_s + \tfrac{1}{2}(F_1 + F_1^*, S_1)_p \tag{23.30}$$

Here, the subscript s denotes an average over ℓ' and the subscript p the quantity minus this average value.

We eliminate ℓ' from F_2^* by choosing

$$F_2^* = \frac{1}{2}(F_1 + F_1^*, S_1)_s = \frac{1}{4\pi}\int_0^{2\pi} (F_1 + F_1^*, S_1)\,\mathrm{d}\ell' \tag{23.31}$$

For S_2 we obtain

$$(F_0, S_2) = -\tfrac{1}{2}(F_1 + F_1^*, S_1)_p \tag{23.32}$$

or

$$\frac{\mathrm{d}F_0}{\mathrm{d}L'}\frac{\partial S_2}{\partial \ell'} = -\frac{1}{2}(F_1 + F_1^*, S_1)_p \tag{23.33}$$

Then

$$S_2 = -\frac{1}{2}\left(\frac{\mathrm{d}F_0}{\mathrm{d}L'}\right)^{-1} \int (F_1 + F_1^*, S_1)_p\,\mathrm{d}\ell' \tag{23.34}$$

To find F_3^* and S_3, we use Eq. (23.13d). If we omit F_3, we have

$$F_3^* = (F_0, S_3) + M_{3s} + M_{3p} \tag{23.35}$$

where

$$M_3 = \tfrac{1}{2}(F_1 + F_1^*, S_2) + \tfrac{1}{2}(F_2 + F_2^*, S_1) + \tfrac{1}{12}((F_1 - F_1^*, S_1), S_1) \tag{23.36}$$

Then

$$M_{3s} = \frac{1}{2\pi}\int_0^{2\pi} M_3\,\mathrm{d}\ell' \tag{23.37a}$$

$$M_{3p} = M_3 - M_{3s} \tag{23.37b}$$

We choose

$$F_3^* = M_{3s} \tag{23.38a}$$

$$(F_0, S_3) = \frac{\mathrm{d}F_0}{\mathrm{d}L'}\frac{\partial S_3}{\partial \ell'} = -M_{3p} \tag{23.38b}$$

$$S_3 = -\frac{1}{2}\left(\frac{\mathrm{d}F_0}{\mathrm{d}L'}\right)^{-1} \int M_{3p}\,\mathrm{d}\ell' \tag{23.38c}$$

We could go on and find all the F_k^* and S_k in the same way. Like F_1^*, F_2^*, F_3^*, S_1, S_2, and S_3, they are all independent of ℓ' and h'. Since $F^* = \Sigma F_k^*$, we have

$$\frac{\mathrm{d}L'}{\mathrm{d}t} = \frac{\partial F^*}{\partial \ell'} = 0 \qquad L' = \text{const} \tag{23.39}$$

$$\frac{\mathrm{d}H'}{\mathrm{d}t} = \frac{\partial F^*}{\partial h'} = 0 \qquad H' = \text{const} \tag{23.40}$$

From the S_k, we find S from $\varepsilon S = \Sigma S_k$, where $\varepsilon = J_2$. From the Lie series of Chapter 22, with the (x, y) as unprimed Delaunay variables and the (ξ, η) as primed Delaunay variables,

$$f(x, y) = f(\xi, \eta) + \sum_{n=1}^{\infty} \frac{\varepsilon^n}{n!} D_s^n f(\xi, \eta) \tag{23.41}$$

Let us work this out for H.

$$H = H' + \sum_{n=1}^{\infty} \frac{\varepsilon^n}{n!} D_s^n H' \tag{23.42}$$

Here

$$D_s H' = (H', S) = \frac{\partial S}{\partial h'} = 0 \tag{23.43a}$$

$$D_s^n H' = 0 \qquad (n > 1) \tag{23.43b}$$

Thus

$$H = H' \tag{23.44}$$

Before going on to the next topic, let us compare results with Brouwer.[2]

V. Comparison with Brouwer's Theory

We have already shown that the Lie and Brouwer methods yield the same results through order J_2 in the splitting off of short periodic terms. Next, we shall show that they yield the same results through order J_2^2 for F_2^* and $L - L'$. The reader may also wish to show that $G - G'$, $\ell - \ell'$, $g - g'$, and $h - h'$, computed by either method, also agree to this order.

Use the subscripts L and B for Lie and Brouwer. From Chapter 19

$$S_{1\mathrm{B}} = -\left(\frac{\mathrm{d}F_0}{\mathrm{d}L'}\right)^{-1} \int F_{1p}\,\mathrm{d}\ell = \psi(L', G', H', \ell, g) \tag{23.45}$$

From this chapter

$$S_{1\mathrm{L}} = \psi(L', G', H', \ell', g') \tag{23.46}$$

the same expression with ℓ and g replaced by ℓ' and g'. Here $\psi(\ell, g)$ and $\psi(\ell', g')$ are both of order J_2, differing by a quantity of order J_2^2.

Comparison of F_2^* by Both Methods

Brouwer's

$$F_{2\mathrm{B}}^* = \bar{N} \tag{23.47}$$

where

$$N = \frac{1}{2}\frac{d^2 F_0}{\mathrm{d}L'^2}\left(\frac{\partial S_1}{\partial \ell}\right)^2 + \frac{\partial F_1}{\partial L'}\frac{\partial S_1}{\partial \ell} + \frac{\partial F_1}{\partial G'}\frac{\partial S_1}{\partial g} \tag{23.47a}$$

Here, $F_0 = \mu^2/(2L'^2)$, and F_1 and S_1 are functions of L', G', H', ℓ, and g. Through order J_2^2, N will be unchanged if we replace ℓ and g in F_1, S_1 and N by ℓ' and g'. Also, through order J_2^2,

$$\bar{N} = \frac{1}{2\pi}\int_0^{2\pi} N\,\mathrm{d}\ell = \frac{1}{2\pi}\int_0^{2\pi} N\,\mathrm{d}\ell' \tag{23.47b}$$

To find $F_{2\mathrm{L}}^*$, use Eq. (23.31):

$$F_{2\mathrm{L}}^* = \frac{1}{4\pi}\int_0^{2\pi} (F_1 + F_1^*, S_1)\,\mathrm{d}\ell' \tag{23.48}$$

where

$$F_1^* = F_1^*(L', G', H') \tag{23.49}$$

given by Eq. (23.26), and

$$F_1 = F_1(L', G', H', \ell', g') \tag{23.50}$$

given by Eqs. (23.20), (23.26), and (23.27). By Eq. (23.49)

$$(F_1^*, S_1) = \frac{\partial F_1^*}{\partial L'}\frac{\partial S_1}{\partial \ell'} + \frac{\partial F_1^*}{\partial G'}\frac{\partial S_1}{\partial g'} \tag{23.51}$$

By Eq. (23.50)

$$(F_1, S_1) = \frac{\partial F_1}{\partial L'}\frac{\partial S_1}{\partial \ell'} + \frac{\partial F_1}{\partial G'}\frac{\partial S_1}{\partial g'} - \frac{\partial F_1}{\partial \ell'}\frac{\partial S_1}{\partial L'} - \frac{\partial F_1}{\partial g'}\frac{\partial S_1}{\partial G'} \tag{23.52}$$

Using $F_1^* = F_1 - F_{1p}$, we find for half the sum of Eqs. (23.51) and (23.52)

$$\begin{aligned}\tfrac{1}{2}(F_1 + F_1^*, S_1) = {}& \frac{\partial F_1}{\partial L'}\frac{\partial S_1}{\partial \ell'} + \frac{\partial F_1}{\partial G'}\frac{\partial S_1}{\partial g'} - \frac{1}{2}\frac{\partial F_{1p}}{\partial L'}\frac{\partial S_1}{\partial \ell'} \\ & - \frac{1}{2}\frac{\partial F_{1p}}{\partial G'}\frac{\partial S_1}{\partial g'} - \frac{1}{2}\frac{\partial F_1}{\partial \ell'}\frac{\partial S_1}{\partial L'} - \frac{1}{2}\frac{\partial F_1}{\partial g'}\frac{\partial S_1}{\partial G'}\end{aligned} \tag{23.53}$$

By Eq. (23.28)

$$(F_0, S_1) = \frac{\mathrm{d}F_0}{\mathrm{d}L'}\frac{\partial S_1}{\partial \ell'} = -F_{1p}$$

so that

$$\frac{\mathrm{d}F_{1p}}{\mathrm{d}L'} = -\frac{d^2F_0}{\mathrm{d}L'^2}\frac{\partial S_1}{\partial \ell'} - \frac{\mathrm{d}F_0}{\mathrm{d}L'}\frac{\partial^2 S_1}{\partial L'\partial \ell'} \tag{23.54}$$

$$\frac{\mathrm{d}F_{1p}}{\mathrm{d}G'} = -\frac{\mathrm{d}F_0}{\mathrm{d}L'}\frac{\partial^2 S_1}{\partial G'\partial \ell'} \tag{23.55}$$

Insertion of Eqs. (23.54) and (23.55) into Eq. (23.53) gives

$$\tfrac{1}{2}(F_1 + F_1^*, S_1) = N' + Q \tag{23.56}$$

where N' is the same as N in Eq. (23.47a), except that ℓ is replaced by ℓ' and g by g'. Also

$$Q = \frac{1}{2}\frac{\mathrm{d}F_0}{\mathrm{d}L'}\frac{\partial^2 S_1}{\partial L'\partial \ell'}\frac{\partial S_1}{\partial \ell'} + \frac{1}{2}\frac{\mathrm{d}F_0}{\mathrm{d}L'}\frac{\partial^2 S_1}{\partial G'\partial \ell'}\frac{\partial S_1}{\partial g'} - \frac{1}{2}\frac{\partial F_1}{\partial \ell'}\frac{\partial S_1}{\partial L'} - \frac{1}{2}\frac{\partial F_1}{\partial g'}\frac{\partial S_1}{\partial G'} \tag{23.57}$$

Using

$$\frac{\mathrm{d}F_0}{\mathrm{d}L'}\frac{\partial S_1}{\partial \ell'} = -F_{1p} \tag{23.28}$$

we find

$$\frac{\partial F_1}{\partial \ell'} = \frac{\partial F_{1p}}{\partial \ell'} = -\frac{\mathrm{d}F_0}{\mathrm{d}L}\frac{\partial^2 S_1}{\partial \ell'^2} \tag{23.58}$$

$$\frac{\partial F_1}{\partial g'} = \frac{\partial F_{1p}}{\partial g'} = -\frac{\mathrm{d}F_0}{\mathrm{d}L'}\frac{\partial^2 S_1}{\partial g'\partial \ell'} \tag{23.59}$$

From Eqs. (23.57)–(23.59), we find

$$Q = \frac{1}{2}\frac{\mathrm{d}F_0}{\mathrm{d}L'}\left(\frac{\partial^2 S_1}{\partial L'\partial \ell'}\frac{\partial S_1}{\partial \ell'} + \frac{\partial^2 S_1}{\partial G'\partial \ell'}\frac{\partial S_1}{\partial g'} + \frac{\partial^2 S_1}{\partial \ell'^2}\frac{\partial S_1}{\partial L'} + \frac{\partial^2 S_1}{\partial g'\partial \ell'}\frac{\partial S_1}{\partial G'}\right) \tag{23.60}$$

or

$$Q = \frac{1}{2}\frac{\mathrm{d}F_0}{\mathrm{d}L'}\frac{\partial}{\partial \ell'}\left(\frac{\partial S_1}{\partial L'}\frac{\partial S_1}{\partial \ell'} + \frac{\partial S_1}{\partial G'}\frac{\partial S_1}{\partial g'}\right) \tag{23.61}$$

Now $\partial S_1/\partial \ell'$ is purely short periodic in ℓ', and S_1 is the sum of a short periodic term and a constant by Eq. (23.28). Thus

$$\bar{Q} = \frac{1}{4\pi}\frac{\mathrm{d}F_0}{\mathrm{d}L'}\left(\frac{\partial S_1}{\partial L'}\frac{\partial S_1}{\partial \ell'} + \frac{\partial S_1}{\partial G'}\frac{\partial S_1}{\partial g'}\right)_{\ell'=0}^{2\pi} = 0 \tag{23.62}$$

By Eqs. (23.48) and (23.56)

$$F_2^* = \bar{N}' + \bar{Q} = \bar{N}' \tag{23.63}$$

using Eq. (23.62). However,

$$F_{2\mathrm{B}}^* = \bar{N} \tag{23.47}$$

Here, $\bar{N} - N = O(j_2^3)$, so that

$$F_2^* = F_{2B}^* \tag{23.63a}$$

through order J_2^2. Through this order, the Lie series and Brouwer's method yield the same transformed Hamiltonian.

Comparison of $L - L'$ by Both Methods

By the Lie method

$$L - L' = \sum_{n=1}^{\infty} \frac{\varepsilon^n}{n!} D_s^n L' \tag{23.64}$$

Refer back to Eqs. (22.10) to see that

$$\frac{\varepsilon^n}{n!} D_s^n L' = \frac{1}{n!}(\cdots(((F_k, S_1 + S_2 + \cdots), S_1 + S_2 + \cdots), S_1 + S_2 + \cdots)\cdots) \tag{23.65}$$

an n-fold nest of Poisson brackets. Then

$$L - L' = \frac{\partial S_1}{\partial \ell'} + \frac{\partial S_2}{\partial \ell'} + \frac{1}{2}\left(\frac{\partial S_1}{\partial \ell'}, S_1\right) + O(\varepsilon^3) \tag{23.66}$$

where the S_1 and S_2 are those of the Lie method. Now, write out the indicated Poisson bracket in Eq. (23.66). Then

$$(L - L')_L = \frac{\partial S_1}{\partial \ell'} + \frac{\partial S_2}{\partial \ell'} + \frac{1}{2}\left[\frac{\partial^2 S_1}{\partial L' \partial \ell'}\frac{\partial S_1}{\partial \ell'} + \frac{\partial^2 S_1}{\partial G' \partial \ell'}\frac{\partial S_1}{\partial g'} + \frac{\partial^2 S_1}{\partial \ell'^2}\frac{\partial S_1}{\partial L'} + \frac{\partial^2 S_1}{\partial g' \partial \ell'}\frac{\partial S_1}{\partial G'}\right] + O\left(J_2^3\right) \tag{23.67}$$

Next, we must express $L - L'$ by the Brouwer method in terms of the same variables:

$$(L - L')_B = \frac{\partial S_{1B}}{\partial \ell} + \frac{\partial S_{2B}}{\partial \ell} + O\left(J_2^3\right) \tag{23.68}$$

If we write the Lie S_1 as

$$S_1 = \psi(L', G', H', \ell', g') \tag{23.69}$$

the Brouwer

$$S_{1B} = \psi(L', G', H', \ell, g) \tag{23.70}$$

For short

$$S_1 = \psi(\ell', g') \qquad S_{1B} = \psi(\ell, g) \tag{23.71}$$

Then

$$S_{1B} = \psi(\ell' + \ell - \ell', g' + g - g') \tag{23.72}$$

Expand this in Taylor's series and use

$$\ell - \ell' = -\frac{\partial S_{1B}}{\partial L'} + O\left(J_2^2\right) \qquad g - g' = -\frac{\partial S_{1B}}{\partial G'} + O\left(J_2^2\right) \tag{23.73}$$

Then

$$S_{1B} = \psi(\ell', g') - \frac{\partial \psi}{\partial \ell'}\frac{\partial S_{1B}}{\partial L'} - \frac{\partial \psi}{\partial g'}\frac{\partial S_{1B}}{\partial G'} + O\left(J_2^3\right) \tag{23.74}$$

To the same accuracy, this may be written

$$S_{1B} = S_1 - \frac{\partial S_1}{\partial \ell'}\frac{\partial S_1}{\partial L'} - \frac{\partial S_1}{\partial g'}\frac{\partial S_1}{\partial G'} + O\left(J_2^3\right) \tag{23.75}$$

Now, since $S_{1B} = \psi(\ell, g)$,

$$\frac{\partial S_{1B}}{\partial \ell} = \frac{\partial S_{1B}}{\partial \ell'}\frac{\partial \ell'}{\partial \ell} + \frac{\partial S_{1B}}{\partial g'}\frac{\partial g'}{\partial \ell} \tag{23.76}$$

By Eqs. (23.73) and (23.75)

$$\frac{\partial \ell'}{\partial \ell} = 1 + \frac{\partial^2 S_1}{\partial \ell' \partial L'} + O\left(J_2^2\right) \qquad \frac{\partial g'}{\partial \ell} = \frac{\partial^2 S_1}{\partial G' \partial \ell'} + O\left(J_2^2\right) \tag{23.77}$$

If we insert Eqs. (23.77) into Eq. (23.76) and use Eq. (23.75) for S_{1B}, we find through order J_2^2

$$\frac{\partial S_{1B}}{\partial \ell} = \frac{\partial S_1}{\partial \ell'} - \frac{\partial^2 S_1}{\partial \ell'^2}\frac{\partial S_1}{\partial L'} - \frac{\partial^2 S_1}{\partial g' \partial \ell'}\frac{\partial S_1}{\partial G'} + O\left(J_2^3\right) \tag{23.78}$$

In Eq. (23.68), we also need $\partial S_{2B}/\partial \ell$, which we have to compare with $\partial S_2/\partial \ell'$ of the Lie theory. First consider S_2. By Eq. (23.33)

$$\frac{\mathrm{d}F_0}{\mathrm{d}L'}\frac{\partial S_2}{\partial \ell'} = -\frac{1}{2}(F_1 + F_1^*, S_1)_p \tag{23.33}$$

By Eqs. (23.56) and (23.33)

$$\frac{\mathrm{d}F_0}{\mathrm{d}L'}\frac{\partial S_2}{\partial \ell'} = -N_p' - Q_p = -N_p' - Q \tag{23.79}$$

since $\bar{Q} = 0$.

Now consider S_{2B}. From Chapter 19

$$\frac{\mathrm{d}F_0}{\mathrm{d}L'}\frac{\partial S_{2B}}{\partial \ell} = -N_p \tag{23.80}$$

Through order J_2^2, we can write this as

$$\frac{\mathrm{d}F_0}{\mathrm{d}L'}\frac{\partial S_{2B}}{\partial \ell'} = -N_p' \tag{23.80a}$$

If

$$\Delta S_2 = S_2 - S_{2B} \tag{23.81}$$

it follows from Eqs. (23.79) and (23.80a) that

$$\frac{\mathrm{d}F_0}{\mathrm{d}L'}\frac{\partial \Delta S_2}{\partial \ell'} = -Q \tag{23.82}$$

By Eq. (23.61)

$$Q = \frac{1}{2}\frac{\mathrm{d}F_0}{\mathrm{d}L'}\frac{\partial}{\partial \ell'}\left(\frac{\partial S_1}{\partial L'}\frac{\partial S_1}{\partial \ell'} + \frac{\partial S_1}{\partial G'}\frac{\partial S_1}{\partial g'}\right) \tag{23.61}$$

and the second-order term in Eq. (23.68) is

$$\frac{\partial \Delta S_2}{\partial \ell'} = -\frac{1}{2}\frac{\partial}{\partial \ell'}\left(\frac{\partial S_1}{\partial L'}\frac{\partial S_1}{\partial \ell'} + \frac{\partial S_1}{\partial G'}\frac{\partial S_1}{\partial g'}\right) \tag{23.83}$$

Thus

$$\begin{aligned}\frac{\partial S_2}{\partial \ell'} &= \frac{\partial S_{2\mathrm{B}}}{\partial \ell'} - \frac{1}{2}\frac{\partial}{\partial \ell'}\left(\frac{\partial S_1}{\partial L'}\frac{\partial S_1}{\partial \ell'} + \frac{\partial S_1}{\partial G'}\frac{\partial S_1}{\partial g'}\right)\\ &= \frac{\partial S_{2\mathrm{B}}}{\partial \ell} - \frac{1}{2}\frac{\partial}{\partial \ell'}\left(\frac{\partial S_1}{\partial L'}\frac{\partial S_1}{\partial \ell'} + \frac{\partial S_1}{\partial G'}\frac{\partial S_1}{\partial g'}\right) + O\left(J_2^3\right)\end{aligned} \tag{23.84}$$

$$\frac{\partial S_{2\mathrm{B}}}{\partial \ell} = \frac{\partial S_2}{\partial \ell'} + \frac{1}{2}\frac{\partial}{\partial \ell'}\left(\frac{\partial S_1}{\partial L'}\frac{\partial S_1}{\partial \ell'} + \frac{\partial S_1}{\partial G'}\frac{\partial S_1}{\partial g'}\right) + O\left(J_2^3\right) \tag{23.85}$$

By Eqs. (23.78), (23.68), and (23.85), we find

$$\begin{aligned}(L - L')_{\mathrm{B}} = {}& \frac{\partial S_1}{\partial \ell'} - \frac{\partial^2 S_1}{\partial \ell'^2}\frac{\partial S_1}{\partial L'} - \frac{\partial^2 S_1}{\partial g'\partial \ell'}\frac{\partial S_1}{\partial G'}\\ &+ \frac{\partial S_2}{\partial \ell'} + \frac{1}{2}\frac{\partial}{\partial \ell'}\left(\frac{\partial S_1}{\partial L'}\frac{\partial S_1}{\partial \ell'} + \frac{\partial S_1}{\partial G'}\frac{\partial S_1}{\partial g'}\right) + O\left(J_2^3\right)\end{aligned}$$

$$\begin{aligned}(L - L')_{\mathrm{B}} = {}& \frac{\partial S_1}{\partial \ell'} + \frac{\partial S_2}{\partial \ell'} + \frac{1}{2}\left[\frac{\partial^2 S_1}{\partial L'\partial \ell'}\frac{\partial S_1}{\partial \ell'} + \frac{\partial^2 S_1}{\partial G'\partial \ell'}\frac{\partial S_1}{\partial g'}\right.\\ &\left. + \frac{\partial^2 S_1}{\partial \ell'^2}\frac{\partial S_1}{\partial L'} + \frac{\partial^2 S_1}{\partial g'\partial \ell'}\frac{\partial S_1}{\partial G'}\right] + O\left(J_2^3\right)\end{aligned} \tag{23.86}$$

which is the same as Eq. (23.67) for $(L - L')_L$. This is what we set out to prove.

VI. A Second Lie Transformation

For an artificial satellite, with zonal harmonics only in the potential, we now have as Hamiltonian

$$F^* = F_0^*(L') + F_1^*(L', G', H') + F_2^*(L', G', H', g') + F_3^*(L', G', H', g') \tag{23.87}$$

where $L' = L$ and $H' = H$. The variable ℓ' has been eliminated.

If we can find a Lie transformation that will make

$$F^{**} = F^{**}(L'', G'', H'') \tag{23.88}$$

the problem will be solved. If so, we shall know the differences between the singly and doubly primed variables, and L'', G'', H'' will be constants that will serve as mean orbital elements. Also, we shall have

$$\frac{d\ell''}{dt} = -\frac{\partial F^{**}}{\partial L''} \qquad \frac{dg''}{dt} = -\frac{\partial F^{**}}{\partial G''} \qquad \frac{dh''}{dt} = -\frac{\partial F^{**}}{\partial H''} \tag{23.89}$$

so that

$$\begin{aligned} \ell'' &= \ell''_0 - \frac{\partial F^{**}}{\partial L'} t \\ g'' &= g''_0 - \frac{\partial F^{**}}{\partial G'} t \\ h'' &= h''_0 - \frac{\partial F^{**}}{\partial H} t \end{aligned} \tag{23.90}$$

Then, ℓ''_0, g''_0, h''_0 will be the remaining mean elements to be determined by observation.

With

$$\begin{aligned} (\xi, \eta) &= (L', G', H', \ell', g', h') \\ (q, p) &= (L'', G'', H'', \ell'', g'', h'') \end{aligned} \tag{23.91}$$

we now perform a Lie series transformation from (ξ, η) to (q, p). Because F^* is time independent, if we use a generating function $S^*(q, p, \varepsilon)$ that is time independent, we shall obtain

$$F^* = \Sigma F_k^*(\xi, \eta) = \Sigma F_k^{**}(q, p) = F_k^{**}(q, p) = \text{const} \tag{23.92}$$

Apply

$$f(\xi, \eta) = \sum_{n=0}^{\infty} \frac{\varepsilon^n}{n!} D_{s^*}^n f(q, p) \tag{23.93}$$

to F^{**}. Then

$$\sum_{n=0}^{\infty} \frac{\varepsilon^n}{n!} \Sigma_k D_{s^*}^n F_k^*(q, p) = \Sigma_k F_k^{**}(q, p) = F_0^{**} + F_1^{**} + F_2^{**} + F_3^{**} + \cdots \tag{23.94}$$

Now apply Eq. (23.93) to $F_0^*(\xi, \eta) = F_0^*(L') = \mu^2/(2L'^2)$. It yields

$$F_0^*(L') = F_0^*(L'') + \sum_{n=0}^{\infty} \frac{\varepsilon^n}{n!} D_{s^*}^n F_0^*(L'') \tag{23.95}$$

Here

$$D_{s^*} F_0^*(L'') = (F_0^*, S^*) = \frac{dF_0^*}{dL''} \frac{\partial S^*}{\partial \ell''} \tag{23.96}$$

Because F^* depends only on L', G', H', and g', we need S^* to depend only on L'', G'', H'', and g''. Thus

$$S^* = S^*(L'', G'', H'', g'') \tag{23.97}$$

Then

$$D_{s^*} F_0^*(L'') = 0$$
$$D_{s^*}^n F_0^*(L'') = 0 \qquad (n > 1) \tag{23.98}$$

As a result, all the terms in Eq. (23.94) involving F_0^* disappear, except F_0^* itself.

It is also convenient at this point to show that $H' = H''$. To do so, apply Eq. (23.93) to H'. Then

$$H' = H'' + \sum_{n=1}^{\infty} \frac{\varepsilon^n}{n!} D_{s^*}^n H'' \tag{23.99}$$

but

$$D_{s^*} H'' = (H'', S^*) = \frac{\partial H''}{\partial H''} \frac{\partial S^*}{\partial h''} = 0 \tag{23.100a}$$

by Eq. (23.97). Then

$$D_{s^*}^n H'' = 0 \qquad (n > 1) \tag{23.100b}$$

Thus, by Eqs. (23.99)–(23.100b)

$$H' = H'' \tag{23.101}$$

We may now rewrite Eq. (23.94) as

$$F_0^* + F_1^* + F_2^* + F_3^* + \cdots + \varepsilon D_{s^*} F_1^* + \varepsilon D_{s^*} F_2^* + \cdots + \frac{\varepsilon^2}{2} D_{s^*}^2 F_1^*$$
$$= F_0^{**} + F_1^{**} + F_2^{**} + F_3^{**} + \cdots \tag{23.102}$$

Then

$$F_0^{**} = F_0^*(L'') = \mu^2/(2L''^2) \tag{23.103}$$

By Eq. (23.26), F_1^* depends only on L', G', and H'. Thus

$$F_1^{**} = F_1^*(L'', G'', H'') \tag{23.104}$$

independent of ℓ'', g'', and h''.

Now choose S_1^*, S_2^*, etc., so that

$$\varepsilon S^* = \Sigma_k S_k^* \tag{23.105}$$

where S_k^* has ε^k as a factor. We then find

$$F_2^* + F_3^* + \cdots + (F_1^*, \varepsilon S^*) + (F_2^*, \varepsilon S^*) + \cdots$$
$$+ \tfrac{1}{2}((F_1^*, \varepsilon S^*), \varepsilon S^*) = F_2^{**} + F_3^{**} + \cdots \tag{23.106}$$

This becomes

$$F_2^* + F_3^* + (F_1^*, S_1^*) + (F_1^*, S_2^*) + (F_2^*, S_1^*)$$
$$+ \tfrac{1}{2}((F_1^*, S_1^*), S_1^*) + O(\varepsilon^4) = F_2^{**} + F_3^{**} + \cdots \tag{23.107}$$

Thus

$$F_2^{**} = F_2^* + (F_1^*, S_1^*) \tag{23.108}$$

$$F_3^{**} = F_3^* + (F_1^*, S_2^*) + (F_2^*, S_1^*) + \tfrac{1}{2}((F_1^*, S_1^*), S_1^*) \tag{23.109}$$

or

$$F_3^{**} = F_3^* + (F_1^*, S_2^*) + \tfrac{1}{2}((F_2^* + F_2^{**}, S_1^*), S_1^*) \tag{23.110}$$

Also

$$F_1^{**} = F_1^*(L'', G'', H'') \tag{23.104}$$

Because we shall choose F^{**} to be $F^{**}(L'', G'', H'')$, independent of ℓ'', g'', and h'', it follows that L'', G'', H'' will all be constant. Thus, F_0^{**} and F_1^{**} will both be constants of the motion.

In Eq. (23.108), write

$$F_2^* = F_{2s}^* + F_{2p}^* \tag{23.111}$$

where the subscript s means an average over g''. That is

$$F_{2s}^* = \frac{1}{2\pi}\int_0^{2\pi} F_2^*(L'', G'', H'', g'')\,\mathrm{d}g'' \tag{23.112}$$

Then

$$F_{2p}^* = F_2^*(L'', G'', H'', g'') - F_{2s}^*(L'', G'', H'') \tag{23.113}$$

and by Eq. (23.108)

$$(F_1^*, S_1^*) + F_{2s}^* + F_{2p}^* = F_2^{**} \tag{23.114}$$

To eliminate g'' from the Hamiltonian, choose

$$F_2^{**} = F_{2s}^* \tag{23.115}$$

Then

$$(F_1^*, S_1^*) = -F_{2p}^* \tag{23.116}$$

Because $F_1^* = F_1^*(L'', G'', H'', g'')$, this becomes

$$\frac{\partial F_1^*}{\partial G''}\frac{\partial S_1^*}{\partial g''} = -F_{2p}^* \tag{23.117}$$

so that

$$S_1^* = -\left(\frac{\partial F_1^*}{\partial G''}\right)^{-1}\int F_{2p}^*(L'', G'', H'', g'')\,\mathrm{d}g'' \tag{23.118}$$

Here, we have been following Brouwer's procedure, so that the results must agree with Brouwer's for the first-order long periodic terms.

Next, consider Eq. (23.110). We can write it as

$$\begin{aligned} F_{3s}^* + F_{3p}^* + (F_1^*, S_2^*) + \tfrac{1}{2}((F_2^* + F_2^{**}, S_1^*), S_1^*)_s \\ + \tfrac{1}{2}((F_2^* + F_2^{**}, S_1^*), S_1^*)_p = F_3^{**} \end{aligned} \tag{23.119}$$

Now choose

$$F_3^{**} = F_{3s}^* + \tfrac{1}{2}((F_2^* + F_2^{**}, S_1^*), S_1^*)_s \tag{23.120}$$

Then

$$(F_1^*, S_2^*) = -F_{3p}^* - \tfrac{1}{2}((F_2^* + F_2^{**}, S_1^*), S_1^*)_p \tag{23.121}$$

so that

$$S_2^* = -\left(\frac{\partial F_1^*}{\partial G''}\right)^{-1} \int \left[F_{3p}^* - \frac{1}{2}((F_2^* + F_2^{**}, S_1^*), S_1^*)_p\right] \mathrm{d}g'' \tag{23.122}$$

If f is any of L, G, H, ℓ, g, h, then

$$f' = f'' + \sum_{n=1}^{\infty} \frac{\varepsilon^n}{n!} D_{s^*}^n f'' \tag{23.123}$$

or

$$\begin{aligned} f' - f'' &= (f'', S_1^* + S_2^* + \cdots) + \tfrac{1}{2}((f'', S_1^* + S_2^*), S_1^* + S_2^*) \\ &\quad + \tfrac{1}{6}(((f'', S_1^* + S_2^*), S_1^* + S_2^*), S_1^* + S_2^*) + \cdots \end{aligned} \tag{23.124}$$

This gives

$$L' = L'' \qquad H' = H''$$

as we have already seen. Knowing the doubly primed Delaunay variables, we can find the singly primed ones.

We can find the doubly primed variables by Eqs. (23.90), since we now have $F^{**} = \Sigma F_k^{**}$. Given the mean elements L', G'', H, ℓ_0'', g_0'', and h_0'', we can work back to L, G, H, ℓ, g, and h at a given time t, then to the Keplerian elements, and finally to the rectangular coordinates and velocities.

References

[1]Hori, G., *Publications of the Astronomical Society of Japan,* Vol. 18, 1966, pp. 287–296.

[2]Brouwer, D., "Solution of Problem of Artificial Satellite Theory Without Drag," *Astronomical Journal,* Vol. 64, No. 9, 1959, pp. 378–397.

Chapter 24

The General Three-Body Problem

I. Introduction

THE general problem of the motion of three particles, moving in response to their gravitational interactions, cannot be solved in closed form. However, there are certain integrals of motion that can be written down, and there are certain stationary solutions that we shall derive. If one of the particles has a mass that is negligible compared with the other masses, a good deal more can be said about the motion. This problem, the restricted three-body problem, will be treated in Chapter 25.

II. Formulation of the General Three-Body Problem

Let the three particles with masses m_1, m_2, and m_3 have position vectors $\boldsymbol{r}_1, \boldsymbol{r}_2, \boldsymbol{r}_3$ in an inertial system $Oxyz$ as shown in Fig. 24.1. Their separation vectors are

$$\boldsymbol{\rho}_{12} = \boldsymbol{r}_2 - \boldsymbol{r}_1 \tag{24.1a}$$

$$\boldsymbol{\rho}_{23} = \boldsymbol{r}_3 - \boldsymbol{r}_2 \tag{24.1b}$$

$$\boldsymbol{\rho}_{31} = \boldsymbol{r}_1 - \boldsymbol{r}_3 \tag{24.1c}$$

The equations of motion are

$$\ddot{\boldsymbol{r}}_1 = \frac{Gm_2}{\rho_{12}^3}\boldsymbol{\rho}_{12} - \frac{Gm_3}{\rho_{31}^3}\boldsymbol{\rho}_{31} \tag{24.2a}$$

$$\ddot{\boldsymbol{r}}_2 = \frac{Gm_3}{\rho_{23}^3}\boldsymbol{\rho}_{23} - \frac{Gm_1}{\rho_{12}^3}\boldsymbol{\rho}_{12} \tag{24.2b}$$

$$\ddot{\boldsymbol{r}}_3 = \frac{Gm_1}{\rho_{31}^3}\boldsymbol{\rho}_{31} - \frac{Gm_2}{\rho_{23}^3}\boldsymbol{\rho}_{23} \tag{24.2c}$$

where G is the gravitational constant.

III. Momentum Integrals

Multiply Eqs. (24.2), respectively, by m_1, m_2, and m_3, and add the results. We find

$$m_1\ddot{\boldsymbol{r}}_1 + m_2\ddot{\boldsymbol{r}}_2 + m_3\ddot{\boldsymbol{r}}_3 = \boldsymbol{0} \tag{24.3}$$

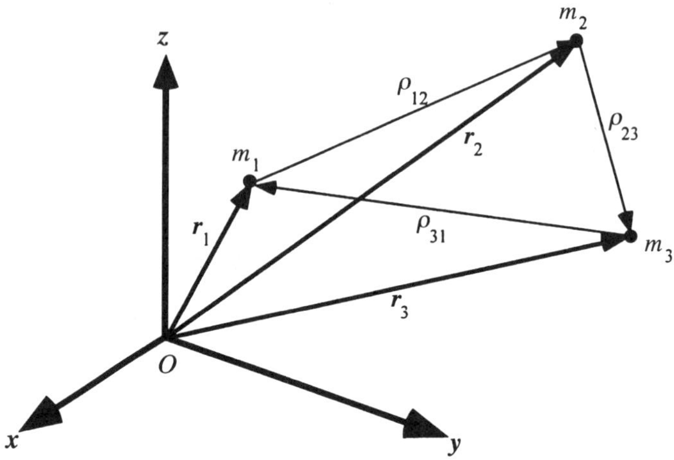

Fig. 24.1 Formulation of the general three-body problem.

Then

$$m_1\dot{\boldsymbol{r}}_1 + m_2\dot{\boldsymbol{r}}_2 + m_3\dot{\boldsymbol{r}}_3 = \mathbf{c}_1 \tag{24.4}$$

and

$$M\boldsymbol{R} \equiv m_1\boldsymbol{r}_1 + m_2\boldsymbol{r}_2 + m_3\boldsymbol{r}_3 = \boldsymbol{c}_1 t + \boldsymbol{c}_2 \tag{24.5}$$

where

$$M \equiv m_1 + m_2 + m_3 \tag{24.6}$$

and where $\boldsymbol{R}$ is the position vector of the center of mass of the three particles. Equations (24.4) and (24.5), when expressed in terms of rectangular components, yield six integrals.

IV. Angular Momentum

The total angular momentum is given by

$$\boldsymbol{L} = \Sigma_i \boldsymbol{r}_i \times (m_i \dot{\boldsymbol{r}}_i) \tag{24.7}$$

Then

$$\frac{\mathrm{d}\boldsymbol{L}}{\mathrm{d}t} = \Sigma_i \boldsymbol{r}_i \times (m_i \ddot{\boldsymbol{r}}_i) \tag{24.8}$$

From Eqs. (24.2) and (24.8), it follows that

$$G^{-1}\frac{\mathrm{d}\boldsymbol{L}}{\mathrm{d}t} = \frac{m_1 m_2}{\rho_{12}^3}(\boldsymbol{r}_1 - \boldsymbol{r}_2) \times \boldsymbol{\rho}_{12} + \frac{m_2 m_3}{\rho_{23}^3}(\boldsymbol{r}_2 - \boldsymbol{r}_3) \times \boldsymbol{\rho}_{23} + \frac{m_3 m_1}{\rho_{31}^3}(\boldsymbol{r}_3 - \boldsymbol{r}_1) \times \boldsymbol{\rho}_{31} \tag{24.9}$$

Then by Eqs. (24.9) and (24.1)

$$\frac{\mathrm{d}\boldsymbol{L}}{\mathrm{d}t} = \mathbf{0} \qquad \boldsymbol{L} = \text{const vector} \tag{24.10}$$

This vector equation yields three integrals.

V. Energy

In Eqs. (24.2), form the scalar product of each $\dot{\boldsymbol{r}}_i$ by $m_i\ddot{\boldsymbol{r}}_i$ and add the results. The sum of the left sides becomes

$$\Sigma_i m_i \dot{\boldsymbol{r}}_i \cdot \ddot{\boldsymbol{r}}_i = \frac{\mathrm{d}T}{\mathrm{d}t} \tag{24.11}$$

where the kinetic energy T is

$$T = \frac{1}{2}\Sigma_i m_i \dot{\boldsymbol{r}}_i^2 \tag{24.11a}$$

The sum R.S. of the right sides becomes

$$\text{R.S.} = G\left[\frac{m_1 m_2}{\rho_{12}^3}(\dot{\boldsymbol{r}}_1 - \dot{\boldsymbol{r}}_2)\cdot\boldsymbol{\rho}_{12} + \frac{m_2 m_3}{\rho_{23}^3}(\dot{\boldsymbol{r}}_2 - \dot{\boldsymbol{r}}_3)\cdot\boldsymbol{\rho}_{23} + \frac{m_3 m_1}{\rho_{31}^3}(\dot{\boldsymbol{r}}_3 - \dot{\boldsymbol{r}}_1)\cdot\boldsymbol{\rho}_{31}\right] \tag{24.12}$$

However,

$$\dot{\boldsymbol{r}}_1 - \dot{\boldsymbol{r}}_2 = -\dot{\boldsymbol{\rho}}_{12} \qquad \dot{\boldsymbol{r}}_2 - \dot{\boldsymbol{r}}_3 = -\dot{\boldsymbol{\rho}}_{23} \qquad \dot{\boldsymbol{r}}_3 - \dot{\boldsymbol{r}}_1 = -\dot{\boldsymbol{\rho}}_{31} \tag{24.13}$$

so that by Eqs. (24.1)

$$(\dot{\boldsymbol{r}}_1 - \dot{\boldsymbol{r}}_2)\cdot\boldsymbol{\rho}_{12} = -\boldsymbol{\rho}_{12}\cdot\dot{\boldsymbol{\rho}}_{12} = -\rho_{12}\dot{\rho}_{12} \tag{24.14a}$$

$$(\dot{\boldsymbol{r}}_2 - \dot{\boldsymbol{r}}_3)\cdot\boldsymbol{\rho}_{23} = -\boldsymbol{\rho}_{23}\cdot\dot{\boldsymbol{\rho}}_{23} = -\rho_{23}\dot{\rho}_{23} \tag{24.14b}$$

$$(\dot{\boldsymbol{r}}_3 - \dot{\boldsymbol{r}}_1)\cdot\boldsymbol{\rho}_{31} = -\boldsymbol{\rho}_{31}\cdot\dot{\boldsymbol{\rho}}_{31} = -\rho_{31}\dot{\rho}_{31} \tag{24.14c}$$

Then

$$\begin{aligned}\text{R.S.} &= -G\left[\frac{m_1 m_2}{\rho_{12}^3}\rho_{12}\dot{\rho}_{12} + \frac{m_2 m_3}{\rho_{23}^3}\rho_{23}\dot{\rho}_{23} + \frac{m_3 m_1}{\rho_{31}^3}\rho_{31}\dot{\rho}_{31}\right] \\ &= G\frac{d}{\mathrm{d}t}\left[\frac{m_1 m_2}{\rho_{12}} + \frac{m_2 m_3}{\rho_{23}} + \frac{m_3 m_1}{\rho_{31}}\right]\end{aligned} \tag{24.15}$$

or

$$\text{R.S.} = -\frac{\mathrm{d}V}{\mathrm{d}t} \tag{24.16}$$

where the potential energy

$$V = -G\left[\frac{m_1 m_2}{\rho_{12}} + \frac{m_2 m_3}{\rho_{23}} + \frac{m_3 m_1}{\rho_{31}}\right] \tag{24.17}$$

From Eqs. (24.11) and (24.16)

$$\frac{\mathrm{d}T}{\mathrm{d}t} + \frac{\mathrm{d}V}{\mathrm{d}t} = 0 \tag{24.18}$$

so that the total energy

$$T + V = \text{const} \tag{24.19}$$

This gives one more integral, so that we now have 10 integrals of the motion. There is another integral[1] obtained by "elimination of the node," but we shall not consider it in this text.

VI. Stationary Solutions

A stationary solution of Eqs. (24.2) is one in which each particle moves in a circle about the center of the mass, each with the same angular velocity n. We shall show that such solutions, which were discovered by Lagrange, do exist.

For such a solution, with origin at the center of mass,

$$\ddot{\boldsymbol{r}}_i = -n^2\boldsymbol{r}_i \qquad i = 1, 2, 3 \tag{24.20}$$

With origin at the center of mass

$$m_1\boldsymbol{r}_1 + m_2\boldsymbol{r}_2 + m_3\boldsymbol{r}_3 = \boldsymbol{0} \tag{24.21}$$

so that

$$m_2\boldsymbol{r}_2 = -m_1\boldsymbol{r}_1 - m_3\boldsymbol{r}_3 \tag{24.22}$$

Equation (24.2a) becomes

$$-\frac{n^2\boldsymbol{r}_1}{G} = \frac{m_2}{\rho_{12}^3}(\boldsymbol{r}_2 - \boldsymbol{r}_1) - \frac{m_3}{\rho_{31}^3}(\boldsymbol{r}_1 - \boldsymbol{r}_3) \tag{24.23}$$

From Eqs. (24.23) and (24.22)

$$-\frac{n^2\boldsymbol{r}_1}{G} = \boldsymbol{r}_1\left(-\frac{m_1}{\rho_{12}^3} - \frac{m_2}{\rho_{12}^3} - \frac{m_3}{\rho_{31}^3}\right) + \boldsymbol{r}_3\left(-\frac{m_3}{\rho_{12}^3} + \frac{m_3}{\rho_{31}^3}\right) \tag{24.24}$$

Apply $\boldsymbol{r}_1 \times$ to Eq. (24.24). The result is

$$m_3\boldsymbol{r}_1 \times \boldsymbol{r}_3\left(\frac{1}{\rho_{31}^3} - \frac{1}{\rho_{12}^3}\right) = \boldsymbol{0} \tag{24.25}$$

The assumed solution requires either that

$$\boldsymbol{r}_1 \times \boldsymbol{r}_3 = \boldsymbol{0} \tag{24.26a}$$

or

$$\rho_{12} = \rho_{31} \tag{24.26b}$$

Next, apply a similar procedure to Eq. (24.2b). In place of Eq. (24.23), we find an equation that can be obtained by the cyclic permutation $1 \to 2, 2 \to 3, 3 \to 1$. The result is

$$-\frac{n^2\boldsymbol{r}_2}{G} = \frac{m_3}{\rho_{23}^3}(\boldsymbol{r}_3 - \boldsymbol{r}_2) - \frac{m_1}{\rho_{12}^3}(\boldsymbol{r}_2 - \boldsymbol{r}_1) \tag{24.27}$$

From Eq. (24.21)

$$m_1\boldsymbol{r}_1 = -m_2\boldsymbol{r}_2 - m_3\boldsymbol{r}_3 \tag{24.28}$$

Insert Eq. (24.28) into Eq. (24.27). The result is

$$-\frac{n^2\boldsymbol{r}_2}{G} = \boldsymbol{r}_2\left(-\frac{m_1}{\rho_{12}^3} - \frac{m_3}{\rho_{23}^3} - \frac{m_2}{\rho_{12}^3}\right) + \boldsymbol{r}_3\left(\frac{m_3}{\rho_{23}^3} - \frac{m_3}{\rho_{12}^3}\right) \tag{24.29}$$

Apply $\boldsymbol{r}_2 \times$ to Eq. (24.29). The result is

$$m_3 \boldsymbol{r}_2 \times \boldsymbol{r}_3 \left(\frac{1}{\rho_{23}^3} - \frac{1}{\rho_{12}^3} \right) = \boldsymbol{0} \tag{24.30}$$

Then, either

$$\boldsymbol{r}_2 \times \boldsymbol{r}_3 = \boldsymbol{0} \tag{24.31a}$$

or

$$\rho_{23} = \rho_{12} \tag{24.31b}$$

To summarize Eqs. (24.26) and (24.31): The following conclusions are both necessary conditions for the stationary solution (24.20).

$$\boldsymbol{r}_1 \times \boldsymbol{r}_3 = \boldsymbol{0} \qquad \text{or} \qquad \rho_{12} = \rho_{31} \tag{24.32a}$$

$$\boldsymbol{r}_2 \times \boldsymbol{r}_3 = \boldsymbol{0} \qquad \text{or} \qquad \rho_{23} = \rho_{12} \tag{24.32b}$$

Now, for example, $\boldsymbol{r}_1$ is the vector from the center of mass to particle 1. If either of the vector products $\boldsymbol{r}_1 \times \boldsymbol{r}_3$ or $\boldsymbol{r}_2 \times \boldsymbol{r}_3$ vanishes, but not the other, then two of the particles lie on a straight line containing the center of mass, and the third lies off this straight line. This result is impossible, as the center of mass of all three particles is the center of mass, e.g., of m_2 and a particle of mass $m_1 + m_3$ at the center of mass of m_1 and m_3. Thus, both vector products must vanish or neither can vanish.

If neither vanishes, we must accept the alternative conditions in Eqs. (24.32a) and (24.32b), which lead to $\rho_{12} = \rho_{31} = \rho_{23}$, i.e., to an equilateral triangle solution. If both vanish, then all three particles are collinear.

VII. The Triangular Stationary Solution

If $\rho_{12} = \rho_{31} = \rho_{23} = \rho =$ const, there is a stationary solution for which

$$n^2 \rho^3 = G(m_1 + m_2 + m_3) \tag{24.33}$$

To show this, apply Eqs. (24.2), with the origin at the center of mass, so that

$$m_1 \boldsymbol{r}_1 + m_2 \boldsymbol{r}_2 + m_3 \boldsymbol{r}_3 = \boldsymbol{0} \tag{24.34}$$

and put

$$\rho_{12} = \rho_{31} = \rho_{23} = \rho \tag{24.35}$$

Equations (24.2) become

$$\frac{\rho^3}{G} \ddot{\boldsymbol{r}}_1 = m_2(\boldsymbol{r}_2 - \boldsymbol{r}_1) - m_3(\boldsymbol{r}_1 - \boldsymbol{r}_3) \tag{24.36a}$$

$$\frac{\rho^3}{G} \ddot{\boldsymbol{r}}_2 = m_3(\boldsymbol{r}_3 - \boldsymbol{r}_2) - m_1(\boldsymbol{r}_2 - \boldsymbol{r}_1) \tag{24.36b}$$

$$\frac{\rho^3}{G} \ddot{\boldsymbol{r}}_3 = m_1(\boldsymbol{r}_1 - \boldsymbol{r}_3) - m_2(\boldsymbol{r}_3 - \boldsymbol{r}_2) \tag{24.36c}$$

since

$$\boldsymbol{\rho}_{ij} = \boldsymbol{r}_j - \boldsymbol{r}_i \tag{24.37}$$

Apply Eq. (24.34) to the right sides of Eqs. (24.36). We find

$$\frac{\rho^3}{G}\ddot{\boldsymbol{r}}_1 = -(m_1 + m_2 + m_3)\boldsymbol{r}_1 \tag{24.38a}$$

$$\frac{\rho^3}{G}\ddot{\boldsymbol{r}}_2 = -(m_1 + m_2 + m_3)\boldsymbol{r}_2 \tag{24.38b}$$

$$\frac{\rho^3}{G}\ddot{\boldsymbol{r}}_3 = -(m_1 + m_2 + m_3)\boldsymbol{r}_3 \tag{24.38c}$$

Thus,

$$\ddot{\boldsymbol{r}}_i = -\frac{GM}{\rho^3}\boldsymbol{r}_i \qquad i = 1, 2, 3 \tag{24.39}$$

where

$$M \equiv m_1 + m_2 + m_3 \tag{24.40}$$

or

$$\ddot{\boldsymbol{r}}_i = -n^2 \boldsymbol{r}_i \tag{24.41}$$

which is the equation for a stationary solution, with

$$n^2\rho^3 = G(m_1 + m_2 + m_3) \tag{24.42}$$

If one of the masses vanishes, Eq. (24.42) becomes the usual equation of the two-body problem

$$n^2a^3 = G(m_1 + m_2) \tag{24.43}$$

corresponding to Kepler's third law.

VIII. The Collinear Stationary Solution

We saw from Eqs. (24.32) that the vanishing of both vector products $\boldsymbol{r}_1 \times \boldsymbol{r}_3$ and $\boldsymbol{r}_2 \times \boldsymbol{r}_3$ was one of the possible choices among necessary conditions for a stationary solution. Let us now assume this vanishing. Then $\boldsymbol{r}_1, \boldsymbol{r}_2$, and $\boldsymbol{r}_3$ are all parallel, with the result that the right sides of Eqs. (24.2) are all parallel to any of these three vectors. It is appropriate to place

$$\begin{aligned} \ddot{\boldsymbol{r}}_1 &= \lambda_1 \boldsymbol{r}_1 \\ \ddot{\boldsymbol{r}}_2 &= \lambda_2 \boldsymbol{r}_2 \\ \ddot{\boldsymbol{r}}_3 &= \lambda_3 \boldsymbol{r}_3 \end{aligned} \tag{24.44}$$

but we need more for a stationary solution. Indeed, we need

$$\lambda_1 = \lambda_2 = \lambda_3 = -n^2 \tag{24.45}$$

so that the three particles will stay on a line rotating with angular velocity n.

Let us now see if this is possible. For Eqs. (24.2), (24.44), and (24.45) to hold, we need to replace $\ddot{\boldsymbol{r}}_i$ by $-n^2\boldsymbol{r}_i$ in Eqs. (24.2). Do so and let $\boldsymbol{i}$ be a unit vector

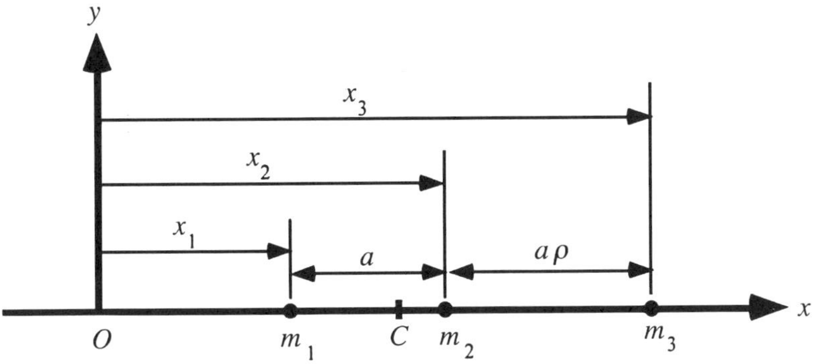

Fig. 24.2 Collinear stationary solution.

pointing along the line of collineation. Call

$$\boldsymbol{r}_k \cdot \boldsymbol{i} = x_k \qquad k = 1, 2, 3 \tag{24.46}$$

Then Eqs. (24.2) become

$$-n^2 x_1 = \frac{Gm_2}{\rho_{12}^3}(x_2 - x_1) - \frac{Gm_3}{\rho_{31}^3}(x_1 - x_3) \tag{24.47a}$$

$$-n^2 x_2 = \frac{Gm_3}{\rho_{23}^3}(x_3 - x_2) - \frac{Gm_1}{\rho_{12}^3}(x_2 - x_1) \tag{24.47b}$$

$$-n^2 x_3 = \frac{Gm_1}{\rho_{31}^3}(x_1 - x_3) - \frac{Gm_2}{\rho_{23}^3}(x_3 - x_2) \tag{24.47c}$$

Consider Fig. 24.2, where we choose the case where the center of mass C lies between m_1 and m_2. Placing

$$x_2 - x_1 = a = \rho_{12} \tag{24.48a}$$

$$x_3 - x_2 = a\rho = \rho_{23} \tag{24.48b}$$

$$x_3 - x_1 = a(1 + \rho) = \rho_{31} \tag{24.48c}$$

we find

$$-n^2 x_1 = \frac{Gm_2}{a^2} + \frac{Gm_3}{a^2(1+\rho)^2} \tag{24.49a}$$

$$-n^2 x_2 = \frac{Gm_3}{a^2\rho^2} - \frac{Gm_1}{a^2} \tag{24.49b}$$

It is not necessary to use Eq. (24.47c). We may use instead the equation for the center of mass

$$m_1 x_1 + m_2(a + x_1) + m_3(a + a\rho + x_1) = 0 \tag{24.50}$$

Solve Eq. (24.50) for x_1; we have

$$(m_1 + m_2 + m_3)x_1 = -m_2 a - m_3 a(1 + \rho) \tag{24.51}$$

With

$$M \equiv m_1 + m_2 + m_3 \tag{24.52}$$

$$x_1 = -\frac{a}{M}[m_2 + m_3(1+\rho)] \tag{24.53}$$

then

$$x_2 = a + x_1 = \frac{a}{M}[m_1 - m_3\rho] \tag{24.54}$$

Insert Eq. (24.54) into Eq. (24.49b) and solve for n^2:

$$n^2 = \frac{GM}{a^3}\frac{m_1 - m_3/\rho^2}{m_1 - m_3\rho} \tag{24.55}$$

If $m_3 = 0$, this yields the usual two-body equation.

From Eqs. (24.53) and (24.55)

$$n^2 x_1 = -\frac{G}{a^2}\frac{m_1 - m_3/\rho^2}{m_1 - m_3\rho}[m_2 + m_3(1+\rho)] \tag{24.56}$$

but by Eq. (24.49a)

$$n^2 x_1 = -\frac{Gm_2}{a^2} - \frac{Gm_3}{a^2(1+\rho)^2} \tag{24.57}$$

On equating Eqs. (24.56) and (24.57), we obtain an equation for ρ:

$$\frac{m_1 - m_3/\rho^2}{m_1 - m_3\rho}[m_2 + m_3(1+\rho)] = m_2 + \frac{m_3}{(1+\rho)^2} \tag{24.58}$$

which reduces to an identity if $m_3 = 0$. Equation (24.58) reduces to a quintic equation for ρ:

$$m_3 F(\rho) = 0 \tag{24.59}$$

where

$$\begin{aligned} F(\rho) \equiv{}& (m_1 + m_2)\rho^5 + (3m_1 + 2m_2)\rho^4 + (3m_1 + m_2)\rho^3 \\ &- (m_2 + 3m_3)\rho^2 - (2m_2 + 3m_3)\rho - (m_2 + m_3) \end{aligned} \tag{24.60}$$

Then, either $m_3 = 0$ or

$$F(\rho) = 0 \tag{24.61}$$

Here, $F(\rho)$ has only one change of sign for positive ρ, so that by Descartes' rule of signs there cannot be more than one positive root. There is one root, because $F(0) < 0$ and $F(\infty) = +\infty$. By renumbering the particles, we can find two other collinear solutions in the stationary case.

Reference

[1]Whittaker, E. T., *A Treatise on Analytical Dynamics,* 4th ed., Dover, New York, 1944, p. 341.

Chapter 25

The Restricted Three-Body Problem

I. Introduction

LET one of the masses, M_3, be very small compared with the other two, M_1 and M_2. This would be true if M_1 and M_2 are the masses of the sun and Jupiter and M_3 that of a Trojan asteroid or if M_1 and M_2 are the masses of the Earth and the moon and M_3 that of a lunar vehicle or an Earth-moon space station.

Label the masses, so that $M_1 > M_2$ and let

$$\frac{M_2}{M_1 + M_2} \equiv m \qquad \frac{M_1}{M_1 + M_2} = 1 - m \tag{25.1}$$

Then $m < 1/2$. The masses M_1 and M_2 are called the primaries. In the bounded case, each moves in an ellipse about the other or about their center of mass C. We shall consider only the "circular restricted" problem where the primary orbits are circles. Either moves in a circle about the other or about C.

Denote X, Y, Z as the rectangular coordinates of a rotating coordinate system such that M_1 is at $(X_1, 0, 0)$, M_2 at $(X_2, 0, 0)$, and C at $(0, 0, 0)$, with C being at rest with respect to an inertial system (Fig. 25.1). The angular velocity in the circle is the mean motion n, and the separation distance between M_1 and M_2 is M_1M_2. Thus

$$n^2a^3 = G(M_1 + M_2) \qquad a = X_2 - X_1 \tag{25.2}$$

Let $\boldsymbol{R}$ be the position vector of M_3, and let $\boldsymbol{V}$ and $\boldsymbol{A}$ be its velocity and acceleration relative to the rotating system. Then $M_3\boldsymbol{A}$ is the sum of two gravitational forces, a Coriolis force, and a centrifugal force. Thus

$$M_3\boldsymbol{A} = -\frac{GM_1M_3}{R_1^3}\boldsymbol{R}_1 - \frac{GM_2M_3}{R_2^3}\boldsymbol{R}_2 - 2M_3n\boldsymbol{k} \times \boldsymbol{V} - M_3n^2\boldsymbol{k} \times (\boldsymbol{k} \times \boldsymbol{R}) \tag{25.3}$$

Here, $\boldsymbol{R}_1$, $\boldsymbol{R}_2$, and $\boldsymbol{R}$ are the position vectors of M_3 relative to M_1, M_2, and C, and $\boldsymbol{k}$ is a unit vector along CZ, so that the vector angular velocity $\mathbf{n} = n\boldsymbol{k}$.

By Eqs. (25.1)

$$GM_1 = G(M_1 + M_2)(1 - m) \qquad GM_2 = G(M_1 + M_2)m \tag{25.4}$$

Let $\boldsymbol{i}$ and $\boldsymbol{j}$ be unit vectors along CX and CY, and denote a time derivative by a prime. Then

$$\begin{aligned} \boldsymbol{R} &= X\boldsymbol{i} + Y\boldsymbol{j} + Z\boldsymbol{k} \\ \boldsymbol{V} &= X'\boldsymbol{i} + Y'\boldsymbol{j} + Z'\boldsymbol{k} \\ \boldsymbol{A} &= X''\boldsymbol{i} + Y''\boldsymbol{j} + Z''\boldsymbol{k} \end{aligned} \tag{25.5}$$

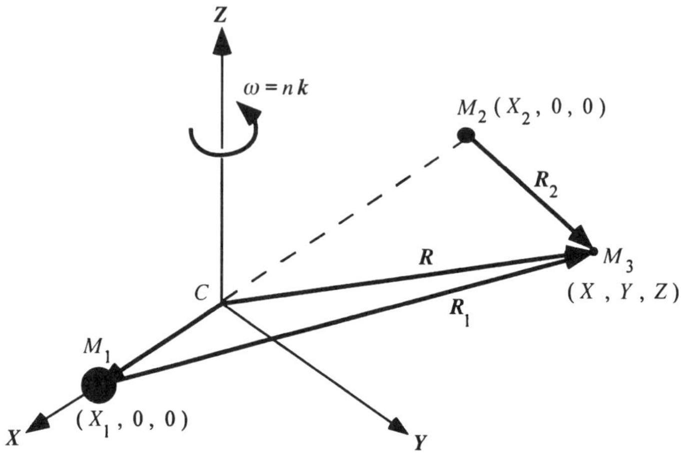

Fig. 25.1 Restricted three-body problem.

$$\begin{aligned} \boldsymbol{R}_1 &= (X - X_1)\boldsymbol{i} + Y\boldsymbol{j} + Z\boldsymbol{k} \\ \boldsymbol{R}_2 &= (X - X_2)\boldsymbol{i} + Y\boldsymbol{j} + Z\boldsymbol{k} \end{aligned} \tag{25.6}$$

$$\boldsymbol{k} \times \boldsymbol{V} = \boldsymbol{k} \times (X'\boldsymbol{i} + Y'\boldsymbol{j} + Z'\boldsymbol{k}) = -Y'\boldsymbol{i} + X'\boldsymbol{j} \tag{25.7}$$

$$\boldsymbol{k} \times (\boldsymbol{k} \times \boldsymbol{R}) = \boldsymbol{k}(\boldsymbol{k} \cdot \boldsymbol{R}) - \boldsymbol{R} = -X\boldsymbol{i} - Y\boldsymbol{j} \tag{25.8}$$

Inserting Eqs. (25.5)–(25.7) into Eq. (25.3) and canceling the M_3, we find

$$\begin{bmatrix} X'' \\ Y'' \\ Z'' \end{bmatrix} = -\frac{GM_1}{R_1^3}\begin{bmatrix} (X - X_1) \\ Y \\ Z \end{bmatrix} - \frac{GM_2}{R_2^3}\begin{bmatrix} (X - X_2) \\ Y \\ Z \end{bmatrix} - 2n\begin{bmatrix} -Y' \\ X' \\ 0 \end{bmatrix} - n^2\begin{bmatrix} -X \\ -Y \\ 0 \end{bmatrix} \tag{25.9}$$

so that

$$\begin{bmatrix} X'' - 2nY' \\ Y'' + 2nX' \\ Z'' \end{bmatrix} = \begin{bmatrix} n^2X - \dfrac{GM_1}{R_1^3}(X - X_1) - \dfrac{GM_2}{R_2^3}(X - X_2) \\ n^2Y - \dfrac{GM_1}{R_1^3}Y - \dfrac{GM_2}{R_2^3}Y \\ -\dfrac{GM_1}{R_1^3}Z - \dfrac{GM_2}{R_2^3}Z \end{bmatrix} \tag{25.10}$$

With $X_2 - X_1 = a$, it is now convenient to use a as the unit of length and $1/n$ as the unit of time. This involves putting $x = X/a, y = Y/a, z = Z/a, R_1 = a\rho_1, R_2 = a\rho_2$, and $\tau = nt$. The length and time units, which depend only on a

and n, become arbitrary by this normalization. Also use Eqs. (25.4) and denote $d/d\tau$ by a superscript dot. Equations (25.10) become

$$\begin{bmatrix} \ddot{x} - 2\dot{y} \\ \ddot{y} + 2\dot{x} \\ \ddot{z} \end{bmatrix} = \begin{bmatrix} x - \dfrac{(1-m)}{\rho_1^3}(x - x_1) - \dfrac{m}{\rho_2^3}(x - x_2) \\ y - \dfrac{(1-m)}{\rho_1^3}y - \dfrac{m}{\rho^3}y \\ -\dfrac{(1-m)}{\rho_1^3}z - \dfrac{m}{\rho_2^3}z \end{bmatrix} \tag{25.11}$$

Next multiply Eqs. (25.11), respectively, by $\dot{x}$, $\dot{y}$, and $\dot{z}$, and form the resulting sum. The result is

$$\dot{x}\ddot{x} + \dot{y}\ddot{y} + \dot{z}\ddot{z} = x\dot{x} + y\dot{y} - \frac{(1-m)}{\rho_1^3}[(x - x_1)\dot{x} + y\dot{y} + z\dot{z}]$$
$$- \frac{m}{\rho_2^3}[(x - x_2)\dot{x} + y\dot{y} + z\dot{z}] \tag{25.12}$$

or

$$\frac{d}{d\tau}\left[\frac{1}{2}(\dot{x}^2 + \dot{y}^2 + \dot{z}^2)\right] = \frac{d}{d\tau}\left[\frac{1}{2}(x^2 + y^2)\right] - \frac{(1-m)}{\rho_1^3}[(x - x_1)\dot{x} + y\dot{y} + z\dot{z}]$$
$$- \frac{m}{\rho_2^3}[(x - x_2)\dot{x} + y\dot{y} + z\dot{z}] \tag{25.13}$$

If we write

$$\boldsymbol{\rho}_1 = (x - x_1)\boldsymbol{i} + y\boldsymbol{j} + z\boldsymbol{k} \tag{25.14a}$$

$$\boldsymbol{\rho}_2 = (x - x_2)\boldsymbol{i} + y\boldsymbol{j} + z\boldsymbol{k} \tag{25.14b}$$

we obtain

$$\boldsymbol{\rho}_1 \cdot \dot{\boldsymbol{\rho}}_1 = (x - x_1)\dot{x} + y\dot{y} + z\dot{z} = \rho_1\dot{\rho}_1$$

$$\boldsymbol{\rho}_2 \cdot \dot{\boldsymbol{\rho}}_2 = (x - x_2)\dot{x} + y\dot{y} + z\dot{z} = \rho_2\dot{\rho}_2$$

$$-\frac{[(x - x_1)\dot{x} + y\dot{y} + z\dot{z}]}{\rho_1^3} = -\frac{\dot{\rho}_1}{\rho_1^2} = \frac{d}{d\tau}\left(\frac{1}{\rho_1}\right)$$

$$-\frac{[(x - x_2)\dot{x} + y\dot{y} + z\dot{z}]}{\rho_2^3} = -\frac{\dot{\rho}_2}{\rho_2^2} = \frac{d}{d\tau}\left(\frac{1}{\rho_2}\right)$$

Then

$$\frac{1}{2}\frac{d}{d\tau}[(\dot{x}^2 + \dot{y}^2 + \dot{z}^2)] = \frac{1}{2}\frac{d}{d\tau}[(x^2 + y^2)] + \frac{d}{d\tau}\left(\frac{1-m}{\rho_1}\right) + \frac{d}{d\tau}\left(\frac{m}{\rho_2}\right) \tag{25.15}$$

or

$$\frac{1}{2}[(\dot{x}^2 + \dot{y}^2 + \dot{z}^2)] - \frac{1}{2}[(x^2 + y^2)] - \frac{1-m}{\rho_1} - \frac{m}{\rho_2} = -2C = \text{const} \tag{25.16}$$

Here C is called the Jacobi constant, and Eq. (25.16) is called the Jacobi integral.

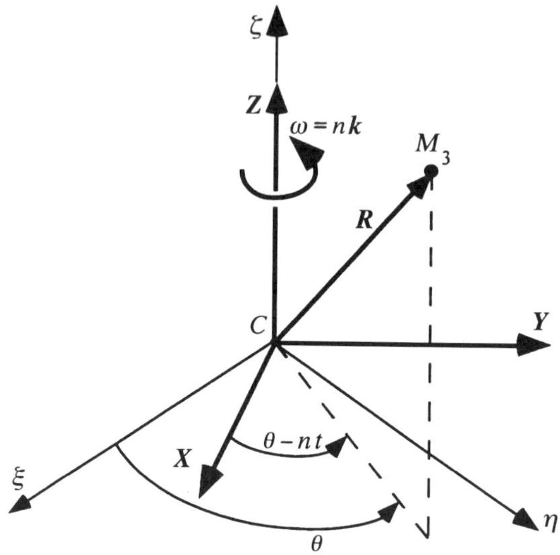

Fig. 25.2 Inertial coordinate system in a rotating coordinate system, with $t = 0$.

It is instructive to derive the Jacobi integral (25.16) by a Hamiltonian method. To do so, we first rewrite Eq. (25.16) in terms of the original variables. The form is

$$\frac{1}{2}\left[\left(\frac{\mathrm{d}X}{\mathrm{d}t}\right)^2 + \left(\frac{\mathrm{d}Y}{\mathrm{d}t}\right)^2 + \left(\frac{\mathrm{d}Z}{\mathrm{d}t}\right)^2\right] - \frac{n^2}{2}(X^2+Y^2) - \frac{GM_1}{R_1} - \frac{GM_2}{R_2} = -2Cn^2a^2 \tag{25.17}$$

To do this derivation, we construct the Hamiltonian in the rotating system and show that it does not depend explicitly on the time. Let ξ, η, ζ be an inertial system of rectangular coordinates, with which $CXYZ$ coincides at time $t = 0$ as shown in Fig. 25.2. The preceding rotating system rotates about CZ with angular velocity n. Then

$$Z = \zeta \tag{25.18}$$

If the projection of $\boldsymbol{R}$ on the plane $\zeta = 0$ makes an angle θ with $C\xi$, it makes an angle $\theta - nt$ with CX. As a complex number, this projection can be written as $\xi + i\eta$ in the inertial system or as $X + iY$ in the rotating system.

If

$$|\xi + i\eta| = |X + iY| = r_p \tag{25.19a}$$

then

$$\xi + i\eta = r_p \varepsilon^{i\theta} \tag{25.19b}$$

$$X + iY = r_p \varepsilon^{i(\theta - nt)} \tag{25.19c}$$

so that

$$\xi + i\eta = (X + iY)\varepsilon^{int} \tag{25.20}$$

With use of a prime for d/dt,

$$\xi' + i\eta' = (X' + iY')\varepsilon^{int} + in(X + iY)\varepsilon^{int}$$

$$\xi' - i\eta' = (X' - iY')\varepsilon^{-int} - in(X - iY)\varepsilon^{-int}$$

$$\xi'^2 + \eta'^2 = X'^2 + Y'^2 + n^2(X^2 + Y^2) + 2n(XY' - YX') \qquad (25.21)$$

The kinetic energy per unit mass

$$T = \frac{1}{2}(\xi'^2 + \eta'^2 + \zeta'^2) = \frac{1}{2}(X'^2 + Y'^2 + Z'^2) + \frac{n^2}{2}(X^2 + Y^2) + n(XY' - YX') \qquad (25.22)$$

Then

$$p_1 = \frac{\partial T}{\partial X'} = X' - nY$$

$$p_2 = \frac{\partial T}{\partial Y'} = Y' + nX$$

$$p_3 = \frac{\partial T}{\partial Y'} = Z'$$

$$\Sigma p\dot{q} = p_1 X' + p_2 Y' + p_3 Z' = X'^2 + Y'^2 + Z'^2 + n(XY' - YX') \qquad (25.23)$$

Lagrangian:

$$L = T - V = \frac{1}{2}(X'^2 + Y'^2 + Z'^2) + \frac{n^2}{2}(X^2 + Y^2) + n(XY' - YX') - V$$

Hamiltonian:

$$H = \Sigma p\dot{q} - L = \frac{1}{2}(X'^2 + Y'^2 + Z'^2) - \frac{n^2}{2}(X^2 + Y^2) + V \qquad (25.24)$$

or

$$H = \frac{1}{2}\left[(p_1 + nY)^2 + (p_2 - nX)^2 + p_3^2\right] - \frac{n^2}{2}(X^2 + Y^2) + V$$

a constant because it does not depend explicitly on the time. However

$$V = -\frac{GM_1}{R_1} - \frac{GM_2}{R_2}$$

and the term containing the p's is simply $(X'^2 + Y'^2 + Z'^2)/2$. Thus

$$\frac{1}{2}(X'^2 + Y'^2 + Z'^2) - \frac{n^2}{2}(X^2 + Y^2) - \frac{GM_1}{R_1} - \frac{GM_2}{R_2} = \text{const} \qquad (25.25)$$

which is the same as Eq. (25.17). This completes the Hamiltonian derivation of the Jacobi integral.

It is also of interest to show that this Jacobi integral is equal to the energy minus the product of the angular velocity and the z component of the angular momentum.

By Eq. (25.22), the energy is

$$W = \frac{1}{2}(X'^2 + Y'^2 + Z'^2) + \frac{n^2}{2}(X^2 + Y^2) + n(XY' - YX') + V \tag{25.26}$$

We first show that the z component of angular momentum, viz.,

$$L_z = \xi\eta' - \eta\xi' = XY' - YX' + n(X^2 + Y^2) \tag{25.27}$$

To do so, use

$$\begin{aligned} \dot{\boldsymbol{r}} &= \boldsymbol{V} + \omega \times \boldsymbol{r} = X'\boldsymbol{i} + Y'\boldsymbol{j} + Z'\boldsymbol{k} + n\boldsymbol{k} \times (X\boldsymbol{i} + Y\boldsymbol{j} + Z\boldsymbol{k}) \\ &= \boldsymbol{V} + n(-Y\boldsymbol{i} + X\boldsymbol{j}) \end{aligned}$$

$$\boldsymbol{L} = \boldsymbol{r} \times \dot{\boldsymbol{r}} = \boldsymbol{r} \times \boldsymbol{V} + n(X\boldsymbol{i} + Y\boldsymbol{j} + Z\boldsymbol{k}) \times (-Y\boldsymbol{i} + X\boldsymbol{j})$$

On taking the z component of each side, we obtain Eq. (25.27). Then

$$n(\xi\eta' - \eta\xi') = n(XY' - YX') + n^2(X^2 + Y^2) \tag{25.27a}$$

On subtracting Eq. (25.27a) from Eq. (25.26), we obtain the Jacobi integral (25.25).

II. Zero-Velocity Curves

Examination of Eq. (25.16) suggests introducing the function

$$U(x, y, z) = \frac{1}{2}[(x^2 + y^2)] + \frac{1 - m}{\rho_1} + \frac{m}{\rho_2} \tag{25.28}$$

By Eq. (25.14)

$$\rho_1^2 = (x - x_1)^2 + y^2 + z^2$$

$$\rho_2^2 = (x - x_2)^2 + y^2 + z^2$$

so that

$$\frac{\partial U}{\partial x} = x - \frac{(1 - m)}{\rho_1^3}(x - x_1) - \frac{m}{\rho_2^3}(x - x_2) \tag{25.29a}$$

$$\frac{\partial U}{\partial y} = y - \frac{(1 - m)}{\rho_1^3}y - \frac{m}{\rho_2^3}y \tag{25.29b}$$

$$\frac{\partial U}{\partial z} = -\frac{(1 - m)}{\rho_1^3}z - \frac{m}{\rho_2^3}z \tag{25.29c}$$

Equations (25.11) then become

$$\begin{bmatrix} \ddot{x} - 2\dot{y} \\ \ddot{y} + 2\dot{x} \\ \ddot{z} \end{bmatrix} = \begin{bmatrix} \dfrac{\partial U}{\partial x} \\ \dfrac{\partial U}{\partial y} \\ \dfrac{\partial U}{\partial z} \end{bmatrix} \tag{25.30}$$

On multiplying Eq. (25.30), respectively, by $\dot{x}$, $\dot{y}$, $\dot{z}$ and adding the results, we find Eq. (25.16) again. If we put

$$v^2 = \dot{x}^2 + \dot{y}^2 + \dot{z}^2 \tag{25.31}$$

we obtain

$$\frac{1}{2}\frac{\mathrm{d}v^2}{\mathrm{d}\tau} - \frac{\mathrm{d}U}{\mathrm{d}\tau} = 0 \tag{25.32}$$

so that

$$v^2 = 2U - C \tag{25.33}$$

the same as Eq. (25.16). For $2U = C$, there corresponds a zero-velocity curve. By Eq. (25.28), on such a curve, C can be large if ρ_1 or ρ_2 is small or if $x^2 + y^2$ is large. In such a case, real motion can take place only inside small ovals enclosing each primary or outside the circle $x^2 + y^2 = C$. As we vary C, we find a pattern of regions where motion is permitted or forbidden. Diagrams are given by Pollard[1] and Brouwer and Clemence.[2] They are used to develop the concept of double points, where two of the curves touch each other. This concept in turn is used to find the equilibrium points. Since there is a simpler method of finding the equilibrium points, we shall not draw such diagrams but proceed in a different way.

III. Equilibrium Points

Return to Eqs. (25.11). Suppose that the first and second derivatives of x, y, z with respect to τ all vanish initially. Equations (25.11) show that all the higher derivatives of x, y, z vanish initially and, thus, vanish for all values of τ by Taylor's theorem. [Remember that $\rho_1^2 = (x - x_1)^2 + y^2 + z^2$ and $\rho_2^2 = (x - x_2)^2 + y^2 + z^2$.] Thus, if $\dot{x}$, $\dot{y}$, $\dot{z}$, $\ddot{x}$, $\ddot{y}$, $\ddot{z}$ all vanish initially, then $z = 0$ for all τ, and x and y remain constant, so that the points of equilibrium satisfy

$$\frac{\partial U}{\partial x} = x - \frac{(1-m)}{\rho_1^3}(x - x_1) - \frac{m}{\rho_2^3}(x - x_2) = 0 \tag{25.34a}$$

$$\frac{\partial U}{\partial y} = y - \frac{(1-m)}{\rho_1^3}y - \frac{m}{\rho_2^3}y = 0 \tag{25.34b}$$

$$\frac{\partial U}{\partial z} = z = 0 \tag{25.34c}$$

for all values of τ.

The Triangular Points of Lagrange

For equilibrium points with $y \neq 0$, Eq. (25.34b) gives

$$1 - \frac{(1-m)}{\rho_1^3} - \frac{m}{\rho_2^3} = 0 \tag{25.35}$$

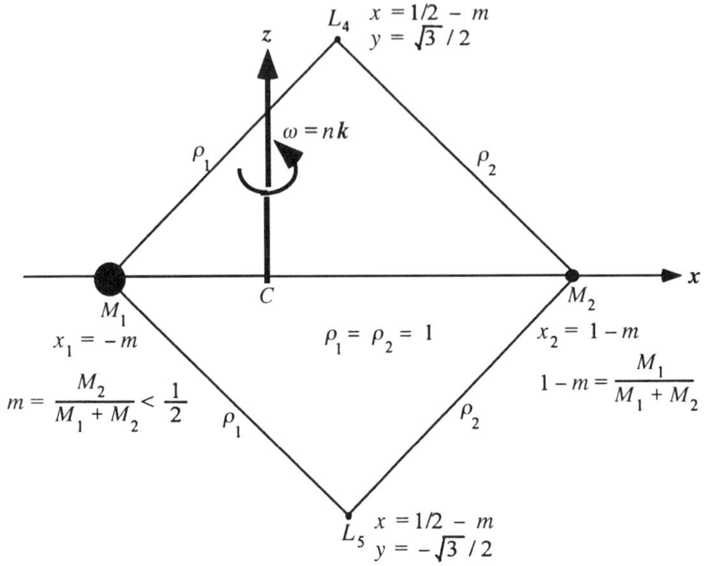

Fig. 25.3 Triangular equilibrium points ($y \neq 0$, $z = 0$) with M_1 and M_2 as primary masses, M_3 at equilibrium points, and $M_1 > M_2 \gg M_3$.

From Eqs. (25.34a) and (25.35)

$$-x\left(1 - \frac{(1-m)}{\rho_1^3} - \frac{m}{\rho_2^3}\right) = \frac{(1-m)x_1}{\rho_1^3} + \frac{mx_2}{\rho_2^3} = 0 \tag{25.36}$$

but, by the property of the center of mass,

$$(1-m)x_1 + mx_2 = 0 \tag{25.37}$$

Also

$$x_2 - x_1 = 1$$

so that

$$x_1 = -m \qquad x_2 = 1 - m \tag{25.38}$$

Insert Eqs. (25.38) into Eq. (25.36). Then, $m(1-m)(\rho_1^{-3} - \rho_2^{-3}) = 0$, so that $\rho_1 = \rho_2$. On inserting $\rho_1 = \rho_2$ into Eq. (25.35), we find

$$\rho_1 = \rho_2 = 1 \tag{25.39}$$

Thus, there are equilibrium points (L_4 and L_5) at the vertices of an equilateral triangle (as shown in Fig. 25.3), which are called the Lagrange triangular points.

The Collinear Equilibrium Points

In Eq. (25.34a), insert $y = z = 0$, $x_1 = -m$, and $x_2 = 1 - m$. Then

$$\rho_1^2 = (x+m)^2 \qquad \rho_2^2 = (x+m-1)^2$$

and

$$f(x) = x - \frac{(1-m)(x+m)}{|x+m|^3} - \frac{m(x+m-1)}{|x+m-1|^3} = 0 \tag{25.40}$$

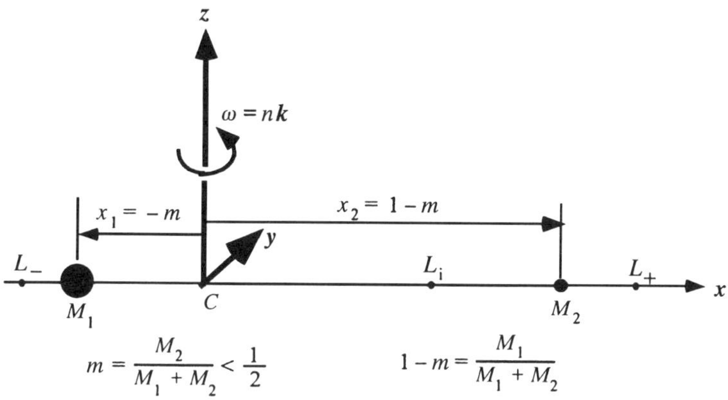

Fig. 25.4 Collinear equilibrium points ($y = z = 0$) with M_1 and M_2 as primary masses, M_3 at equilibrium points, and $M_1 > M_2 \gg M_3$.

This is the equation for equilibrium points ($L_{\rm i}$, L_+, and L_- that some authors denote, respectively, as L_1, L_2, and L_3) on the x axis joining the primaries (M_1 and M_2) as shown in Fig. 25.4. For the Earth-moon-space station system, the equilibrium points are depicted in Fig. 25.5.

Case 1: $x < -m$

In this case

$$
\begin{aligned}
x + m &< 0 \qquad & x + m - 1 &< -1 \\
x + m &= -|x + m| \qquad & x + m - 1 &= -|x + m - 1|
\end{aligned}
\tag{25.41}
$$

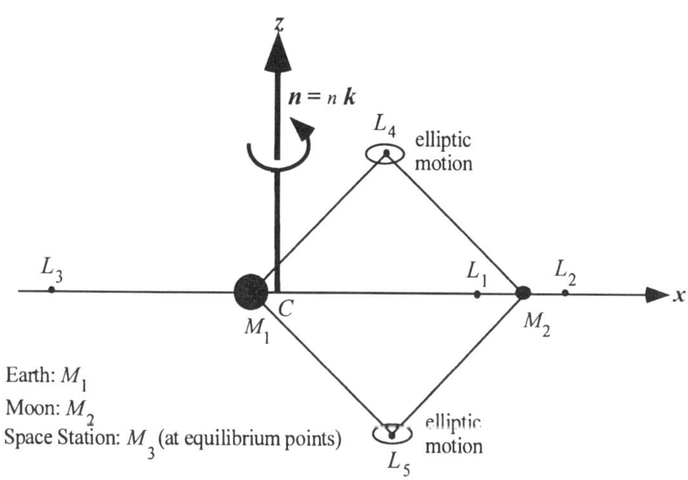

Fig. 25.5 The triangular and collinear equilibrium points (L_1, L_2, L_3, L_4, and L_5) for the Earth-moon-space station system.

By Eqs. (25.40) and (25.41)

$$f(x) = x + \frac{1-m}{(x+m)^2} + \frac{m}{(x+m-1)^2} \tag{25.42a}$$

$$f'(x) = 1 - \frac{2(1-m)}{(x+m)^3} - \frac{2m}{(x+m-1)^3} > 0 \tag{25.42b}$$

the sign being positive by Eqs. (25.41).

Also

$$f(-\infty) = -\infty \qquad f(-m) = +\infty$$

Because $f'(x) > 0$ for $x < -m$, it follows that the curve of $f(x)$ vs x crosses the x axis once and only once when $x < -m$. Call this zero L_-. In Fig. 25.4, this is the collinear equilibrium point to the left of the larger mass M_1.

Case 2: $-m < x < 1 - m$

Here

$$\begin{aligned} &x + m > 0 & &x + m - 1 < 0 \\ &x + m = |x + m| & &x + m - 1 = -|x + m - 1| \end{aligned} \tag{25.43}$$

By Eqs. (25.40) and (25.43)

$$f(x) = x - \frac{1-m}{(x+m)^2} + \frac{m}{(x+m-1)^2} \tag{25.44a}$$

$$f'(x) = 1 + \frac{2(1-m)}{(x+m)^3} - \frac{2m}{(x+m-1)^3} > 0 \tag{25.44b}$$

Here $f'(x) > 0$ by Eqs. (25.43).

Also

$$f(-m+\varepsilon) = -m + \varepsilon - \frac{1-m}{\varepsilon^2} + \frac{m}{(\varepsilon - 1)^2} \to -\infty \qquad \text{as} \qquad \varepsilon \to 0$$

$$f(1-m-\varepsilon) = 1 - m - \varepsilon - \frac{1-m}{(1-\varepsilon)^2} + \frac{m}{\varepsilon^2} \to +\infty \qquad \text{as} \qquad \varepsilon \to 0$$

In this interval between the primaries, $f(x)$ starts out at $-\infty$ and goes to $+\infty$, always increasing. There is one and only one equilibrium point between the primaries; call it L_i (intermediate).

Case 3: $x > 1 - m$

This is to the right of the smaller mass in Fig. 25.4. Here

$$\begin{aligned} &x + m > 1 & &x + m - 1 > 0 \\ &x + m = |x + m| & &x + m - 1 = |x + m - 1| \end{aligned} \tag{25.45}$$

By Eqs. (25.40) and (25.45)

$$f(x) = x - \frac{1-m}{(x+m)^2} - \frac{m}{(x+m-1)^2} \tag{25.46a}$$

$$f'(x) = 1 + \frac{2(1-m)}{(x+m)^3} + \frac{2m}{(x+m-1)^3} > 1 \tag{25.46b}$$

Also

$$f(1-m-\varepsilon) = 1 - m - \varepsilon - \frac{1-m}{(1-\varepsilon)^2} - \frac{m}{\varepsilon^2} \to -\infty \tag{25.47}$$

as $\varepsilon \to 0$, and $f(\infty) = +\infty$, so that $f(x)$ goes from $-\infty$ at $x = 1-m$ to $+\infty$ at $+\infty$, always increasing. There is one and only one equilibrium point for $x > 1-m$; call it L_+.

Solution for L_+

Put $x = x_2 + \rho = 1 - m + \rho$, so that $\rho = \rho_2$. Insert this into Eq. (25.46a) with $f(x) = 0$. Then

$$1 - m + \rho - \frac{1-m}{(1+\rho)^2} - \frac{m}{\rho^2} = 0 \tag{25.48}$$

Divide by $1-m$ and transpose:

$$1 + \frac{\rho}{1-m} - \frac{1}{(1+\rho)^2} = \frac{m}{1-m}\frac{1}{\rho^2}$$

Use

$$\frac{\rho}{1-m} = \left(1 + \frac{m}{1-m}\right)\rho$$

$$1 + \left(1 + \frac{m}{1-m}\right)\rho - \frac{1}{(1+\rho)^2} = \frac{m}{1-m}\frac{1}{\rho^2}$$

$$\frac{m}{1-m}\left(\frac{1}{\rho^2} - \rho\right) = 1 + \rho - \frac{1}{(1+\rho)^2} = \frac{(1+\rho)^3 - 1}{(1+\rho)^2}$$

$$\frac{m}{1-m}\left(\frac{1-\rho^3}{\rho^2}\right) = \frac{\rho^3 + 3\rho^2 + 3\rho}{(1+\rho)^2} > 0$$

since $m < 1/2$ as defined in the first section of this chapter. Thus, $\rho_2 \equiv \rho < 1$ and

$$\frac{m}{1-m} = \frac{3\rho^3(1 + \rho + \rho^2/3)}{(1+\rho)^2(1-\rho^3)} \tag{25.49}$$

Now put

$$\lambda \equiv \left(\frac{m}{3(1-m)}\right)^{\frac{1}{3}} \tag{25.50}$$

Then

$$\rho^3(1+\rho+\rho^2/3) = \lambda^3(1+\rho)^2(1-\rho^3) \tag{25.51}$$

If M_1 is the sun, M_2 the Earth, and M_3 an Earth-orbiting satellite, then $\rho_2 \equiv \rho$ is small. For small ρ, we have $\rho \approx \lambda$, which suggests use of a series expansion

$$\rho = \lambda(1 + c_1\lambda + c_2\lambda^2 + \cdots) \tag{25.52}$$

Insert Eq. (25.52) into Eq. (25.51) to find

$$1 + (1+3c_1)\lambda + \left(\tfrac{1}{3} + 4c_1 + 3c_2 + 3c_1^2\right)\lambda^2 = 1 + 2\lambda + (1+2c_1)\lambda^2 \tag{25.53}$$

Comparing coefficients, we find

λ:

$$1 + 3c_1 = 2 \qquad \text{or} \qquad c_1 = \tfrac{1}{3}$$

λ^2:

$$\tfrac{1}{3} + 4c_1 + 3c_2 + 3c_1^2 = 1 + 2c_1 \qquad c_2 = -\tfrac{1}{9}$$

Thus

$$\rho = \lambda\left(1 + \tfrac{1}{3}\lambda - \tfrac{1}{9}\lambda^2 + \cdots\right)$$

where $m \equiv M_2/(M_1+M_2)$ and $\lambda \equiv (M_2/(3M_1))^{1/3}$ is given by Eq. (25.50). Thus, for L_+

$$x = 1 - m + \lambda\left(1 + \tfrac{1}{3}\lambda - \tfrac{1}{9}\lambda^2 + \cdots\right) \tag{25.54}$$

Solution for L_i

Put $x = x_2 - \rho = 1 - m - \rho$, so that $\rho = \rho_2$. Here $\rho < 1$. Insert this into Eq. (25.44a) with $f(x) = 0$. Then

$$1 - m - \rho - \frac{1-m}{(1-\rho)^2} + \frac{m}{\rho^2} = 0 \tag{25.55}$$

Proceed as previously described, solving for $m/(1-m)$ in terms of ρ.

$$\frac{m}{1-m} = \frac{3\rho^3(1-\rho+\rho^2/3)}{(1-\rho)^2(1-\rho^3)} \tag{25.56}$$

Again

$$\rho \approx \lambda \equiv \left(\frac{m}{3(1-m)}\right)^{\frac{1}{3}} \tag{25.56a}$$

for small ρ. We could go through the same procedure of expanding in powers of λ; it is easier, however, to solve for ρ in terms of λ by means of a trick, from that for L_+. In neither Eqs. (25.49) nor (25.56) does the ρ^3 in $1-\rho^3$ contribute to the expansion if we stop at λ^3.

For L_+, we had

$$\frac{m}{3(1-m)} = \frac{\rho^3(1+\rho+\rho^2/3)}{(1+\rho)^2} = F(\rho) \tag{25.57}$$

Now in Eq. (25.56), put $\rho = -\rho'$, and put $\lambda = -\lambda'$. Then by Eqs. (25.56) and (25.56a)

$$\frac{\rho'^3(1+\rho'+\rho'^2/3)}{(1+\rho')^2} = -\lambda'^3 \tag{25.58}$$

Equation (25.58) has the same form as Eq. (25.57), with ρ' replacing ρ and λ' replacing λ. Thus, ρ' is the same function of λ' that ρ is of λ, so that

$$\rho' = \lambda'\left(1 + \tfrac{1}{3}\lambda' - \tfrac{1}{9}\lambda'^2 + \cdots\right) \tag{25.59}$$

or

$$\rho = \lambda\left(1 + \tfrac{1}{3}\lambda - \tfrac{1}{9}\lambda^2 + \cdots\right)$$

This is the solution for L_i. With $m \equiv M_2/(M_1+M_2)$ and $\lambda \equiv (M_2/(3M_1))^{1/3}$,

$$x = 1 - m - \lambda\left(1 + \tfrac{1}{3}\lambda - \tfrac{1}{9}\lambda^2 + \cdots\right) \tag{25.60}$$

Solution for L_-

Put $x = x_1 - \rho = -m - \rho$, so that $\rho_1 = \rho$. Here $\rho < 1$. Insert this into Eq. (25.42a) with $f(x) = 0$. Then

$$-m - \rho + \frac{1-m}{\rho^2} + \frac{m}{(1+\rho)^2} = 0 \tag{25.61}$$

Then

$$m\left(\frac{1}{(1+\rho)^2} - 1\right) = \rho - \frac{1-m}{\rho^2}$$

$$\frac{m}{1-m}\left(\frac{1}{(1+\rho)^2} - 1\right) = \frac{\rho}{1-m} - \frac{1}{\rho^2} = \left(1 + \frac{m}{1-m}\right)\rho - \frac{1}{\rho^2}$$

$$\frac{m}{1-m}\left(\frac{1}{(1+\rho)^2} - 1 - \rho\right) = \rho - \frac{1}{\rho^2}$$

$$\frac{m}{1-m}\left(\frac{1-(1+\rho)^3}{(1+\rho)^2}\right) = \rho - \rho^{-2}$$

$$\frac{m}{1-m} = \frac{(\rho-\rho^{-2})(1+\rho)^2}{1-(1+\rho)^3} = \frac{(\rho^3-1)(1+\rho)^2}{\rho^2[1-(1+\rho)^3]} \tag{25.62}$$

We may write this as

$$\frac{m}{1-m} = -\frac{N}{D} \tag{25.63}$$

where if

$$\alpha \equiv \rho - 1 \tag{25.64}$$

$$N = 12\alpha + 24\alpha^2 + 19\alpha^3 + 7\alpha^4 + \alpha^5 \tag{25.65}$$

$$D = 7 + 26\alpha + 37\alpha^2 + 25\alpha^3 + 8\alpha^4 + \alpha^5 \tag{25.66}$$

From Eq. (25.63)

$$m = -\frac{N}{D - N} \tag{25.67}$$

so that

$$-\frac{m}{\alpha} = \frac{12 + 24\alpha + 19\alpha^2 + 7\alpha^3 + \alpha^4}{7 + 14\alpha + 13\alpha^2 + 6\alpha^3 + \alpha^4} \tag{25.68}$$

By the elementary algorithm for division

$$-\frac{m}{\alpha} = \frac{12}{7} - \frac{23}{49}\alpha^2 + O(\alpha^3) \tag{25.69}$$

The first approximation to a solution for α is

$$\alpha = \alpha_0 = -\frac{7}{12}m \tag{25.70}$$

Insert this into the α^2 term in Eq. (25.69). The next approximation is

$$-\frac{m}{\alpha} = \frac{12}{7} - \frac{23}{49}\left(-\frac{7}{12}m\right)^2 = \frac{12}{7}\left[1 - \frac{23}{49}\frac{7}{12}m^2\right]$$

$$\alpha = -\frac{m}{\frac{12}{7}\left[1 - \frac{23}{49}\frac{7}{12}m^2\right]} = -\frac{7m}{12}\left[1 + \frac{23}{49}\frac{7}{12}m^2\right]$$

or

$$\alpha = -\frac{7m}{12}\left[1 + \frac{23}{84}m^2\right] \tag{25.71}$$

$$\rho_1 = \rho = 1 - \frac{7m}{12}\left[1 + \frac{23}{84}m^2\right] \tag{25.72}$$

This is the solution for L_-. With $m \equiv M_2/(M_1 + M_2)$ and $\lambda \equiv (M_2/(3M_1))^{1/3}$,

$$x = -m - 1 + \frac{7m}{12}\left[1 + \frac{23}{84}m^2\right] \tag{25.73}$$

IV. Motion near the Equilibrium Points

Return to Eqs. (25.30)

$$\ddot{x} - 2\dot{y} = U_x \tag{25.30a}$$

$$\ddot{y} + 2\dot{x} = U_y \tag{25.30b}$$

$$\ddot{z} = U_z \tag{25.30c}$$

where the subscript of U denotes a partial derivative and

$$U(x, y, z) = \frac{1}{2}[(x^2 + y^2)] + \frac{1-m}{\rho_1} + \frac{m}{\rho_2} \tag{25.28}$$

Again, Eqs. (25.28) and (25.30) apply only the "circular restricted" problem where the primary orbits are circles. From Eq. (25.28)

$$U_z = -\frac{1-m}{\rho_1^3}z - \frac{m}{\rho_2^3}z \tag{25.74}$$

By Eqs. (25.28) and (25.30c)

$$\ddot{z} + kz = 0 \tag{25.75}$$

where

$$k = \frac{1-m}{\rho_1^3} + \frac{m}{\rho_2^3} \tag{25.76}$$

If ρ_1 and ρ_2 are the distances of an equilibrium point from the primaries, then U_z is the first term in the expansion of U_z in a Taylor's series in the neighborhood of the equilibrium point. Then k is a constant, and by Eq. (25.75)

$$z = b_1 \cos\left(k^{\frac{1}{2}}\tau - \theta\right) \tag{25.77}$$

θ being a constant. Equation (25.77) shows that motion of the orbiter perpendicular to the plane of the primary motion is simple harmonic if the orbiter remains near an equilibrium point. If the equilibrium point is a triangular one, then $\rho_1 = \rho_2 = 1$ and

$$k = 1 - m - m = 1 \tag{25.78}$$

To the approximation considered, the z frequency is the same as that of the primaries.

V. Motion in the Plane of the Primaries

In Eqs. (25.30a) and (25.30b) put

$$x = x_0 + \beta_1$$

$$y = y_0 + \beta_2 \tag{25.79}$$

In the neighborhood of the point x_0, y_0, the Taylor expansion of a function $f(x, y)$ takes the form

$$f(x, y) = f(x_0, y_0) + \beta_1 f_x(x_0, y_0) + \beta_1 f_y(x_0, y_0) + \cdots \tag{25.80}$$

where

$$f_x = \frac{\partial f}{\partial x} \qquad f_y = \frac{\partial f}{\partial y}$$

Then, in Eqs. (25.30a) and (25.30b), by Eqs. (25.79) and (25.80)

$$\begin{aligned}\ddot{\beta}_1 - 2\dot{\beta}_2 &= (U_x)_0 + \beta_1(U_{xx})_0 + \beta_2(U_{xy})_0 + \cdots \\ \ddot{\beta}_2 + 2\dot{\beta}_1 &= (U_y)_0 + \beta_1(U_{yx})_0 + \beta_2(U_{yy})_0 + \cdots\end{aligned} \tag{25.81}$$

Because

$$U = \frac{1}{2}[(x^2 + y^2)] + \frac{1-m}{\rho_1} + \frac{m}{\rho_2} \tag{25.28}$$

we find

$$U_x = x - \frac{(1-m)}{\rho_1^3}(x - x_1) - \frac{m}{\rho_2^3}(x - x_2)$$

$$U_y = y - \frac{(1-m)}{\rho_1^3}y - \frac{m}{\rho_2^3}y$$

$$U_{xx} = 1 - \frac{(1-m)}{\rho_1^3} - \frac{m}{\rho_2^3} + \frac{3(1-m)}{\rho_1^5}(x - x_1)^2 + \frac{3m}{\rho_2^5}(x - x_2)^2 = A$$

$$U_{xy} = U_{yx} = \frac{3(1-m)}{\rho_1^5}(x - x_1)y + \frac{3m}{\rho_2^5}(x - x_2)y = B$$

$$U_{yy} = 1 - \frac{(1-m)}{\rho_1^3} - \frac{m}{\rho_2^3} + \frac{3(1-m)}{\rho_1^5}y^2 + \frac{3m}{\rho_2^5}y^2 = C$$

At the triangular points

$$x - x_1 = \tfrac{1}{2} \qquad x_2 - x = \tfrac{1}{2} \qquad \rho_1 = \rho_2 = 1$$

$$L_4: \quad y = \tfrac{1}{2}\sqrt{3} \qquad L_5: \quad y = -\tfrac{1}{2}\sqrt{3}$$

$$A = \tfrac{3}{4} \tag{25.82a}$$

$$B = \frac{3\sqrt{3}}{4}(1 - 2m) \text{ at } L_4 \qquad B = -\frac{3\sqrt{3}}{4}(1 - 2m) \text{ at } L_5 \tag{25.82b}$$

$$C = 9/4 \tag{25.82c}$$

At the collinear points

$$A = 1 - k + \frac{3(1-m)}{\rho_1^5}(x - x_1)^2 + \frac{3m}{\rho_2^5}(x - x_2)^2$$

$$\rho_1^2 = (x + m)^2 = (x - x_1)^2$$

$$\rho_2^2 = (x + m - 1)^2 = (x - x_2)^2$$

$$A = 1 - k + \frac{3(1-m)}{\rho_1^3} + \frac{3m}{\rho_2^3} = 1 - k + 3k = 1 + 2k \tag{25.82d}$$

$$B = 0 \tag{25.82e}$$

$$C = 1 - \frac{(1-m)}{\rho_1^3} - \frac{m}{\rho_2^3} = 1 - k \tag{25.82f}$$

Since $(U_x)_0 = (U_y)_0 = 0$ at equilibrium points, we find from Eqs. (25.81)

$$\ddot{\beta}_1 - 2\dot{\beta}_2 = A\beta_1 + B\beta_2$$
$$\ddot{\beta}_2 + 2\dot{\beta}_1 = B\beta_1 + C\beta_2 \tag{25.83}$$

If the operator $D \equiv \mathrm{d}/\mathrm{d}\tau$, then

$$(D^2 - A)\beta_1 = (2D + B)\beta_2 \tag{25.84}$$

$$(D^2 - C)\beta_2 = -(2D - B)\beta_1 \tag{25.85}$$

Operate on Eq. (25.84) with $2D - B$. Then

$$(2D - B)(D^2 - A)\beta_1 = (4D^2 - B^2)\beta_2$$

The operators commute, so that

$$(D^2 - A)(2D - B)\beta_1 = (4D^2 - B^2)\beta_2 \tag{25.86}$$

but by Eq. (25.85)

$$-(D^2 - A)(D^2 - C)\beta_2 = (4D^2 - B^2)\beta_2 \tag{25.87}$$

Thus

$$[D^4 + (4 - A - C)D^2 + (AC - B^2)]\beta_2 = 0 \tag{25.88}$$

Now, operate on Eq. (25.85) with $2D + B$ to obtain

$$(2D + B)(D^2 - C)\beta_2 = -(4D^2 - B^2)\beta_1 \tag{25.89}$$

Thus

$$(D^2 - C)(2D + B)\beta_2 = -(4D^2 - B^2)\beta_1$$

Apply Eq. (25.84):

$$(D^2 - C)(D^2 - A)\beta_1 = -(4D^2 - B^2)\beta_1$$

so that

$$[D^4 + (4 - A - C)D^2 + (AC - B^2)]\beta_1 = 0 \tag{25.90}$$

Thus, β_1 and β_2 both satisfy the same fourth-order differential equation

$$D^4 f + (4 - A - C)D^2 f + (AC - B^2) f = 0 \tag{25.91}$$

To solve, place

$$f = \varepsilon^{p\tau}$$

Then

$$p^4 + (4 - A - C)p^2 + AC - B^2 = 0 \tag{25.92}$$

There are four roots, so that the solutions take the form

$$f = \sum_{i=1}^{4} a_i \varepsilon^{p_i \tau} \tag{25.93}$$

If we put

$$q = p^2 \tag{25.94}$$

then

$$q^2 + (4 - A - C)q + AC - B^2 = 0 \tag{25.95}$$

Stability of Motion near the Triangular Points

For these points, we have

$$A = \frac{3}{4} \qquad B = \pm\frac{3\sqrt{3}}{4}(1 - 2m) \qquad C = \frac{9}{4}$$

Thus

$$4 - A - C = 1$$

$$AC - B^2 = \frac{27}{4}m(1 - m) \tag{25.96}$$

so that

$$q^2 + q + \frac{27}{4}m(1 - m) = 0 \tag{25.97}$$

Then

$$q = -\tfrac{1}{2} \pm \tfrac{1}{2}[1 - 27m(1 - m)]^{\frac{1}{2}} \tag{25.98}$$

If $27m(1 - m) < 1$, then q is real and < 0, and all values of p are pure imaginary. This means that the solutions for β_1 and β_2 contain only sines and cosines, with no increasing exponential functions. If

$$27m(1 - m) < 1 \tag{25.99}$$

the orbit never goes to infinity, and the motion is stable—as far as we can tell from the linearized equations. The motion near the triangular points has been proved to be stable even with the nonlinearized equations.[1]

Instability of Motion near the Collinear Points

Lemma: At the collinear points

$$k \equiv \frac{1 - m}{\rho_1^3} + \frac{m}{\rho_2^3} > 1 \tag{25.99a}$$

Proof: Write down Eq. (25.34a) for an equilibrium point

$$x - \frac{(1 - m)}{\rho_1^3}(x - x_1) - \frac{m}{\rho_2^3}(x - x_2) = 0 \tag{25.34a}$$

and use $x_1 = -m$, $x_2 = 1 - m$. At any equilibrium point

$$x - \frac{(1 - m)}{\rho_1^3}(x + m) - \frac{m}{\rho_2^3}(x + m - 1) = 0 \tag{25.100}$$

We also have the identity

$$x \equiv \frac{(1-m)(x+m)}{\rho_1}\rho_1 + \frac{m(x+m-1)}{\rho_2}\rho_2 \tag{25.101}$$

Insert Eq. (25.101) into Eq. (25.100). The result is

$$\frac{(1-m)(x+m)}{\rho_1}\left(\rho_1 - \rho_1^{-2}\right) + \frac{m(x+m-1)}{\rho_2}\left(\rho_2 - \rho_2^{-2}\right) = 0 \tag{25.102}$$

Equation (25.102) holds for all the equilibrium points. To prove Eq. (25.99a) for all the collinear points, we have to treat each one.

At L_-

Here, $x = -m - \rho_1$ and $\rho_2 - \rho_1 = 1$. Thus

$$x + m = -\rho_1$$

$$x + m - 1 = -\rho_1 - 1 = -\rho_2$$

Insert into Eq. (25.102). Then

$$\frac{(1-m)(-\rho_1)}{\rho_1}\left(\rho_1 - \rho_1^{-2}\right) + \frac{m(-\rho_2)}{\rho_2}\left(\rho_2 - \rho_2^{-2}\right) = 0 \tag{25.103}$$

or

$$(1-m)\left(\rho_1 - \rho_1^{-2}\right) + m\left(\rho_2 - \rho_2^{-2}\right) = 0 \tag{25.104}$$

or

$$(1-m)\rho_1 - \frac{(1-m)\rho_1}{\rho_1^3} + m\rho_2 - \frac{m\rho_2}{\rho_2^3} = 0 \tag{25.105}$$

Because $\rho_2 = \rho_1 + 1$, this becomes

$$(1-m)\rho_1 - \frac{(1-m)\rho_1}{\rho_1^3} + m + m\rho_1 - \frac{m}{\rho_2^3} - \frac{m\rho_1}{\rho_2^3} = 0$$

By definition

$$k = \frac{1-m}{\rho_1^3} + \frac{m}{\rho_2^3}$$

Thus

$$(1-m)\rho_1 + m + m\rho_1 - \frac{m}{\rho_2^3} - \rho_1 k = 0 \tag{25.106}$$

or

$$\rho_1(k-1) = m\left(1 - \rho_2^{-3}\right) \tag{25.107}$$

Since for L_-, we have $\rho_2 > 1$, it follows that $k > 1$ for L_-.

At L_i

Here, $x = -m + \rho_1$ and $\rho_2 + \rho_1 = 1$. Thus

$$x + m = \rho_1$$

$$x + m - 1 = \rho_1 - 1 = -\rho_2$$

Insert into Eq. (25.102). Then

$$(1 - m)\left(\rho_1 - \rho_1^{-2}\right) - m\left(\rho_2 - \rho_2^{-2}\right) = 0 \tag{25.108}$$

or

$$\rho_1 - m\rho_1 - \frac{(1-m)\rho_1}{\rho_1^3} - m + m\rho_1 - \frac{m\rho_1}{\rho_2^3} + \frac{m}{\rho_2^3} = 0$$

or

$$\rho_1\left[1 - \frac{(1-m)}{\rho_1^3} - \frac{m}{\rho_2^3}\right] = m\left(1 - \rho_2^{-3}\right)$$

because $\rho_2 = 1 - \rho_1$. However, the bracketed factor on the left is simply $1 - k$. Thus

$$\rho_1(1 - k) = m\left(1 - \rho_2^{-3}\right) \tag{25.109}$$

Since for L_i we have $\rho_2 < 1$, it follows that $k > 1$ for L_i.

At L_+

Here, $x = -m + \rho_1$ and $\rho_2 - \rho_1 = -1$. Thus

$$x + m = \rho_1$$

$$x + m - 1 = \rho_1 - 1 = \rho_2$$

Insert into Eq. (25.102). Then

$$(1 - m)\left(\rho_1 - \rho_1^{-2}\right) + m\left(\rho_2 - \rho_2^{-2}\right) = 0$$

or

$$\rho_1 - m\rho_1 - \frac{(1-m)\rho_1}{\rho_1^3} - m + m\rho_1 - \frac{m\rho_1}{\rho_2^3} + \frac{m}{\rho_2^3} = 0$$

or

$$\rho_1\left[1 - \frac{(1-m)}{\rho_1^3} - \frac{m}{\rho_2^3}\right] = m\left(1 - \rho_2^{-3}\right) \tag{25.110}$$

because $\rho_2 = \rho_1 - 1$. However, the bracketed factor on the left is simply $1 - k$. Thus

$$\rho_1(1 - k) = m\left(1 - \rho_2^{-3}\right)$$

Since for L_+ we have $\rho_2 < 1$, it follows that $k > 1$ for L_+. This completes the proof of the lemma

$$k \equiv \frac{1-m}{\rho_1^3} + \frac{m}{\rho_2^3} > 1$$

for all the collinear equilibrium points.

We have now to investigate the roots of

$$q^2 + (4 - A - C)q + AC - B^2 = 0 \tag{25.95}$$

For the collinear equilibrium points, we found

$$A = 1 + 2k \qquad B = 0 \qquad C = 1 - k$$

in Eqs. (25.82d)–(25.82f). Insertion of these values into Eq. (25.95) yields

$$q^2 + (2 - k)q + (1 + 2k)(1 - k) = 0 \tag{25.111}$$

so that

$$q = -\frac{(2-k)}{2} \pm \frac{1}{2}[(2-k)^2 - 4(1+2k)(1-k)]^{\frac{1}{2}}$$

$$q = -\frac{(2-k)}{2} \pm \frac{1}{2}(9k^2 - 8k)^{\frac{1}{2}}$$

The two roots q_1 and q_2 satisfy

$$2q_1 = k - 2 + (9k^2 - 8k)^{\frac{1}{2}} \tag{25.112}$$

$$2q_2 = k - 2 - (9k^2 - 8k)^{\frac{1}{2}} \tag{25.113}$$

Consider

$$9k^2 - 8k = k^2 + 8k^2 - 8k$$

Because $k > 1$, $8k^2 - 8k > 0$, so that

$$9k^2 - 8k > k^2 > 1$$

or

$$(9k^2 - 8k)^{\frac{1}{2}} > k > 1 \tag{25.113a}$$

By Eqs. (25.112) and (25.113a)

$$2q_1 > 1 - 2 + 1$$

Thus

$$q_1 > 0 \tag{25.113b}$$

For q_2, use Eqs. (25.113) and (25.113a). Then

$$-(9k^2 - 8k)^{\frac{1}{2}} < -k \tag{25.113c}$$

By Eqs. (25.113) and (25.113c)

$$2q_2 < k - 2 - k < -2$$

Thus

$$q_2 < -1 \tag{25.113d}$$

Because $q = p^2$, $q_1 > 0$ gives rise to two real roots of opposite signs, and $q_2 < 0$ gives rise to two imaginary roots.

For the collinear points, the fourth-order equation for β_1 and β_2 leads to solutions of the form

$$\beta_j = c_{j1}\varepsilon^{\lambda_1\tau} + c_{j2}\varepsilon^{-\lambda_2\tau} + c_{j3}\cos(\lambda_3\tau - c_{j4}) \tag{25.113e}$$

Even though the initial conditions may be such as not to bring the positive exponential function $\varepsilon^{\lambda_1\tau}$ into the solution, a small change in the initial conditions can always bring it in. Thus, the motion is unstable near a collinear equilateral point or libration point.

VI. Further Considerations About L_4 and L_5

We have found that the solutions for β_1 and β_2 are of the form $\varepsilon^{p_i\tau}$, where p_i satisfies

$$p^2 = -\tfrac{1}{2} \pm \tfrac{1}{2}[1 - 27m(1-m)]^{1/2} \tag{25.114}$$

Case 1: $1 - 27m(1-m) < 0$

$$b^2 \equiv 27m(1-m) - 1 > 0 \tag{25.115}$$

where we can take $b > 0$. The values of p_1, p_2, p_3, and p_4 are given by

$$p_{1,2}^2 = \tfrac{1}{2}(-1 + ib) = \tfrac{1}{2}(1+b^2)^{\frac{1}{2}}\varepsilon^{i\theta} \tag{25.116}$$

$$p_{1,2} = \pm 2^{-\frac{1}{2}}(1+b^2)^{\frac{1}{4}}\varepsilon^{i\theta/2} \tag{25.117}$$

$$p_{3,4}^2 = \tfrac{1}{2}(-1 - ib) = \tfrac{1}{2}(1+b^2)^{\frac{1}{2}}\varepsilon^{i\phi} \tag{25.118}$$

$$p_{3,4} = \pm 2^{-\frac{1}{2}}(1+b^2)^{\frac{1}{4}}\varepsilon^{i\phi/2} \tag{25.119}$$

From Eq. (25.116)

$$\frac{1}{2}(1+b^2)^{\frac{1}{2}}\cos\theta = -\frac{1}{2} < 0 \qquad \frac{1}{2}(1+b^2)^{\frac{1}{2}}\sin\theta = \frac{b}{2} > 0$$

Thus, $90° < \theta < 180°$ and $45° < \theta/2 < 90°$, so that

$$\cos\frac{\theta}{2} > 0 \tag{25.120}$$

By Eqs. (25.117) and (25.120)

$$\mathrm{Re}(p_1) > 0 \qquad \mathrm{Re}(p_2) < 0 \tag{25.121}$$

From Eq. (25.118)

$$\frac{1}{2}(1+b^2)^{\frac{1}{2}}\cos\phi = -\frac{1}{2} < 0 \qquad \frac{1}{2}(1+b^2)^{\frac{1}{2}}\sin\phi = -\frac{b}{2} < 0$$

Thus, $180° < \phi < 270°$ and $90° < \phi/2 < 135°$, so that

$$\cos\frac{\phi}{2} < 0 \tag{25.122}$$

By Eqs. (25.119) and (25.122)

$$\mathrm{Re}(p_3) < 0 \qquad \mathrm{Re}(p_4) > 0 \tag{25.123}$$

Two of the solutions have positive exponential factors, so that the motion is unstable.

Case 2: $1 > 1 - 27m(1-m) > 0$

By Eq. (25.114), all four values of p are pure imaginary, so that the solutions are all cosines and the motion is stable. Now, consider Eq. (25.114) and put

$$27f(m) \equiv 1 - 27m(1-m) \tag{25.124}$$

Then

$$f(m) = m^2 - m + \frac{1}{27} = \left(m - \frac{1}{2}\right)^2 - \frac{1}{4} + \frac{1}{27}$$

$$f(m) = \left(m - \frac{1}{2}\right)^2 - \frac{23}{108} > 0 \tag{25.125}$$

and

$$\left(m - \frac{1}{2}\right)^2 > \frac{23}{108}$$

or

$$m < 0.03852 \approx 1/26 \tag{25.126}$$

This is a necessary and sufficient condition for stability of motion near the triangular points, at least in the linearized theory. It is satisfied when the primaries are the sun and Jupiter, the sun and the Earth, and the Earth and the moon. Because $m \equiv M_2/(M_1 + M_2)$, we have

sun–Jupiter	$m \approx 1/1000$
sun–Earth	$m \approx 1/300{,}000$
Earth–moon	$m \approx 1/80$

Now let

$$1 - 27m(1-m) \equiv \lambda^2 > 0 \tag{25.127}$$

By Eqs. (25.114) and (25.127), with $\lambda < 1$ for the preceding three combinations of primaries,

$$p^2 = -\frac{1}{2} \pm \frac{\lambda}{2} < 0 \tag{25.128}$$

Let p_1 and p_2 correspond to the + sign and p_3 and p_4 to the − sign. Then

$$p_{1,2}^2 = -\frac{1}{2} + \frac{\lambda}{2} \qquad p_{3,4}^2 = -\frac{1}{2} - \frac{\lambda}{2} \tag{25.129}$$

$$p_1 = i\left(\frac{1-\lambda}{2}\right)^{\frac{1}{2}} \tag{25.130a}$$

$$p_2 = -i\left(\frac{1-\lambda}{2}\right)^{\frac{1}{2}} \tag{25.130b}$$

$$p_3 = i\left(\frac{1+\lambda}{2}\right)^{\frac{1}{2}} \tag{25.130c}$$

$$p_4 = -i\left(\frac{1+\lambda}{2}\right)^{\frac{1}{2}} \tag{25.130d}$$

There are two frequencies ν_1 and ν_2, given by

$$\omega_1 = 2\pi\nu_1 = \left(\frac{1-\lambda}{2}\right)^{\frac{1}{2}} \tag{25.131}$$

$$\omega_2 = 2\pi\nu_2 = \left(\frac{1+\lambda}{2}\right)^{\frac{1}{2}} \tag{25.132}$$

For the sun–Jupiter case, the motion near a triangular equilibrium point is exemplified by a Trojan planet. Here

$$m = 0.00095388 < 0.03852$$

$$\lambda^2 \equiv 1 - 27m(1-m) = 0.974270$$

$$\lambda = 0.987051$$

$$\omega_1 = \left(\frac{1-\lambda}{2}\right)^{\frac{1}{2}} = 0.08046$$

$$\omega_2 = \left(\frac{1+\lambda}{2}\right)^{\frac{1}{2}} = 0.996757$$

If n is the Jupiter mean motion, we have

$$\tau \equiv nt \tag{25.133}$$

Let T_j be Jupiter's actual period, T the actual period of the Trojan, and P its period in τ units. By Eq. (25.133),

$$P \equiv \frac{2\pi}{\omega} = \frac{2\pi}{T_j}T \tag{25.134}$$

so that

$$T = T_j/\omega$$

Corresponding to ω_1 and ω_2, we find

$$T_1 = \frac{T_j}{0.08046} = 12.43\,T_j \tag{25.135}$$

$$T_2 = \frac{T_j}{0.996757} = 1.003254\,T_j \tag{25.136}$$

Because $T_j = 11.862$ tropical years, we find

$$T_1 = 147.4\,\text{years} \tag{25.137}$$

$$T_2 = 11.9\,\text{years} \tag{25.138}$$

These are the periods of the Trojan in the rotating (synodic) system.

VII. Further Considerations About the Collinear Points

The Exponents

Refer back to Eqs. (25.112) and (25.113). If the exponential factors are p_1, p_2, p_3, and p_4, then since $p^2 = q$, we find

$$p_{1,2}^2 = \frac{(k-2)}{2} + \frac{1}{2}(9k^2 - 8k)^{\frac{1}{2}} > 0 \tag{25.139}$$

$$p_{3,4}^2 = \frac{(k-2)}{2} - \frac{1}{2}(9k^2 - 8k)^{\frac{1}{2}} < 0 \tag{25.140}$$

where

$$k \equiv \frac{1-m}{\rho_1^3} + \frac{m}{\rho_2^3} > 1$$

ρ_1 and ρ_2 being evaluated at the equilibrium points. The signs of $p_{1,2}^2$ and $p_{3,4}^2$ are the signs of q_1 in Eq. (25.113b) and ρ_2 in Eq. (25.113d). Then

$$\begin{aligned} p_1 &= a & p_3 &= ib \\ p_2 &= -a & p_4 &= -ib \end{aligned} \tag{25.141}$$

where

$$a^2 = \tfrac{1}{2}\left[(k-2) + (9k^2 - 8k)^{\frac{1}{2}}\right] \tag{25.142}$$

$$b^2 = -\tfrac{1}{2}\left[(k-2) - (9k^2 - 8k)^{\frac{1}{2}}\right] \tag{25.143}$$

and where $a > 0$ and $b > 0$.

Motion in the Primary Plane near a Collinear Equilibrium Point

For this case, $A = 1 + 2k$, $B = 0$, $C = 1 - k$, where

$$k = \frac{1-m}{\rho_1^3} + \frac{m}{\rho_2^3} \tag{25.144}$$

where ρ_1 and ρ_2 are the distances of the equilibrium point from the primaries. Here

$$\begin{aligned} x &= x_0 + \alpha \\ y &= \beta \end{aligned} \tag{25.145}$$

The linearized equations (25.83) take the form

$$\ddot{\alpha} - 2\dot{\beta} = (1 + 2k)\alpha \tag{25.146a}$$

$$\ddot{\beta} + 2\dot{\alpha} = (1 - k)\beta \tag{25.146b}$$

The solution for either α or β is a linear combination of $\varepsilon^{a\tau}$, $\varepsilon^{-a\tau}$, $\varepsilon^{ib\tau}$, and $\varepsilon^{-ib\tau}$, where a and b are given by Eqs. (25.142) and (25.143). A real solution is a linear combination of $\varepsilon^{a\tau}$, $\varepsilon^{-a\tau}$, $\cos b\tau$, and $\sin b\tau$.

Suppose we consider only those orbits that are bounded and periodic. For such orbits, α and β will be linear combinations of $\cos b\tau$ and $\sin b\tau$, expressible as

$$\alpha = k_1 \cos(b\tau + \phi_1) \tag{25.147a}$$

$$\beta = k_2 \sin(b\tau + \phi_2) \tag{25.147b}$$

where the k's and ϕ's are constants. It is understood that the initial conditions are such as to yield zero coefficients for $\varepsilon^{a\tau}$ and $\varepsilon^{-a\tau}$.

The Orbit Is an Ellipse

We shall next show that an orbit that remains near a collinear equilibrium point and that is periodic is an ellipse in the rotating system. To show this, we first insert the expressions (25.147) into Eqs. (25.146), thereby obtaining

$$\begin{aligned} -k_1 b^2 \cos(b\tau + \phi_1) - 2k_2 b \cos(b\tau + \phi_2) \\ = (1 + 2k)k_1 \cos(b\tau + \phi_1) \end{aligned} \tag{25.148a}$$

$$\begin{aligned} -k_2 b^2 \sin(b\tau + \phi_2) - 2k_1 b \sin(b\tau + \phi_1) \\ = (1 - k)k_2 \sin(b\tau + \phi_2) \end{aligned} \tag{25.148b}$$

These equations hold for all values of τ. Let us first put $b\tau = 0$ and $b\tau = \pi/2$ in Eq. (25.148a). The results are

$$-\, k_1 b^2 \cos\phi_1 - 2k_2 b \cos\phi_2 = (1 + 2k)k_1 \cos\phi_1 \tag{25.149a}$$

$$k_1 b^2 \sin\phi_1 + 2k_2 b \sin\phi_2 = -(1 + 2k)k_1 \sin\phi_1 \tag{25.149b}$$

Doing the same in Eq. (25.148b) gives

$$-k_2 b^2 \sin\phi_2 - 2k_1 b \sin\phi_1 = (1 - k)k_2 \sin\phi_2 \tag{25.150a}$$

$$-k_2 b^2 \cos\phi_2 - 2k_1 b \cos\phi_1 = (1 - k)k_2 \cos\phi_2 \tag{25.150b}$$

Multiply Eq. (25.149b) by i, and add the result to Eq. (25.149a) to obtain

$$-\, k_1 b^2 \varepsilon^{-i\phi_1} - 2k_2 b\, \varepsilon^{-i\phi_2} = (1 + 2k)k_1 \varepsilon^{-i\phi_1} \tag{25.151a}$$

$$-\, k_2 b^2 \varepsilon^{i\phi_2} - 2k_1 b \varepsilon^{i\phi_1} = (1 - k)k_2 \varepsilon^{i\phi_2} \tag{25.151b}$$

On multiplying Eq. (25.151a) by $\varepsilon^{i\phi_2}$ and Eq. (25.152b) by $\varepsilon^{-i\phi_1}$, we find that if

$$\phi = \phi_2 - \phi_1 \tag{25.152}$$

$$-k_1 b^2 \varepsilon^{i\phi} - 2k_2 b = (1 + 2k)k_1 \varepsilon^{i\phi} \tag{25.153a}$$

$$-k_2 b^2 \varepsilon^{i\phi} - 2k_1 b = (1 - k)k_2 \varepsilon^{i\phi} \tag{25.153b}$$

or

$$\left[(1 + 2k)k_1 + k_1 b^2\right]\varepsilon^{i\phi} = -2k_2 b \tag{25.154a}$$

$$\left[(1 - k)k_2 + k_2 b^2\right]\varepsilon^{i\phi} = -2k_1 b \tag{25.154b}$$

All quantities in Eqs. (25.154) are manifestly real, except $\varepsilon^{i\phi}$. It follows that

$$\sin\phi = 0 \tag{25.155a}$$

$$\cos\phi = \pm 1 \tag{25.155b}$$

We may choose either sign in Eq. (25.155b). If we choose plus, then $\phi = 0$, and k_2/k_1 comes out minus. Because the signs of k_1 and k_2 in Eqs. (25.147) are arbitrary, we may choose either sign, and then the sine and cosine in Eqs. (25.147) have the same argument. Then

$$\alpha = k_1 \cos(b\tau + \phi_1) \tag{25.156a}$$

$$\beta = k_2 \sin(b\tau + \phi_1) \tag{25.156b}$$

where k_1 and k_2 are opposite in sign. It follows that

$$\frac{\alpha^2}{k_1^2} + \frac{\beta^2}{k_2^2} = 1 \tag{25.157}$$

so that the orbit in the rotating system is an ellipse, with principal axes along the axes of the primary system.

The Amplitudes k_1 and k_2

We have to examine k_2/k_1 to find which axis, α or β, is the major axis and to find the eccentricity of the ellipse. From Eq. (25.154a), with $\phi = 0$

$$-\frac{k_2}{k_1} = \frac{1 + 2k + b^2}{2b} \tag{25.158a}$$

$$-\frac{k_2}{k_1} = \frac{2b}{1 - k + b^2} \tag{25.158b}$$

If we equate Eqs. (25.158a) and (25.158b), we find

$$b^4 + (k - 2)b^2 + (1 + 2k)(1 - k) = 0 \tag{25.159}$$

with solutions

$$b^2 = \tfrac{1}{2}\left[(2 - k) \pm (9k^2 - 8k)^{\frac{1}{2}}\right] \tag{25.160}$$

Because $k > 1$, we have $9k^2 - 8k \equiv 8k^2 - 8k + k^2 > k^2$. Using the minus sign in Eq. (25.160) would yield $b^2 < 0$, so that we must use the plus sign, and then

$$2b^2 = 2 - k + (9k^2 - 8k)^{\frac{1}{2}}$$

in agreement with Eq. (25.143). Thus

$$R \equiv \left|\frac{k_2}{k_1}\right| = -\frac{k_2}{k_1} = \frac{1 + 2k + b^2}{2b} \tag{25.161}$$

Because $k > 1$, this gives

$$2R > b + (3/b) \tag{25.162}$$

The function

$$\psi(b) \equiv b + (3/b) \tag{25.163}$$

has the minimum value $2\sqrt{3}$, so that

$$R > \sqrt{3} \tag{25.164}$$

Now $|k_2/k_1|$ is the ratio of the β axis to the α axis, so that the β axis is the major axis. If e is the eccentricity

$$1 - e^2 = R^{-2} < 1/3$$

and $e^2 > 2/3$, so that $e > (2/3)^{1/2} \approx 0.816$. Of course, $1 - e^2 = R^{-2}$, and either of Eqs. (25.158), along with Eq. (25.143), will yield the eccentricity as an explicit function of

$$k = (1 - m)\rho_1^{-3} + m\rho_2^{-3}$$

a rather complicated function. What is worthy of note is that e depends only on m, ρ_1, and ρ_2, i.e., only on the primary masses and their separation, and not at all on the initial conditions. The initial conditions must be such as to make the motion bounded and periodic.

The Sense of Circulation

If we use polar coordinates η and θ for the displacement of the orbiter from the collinear equilibrium point, then

$$\begin{aligned} \alpha &= \eta \cos\theta \\ \beta &= \eta \sin\theta \end{aligned} \tag{25.165}$$

Therefore,

$$\begin{aligned} \tan\theta &= \beta/\alpha \\ \dot{\theta}\sec^2\theta &= \frac{\alpha\dot{\beta} - \beta\dot{\alpha}}{\alpha^2} \end{aligned} \tag{25.166}$$

If the motion is periodic, Eqs. (25.156) have to be satisfied. Then

$$\alpha = k_1 \cos(b\tau + \phi_1) \tag{25.156a}$$

$$\beta = k_2 \sin(b\tau + \phi_1) \tag{25.156b}$$

$$\dot{\alpha} = -k_1 b \sin(b\tau + \phi_1) \tag{25.167a}$$

$$\dot{\beta} = k_2 b \cos(b\tau + \phi_1) \tag{25.167b}$$

and

$$\alpha\dot{\beta} = bk_1k_2 \cos^2(b\tau + \phi_1)$$

$$\beta\dot{\alpha} = -bk_1k_2 \sin^2(b\tau + \phi_1)$$

so that

$$\alpha\dot{\beta} - \beta\dot{\alpha} = bk_1k_2 \tag{25.168}$$

From Eqs. (25.166) and (25.168)

$$\dot{\theta}\sec^2\theta = \frac{bk_1k_2}{\alpha^2} \tag{25.169}$$

Here, k_1 and k_2 are opposite in sign and $b > 0$. Thus

$$\dot{\theta} < 0 \tag{25.170}$$

The circulation of the orbiter around a collinear equilibrium point, when the motion is bounded and periodic, is retrograde relative to the motion of the primaries around their center of mass.

References

[1]Pollard, H., *Mathematical Introduction to Celestial Mechanics,* Prentice–Hall, Englewood Cliffs, NJ, 1966.

[2]Brouwer, D., and Clemence, G., *Methods of Celestial Mechanics,* Academic Press, New York, 1961, p. 562.

Chapter 26

Staeckel Systems

I. Staeckel's Theorem

SEPARABLE systems occur often in the theory of orbits, and they have all been of the Staeckel type, which we shall now consider.

For an orthogonal coordinate system with metric

$$ds^2 = \sum_{k=1}^{3} A_k^{-1}\, dq_k^2 \tag{26.1}$$

the kinetic energy of a particle of unit mass is

$$T = \frac{1}{2}\sum_{k=1}^{3} A_k^{-1}\dot{q}_k^2 \tag{26.2}$$

The generalized momenta are

$$p_k = \frac{\partial T}{\partial \dot{q}_k} = A_k^{-1}\dot{q}_k \tag{26.3}$$

If the potential energy is

$$V = V(q_1, q_2, q_3)$$

the Hamiltonian is

$$H = \frac{1}{2}\sum_{k=1}^{3} A_k p_k^2 + V \tag{26.4}$$

and the Hamilton–Jacobi equation is

$$\frac{1}{2}\sum_{k=1}^{3} A_k\left(\frac{\partial W}{\partial q_k}\right)^2 + V(q) = \alpha_1 \tag{26.5}$$

Staeckel's theorem states that, the A_k being all positive, the *HJ* equation is separable if and only if there exists a 3×3 matrix (ϕ_{kj}), where ϕ_{kj} depends only on q_k, and a column matrix (ψ_1, ψ_2, ψ_3), where ψ_k depends only on q_k, such that

$$\sum_{k=1}^{3} A_k\phi_{kj}(q_k) = \delta_{1j} \tag{26.6}$$

$$\sum_{k=1}^{3} A_k\psi_k(q_k) = V \tag{26.7}$$

Proof of Necessity: Given a solution of Eq. (26.5), viz.,

$$W = W_1(q_1, \alpha_1, \alpha_2, \alpha_3) + W_2(q_2, \alpha_1, \alpha_2, \alpha_3) + W_3(q_3, \alpha_1, \alpha_2, \alpha_3) \tag{26.8}$$

show that functions $\phi_{kj}(q_k)$ and $\psi_k(q_k)$ exist, satisfying Eqs. (26.6) and (26.7).

In proving this statement, we shall let Eq. (26.8) be a complete integral of Eq. (26.5). This is one depending on three arbitrary constants α_1, α_2, and α_3 with determinant

$$\det\left[\frac{\partial^2 W}{\partial q_k \partial \alpha_j}\right] \neq 0 \tag{26.9}$$

To prove necessity, substitute Eq. (26.8) into Eq. (26.5). Then

$$\frac{1}{2}\sum_{k=1}^{3} A_k\left(\frac{\partial W_k}{\partial q_k}\right)^2 + V(q) = \alpha_1 \tag{26.10}$$

Differentiate Eq. (26.10) successively with respect to α_1, α_2, and α_3 to find

$$\sum_{k=1}^{3} A_k \frac{\partial W_k}{\partial q_k}\frac{\partial^2 W}{\partial q_k \partial \alpha_1} = 1 \tag{26.11a}$$

$$\sum_{k=1}^{3} A_k \frac{\partial W_k}{\partial q_k}\frac{\partial^2 W}{\partial q_k \partial \alpha_2} = 0 \tag{26.11b}$$

$$\sum_{k=1}^{3} A_k \frac{\partial W_k}{\partial q_k}\frac{\partial^2 W}{\partial q_k \partial \alpha_3} = 0 \tag{26.11c}$$

This is a system of linear equations for the A_k's with determinant

$$D = \frac{\partial W_1}{\partial q_1}\frac{\partial W_2}{\partial q_2}\frac{\partial W_3}{\partial q_3}\det\left[\frac{\partial^2 W}{\partial q_k \partial \alpha_j}\right] \neq 0 \tag{26.12}$$

by the hypothesis of the completeness of the integral, so that Eqs. (26.11) are all independent.

The coefficient of each A_k in Eqs. (26.11) is a function only of q_k. Thus, there exist functions $\phi_{kj}(q_k)$ satisfying Eq. (26.6). They are

$$\phi_{k1}(q_k) = \frac{\partial W_k}{\partial q_k}\frac{\partial^2 W}{\partial q_k \partial \alpha_1} \tag{26.13a}$$

$$\phi_{k2}(q_k) = \frac{\partial W_k}{\partial q_k}\frac{\partial^2 W}{\partial q_k \partial \alpha_2} \tag{26.13b}$$

$$\phi_{k3}(q_k) = \frac{\partial W_k}{\partial q_k}\frac{\partial^2 W}{\partial q_k \partial \alpha_3} \tag{26.13c}$$

Next, we have to show that functions $\psi_k(q_k)$ exist, satisfying Eq. (26.7). From Eq. (26.5)

$$V = \alpha_1 - \frac{1}{2}\sum_{k=1}^{3} A_k\left(\frac{\partial W_k}{\partial q_k}\right)^2 \tag{26.14}$$

because $\partial W/\partial q_k = \partial W_k/\partial q_k$. Now, since we have shown that the functions $\phi_{k1}(q_k)$ satisfy Eq. (26.6) with $\delta_{1k} = 1$, we have

$$\alpha_1 = \alpha_1 \sum_{k=1}^{3} A_k \phi_{k1}(q_k) = \sum_{k=1}^{3} A_k \alpha_1 \phi_{k1}(q_k) \tag{26.15}$$

Inserting Eq. (26.15) into Eq. (26.14), we find

$$V = \sum_{k=1}^{3} A_k \left\{ \alpha_1 \phi_{k1}(q_k) - \frac{1}{2}\left(\frac{\partial W_k}{\partial q_k}\right)^2 \right\} \tag{26.16}$$

so that Eq. (26.7) is satisfied, with

$$\psi_k(q_k) = \alpha_1 \phi_{k1}(q_k) - \frac{1}{2}\left(\frac{\partial W_k}{\partial q_k}\right)^2 \tag{26.17}$$

This completes the proof of necessity.

To prove sufficiency, we have to begin with Eqs. (26.6) and (26.7) and show that they lead to the separability of Eq. (26.5). To do so, first insert Eq. (26.7) into Eq. (26.5):

$$\frac{1}{2}\sum_{k=1}^{3} A_k \left(\frac{\partial W_k}{\partial q_k}\right)^2 + \sum_{k=1}^{3} A_k \psi_k(q_k) = \alpha_1 \tag{26.18}$$

Next, from Eq. (26.6)

$$\sum_{k=1}^{3} A_k \phi_{k2}(q_k) = 0 \tag{26.19a}$$

$$\sum_{k=1}^{3} A_k \phi_{k3}(q_k) = 0 \tag{26.19b}$$

Multiply Eq. (26.19a) by an arbitrary constant α_2 and Eq. (26.19b) by an arbitrary constant α_3, add the results to Eq. (26.18) and use Eq. (26.15). We obtain

$$\sum_{k=1}^{3} A_k \left[\frac{1}{2}\left(\frac{\partial W_k}{\partial q_k}\right)^2 + \psi_k(q_k)\right] = \sum_{k=1}^{3} A_k [\alpha_1 \phi_{k1}(q_k) + \alpha_2 \phi_{k2}(q_k) + \alpha_3 \phi_{k3}(q_k)] \tag{26.19c}$$

or

$$\sum_{k=1}^{3} A_k \left[\frac{1}{2}\left(\frac{\partial W_k}{\partial q_k}\right)^2 - \{\alpha_1 \phi_{k1}(q_k) + \alpha_2 \phi_{k2}(q_k) + \alpha_3 \phi_{k3}(q_k) - \psi_k(q_k)\}\right] = 0 \tag{26.19d}$$

Here, ψ_k and the ϕ_{kj}'s depend only on q_k. The *HJ* equation is then satisfied if

$$W = W_1(q_1) + W_2(q_2) + W_3(q_3) \tag{26.20}$$

with

$$\left(\frac{\partial W_k}{\partial q_k}\right)^2 = 2\{\alpha_1 \phi_{k1}(q_k) + \alpha_2 \phi_{k2}(q_k) + \alpha_3 \phi_{k3}(q_k) - \psi_k(q_k)\} \tag{26.21}$$

It is separable. This completes the proof of sufficiency.

II. Staeckel Systems

We define a Staeckel system as a system described by Eq. (26.4) as its Hamiltonian and the auxiliary conditions (26.6) and (26.7). We may simplify this definition.

Let A be the row matrix (A_1, A_2, A_3) and Φ the square matrix $[\phi_{kj}(q_k)]$. Then by Eq. (26.6)

$$A\Phi = (1, 0, 0) \tag{26.22}$$

With the requirement that Φ^{-1} exists, we find

$$A = (1, 0, 0)\Phi^{-1} \tag{26.23}$$

On writing this out, we find

$$(A_1, A_2, A_3) = \left(\Phi_{11}^{-1}, \Phi_{12}^{-1}, \Phi_{13}^{-1}\right) \tag{26.24}$$

or

$$A_k = (\Phi^{-1})_{1k} \tag{26.25}$$

Now, if Φ is a 3×3 square matrix and x is a column matrix of three elements, then

$$\Phi x = y \tag{26.25a}$$

is also a column matrix of three elements. Equation (26.25a) is a set of three linear equations for the x's. Solution of Eq. (26.25a) by Kramer's rule gives

$$x_1 = \Delta^{-1}(y_1 M_{11} + y_2 M_{21} + y_3 M_{31}) \tag{26.25b}$$

where Δ is the determinant of Φ and M_{k1} is the cofactor of Φ_{k1} in Φ. From Eq. (26.25a), we can also write

$$x = \Phi^{-1} y \tag{26.25c}$$

so that

$$x_1 = \Phi_{11}^{-1} y_1 + \Phi_{12}^{-1} y_2 + \Phi_{13}^{-1} y_3 \tag{26.25d}$$

Comparison of Eqs. (26.25d) and (26.25b) shows that

$$(\Phi^{-1})_{1k} = \frac{M_{k1}}{\Delta} \tag{26.25e}$$

We may now redefine a Staeckel system as an orthogonal system with Hamiltonian

$$H = \frac{1}{2}\sum_{k=1}^{3} A_k(q_1, q_2, q_3) p_k^2 + V \tag{26.26}$$

where there exist functions $\phi_{kj}(q_k)$ and $\psi_k(q_k)$ such that

$$A_k = (\Phi^{-1})_{1k} \tag{26.27}$$

$$V = \sum_{k=1}^{3} A_k \psi_k(q_k) \tag{26.28}$$

where

$$\Phi \equiv [\phi_{kj}(q_k)] \qquad \Delta = \det \Phi \neq 0$$

and

$$(\Phi^{-1})_{1k} = \frac{M_{k1}}{\Delta} \tag{26.29}$$

M_{k1} being the cofactor of ϕ_{k1} in Δ.

We can now write the Hamiltonian (26.26) as

$$H = \sum_{k=1}^{3} (\Phi^{-1})_{1k} \left(\frac{1}{2} p_k^2 + \psi_k(q_k) \right) = \alpha_1 \tag{26.30}$$

or

$$\left(\Phi^{-1} \begin{bmatrix} \frac{1}{2} p_1^2 + \psi_1(q_1) \\ \frac{1}{2} p_2^2 + \psi_2(q_2) \\ \frac{1}{2} p_3^2 + \psi_3(q_3) \end{bmatrix} \right)_1 = H = \alpha_1 \tag{26.31}$$

This is satisfied if

$$\Phi^{-1} \begin{bmatrix} \frac{1}{2} p_1^2 + \psi_1(q_1) \\ \frac{1}{2} p_2^2 + \psi_2(q_2) \\ \frac{1}{2} p_3^2 + \psi_3(q_3) \end{bmatrix} = \begin{bmatrix} \alpha_1 \\ \alpha_2 \\ \alpha_3 \end{bmatrix} \tag{26.32}$$

where α_2 and α_3 are arbitrary constants. This gives

$$\frac{1}{2} p_k^2 + \psi_k(q_k) = \left(\Phi \begin{bmatrix} \alpha_1 \\ \alpha_2 \\ \alpha_3 \end{bmatrix} \right)_k \tag{26.33}$$

or

$$\frac{1}{2} p_k^2 + \psi_k(q_k) = \sum_{i=1}^{3} \phi_{ki}(q_k) \alpha_i \tag{26.34}$$

but Eq. (26.34) leads to separability at once. Thus, in Eq. (26.26), with V given by Eq. (26.28), the condition

$$A_k = (\Phi^{-1})_{1k} \tag{26.35}$$

where the elements of Φ are $\phi_{kj}(q_k)$, is necessary and sufficient that the *HJ* equation (26.18) be separable. This condition (26.35) is called the Staeckel condition.

III. The Staeckel Integrals

From Eq. (26.32) we have

$$\frac{1}{2} \sum_{j=1}^{3} (\Phi^{-1})_{kj} p_j^2 + \sum_{j=1}^{3} (\Phi^{-1})_{kj} \psi_j(q_j) = \alpha_k \qquad k = 1, 2, 3 \tag{26.36}$$

For each value of k, Eq. (26.36) gives an integral of the motion. If we multiply Eq. (26.32) by Φ on the left, we obtain

$$\begin{bmatrix} \frac{1}{2}p_1^2 + \psi_1(q_1) \\ \frac{1}{2}p_2^2 + \psi_2(q_2) \\ \frac{1}{2}p_3^2 + \psi_3(q_3) \end{bmatrix} = \Phi \begin{bmatrix} \alpha_1 \\ \alpha_2 \\ \alpha_3 \end{bmatrix} \tag{26.37}$$

leading to equations for the p's, viz.,

$$p_k^2 = -2\psi_k(q_k) + 2\sum_{j=1}^{3} \Phi_{kj}(q_k)\alpha_j \tag{26.38}$$

IV. An Example: The Kepler Problem

By Chapter 6, if θ is the latitude, the Hamiltonian for the Kepler problem is

$$H = \frac{1}{2}\left(p_r^2 + \frac{p_\theta^2}{r^2} + \frac{p_\phi^2}{r^2\cos^2\theta}\right) - \frac{\mu}{r} = \alpha_1 \tag{26.39}$$

To agree with the notation of this chapter, one has to replace the α_2 and α_3 of Chapter 6 as follows:

$$\begin{aligned} \alpha_2^2 &\to 2\alpha_2 \\ \alpha_3^2 &\to 2\alpha_3 \end{aligned} \tag{26.40}$$

If $p_r = p_1$, $p_\theta = p_2$, $p_\phi = p_3$, the equations of Chapter 6 become

$$p_1^2 = r^{-2}(-2\alpha_2 + 2\mu r + 2\alpha_1 r^2) \tag{26.41a}$$

$$p_2^2 = 2\alpha_2 - 2\alpha_3 \sec^2\theta \tag{26.41b}$$

$$p_3^2 = 2\alpha_3 \tag{26.41c}$$

If we compare Eqs. (26.41) with Eq. (26.38), we obtain

$$\psi_1 = -\mu/r \qquad \psi_2 = 0 \qquad \psi_3 = 0 \tag{26.42}$$

$$\Phi = \begin{bmatrix} 1 & -r^{-2} & 0 \\ 0 & 1 & -\sec^2\theta \\ 0 & 0 & 1 \end{bmatrix} \tag{26.43}$$

Because $\Delta \equiv \det \Phi = 1$, we have

$$(\Phi^{-1})_{jk} = M_{kj}/\Delta = M_{kj} \tag{26.44}$$

Thus

$$\Phi^{-1} = \begin{bmatrix} 1 & r^{-2} & r^{-2}\sec^2\theta \\ 0 & 1 & \sec^2\theta \\ 0 & 0 & 1 \end{bmatrix} \tag{26.45}$$

If we solve Eqs. (26.41) for the α's, we find

$$\begin{aligned}\alpha_1 &= \frac{1}{2}p_1^2 + \frac{1}{2r^2}p_2^2 + \frac{\sec^2\theta}{2r^2}p_3^2 - \frac{\mu}{r}\\ \alpha_2 &= \frac{1}{2}p_2^2 + \frac{\sec^2\theta}{2}p_3^2\\ \alpha_3 &= \frac{1}{2}p_3^2\end{aligned} \tag{26.46}$$

On writing out Eq. (26.36), with use of Eqs. (26.42), we obtain

$$\begin{aligned}\alpha_1 &= \frac{1}{2}\Phi_{11}^{-1}p_1^2 + \frac{1}{2}\Phi_{12}^{-1}p_2^2 + \frac{1}{2}\Phi_{13}^{-1}p_3^2 - \frac{\mu}{r}\Phi_{11}^{-1}\\ \alpha_2 &= \frac{1}{2}\Phi_{21}^{-1}p_1^2 + \frac{1}{2}\Phi_{22}^{-1}p_2^2 + \frac{1}{2}\Phi_{23}^{-1}p_3^2 - \frac{\mu}{r}\Phi_{21}^{-1}\\ \alpha_3 &= \frac{1}{2}\Phi_{31}^{-1}p_1^2 + \frac{1}{2}\Phi_{32}^{-1}p_2^2 + \frac{1}{2}\Phi_{33}^{-1}p_3^2 - \frac{\mu}{r}\Phi_{31}^{-1}\end{aligned} \tag{26.47}$$

Comparison of Eqs. (26.46) and (26.47) yields

$$\Phi^{-1} = \begin{bmatrix} 1 & r^{-2} & r^{-2}\sec^2\theta \\ 0 & 1 & \sec^2\theta \\ 0 & 0 & 1 \end{bmatrix} \tag{26.48}$$

in agreement with Eq. (26.45).

V. General Remarks About Separable Systems

References 1 and 2 illustrated that all the separable cases of particle motion in Euclidean space are Staeckelian or reducible to Staeckelian by a point transformation. The qualification is easily explained. In oblique coordinates, the motion of a projectile in a uniform field is not Staeckelian but is reducible to such by a point transformation to rectangular coordinates.

The list of the 11 possible coordinate systems for separability of particle motion in Euclidean space is[3]: rectangular, spherical, cylindrical, parabolic, prolate spheroidal, oblate spheroidal, parabolic cylindrical, conical, elliptic cylindrical, paraboloidal, and ellipsoidal.

Systems may be classified as follows: 1) Staeckelian and Euclidean (Kepler problem); 2) Staeckelian and non-Euclidean (spherical pendulum and particle in a parabolic bowl); 3) Separable and non-Euclidean, but not Staeckelian (symmetric top, with one point fixed, in a uniform field; the cross-product terms in the momenta making it non-Staeckelian); 4) Euclidean, if properly scaled, but not separable (three-body problem); and 5) non-Euclidean and nonseparable (asymmetric top) (see Fig. 26.1).

VI. Motion According to $\dot{x}_2 = F(x)$

This section is a necessary preliminary to the next one on conditionally periodic systems; for more details see Ref. 4.

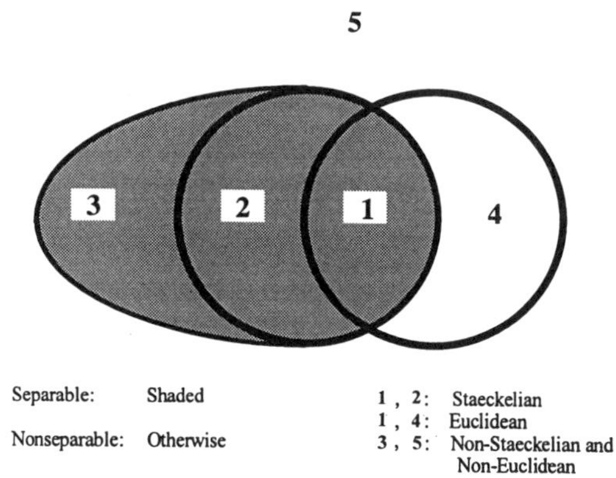

Fig. 26.1 A set-theoretical diagram.

Suppose a particle with coordinate x moves according to

$$\dot{x}^2 = F(x) \tag{26.49}$$

If $F(x)$ has a zero at $x = a$, it may be a simple zero, so that

$$F(x) = (a - x)\psi(x) \tag{26.50}$$

where $\psi(x)$ has no factor $a - x$. (It happens to be convenient here to write $a - x$ rather than $x - a$.) It may also be a multiple zero, the curve $y = F(x)$ being tangent to the x axis at $x = a$. Then

$$F(x) = (a - x)^s \psi(x) \tag{26.51}$$

where $s > 1$. For values of x close to a, we can see what is happening by taking ψ to be a constant k^2. Then

$$F(x) = k^2(a - x)^s \tag{26.52}$$

and by Eqs. (26.49) and (26.52)

$$\dot{x} = k(a - x)^{s/2} \tag{26.53}$$

where it is convenient to choose the plus sign, in order to consider motion from $x = a - \eta$ to $x = a$. Thus

$$\mathrm{d}t = k^{-1}(a - x)^{-s/2}\,\mathrm{d}x \tag{26.54}$$

The time Δt for passage from $x = a - \eta$ to $x = a$ is

$$\Delta t = k^{-1} \int_{a-\eta}^{a} (a - x)^{-s/2}\,\mathrm{d}x \tag{26.55}$$

If $u = a - x$, then

$$\Delta t = k^{-1} \int_0^{\eta} u^{-s/2}\,\mathrm{d}u \tag{26.56}$$

If $s = 1$, this becomes

$$\Delta t = 2k^{-1}\eta^{\frac{1}{2}} \tag{26.57}$$

but it diverges for $s > 1$. Thus, any zero at $x = a$ leads to accessibility of the particle to $x = a$ only in an infinite time, unless it is a simple zero, for which $s = 1$. A simple zero can be reached in a finite time.

If we are dealing with Staeckel systems by Hamiltonian methods, we have to deal with equations such as $p_k^2 = F(q_k)$, where F depends only on the single coordinate q_k, and p_k will always be proportional to $\dot{q}_k$. Near a zero of $F(q_k)$, we shall have, approximately, that

$$\dot{q}_k^2 = k_1 F(q_k) \tag{26.58}$$

where $k_1 > 0$. The preceding analysis showed that any zero of $F(q_k)$ must be a simple zero if it can be reached in finite time. Furthermore, if q_k oscillates between two values a and b, it follows that the necessary form for $F(q_k)$ is

$$F(q_k) = (q_k - a)(b - q_k)\psi(q_k) \tag{26.59}$$

where $\psi(q_k)$ has no zeros and

$$a \leq q_k \leq b \tag{26.60}$$

VII. Conditionally Periodic Staeckel Systems

The physical pendulum is a simple Staeckel system. It can have three types of motion. It may move as in a clock, back and forth from an angle $-\theta_m$ to $+\theta_m$; this is libration. It may have enough energy to keep going in a circle; this is circulation. Finally, it may have just enough energy to approach $\theta = 180°$ in an infinite time; this is asymptotic motion.

A bounded Staeckel system can, in general, have q's that vary in all three ways. If it has only circulational and librational coordinates, it is called conditionally periodic.

Circulational Coordinates

A coordinate q_k is circulational if all these conditions hold:

1) If it is an angle.

2) If $p_k^2 = F(q_k)$, with $p_k > 0$ for $\dot{q}_k > 0$ and $p_k < 0$ for $\dot{q}_k < 0$.

3) If F_k is so bounded that there exist constants c_{1k} and c_{2k} satisfying $c_{2k} \geq F_k(q_k) \geq c_{1k} > 0$.

4) If $F_k(q_k + 2\pi) = F_k(q_k)$.

Note that $c_{1k} > 0$ rules out asymptotic motion and that the condition 4 may be either periodicity or constancy. For an artificial satellite, for example, $p_\phi^2 = \text{const}$ if the potential V does not depend on ϕ, the right ascension.

From the preceding conditions

$$p_k \geq (c_{1k})^{\frac{1}{2}} \qquad \text{if} \qquad \dot{q}_k > 0$$

$$p_k \leq -(c_{1k})^{\frac{1}{2}} \qquad \text{if} \qquad \dot{q}_k < 0$$

In either case

$$v_k \equiv \int_{q_{k0}}^{q_k} p_k^{-1}\, \mathrm{d}q_k \tag{26.61}$$

in a single-valued differentiable (SVD) function of q_k, with dv_k/dq_k existing and positive for all q_k. It is a monotonically increasing function of q_k. Conversely, q_k is a SVD function of v_k.

Librational Coordinates

A librational coordinate is one that fluctuates back and forth between values a_k and b_k. From the previous section, this means that $p_k^2 = F(q_k)$ has zeros only at a_k and b_k and that they are simple zeros. These facts lead to the following specification: q_k is librational if there exist constants a_k, b_k, c_{1k}, c_{2k}, and a function $G_k(q_k)$ such that

$$c_{2k} \geq G_k(q_k) \geq c_{1k} > 0 \qquad \text{for} \qquad a_k \leq q_k \leq b_k \tag{26.61a}$$

with $a_k \leq q_k(0) \leq b_k$, where $q_k(0)$ is the initial value of q_k as a function of time t and where

$$p_k^2 = (q_k - a_k)(b_k - q_k)G_k(q_k) \tag{26.62}$$

It may be difficult to find $G_k(q_k)$. Consider the Kepler problem with

$$p_\theta^2 = \alpha_2^2 - \alpha_3^2 \sec^2\theta \tag{26.63}$$

where θ is the latitude. The inclination I is given by

$$\cos I = \alpha_3/\alpha_2 \tag{26.63a}$$

so that

$$p_\theta^2 = \alpha_2^2(1 - \cos^2 I \sec^2\theta) = F(\theta) \tag{26.64}$$

For direct orbits, $F(\theta)$ has zeros for $\theta = \pm I$. To show that Eq. (26.64) can be put into the form of Eq. (26.62), note that

$$\frac{dF(\theta)}{d\theta} = -2\alpha_2^2 \cos^2 I \sec^2\theta \tan\theta \tag{26.65}$$

For

$$\frac{dF(\theta)}{d\theta} = \mp 2\alpha_2^2 \tan I \tag{26.66}$$

Thus, for $I \neq 0$, the derivative does not vanish at the zeros of $F(\theta)$, so that these zeros are simple zeros. This completes the proof that Eq. (26.64) can be expressed in the form of Eq. (26.62), which becomes

$$p_\theta^2 = (I^2 - \theta^2)G(\theta) \tag{26.67}$$

but it does not find the upper and lower limits on $G(\theta)$.

By Eq. (26.63a)

$$\alpha_2^2 = \alpha_3^2 \sec^2 I \tag{26.68}$$

Then by Eqs. (26.64) and (26.68)

$$p_\theta^2 = \alpha_3^2(\sec^2 I - \sec^2\theta) = \alpha_3^2(\tan^2 I - \tan^2\theta) \tag{26.69}$$

By Eqs. (26.67) and (26.69)

$$G(\theta) = \alpha_3^2 \frac{\tan^2 I - \tan^2 \theta}{I^2 - \theta^2} \tag{26.70}$$

an even function of θ. To investigate its behavior, we need consider only the range

$$0 < \theta < I < \pi/2 \tag{26.70a}$$

Note that

$$G(0) = \left(\alpha_3^2 \tan^2 I / I^2\right) \tag{26.71}$$

At $\theta = I$, $G(\theta)$ takes the form 0/0, but by L'Hospital's rule

$$G(I) = \alpha_3^2 \frac{\tan I}{I} \sec^2 I \tag{26.72}$$

One suspects that $G(0)$ and $G(I)$ are the lower and upper limits of $G(\theta)$. To verify that $G(0)$ is the lower limit, form

$$\frac{\tan^2 I - \tan^2 \theta}{I^2 - \theta^2} - \frac{\tan^2 I}{I^2} = \frac{I^2 \theta^2}{I^2(I^2 - \theta^2)} \left(\frac{\tan^2 I}{I^2} - \frac{\tan^2 \theta}{\theta^2} \right) \tag{26.73}$$

Now, from Pierce's integral tables[5]

$$\frac{\tan x}{x} = 1 + \frac{x^2}{3} + \frac{2x^2}{15} + \cdots \qquad (x^2 < \pi^2/4) \tag{26.74}$$

so that for $0 < \theta < I < \pi/2$

$$\frac{\tan^2 I}{I^2} \geq \frac{\tan^2 \theta}{\theta^2} \tag{26.75}$$

Therefore, by Eqs. (26.70) and (26.75)

$$G(\theta) \geq \frac{\alpha_3^2 \tan^2 I}{I^2} \tag{26.76}$$

and $G(0)$ is the lower limit.

For the upper limit, write

$$\frac{\tan^2 I - \tan^2 \theta}{I^2 - \theta^2} = \frac{\tan I + \tan \theta}{I + \theta} \frac{\tan I - \tan \theta}{I - \theta} \tag{26.77}$$

Compare $(\tan I)/I$ with the first factor on the right and $\sec^2 I$ with the second factor.

$$\frac{\tan I}{I} - \frac{\tan I + \tan \theta}{I + \theta} = \frac{\theta \tan I - I \tan \theta}{I(I + \theta)} = \frac{\theta}{(I + \theta)} \left(\frac{\tan I}{I} - \frac{\tan \theta}{\theta} \right) \geq 0 \tag{26.78}$$

for $0 < \theta < I < \pi/2$.

Now

$$\tan I - \tan \theta = \frac{\sin I \cos \theta - \cos I \sin \theta}{\cos I \cos \theta} \tag{26.79}$$

$$f \equiv \sec^2 I - \frac{\tan I - \tan \theta}{I - \theta} = \sec^2 I - \frac{\sin(I - \theta)}{(I - \theta) \cos I \cos \theta} \tag{26.80}$$

For $0 < \theta < I < \pi/2$

$$\frac{\sin(I-\theta)}{I-\theta} < 1$$

Thus

$$f \geq \sec^2 I - \sec I \sec\theta \geq \sec I(\sec I - \sec\theta) \geq 0$$

so that

$$\frac{\tan I - \tan\theta}{I-\theta} \leq \sec^2 I \tag{26.81}$$

From Eqs. (26.77), (26.78), and (26.81)

$$\frac{\tan^2 I - \tan^2\theta}{I^2-\theta^2} \leq \frac{\tan I}{I}\sec^2 I \tag{26.82}$$

Thus, $G(I)$ is the upper limit of $G(\theta)$.

We now define v_k as before, viz.,

$$v_k \equiv \int_{q_{k0}}^{q_k} p_k^{-1}\,\mathrm{d}q_k \tag{26.83}$$

From Eq. (26.62)

$$v_k \equiv \int_{q_{k0}}^{q_k} \pm[(q_k-a_k)(b_k-q_k)G_k(q_k)]^{-\frac{1}{2}}\,\mathrm{d}q_k \tag{26.84}$$

Because $p_k > 0$ for $\dot{q}_k > 0$ and $p_k < 0$ for $\dot{q}_k < 0$, the upper sign goes with $dq_k > 0$ and the lower with $dq_k < 0$. To show also in this case that v_k is a SVD function of q_k, introduce a uniformizing variable E_k, such that $\dot{E}_k > 0$ for all q_k and

$$2q_k = a_k + b_k + (a_k - b_k)\cos E_k \tag{26.85}$$

This gives maximum $q_k = b_k$ for $\cos E_k = -1$ and minimum $q_k = a_k$ for $\cos E_k = 1$. This definition of E_k covers all values of q_k in the interval $a_k \leq q_k \leq b_k$.

$$2(q_k - a_k) = (b_k - a_k)(1 - \cos E_k) \tag{26.86}$$

$$2(b_k - q_k) = (b_k - a_k)(1 + \cos E_k) \tag{26.87}$$

$$4(q_k - a_k)(b_k - q_k) = (b_k - a_k)^2 \sin^2 E_k \tag{26.88}$$

$$[(q_k - a_k)(b_k - q_k)]^{-\frac{1}{2}} = 2(b_k - a_k)^{-1}|\sin E_k|^{-1} \tag{26.89}$$

From Eq. (26.85)

$$\mathrm{d}q_k = \tfrac{1}{2}(b_k - a_k)\sin E_k\,\mathrm{d}E_k \tag{26.90}$$

so that

$$\pm[(q_k - a_k)(b_k - q_k)]^{-\frac{1}{2}}\,\mathrm{d}q_k = \pm\frac{\sin E_k\,\mathrm{d}E_k}{|\sin E_k|} \tag{26.91}$$

We saw that in Eq. (26.84) the upper sign goes with $dq_k > 0$ and the lower with $dq_k < 0$. By Eq. (26.90), because $\mathrm{d}E_k > 0$ for all q_k, it follows that $\sin E_k > 0$

for the upper sign and $\sin E_k < 0$ for the lower sign. Thus, Eq. (26.91) becomes

$$\pm[(q_k - a_k)(b_k - q_k)]^{-\frac{1}{2}}\,\mathrm{d}q_k = \mathrm{d}E_k \tag{26.92}$$

Insertion of this into Eq. (26.84) gives

$$v_k \equiv \int_{E_{k0}}^{E_k} G_k^{-\frac{1}{2}}\,\mathrm{d}E_k \tag{26.93}$$

Because $c_{2k} \geq G_k(q_k) \geq c_{1k} > 0$, it follows that v_k is a SVD function of E_k, monotonically increasing with E_k. This means E_k is a SVD function of v_k. From Eq. (26.85), q_k is a SVD function of E_k. Thus, q_k is a SVD function of v_k.

Summary

In a conditionally periodic system, each coordinate is a single-valued differentiable function of

$$v_k \equiv \int_{q_{k0}}^{q_k} p_k^{-1}\,\mathrm{d}q_k \tag{26.94}$$

VIII. Action and Angle Variables

Before we can go further with conditionally periodic systems, we need to introduce a new set of canonical variables, called action and angle variables.

We first define a single cycle of q_k as an increase of 2π if q_k is circulational and as a round-trip from a_k to b_k if q_k is librational. A small circle on an integral sign will denote an integral over one cycle.

The following quantities J_k are called action variables:

$$J_k = \oint p_k\,\mathrm{d}q_k \qquad k = 1, 2, 3 \tag{26.95}$$

By Eq. (26.38), we have for a Staeckel system

$$p_k = \pm\left[-2\psi_k(q_k) + 2\sum_{i=1}^{3}\Phi_{ki}(q_k)\alpha_i\right]^{\frac{1}{2}} \tag{26.96}$$

Thus

$$J_k = \oint \pm\left[-2\psi_k(q_k) + 2\sum_{i=1}^{3}\Phi_{ki}(q_k)\alpha_i\right]^{\frac{1}{2}}\mathrm{d}q_k \tag{26.97}$$

so that

$$J_k = J_k(\alpha_1, \alpha_2, \alpha_3) \tag{26.98}$$

The J_k's are functions of the α's. If we express the α's as functions of the J's, then the *HJ* function W becomes expressible as

$$W = W(q, J) \tag{26.99}$$

Let this be the generating function for a canonical transformation, where the q's are the "old" coordinates and the J's are the "new" momenta. If we denote the "new" coordinates by w_k $(k = 1, 2, 3)$, then

$$p_k = \frac{\partial W(q, J)}{\partial q_k} \tag{26.100a}$$

$$w_k = \frac{\partial W(q, J)}{\partial J_k} \tag{26.100b}$$

Here, the w's are called angle variables, and the J's are the action variables. They are canonical with respect to the Hamiltonian, which may now be expressed as

$$H = \alpha_1(J_1, J_2, J_3) \tag{26.101}$$

and

$$\dot{J}_k = -\frac{\partial H}{\partial w_k} = -\frac{\partial \alpha_1}{\partial w_k} = 0 \tag{26.102a}$$

$$\dot{w}_k = \frac{\partial \alpha_1(J_1, J_2, J_3)}{\partial J_k} \tag{26.102b}$$

By Eq. (26.102a), the J's are constants, so that by Eq. (26.102b)

$$\dot{w}_k = \text{const} = \nu_k \tag{26.103}$$

Thus

$$w_k = \nu_k t + \delta_k \tag{26.104}$$

The new set of canonical variables has J's as constants and w's as linear functions of the time. Here

$$\nu_k = \frac{\partial \alpha_1(J_1, J_2, J_3)}{\partial J_k} \tag{26.105}$$

is called the kth fundamental frequency. In a general Staeckel system, a given coordinate q_k may go through successive cycles in different times. It is one of the main points of this chapter, however, to show that the mean frequency of each q_k of a conditionally periodic Staeckel system is equal to the fundamental frequency ν_k.

IX. Keplerian Action Variables

The Keplerian example will help to clarify our ideas. For simplicity, use the α's of Chapter 6. Then

$$p_r = r^{-1}\left(-\alpha_2^2 + 2\mu r + 2\alpha_1 r^2\right)^{\frac{1}{2}} \tag{26.106a}$$

$$p_\theta = \left(\alpha_2^2 - \alpha_3^2 \sec^2\theta\right)^{\frac{1}{2}} \tag{26.106b}$$

$$p_\phi = \alpha_3 \tag{26.106c}$$

Then

$$J_1 = \oint p_r \,\mathrm{d}r = 2\int_{r_1}^{r_2} r^{-1}\left(-\alpha_2^2 + 2\mu r + 2\alpha_1 r^2\right)^{\frac{1}{2}} \mathrm{d}r$$

$$= 2\int_{r_1}^{r_2} \frac{r^{-1}\left(-\alpha_2^2 + 2\mu r + 2\alpha_1 r^2\right)}{\left(-\alpha_2^2 + 2\mu r + 2\alpha_1 r^2\right)^{\frac{1}{2}}} \mathrm{d}r \tag{26.107a}$$

where

$$r_1 = a(1-e) \qquad r_2 = a(1+e)$$
$$a = -\frac{\mu}{2\alpha_1} > 0 \qquad e = \left(1 + \frac{2\alpha_1\alpha_2^2}{\mu^2}\right)^{\frac{1}{2}} < 1 \tag{26.107b}$$

Write the denominator in Eq. (26.107a) as $[-2\alpha_1(r-r_1)(r_2-r)]^{\frac{1}{2}}$. Then

$$J_1 = 2(-2\alpha_1)^{-\frac{1}{2}} \int_{r_1}^{r_2} \frac{r^{-1}\left(-\alpha_2^2 + 2\mu r + 2\alpha_1 r^2\right)}{[(r-r_1)(r_2-r)]^{\frac{1}{2}}} \mathrm{d}r \tag{26.108}$$

$$= 2(-2\alpha_1)^{-\frac{1}{2}} \left\{ -\alpha_2^2 \int_{r_1}^{r_2} \frac{r^{-1}\,\mathrm{d}r}{[(r-r_1)(r_2-r)]^{\frac{1}{2}}} + 2\mu \int_{r_1}^{r_2} \frac{\mathrm{d}r}{[(r-r_1)(r_2-r)]^{\frac{1}{2}}} \right.$$
$$\left. + \, 2\alpha_1 \int_{r_1}^{r_2} \frac{r\,\mathrm{d}r}{[(r-r_1)(r_2-r)]^{\frac{1}{2}}} \right\} \tag{26.109}$$

If we place

$$r = \frac{a(1-e^2)}{1+e\cos f} \qquad \mathrm{d}r = \frac{a(1-e^2)e\sin f}{(1+e\cos f)^2}\,\mathrm{d}f \tag{26.110}$$

in the first integral, along with Eq. (26.107b), we obtain

$$r^{-1}\,\mathrm{d}r = \frac{e\sin f}{1+e\cos f}\,\mathrm{d}f \qquad [(r-r_1)(r_2-r)]^{\frac{1}{2}} = \frac{ae(1-e^2)^{\frac{1}{2}}|\sin f|}{1+e\cos f}$$

and

$$\frac{r^{-1}\,\mathrm{d}r}{[(r-r_1)(r_2-r)]^{\frac{1}{2}}} = \frac{1}{a(1-e^2)^{\frac{1}{2}}}\frac{\sin f}{|\sin f|}\,\mathrm{d}f = \frac{\mathrm{d}f}{a(1-e^2)^{\frac{1}{2}}}$$

since $\sin f = |\sin f|$ as r increases from r_1 to r_2. Then

$$\int_{r_1}^{r_2} \frac{r^{-1}\,\mathrm{d}r}{[(r-r_1)(r_2-r)]^{\frac{1}{2}}} = \frac{\pi}{a(1-e^2)^{\frac{1}{2}}} \tag{26.111}$$

In the next two integrals in Eq. (26.109), use

$$r = a(1-e\cos E)$$
$$\mathrm{d}r = ae\sin E\,\mathrm{d}E \tag{26.112}$$

Then

$$\frac{dr}{[(r-r_1)(r_2-r)]^{\frac{1}{2}}} = \frac{\sin E}{|\sin E|} \, dE = dE \tag{26.113}$$

$$\frac{r\,dr}{[(r-r_1)(r_2-r)]^{\frac{1}{2}}} = a(1-e\cos E)\,dE \tag{26.114}$$

The second and third integrals in Eq. (26.109) become π and πa, so that

$$J_1 = 2(-2\alpha_1)^{-\frac{1}{2}}\left\{-\alpha_2^2\frac{\pi}{a(1-e^2)^{\frac{1}{2}}} + 2\mu\pi + 2\alpha_1\pi a\right\} \tag{26.115}$$

Using $a = -\mu/(2\alpha_1)$ and $1-e^2 = -2\alpha_1\alpha_2^2/\mu^2$, we find

$$J_1 = 2(-2\alpha_1)^{-\frac{1}{2}}\{-\pi\alpha_2(-2\alpha_1)^{\frac{1}{2}} + \mu\pi\}$$

or

$$J_1 = -2\pi\alpha_2 + 2\pi\mu(-2\alpha_1)^{-\frac{1}{2}} \tag{26.115a}$$

Next

$$J_2 = \oint p_\theta \, d\theta = 2\int_{\theta_{\min}}^{\theta_{\max}} p_\theta \, d\theta = 4\int_0^{\theta_{\max}} p_\theta \, d\theta \tag{26.116}$$

since $\theta_{\max} = -\theta_{\min}$ and p_θ is even in θ. Here, $\theta_{\max}$ is given by

$$\cos\theta_{\max} = |\alpha_3|/\alpha_2$$

$$\theta_{\max} = I \qquad \text{for direct orbits}$$

$$\theta_{\max} = \pi - I \qquad \text{for retrograde orbits}$$

Abbreviate $\theta_{\max}$ to θ_m and use

$$p_\theta = \left(\alpha_2^2 - \alpha_3^2\sec^2\theta\right)^{\frac{1}{2}}$$

so that

$$J_2 = 4\int_0^{\theta_m}\left(\alpha_2^2 - \alpha_3^2\sec^2\theta\right)^{\frac{1}{2}} d\theta \tag{26.117}$$

$$= 4\int_0^{\theta_m}\frac{\alpha_2^2 - \alpha_3^2\sec^2\theta}{\left(\alpha_2^2 - \alpha_3^2\sec^2\theta\right)^{\frac{1}{2}}} d\theta \tag{26.117a}$$

$$= 4\left(\alpha_2^2 N_1 - \alpha_3^2 N_2\right) \tag{26.117b}$$

where

$$N_1 = \int_0^{\theta_m}\left(\alpha_2^2 - \alpha_3^2\sec^2\theta\right)^{-\frac{1}{2}} d\theta \tag{26.117c}$$

$$N_2 = \int_0^{\theta_m}\left(\alpha_2^2 - \alpha_3^2\sec^2\theta\right)^{-\frac{1}{2}}\sec^2\theta \, d\theta \tag{26.117d}$$

Now

$$N_1 = \int_0^{\theta_m} \frac{\cos\theta \, d\theta}{\left(\alpha_2^2 \cos^2\theta - \alpha_3^2\right)^{\frac{1}{2}}} = \int_0^{\theta_m} \frac{\cos\theta \, d\theta}{\left(\alpha_2^2 - \alpha_3^2 - \alpha_2^2 \sin^2\theta\right)^{\frac{1}{2}}}$$

$$N_1 = \left(\alpha_2^2 - \alpha_3^2\right)^{-\frac{1}{2}} \int_0^{\theta_m} \left(1 - \frac{\sin^2\theta}{\sin^2 I}\right)^{-\frac{1}{2}} \cos\theta \, d\theta \tag{26.118}$$

since

$$\cos I = \frac{\alpha_3}{\alpha_2} \qquad \sin^2 I = \frac{\alpha_2^2 - \alpha_3^2}{\alpha_2^2} \tag{26.118a}$$

Put

$$u = \frac{\sin\theta}{\sin I} \qquad du = \frac{\cos\theta \, d\theta}{\sin I} \tag{26.118b}$$

and $u = 1$ when $\theta = \theta_m$. Thus

$$N_1 = \left(\alpha_2^2 - \alpha_3^2\right)^{-\frac{1}{2}} \sin I \int_0^1 (1 - u^2)^{-\frac{1}{2}} \, du$$

$$N_1 = (\pi/2)\left(\alpha_2^2 - \alpha_3^2\right)^{-\frac{1}{2}} \sin I \tag{26.119}$$

but

$$\sin I = \frac{\left(\alpha_2^2 - \alpha_3^2\right)^{\frac{1}{2}}}{\alpha_2}$$

by Eq. (26.118a), so that

$$N_1 = \pi/2\alpha_2 \tag{26.120}$$

For N_2, put $v = \tan\theta$ in Eq. (25.117d). Then

$$N_2 = \int_0^{\theta_m} \left(\alpha_2^2 - \alpha_3^2 - \alpha_3^2 v^2\right)^{-\frac{1}{2}} \, dv \tag{26.121}$$

Because

$$p_\theta^2 = \alpha_2^2 - \alpha_3^2 \sec^2\theta$$

we have

$$\sec^2\theta_m = \alpha_2^2/\alpha_3^2$$

$$\tan^2\theta_m = \frac{\alpha_2^2 - \alpha_3^2}{\alpha_3^2}$$

so that

$$v_m = |\alpha_3|^{-1}\left(\alpha_2^2 - \alpha_3^2\right)^{\frac{1}{2}} \tag{26.122}$$

From Eq. (26.121)

$$N_2 = (\alpha_2^2 - \alpha_3^2)^{-\frac{1}{2}} \int_0^{v_m} \left(1 - \frac{v^2}{v_m^2}\right)^{-\frac{1}{2}} dv \tag{26.123}$$

Next, put

$$v = v_m \eta \tag{26.124}$$

Then

$$\begin{aligned} N_2 &= (\alpha_2^2 - \alpha_3^2)^{-\frac{1}{2}} v_m \int_0^1 (1 - \eta^2)^{-\frac{1}{2}} d\eta \\ &= \pi/2(\alpha_2^2 - \alpha_3^2)^{-\frac{1}{2}} v_m \end{aligned} \tag{26.125}$$

or

$$N_2 = \pi/2|\alpha_3| \tag{26.126}$$

By Eqs. (26.117b), (26.119), and (26.126)

$$J_2 = 4\left(\alpha_2^2 \frac{\pi}{2\alpha_2} - \alpha_3^2 \frac{\pi}{2|\alpha_3|}\right) = 2\pi(\alpha_2 - |\alpha_3|) \tag{26.127}$$

Also

$$J_3 = 2\pi\alpha_3 \tag{26.128}$$

Adding J_1 and J_2, we find

$$J_1 + J_2 = 2\pi\mu(-2\alpha_1)^{-\frac{1}{2}} - 2\pi|\alpha_3| \tag{26.129}$$

and

$$J_1 + J_2 + J_3 \operatorname{sgn}\alpha_3 = 2\pi\mu(-2\alpha_1)^{-\frac{1}{2}} \tag{26.130}$$

Thus

$$\alpha_1 = -\frac{2\pi^2\mu^2}{(J_1 + J_2 + J_3 \operatorname{sgn}\alpha_3)^2} \tag{26.131}$$

The fundamental frequencies are

$$\nu_1 = \frac{\partial\alpha_1(J_1, J_2, J_3)}{\partial J_1} = \frac{4\pi^2\mu^2}{(J_1 + J_2 + J_3 \operatorname{sgn}\alpha_3)^3} \tag{26.132}$$

$$\nu_2 = \frac{\partial\alpha_1(J_1, J_2, J_3)}{\partial J_2} = \nu_1 \tag{26.133}$$

$$\nu_3 = \frac{\partial\alpha_1(J_1, J_2, J_3)}{\partial J_3} = \nu_1 \operatorname{sgn}\alpha_3 \tag{26.134}$$

From Eq. (26.130)

$$(J_1 + J_2 + J_3 \operatorname{sgn}\alpha_3)^{-3} = \frac{(-2\alpha_1)^{\frac{3}{2}}}{8\pi^3\mu^3} \tag{26.135}$$

and from Eqs. (26.132) and (26.135)

$$\nu_1 = \frac{1}{2\pi\mu}(-2\alpha_1)^{\frac{3}{2}}$$

or

$$\nu_1 = \frac{1}{2\pi\mu}\left(\frac{\mu}{a}\right)^{\frac{3}{2}}$$

$$2\pi\nu_1 = \mu^{\frac{1}{2}}a^{-\frac{3}{2}} = n \tag{26.136}$$

the mean motion. Thus, $2\pi\nu_1 = 2\pi\nu_2 = 2\pi|\nu_3| = n$.

X. Conditionally Periodic Staeckel Systems, Continued

We saw in Sec. VII that each

$$v_k \equiv \int_{q_{k0}}^{q_k} p_k^{-1}\,\mathrm{d}q_k \tag{26.137}$$

is a SVD function of q_k and conversely that each q_k is a SVD function of v_k. We now proceed to prove the theorem that the mean frequency of each q_k is equal to the fundamental frequency ν_k. To do so, we must know how the v_k's behave as functions of the angle variables w_k. Such knowledge will tell us how the q_k's behave as functions of the w_k's.

We begin with

$$J_k = \oint p_k\,\mathrm{d}q_k \qquad k = 1, 2, 3$$

$$w_k = \frac{\partial W(q, J)}{\partial J_k}$$

$$\dot{w}_k = \frac{\partial\alpha_1(J_1, J_2, J_3)}{\partial J_k} = \nu_k = \text{const} \tag{26.137a}$$

The Jacobi β's are given by

$$t + \beta_1 = \frac{\partial W(q, \alpha)}{\partial\alpha_1}$$

$$\beta_2 = \frac{\partial W(q, \alpha)}{\partial\alpha_2} \tag{26.138}$$

$$\beta_3 = \frac{\partial W(q, \alpha)}{\partial\alpha_3}$$

where W is the separated solution for the *HJ* equation for the Staeckel system. These equations can be inverted to give the q's as functions of t. We shall not follow that usual procedure here, but that possibility is mentioned only to show that q's can be expressed as functions of the w's, because the w's are linear functions of t.

Let us put

$$\beta_i + t\delta_{1i} \equiv B_i \tag{26.139}$$

By Eqs. (26.138) and (26.139)

$$B_i = \frac{\partial W(q, \alpha)}{\partial \alpha_i} \tag{26.140}$$

If we now introduce the J's and the w's, we can write

$$B_i = \sum_{k=1}^{3} \frac{\partial W(q, J)}{\partial J_k} \frac{\partial J_k}{\partial \alpha_i} \qquad i = 1, 2, 3 \tag{26.141}$$

[It is of course to be understood that $W(q, J)$ does not have the same functional form in the q's and J's as does $W(q, \alpha)$ in the q's and α's.]

If we put

$$\omega_{ki} \equiv \frac{\partial J_k}{\partial \alpha_i} \tag{26.142}$$

then

$$[\omega_{ki}] \equiv \Omega \tag{26.143}$$

is a square matrix. From Eqs. (26.141), (26.142), and (26.137a), we find

$$B_i = \sum_{k=1}^{3} w_k \omega_{ki} \tag{26.144}$$

The differential of B_i in terms of the dw's is

$$\mathrm{d}B_i = \sum_{k=1}^{3} \mathrm{d}w_k \omega_{ki} \tag{26.145}$$

or

$$\mathrm{d}B_i = \sum_{k=1}^{3} [(\mathrm{d}w)^T \Omega]_i \tag{26.145a}$$

where

$$(\mathrm{d}w)^T = (\mathrm{d}w_1, \mathrm{d}w_2, \mathrm{d}w_3) \tag{26.146}$$

is a row matrix, the transpose of the column matrix of $\mathrm{d}w_1$, $\mathrm{d}w_2$, and $\mathrm{d}w_3$.

We can now express $\mathrm{d}B_i$ in another way. Because

$$W = \sum_{k=1}^{3} W_k(q_k) \tag{26.147}$$

and

$$p_k = \frac{\partial W}{\partial q_k} = \frac{\partial W_k}{\partial q_k} \tag{26.148}$$

we have

$$W = \sum_{k=1}^{3} \int_{q_{k0}}^{q_k} p_k \, \mathrm{d}q_k \tag{26.149}$$

From Eqs. (26.140) and (26.149)

$$B_i = \frac{\partial W(q, \alpha)}{\partial \alpha_i} = \sum_{k=1}^{3} \int_{q_{k0}}^{q_k} \frac{\partial p_k(q, \alpha)}{\partial \alpha_i} \, \mathrm{d}q_k \tag{26.150}$$

The differential of B_i in terms of the dq's is given by dropping the integral signs in Eq. (26.150), so that

$$\mathrm{d}B_i = \sum_{k=1}^{3} \frac{\partial p_k(q, \alpha)}{\partial \alpha_i} \mathrm{d}q_k \tag{26.151}$$

or

$$\mathrm{d}B_i = \frac{1}{2} \sum_{k=1}^{3} \frac{\partial p_k^2}{\partial \alpha_i} \frac{dq_k}{p_k} = \frac{1}{2} \sum_{k=1}^{3} \frac{\partial p_k^2}{\partial \alpha_i} \mathrm{d}v_k \tag{26.152}$$

with use of

$$\frac{\mathrm{d}q_k}{p_k} = \mathrm{d}v_k$$

from Eq. (26.137).

Now, by Eq. (26.38)

$$p_k^2 = -2\psi_k(q_k) + 2 \sum_{j=1}^{3} \Phi_{kj}(q_k)\alpha_j \tag{26.38}$$

so that

$$\frac{\partial p_k^2}{\partial \alpha_i} = 2\Phi_{kj}(q_k) \tag{26.153}$$

By Eqs. (26.152) and (26.153),

$$\mathrm{d}B_i = \sum_{k=1}^{3} \mathrm{d}v_k \Phi_{kj}(q_k) = [(\mathrm{d}v)^T \Phi]_i \tag{26.154}$$

where

$$(\mathrm{d}v)^T = (\mathrm{d}v_1, \mathrm{d}v_2, \mathrm{d}v_3)$$

a row matrix. Comparison of Eqs. (26.145a) and (26.154) shows that

$$(\mathrm{d}v)^T \Phi = (\mathrm{d}w)^T \Omega \tag{26.155}$$

Because Φ is a nonsingular matrix for a Staeckel system, Φ^{-1} exists, so that

$$(\mathrm{d}v)^T = (\mathrm{d}w)^T \Omega \Phi^{-1} \tag{26.156}$$

or

$$\mathrm{d}v_i = \sum_{k=1}^{3} \mathrm{d}w_k (\Omega \Phi^{-1})_{ki} \tag{26.157}$$

Thus

$$\frac{\partial v_i}{\partial w_k} = (\Omega \Phi^{-1})_{ki} \tag{26.158}$$

Before we can draw any conclusions from Eq. (26.158), we need to show that the domain Q of the q's for which all the p's are real corresponds exactly to w space, which is the space of all real numbers. This means that if all the p's are real, so are all the w's and vice versa. To show this, note that

$$p_k^2 = -2\psi_k(q_k) + 2\sum_{i=1}^{3} \Phi_{ki}(q_k)\alpha_i \tag{26.38}$$

and

$$W = \sum_{k=1}^{3} \int_{q_{k0}}^{q_k} p_k \, \mathrm{d}q_k \tag{26.149}$$

Then

$$w_j = \frac{\partial W(q, J)}{\partial J_j} = \sum_{k=1}^{3} \int_{q_{k0}}^{q_k} \frac{\partial p_k^2}{\partial J_j} \frac{\mathrm{d}q_k}{2p_k} \tag{26.159}$$

However, by Eq. (26.38)

$$\frac{\partial p_k^2}{\partial J_j} = 2\sum_{i=1}^{3} \Phi_{ki}(q_k)\frac{\partial \alpha_i}{\partial J_j} \tag{26.159a}$$

Thus

$$w_j = \sum_{k=1}^{3} \int_{q_{k0}}^{q_k} \frac{1}{p_k} \sum_{i=1}^{3} \Phi_{ki}(q_k)\frac{\partial \alpha_i}{\partial J_j} \, \mathrm{d}q_k \tag{26.159b}$$

This shows that if all the p's are real, so are all the w's. Also, from Eq. (26.159b)

$$\frac{\partial w_j}{\partial q_k} = \frac{1}{p_k} \sum_{i=1}^{3} \Phi_{ki}(q_k)\frac{\partial \alpha_i}{\partial J_j} \tag{26.160}$$

Here the sum is real. Also, if all the w's are real, so are all the $\partial w_j/\partial q_k$; then all the p's are real. Thus, the domain Q corresponds exactly to the set of all possible values for the w's. From this fact and from Eq. (26.158), we have the result that each v_k is a SVD function of the w's. However, we just showed that each q_k is a SVD function of the corresponding v_k. Thus, for a conditionally periodic Staeckel system, each q_k is a SVD function of the w's:

$$q_k = f_k(w_1, w_2, w_3) \tag{26.161}$$

with f_k single valued and differentiable.

Periodic Properties of $q_k = f_k(w_1, w_2, w_3)$

Because

$$W = \sum_{i=1}^{3} \int_{q_{i0}}^{q_i} p_i(q, J) \, \mathrm{d}q_i \tag{26.162}$$

we have

$$w_k = \frac{\partial W(q, J)}{\partial J_k} = \sum_{i=1}^{3} \int_{q_{i0}}^{q_i} \frac{\partial p_i(q, J)}{\partial J_K} \, \mathrm{d}q_i \tag{26.163}$$

The change of w_k with changes in the q's alone is given by

$$dw_k = \sum_{i=1}^{3} \frac{\partial p_i(q, J)}{\partial J_k} dq_i \tag{26.164}$$

Mathematically, we may let each coordinate q_i go through an integer number of cycles. [Such an event may not be possible physically, but we are concerned here only with the mathematical structure and behavior of the functions $f_k(w_1, w_2, w_3)$.] What happens to the w's? If each coordinate q_i goes through m_i cycles, then by Eq. (26.164)

$$\begin{aligned}
\Delta w_k &= \sum_{i=1}^{3} \int_{m_i \text{ cycle}} \frac{\partial p_i}{\partial J_k} dq_i = \sum_{i=1}^{3} m_i \oint \frac{\partial p_i}{\partial J_k} dq_i \\
&= \frac{\partial}{\partial J_k} \sum_{i=1}^{3} m_i \oint p_i \, dq_i \\
&= \frac{\partial}{\partial J_k} \sum_{i=1}^{3} m_i J_i \\
&= m_k
\end{aligned} \tag{26.165}$$

If each of the functions $q_k = f_k(w_1, w_2, w_3)$ goes through m_k cycles, each w_k increases by the integer m_k.

This means that if, initially,

$$q_k = f_k(w_1, w_2, w_3) \tag{26.166}$$

then if each q_k goes through an integer number m_k of cycles, the resulting q's will be given by

$$q_k' = f_k(w_1 + m_1, w_2 + m_2, w_3 + m_3) \tag{26.167}$$

Now consider the inverse problem, where each w_k increases by m_k. What happens to the q? If we begin with Eq. (26.166), the resulting q's will be

$$q_k'' = f_k(w_1 + m_1, w_2 + m_2, w_3 + m_3) \tag{26.168}$$

This is the same as Eq. (26.167), since the functions f_k are single valued. In Eq. (26.167), however, the librational q's are unchanged from Eq. (26.166), and each of the circulational q's has increased by $2\pi m_k$. That is, each of the circulational q's has gone through m_k cycles, and each of the librational q's has gone through an integer number τ_k of cycles. By Eqs. (26.165), Δw_k for a librational coordinate equals τ_k, but, by the hypothesis of the inverse problem, $\Delta w_k = m_k$, so that $\tau_k = m_k$. Thus, for either type of coordinate, if the corresponding $\Delta w_k = m_k$, that coordinate has gone through m_k cycles.

The Mean Frequencies

If, in a time interval T, the number of complete cycles passed through by any coordinate is N_k, the corresponding mean frequency is defined by

$$n_k = \lim_{T \to \infty} (N_k / T) \tag{26.169}$$

if the limit exists.

We now wish to show that

$$n_k = \nu_k = \frac{\partial \alpha_1}{\partial J_k} \tag{26.170}$$

for any conditionally periodic Staeckel system. To do so, note that if ν_1, ν_2, ν_3 are all commensurable, there exists a ν_0 and positive integers m_1, m_2, m_3 such that

$$\nu_k = m_k \nu_0 \qquad k = 1, 2, 3 \tag{26.171}$$

Here, we may choose ν_0 to be the greatest common divisor of the ν's. From

$$w_k = \nu_k t + \delta_k \tag{26.172}$$

and Eq. (26.171), we may write

$$w_k = m_k \nu_0 t + \delta_k \tag{26.173}$$

In the time interval $\tau \equiv 1/\nu_0$, we have

$$\Delta w_k = m_k \tag{26.174}$$

By the result for the preceding inverse problem, each q_k goes through exactly m_k cycles in this time, so that the motion is truly periodic, with period $\tau \equiv 1/\nu_0$. In the time interval

$$T = \mu\tau + \varepsilon \tag{26.175}$$

where μ is an integer and ε a positive proper fraction of the period $1/\nu_0$, the number of complete cycles passed through by q_k is μm_k. Then

$$n_k = \lim_{T\to\infty} \frac{\mu m_k}{\tau} = \lim_{\mu\to\infty} \frac{\mu m_k}{\mu\tau + \varepsilon} = \frac{m_k}{\tau} = m_k \nu_0 = \nu_k \tag{26.176}$$

This completes the proof of the theorem for the commensurable case.

The incommensurable case is treated in Ref. 6. Physically, one cannot distinguish between a rational number and an irrational number, so that the preceding result should hold in all cases of bounded Staeckel systems without asymptotic coordinates. The main reason for giving the proof of the incommensurable case is to verify the correctness of the mathematical formulation.

References

[1]Weinacht, J., *Mathematische Annalen,* Vol. 91, 1924, pp. 279–299.

[2]Eisenhart, L. P., *Annals of Mathematics,* Vol. 35, 1934, pp. 284–305.

[3]Morse, P. M., and Feshbach, H., *Methods of Theoretical Physics,* McGraw–Hill, New York, 1953, pp. 494, 509, 655.

[4]Pars, L. A., *A Treatise on Analytical Dynamics,* Wiley, New York, 1965, p. 4.

[5]Pierce, B. O., *A Short Table of Integrals,* Ginn and Co., Boston, 1910, p. 91.

[6]Vinti, J. P., "Mean Motions in Conditionally Periodic Separable Systems," *Journal of Research of the National Bureau of Standards,* Vol. 65B, April–June 1961, pp. 131–135.

Appendix A

Coordinate Systems and Coordinate Transformations

FOR theories of satellite orbits and ballistic trajectories, appropriate coordinates are rectangular, spherical, and oblate spheroidal. In the appendices, the Earth-centered inertial (ECI) coordinate system is the rectangular coordinate system unless otherwise specified. Special perturbations methods solve the equations of motion by numerical integration. Traditional methods of general perturbations seek the solution of the equations of motion by series expansion and term-by-term analytic integration of the disturbed acceleration. Brouwer's method, which is a traditional general perturbations method, performs contact transformations on the Delaunay variables. Vinti's method, which is not a traditional general perturbations method, solves the Hamilton–Jacobi equation in the oblate spheroidal coordinate system.

If the position and velocity vectors of a satellite or a ballistic object can be computed at any time, then the equations of motion for the object are essentially solved. The algebraic approach is to determine six integration constants of motion. The geometrical approach is to draw a figure of the desired coordinate systems and then deduce the position and velocity vectors from it. This is simple for the position vector. The velocity vector is obtained from the three-dimensional metric as in theoretical physics.

To describe a coordinate system, the origin of the coordinate system must be defined first. In trajectory mechanics, the center of mass must be defined with respect to a coordinate system in which the trajectories are described. In other words, the geometrical and physical principles must be clearly defined before a theory can be developed. Vinti[1] gives the general theory and physical principles for inclusion of the third zonal harmonic J_3 of a planet's gravitational potential in an accurate reference orbit of an artificial satellite. Here, we provide a few figures to supplement his geometrical interpretation of the coordinate systems and coordinate transformations. This may help the reader to visualize the coordinate systems and physical principles underlying the Vinti spheroidal method. Finally, if the translation and rotation between two coordinate systems can be depicted in a figure, then the figure can be a useful aid in deriving the coordinate transformation between two coordinate systems.

I. Coordinate Systems

Spherical Coordinate System

The center of mass of the Earth is always at the origin of the ECI coordinate system. If the Earth were a perfect sphere and the motion of an object unperturbed,

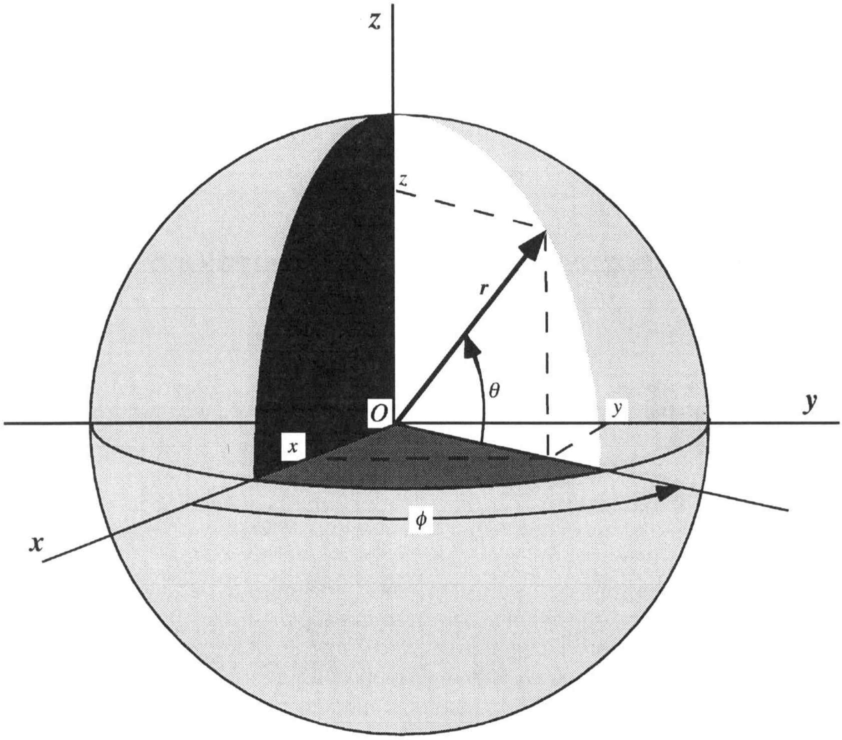

Fig. A.1 An ECI position vector expressed in terms of the spherical coordinates.

then the solution of the equations of motion would be Keplerian. In Fig. A.1, x, y, z are the rectangular coordinates of the ECI coordinate system, and r, θ, ϕ are those of the spherical coordinate system. The equation of the sphere of radius r can be expressed in the form

$$\frac{x^2}{r^2} + \frac{y^2}{r^2} + \frac{z^2}{r^2} = 1 \tag{A.1}$$

and the ECI position vector can be expressed in terms of r, θ, ϕ as

$$\boldsymbol{r} = \begin{pmatrix} x \\ y \\ z \end{pmatrix} = \begin{pmatrix} r\cos\theta\cos\phi \\ r\cos\theta\sin\phi \\ r\sin\theta \end{pmatrix} \tag{A.2}$$

The metric $(\Delta s)^2$, which is the square of the magnitude of the differential position vector Δs, is given by

$$(\Delta s)^2 = (\Delta r)^2 + (r\,\Delta\theta)^2 + (r\cos\theta\,\Delta\phi)^2 \tag{A.3}$$

Dividing Eq. (A.3) by $(\Delta t)^2$ and taking to the limit $\Delta t \to 0$, we find

$$\dot{s}^2 = \dot{r}^2 + (r\dot{\theta})^2 + (r\cos\theta\dot{\phi})^2 \tag{A.4}$$

The three terms on the right side of Eq. (A.4) are the square of the components of the velocity vector in the spherical coordinate system that are identical to the

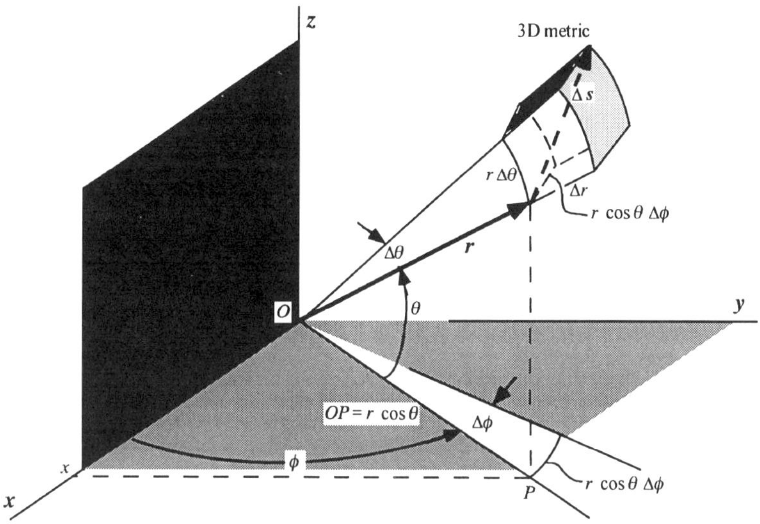

Fig. A.2 An ECI differential position vector expressed in terms of the spherical coordinates.

square of the components of the three-dimensional metric divided by $(\Delta t)^2$ as shown in Fig. A.2. The kinetic energy and the momenta in the spherical coordinate system can be derived, and the Kepler problem can be solved by the Hamilton–Jacobi procedure as shown in Chapter 6. After the position and velocity vectors at any time are solved in the spherical coordinate system, they are required to be transformed back to the ECI coordinate system. Taking the derivative of Eq. (A.2) with respect to time, the ECI velocity vector becomes

$$\dot{\boldsymbol{r}} = \begin{pmatrix} \dot{x} \\ \dot{y} \\ \dot{z} \end{pmatrix} = \begin{pmatrix} \dot{r}\cos\theta\cos\phi - r\dot{\theta}\sin\theta\cos\phi - r\cos\theta\dot{\phi}\sin\phi \\ \dot{r}\cos\theta\sin\phi - r\dot{\theta}\sin\theta\sin\phi + r\cos\theta\dot{\phi}\cos\phi \\ \dot{r}\sin\theta + r\dot{\theta}\cos\theta \end{pmatrix} \tag{A.5}$$

Oblate Spheroidal Coordinate System

The oblate spheroidal coordinates ρ, η, ϕ are depicted in Figs. A.3 and A.4. The ρ coordinate describes the surface of an oblate spheroid, and $\rho \geq 0$ everywhere for real motion. The η coordinate describes the surface of a hyperboloid of one sheet, and $0 \leq \eta \leq 1$ for $z \geq 0$ and $-1 \leq \eta < 0$ for $z < 0$. (The sheet is a surface of revolution, not a solid object.) The ϕ coordinate is the plane through the polar z axis. For clarity, we first use the 1959 Vinti potential models such that the origin of the ECI coordinate system coincides with the origin of the oblate spheroidal coordinate system. The equations of an oblate spheroid and a hyperboloid of one sheet can be expressed, respectively, in the form

$$\frac{x^2 + y^2}{a^2} + \frac{z^2}{b^2} - 1 \tag{A.6}$$

$$\frac{x^2 + y^2}{A^2} - \frac{z^2}{B^2} = 1 \tag{A.7}$$

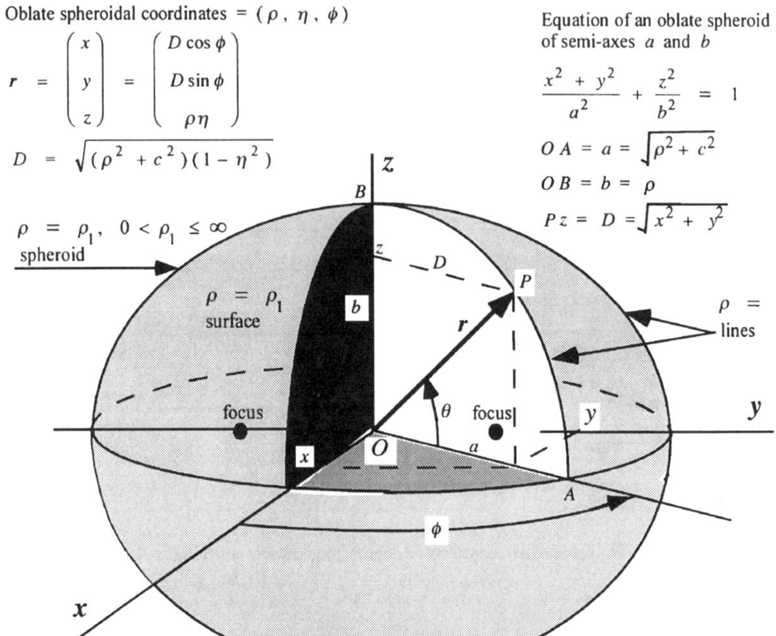

Fig. A.3 An ECI position vector expressed in terms of the oblate spheroidal coordinates with respect to an oblate spheroid.

The ECI position vector can be expressed in terms of ρ, η, ϕ as

$$\boldsymbol{r} = \begin{pmatrix} x \\ y \\ z \end{pmatrix} = \begin{pmatrix} \sqrt{(\rho^2 + c^2)(1 - \eta^2)} \cos\phi \\ \sqrt{(\rho^2 + c^2)(1 - \eta^2)} \sin\phi \\ \rho\eta \end{pmatrix} \tag{A.8}$$

where $c^2 = r_e^2 J_2$ and r_e is the Earth equatorial radius. Now, for the 1966 Vinti potential models, we have $c^2 = r_e^2 J_2[1 - J_3^2/(4J_2^3)]$. The constant, $c(\approx 210\,\text{km})$ as shown in Fig. A.4, is the radius of a focal circle in the spheroidal equatorial plane (see also Chapter 8, Sec. II). The portion of the equatorial plane inside the focal circle is the surface $\rho = 0$, while the portion outside is the surface $\eta = 0$. Note that for large r, $\rho \approx r$ and $\eta \approx \sin\theta$. The magnitudes of the spheroidal coordinates ρ, η, ϕ are bounded by $\rho > 0, 1 \geq \eta \geq 1, 2\pi > \phi \geq 0$ for real motion. We shall revisit the focal circle later in the physical principles of this appendix.

The intersection of the two surfaces of revolution (an oblate spheroid surface and a hyperboloid surface) is depicted in Fig. A.5. The foci belong to both the oblate spheroid and hyperboloid. Note that the 1966 Vinti potential model requires the origin of the oblate spheroidal coordinate system be shifted to a negative distance $\delta(\approx 7\,\text{km})$ along the Earth axis of rotation (z axis); thus, z is replaced by $z + \delta$ in Eqs. (A.6)–(A.8). For example, the δ in Eq. (A.8) can be rearranged, and then the third component becomes $\rho\eta - \delta$. Physically, the origin of the ECI coordinate

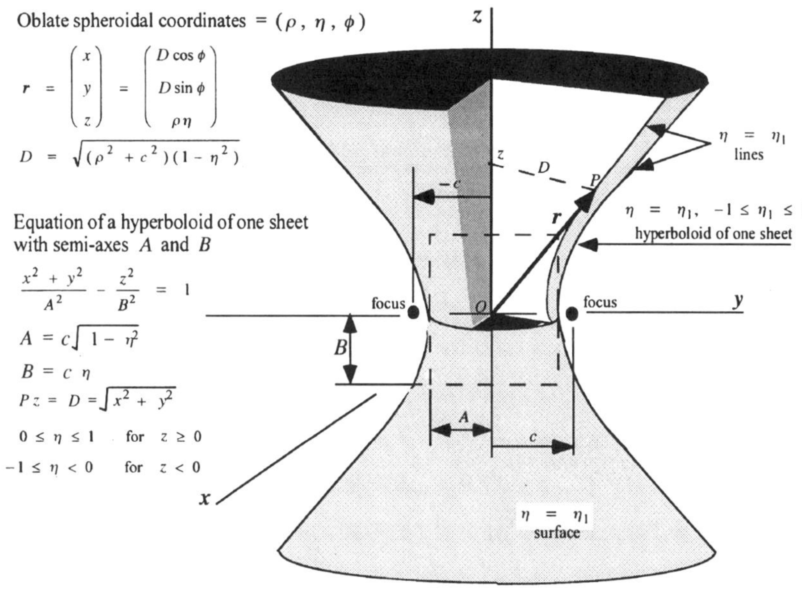

Fig. A.4 An ECI position vector expressed in terms of the oblate spheroidal coordinates with respect to a hyperboloid of one sheet.

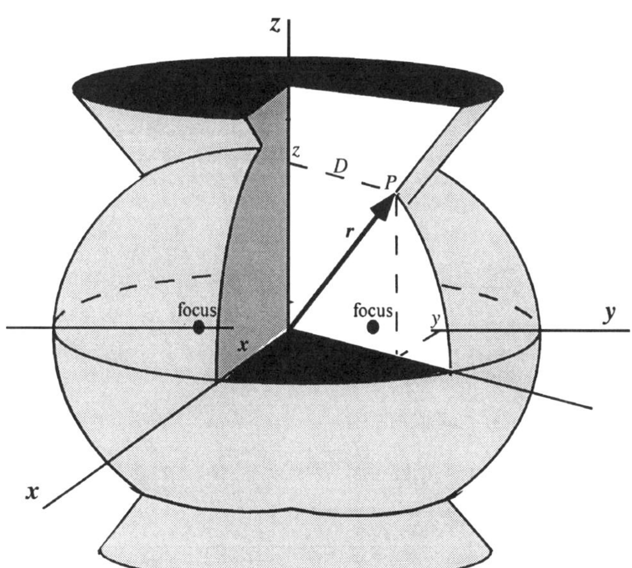

Fig. A.5 An ECI position vector in the oblate spheroidal coordinates is a point on the intersection of an oblate spheroid surface and a hyperboloid surface.

system is approximately 7 km north of the origin of the oblate spheroidal coordinate system.

The metric $(\Delta s)^2$, which is the square of the magnitude of the differential position vector $\Delta \boldsymbol{s}$, is given by

$$(\Delta s)^2 = (h_1 \Delta\rho)^2 + (h_2 \Delta\eta)^2 + (h_3 \Delta\phi)^2 \tag{A.9}$$

where the coefficients h_1, h_2, and h_3 can be derived from

$$\begin{aligned} h_1^2 &= \left(\frac{\partial x}{\partial \rho}\right)^2 + \left(\frac{\partial y}{\partial \rho}\right)^2 + \left(\frac{\partial z}{\partial \rho}\right)^2 = \frac{\rho^2 + c^2\eta^2}{(\rho^2 + c^2)} \\ h_2^2 &= \left(\frac{\partial x}{\partial \eta}\right)^2 + \left(\frac{\partial y}{\partial \eta}\right)^2 + \left(\frac{\partial z}{\partial \eta}\right)^2 = \frac{\rho^2 + c^2\eta^2}{(1 - \eta^2)} \\ h_3^2 &= \left(\frac{\partial x}{\partial \phi}\right)^2 + \left(\frac{\partial y}{\partial \phi}\right)^2 + \left(\frac{\partial z}{\partial \phi}\right)^2 = (\rho^2 + c^2)(1 - \eta^2) \end{aligned} \tag{A.10}$$

as shown by Ref. 2. Dividing Eq. (A.9) by $(\Delta t)^2$ and taking to the limit $\Delta t \to 0$, we find

$$\dot{s}^2 = (h_1 \dot{\rho})^2 + (h_2 \dot{\eta})^2 + (h_3 \dot{\phi})^2 \tag{A.11}$$

The three terms on the right side of Eq. (A.11) are the square of the components of the velocity vector in the oblate spheroidal coordinate system, which are identical to the square of the components of the three-dimensional metric divided by $(\Delta t)^2$ as shown in Fig. A.6. Thus, the kinetic energy and the momenta in the oblate spheroidal coordinate system can be derived, and the Kepler problem can be solved by the Hamilton–Jacobi procedure as described in Chapter 8. After the position and velocity vectors at any time are determined in the oblate spheroidal coordinate system, they are required to be transformed back to the ECI coordinate system. Taking the derivative of Eq. (A.8) with respect to time, the ECI velocity vector becomes

$$\dot{\boldsymbol{r}} = \begin{pmatrix} \dot{x} \\ \dot{y} \\ \dot{z} \end{pmatrix} = \begin{pmatrix} \dot{D}\cos\phi - \dot{\phi} D \sin\phi \\ \dot{D}\sin\phi - \dot{\phi} D \cos\phi \\ \dot{\rho}\eta + \rho\dot{\eta} \end{pmatrix} \tag{A.12}$$

where

$$D = \sqrt{(\rho^2 + c^2)(1 - \eta^2)}$$

$$\dot{D} = [\rho\dot{\rho}(1 - \eta^2) - \eta\dot{\eta}(\rho^2 + c^2)]/D$$

Physical Principles

Vinti[1] gives the general theory and physical principles for inclusion of the third zonal harmonic J_3 of a planet's gravitational potential in an accurate reference orbit of an artificial satellite. Traditional methods of general perturbations seek to develop a perturbed Keplerian orbit, and, therefore, the reference orbit is Keplerian. Vinti's reference orbit is not Keplerian. Vinti[3] indicates that his accurate reference

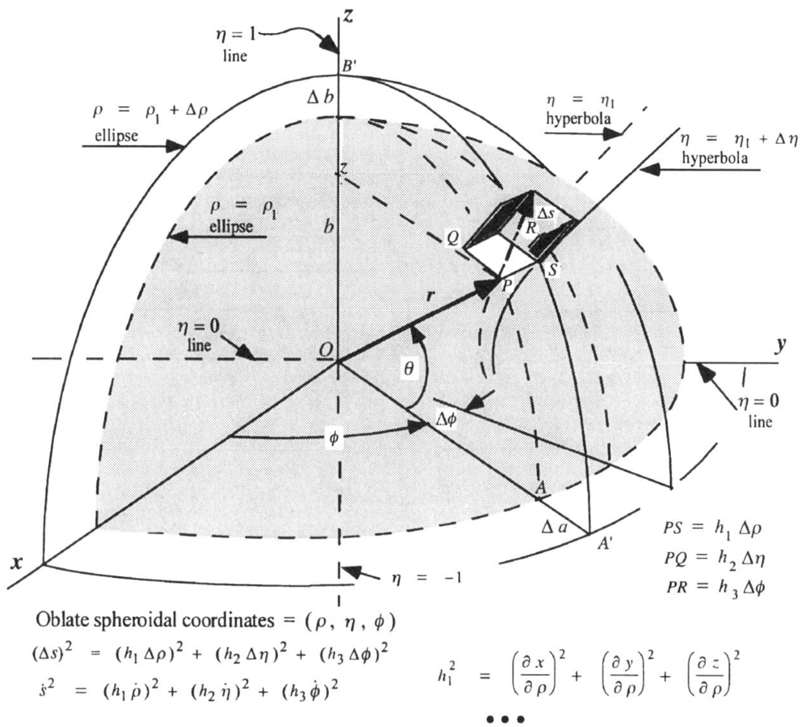

Fig. A.6 An ECI differential position vector expressed in terms of the oblate spheroidal coordinates.

orbit is a Newtonian approximation to the general relativistic orbit since the potential has a vanishing Laplacian. The Vinti spheroidal method, which is developed from the Hamilton–Jacobian formulation of Newtonian mechanics, belongs to a separate class of methods of general perturbations.

The physical significance of δ or the translation of the origin of the spheroidal coordinate system verifies the motion of Earth satellites in some "equatorial" and polar orbits. The underlying physical principles should be of great interest in the fields of orbital and celestial mechanics. However, for the general reader, this translation of the origin may raise the disturbing question: Where is the mass center after the translation? Here, we shall answer this question and re-emphasize several basic concepts of coordinate systems for the Vinti spheroidal method.

Figure A.7 results if we could discard the top half of the sphere in Fig. A.1. The mass center is at the origin O of the spherical coordinate system. When the oblate spheroidal coordinate system degenerates into the spherical coordinate system, the foci coincide at the mass center. A satellite trajectory described by the position vector $\boldsymbol{r}$ is Keplerian if the force acting on the satellite is due only to the gravitational potential $-\mu/r$. Using this potential and putting $c = 0$ and $\delta = 0$, a Vinti trajectory degenerates into a Keplerian trajectory.

Figure A.8 results if we could discard the top half of the oblate spheroid in Fig. A.3. The mass center is at the origin O of the ECI coordinate system and is

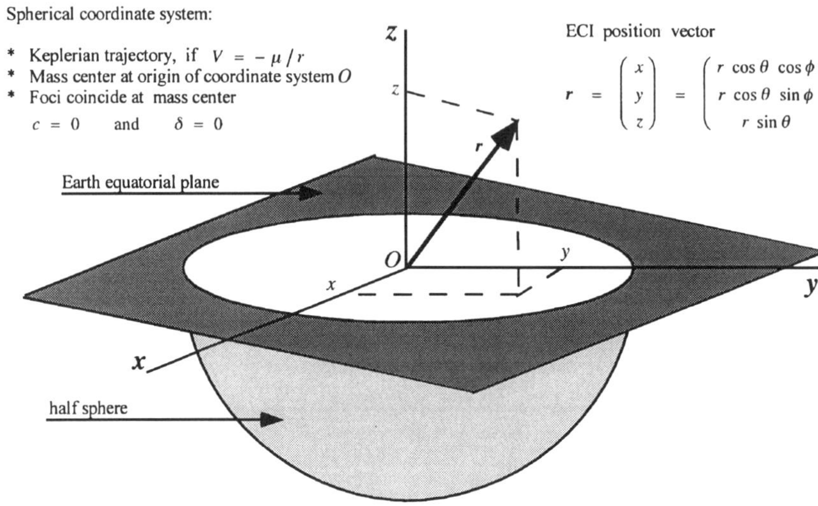

Fig. A.7 The basic concept and coordinate system of a Keplerian trajectory.

identical to that of the oblate spheroidal coordinate system. A satellite trajectory described by the position vector $\boldsymbol{r}$ is a Vinti trajectory if the force acting on the satellite is due only to the gravitational potential $V = -\mu\rho(\rho^2 + c^2\eta^2)^{-1}$. This 1959 Vinti potential model requires that $c^2 = r_e^2 J_2$ and $\delta = 0$. Physically, the foci of the oblate spheroid are at a distance $\pm c$ km from the mass center, and there is no translation of the oblate spheroidal coordinate system. The spheroidal equatorial plane, which is perpendicular to the polar z axis, passes through the center of mass at O and is a plane of symmetry of the 1959 Vinti potential V.

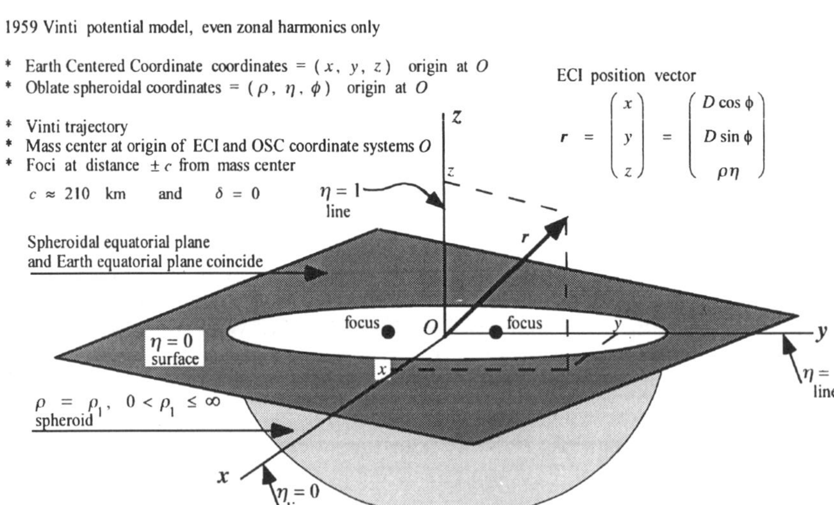

Fig. A.8 The basic concept and coordinate system of the 1959 Vinti trajectory.

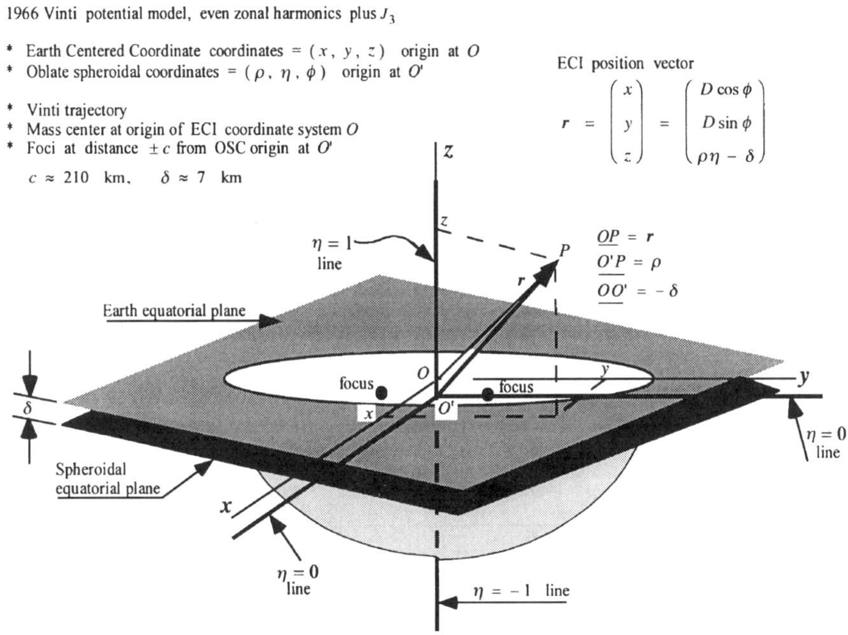

Fig. A.9 The basic concept and coordinate system of the 1966 Vinti trajectory.

Figure A.9 results if we again discard the top half of the oblate spheroid in Fig. A.3. The mass center is still at the origin O of the ECI coordinate system, while the origin of the oblate spheroidal coordinate system is translated a negative distance to O' along the Earth rotational z axis. Vinti assumes that the z axis is also the axis of symmetry; i.e., the small wobbling motion of the polar z axis is neglected. A satellite trajectory described by the position vector $\boldsymbol{r}$ is a Vinti trajectory if the force acting on the satellite is due only to the gravitational potential $V = -\mu(\rho + \delta\eta)(\rho^2 + c^2\eta^2)^{-1}$. This 1966 Vinti potential model requires that $c^2 = r_e^2 J_2[1 - J_3^2/(4J_2^3)]$ and $\delta = -r_e J_3/(2J_2)$. Physically, the foci of the oblate spheroid are at a distance $\pm c$ km from the origin O' of the oblate spheroidal coordinate system. The inclusion of J_3 reduces the original c of the 1959 potential model by 2 parts in 1000, and c is still approximately 210 km from the origin O' of the oblate spheroidal coordinate system. The ECI coordinate system is inertial or fixed with respect to some stars at a reference date; i.e., the J2000 coordinate system for orbital mechanics and the FK5 coordinate system for celestial mechanics; they are identical and referenced to exactly noon on Jan. 1, 2000. Naturally, the origin of the oblate spheroidal coordinate system must be translated, and the distance δ is approximately -7 km, taken positive northward, from the origin O of the ECI coordinate system. The spheroidal equatorial plane, which is perpendicular to the polar z axis, does *not* pass through the center of mass at O and is *not* a plane of symmetry of the 1966 Vinti potential V.

Figure A.10 results if we could discard the portion of the oblate spheroid above the spheroidal equatorial plane. The mass center, which is at the origin

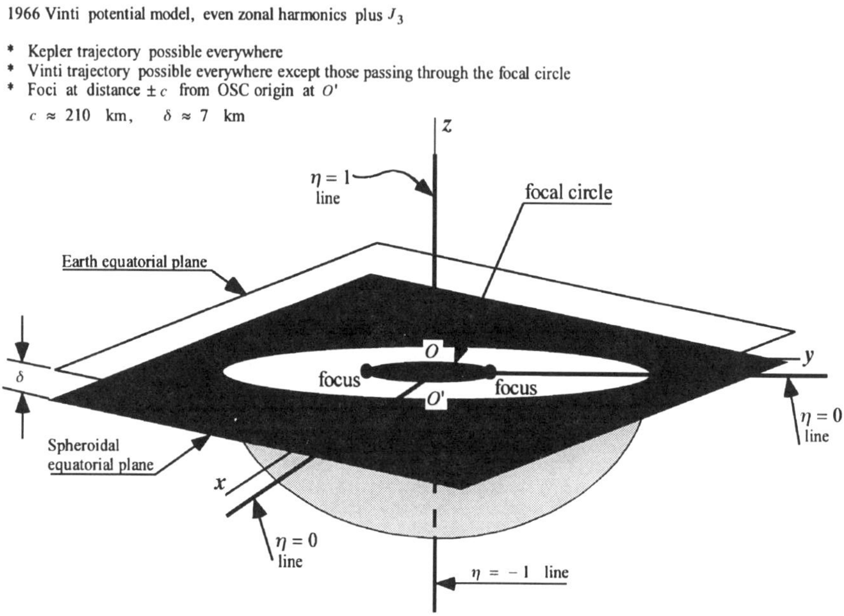

Fig. A.10 The focal circle or the forbidden zone of the Vinti spheroidal method.

O of the ECI coordinate system, is still the attraction center for any real motion. Any real trajectory or its extension must pass through the Earth equatorial plane. The focal circle, whose radius is c, lies on the spheroidal equatorial plane. Only when δ vanishes, the focal circle lies on the equatorial planes of both the ECI and oblate spheroidal coordinate systems. Very few real trajectories pass through the focal circle, but for a rocket shooting straight up with an eccentricity very close to unity, it may. Thus, this type of trajectory exists. If the extension of a real trajectory passes the focal circle or the forbidden zone of the Vinti spheroidal method, then an analytic Vinti representation of this trajectory does not exist.

Figure A.11 depicts a satellite orbit and its two foci. These are not the same foci as described previously in the oblate spheroidal coordinate system. The center of mass is, of course, at the origin of the ECI coordinate system O, and it coincides with the satellite orbit focus F1. We did not specifically name the foci of the oblate spheroidal coordinate system to avoid confusion with the trajectory foci (F1 and F2) of the ECI coordinate system.

In Fig. A.12, the ballistic object is an exo-atmospheric interceptor, which has an apogee altitude of approximately 150 km above the surface of the Earth. The hypothetical perigee of this ballistic object passes inside the Vinti focal circle or the Vinti forbidden zone, and no analytic Vinti representation of this trajectory is possible. In this case, a Keplerian or numerically integrated solution must be used. This is a very special case because there is no analytic Vinti trajectory even though the motion of the physical object is real. Since this type of trajectory is very short, a Keplerian trajectory is used in the **vinti6** computer routine to circumvent this special case.

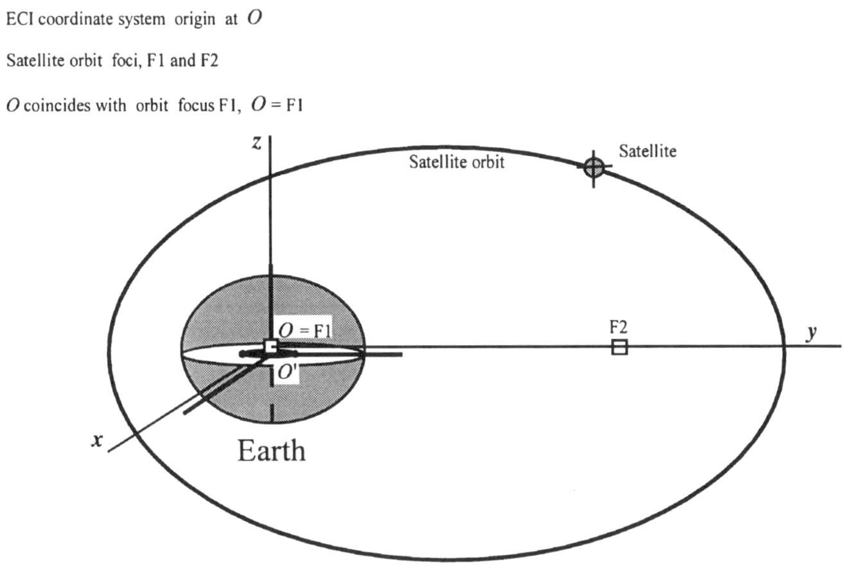

Fig. A.11 The satellite orbit focus F1 at the origin of the ECI origin and mass center.

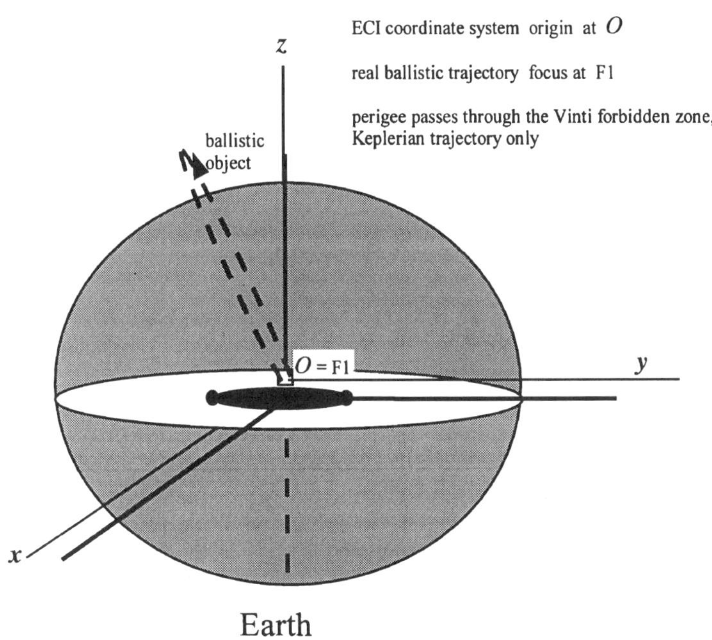

Fig. A.12 Real ballistic trajectory that passes the Vinti forbidden zone.

II. Coordinate Transformations

The state vectors (position and velocity vectors) in the ECI and oblate spheroidal coordinate (OSC) systems are defined as

$$\boldsymbol{x} = \begin{pmatrix} x \\ y \\ z \\ \dot{x} \\ \dot{y} \\ \dot{z} \end{pmatrix} \qquad \boldsymbol{X} = \begin{pmatrix} \rho \\ \eta \\ \phi \\ \dot{\rho} \\ \dot{\eta} \\ \dot{\phi} \end{pmatrix}$$

If the ECI state vector $\boldsymbol{x}$ is given, then the OSC state vector can be obtained from

$$\boldsymbol{X} = \begin{pmatrix} \rho \\ \eta \\ \phi \\ \dot{\rho} \\ \dot{\eta} \\ \dot{\phi} \end{pmatrix} = \begin{pmatrix} [0.5d + 0.5\sqrt{d^2 + 4c^2(z+\delta)^2}]^{\frac{1}{2}} \\ (z+\delta)/\rho \\ \tan^{-1}(y/x) \\ \sqrt{F}/(\rho^2 + c^2\eta^2) \\ \sqrt{G}/(\rho^2 + c^2\eta^2) \\ (x\dot{y} + \dot{x}y)/D^2 \end{pmatrix} \tag{A.13}$$

where

$$r^2 = x^2 + y^2 + z^2$$

$$r\dot{r} = x\dot{x} + y\dot{y} + z\dot{z}$$

$$d = (r^2 - c^2) + \delta(2z + \delta)$$

$$\sqrt{F} = \rho r\dot{r} + (c^2\eta + \delta\rho)\dot{z}$$

$$\sqrt{G} = -\eta r\dot{r} - (\delta\eta - \rho)\dot{z}$$

$$D = \sqrt{(\rho^2 + c^2)(1 - \eta^2)}$$

Note that for real motion, $F \geq 0$ and $G \geq 0$ with $\rho \geq 0$ and $-1 \leq \eta \leq 1$. The case of $\rho = 0$ implies that a trajectory or its extension passes inside or on the focal circle.

If the OSC state vector $\boldsymbol{X}$ is given, then the ECI state vector $\boldsymbol{x}$ can be obtained from

$$\boldsymbol{x} = \begin{pmatrix} x \\ y \\ z \\ \dot{x} \\ \dot{y} \\ \dot{z} \end{pmatrix} = \begin{pmatrix} D\cos\phi \\ D\sin\phi \\ \rho\eta - \delta \\ \dot{D}\cos\phi - \dot{\phi}D\sin\phi \\ \dot{D}\sin\phi + \dot{\phi}D\cos\phi \\ \dot{\rho}\eta + \rho\dot{\eta} \end{pmatrix} \tag{A.14}$$

where

$$\dot{D} = [\rho\dot{\rho}(1 - \eta^2) - \eta\dot{\eta}(\rho^2 + c^2)]/D$$

The input and output of a Kepler or Vinti algorithm are the ECI state vectors, $\boldsymbol{x}(t_0)$ at the initial time t_0 and $\boldsymbol{x}(t)$ at the final time t. Equation (A.13) is used to transform the given ECI state vector $\boldsymbol{x}(t_0)$ to the OSC form at time t_0. The Vinti spheroidal method solves the equations of motion in the OSC system, giving the OSC state vector $X(t)$ at time t. Equation (A.14) is used to transform the OSC state vector back to the ECI form $\boldsymbol{x}(t)$ at time t.

An ECI state vector $\boldsymbol{x}$ consisting of the position and velocity vectors can be transformed to an arbitrary set of osculating orbital elements $(a, e, I, \Omega, \omega, M)$. This coordinate transformation involves only instantaneous conversion since time is not changed. The osculating orbital elements $(a, e, I, \Omega, \omega, M)$, which are the classical orbital elements, are different from the mean orbital elements as used for the input of the simplified general perturbations (SGP) algorithms. Given an initial ECI state vector, an initialization procedure to convert from the initial ECI state vector to the SGP mean elements is necessary. The output of an SGP algorithm is in the ECI form, and, therefore, no conversion is needed at the final time.

For the sake of completeness, we provide a set of computer routines for the conversion of osculating elements to SGP mean elements. These SGP routines, which were downloaded via the Internet from a computer at the U.S. Air Force Institute of Technology, have been slightly modified to achieve true double precision computation. The conversion routines are developed from the epoch point conversion method of Walter.[4] A more robust epoch point conversion method is developed by Der and Danchick.[5] Since the conversion is instantaneous, Izsak's method, described by Uphoff, is also applicable.[6] Differential correction methods for the determination of mean orbital elements are expensive and unnecessary for this purpose.

References

[1] Vinti, J. P., "Invariant Properties of the Spheroidal Potential of an Oblate Planet," *Journal of Research of the National Bureau of Standards,* Vol. 70B, No. 1, Jan.–March 1966, pp. 1–16.

[2] Morse, P. M., and Feshbach, H., *Methods of Theoretical Physics,* McGraw–Hill, New York, 1953.

[3] Vinti, J. P., "The Spheroidal Method in the Theory of the Orbit of an Artificial Satellite," *Satellitenmechanik Im Vakuum Und In Der Atmosphäre,* 7 Himmelsmechanik, July 1964.

[4] Walter, H. G., "Conversion of Osculating Orbital Elements into Mean Elements," *Astronomical Journal,* Vol. 72, No. 8, Oct. 1967.

[5] Der, G. J., and Danchick, R., "Conversion of Osculating Orbital Elements to Mean Orbital Elements," Goddard Flight Mechanics Symposium, May, 1996.

[6] Uphoff, C., "Conversion Between Mean and Osculating Elements," JPL IOM 312/85.2-927, GSFC/NASA, Greenbelt, MD, Jan. 23, 1985.

Appendix B

Vinti Spheroidal Method Computational Procedure and Trajectory Propagators

THE algorithmic implementation of a theory and the code realization (or computer routine) of the algorithm are different aspects of applying the theory to produce useful numbers. Conceptually, the theory of the Vinti spheroidal method for satellite orbits and ballistic trajectories can be reduced to three steps:

1) Transform the given ECI state vector $\boldsymbol{x}(t_i)$ to the oblate spheroidal form at time t_i.

2) Solve the kinematical equations for the oblate spheroidal state vector $\boldsymbol{X}(t_f)$ at time t_f.

3) Transform the oblate spheroidal state vector $\boldsymbol{X}(t_f)$ back to the ECI state vector $\boldsymbol{x}(t_f)$ at time t_f.

The six computer routines provided in this appendix were developed on the basis of the preceding theory. However, the six algorithms differ in the second step.

An algorithm may be defined as a set of equations arranged in their order of execution or a cookbook format. Vinti[1] provided the algorithms for both his 1959 and 1966 potential models. With the exception of the **vinti3** and **vinti4** routines of Bonavito[2] and Lang,[3] none of the other four computer routines follow the original Vinti algorithms. A computer routine is written in lower case **bold** characters, i.e., **vinti3**. In Chapter 8, the equations used in the algorithm for the 1959 Vinti model are described. The extension of the algorithm to the 1966 Vinti model requires only a few more equations and additional terms, but the order of execution is the same. The first three algorithms use classical orbital elements, while the last three use universal variables. The less obvious difference between the algorithms is the use of elliptic integrals; Vinti tried to avoid them, but Getchell[3] showed their advantages.

The implementation of an algorithm results in the computational procedure of a computer routine. Even using the same algorithm, a computational procedure can be different if the code developer chooses to implement a subset of equations in the algorithm using his own favorite method or changes the order of execution of a subset of equations to avoid a singularity. In this appendix, we shall describe only the computational procedures in the **vinti3** and **vinti6** routines, which are based primarily on Refs. 2, 4, works of Bonavito,[2] Getchell,[4] Monuki,[5] and Der.[6] The critical equations are listed without proof, but the interested reader can find the answers in this text, Bonavito,[2] and Getchell.[4] The computational procedures used in **vinti3** and **vinti6** are depicted in Fig. B.1.

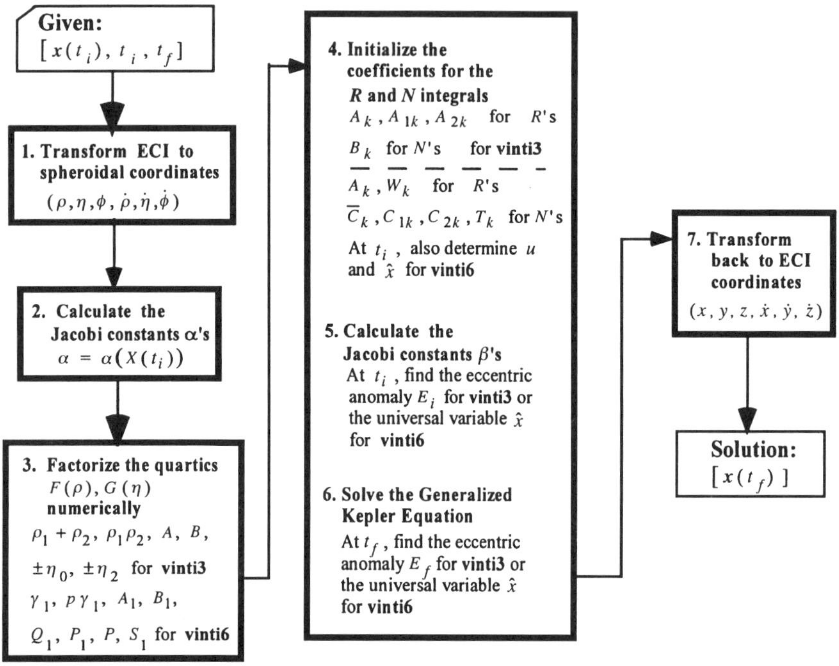

Fig. B.1 The Vinti computational procedures used in vinti3 and vinti6.

I. The Kepler Problem

Given the initial ECI state vector (position and velocity vectors) $\boldsymbol{x}(t_i)$, the initial time t_i, and the final time t_f, find the final ECI state vector $\boldsymbol{x}(t_f)$; units are kilometers and seconds.

II. Given Constants

The following set of constants are used for calculating trajectories about the planet Earth. However, if the four constants—geocentric gravitational constant μ, the equatorial radius r_e, and the zonal gravitational harmonics J_2 and J_3—are replaced by those of another solar system planet, then the Vinti routines compute trajectories in the same way without any change in the algorithms. The Vinti routines, which are applicable to orbital and celestial mechanics, remove all doubts that Vinti's works are not just theory. This book provides not just a complete theoretical treatment of these fields, but the computer source code to prove that Vinti's theory is years ahead of its time.

$\mu = 3.986005 \times 10^5$ km^3/s^2 (the gravitational constant in **vinti3**)
$= 1.0$ (the normalized gravitational constant in **vinti6**)
$r_e = 6378.137$ km (Earth equatorial radius in **vinti3**)
$= 1.0$ (normalized Earth equatorial radius in **vinti6**)
$J_2 = 1082.62999 \times 10^{-6}$

$J_3 = -2.53215 \times 10^{-6}$
$J_4 = -1.61099 \times 10^{-6}$
$c^2 = r_e^2 J_2\left(1 - \frac{J_3^2}{4J_2^3}\right) \text{ km}^2$
$\delta = -\frac{r_e J_3}{2J_2} \text{ km}$

III. The vinti3 Computation Procedure

1) Transform the given ECI state vector $\boldsymbol{x}(t_i)$ to the oblate spheroidal form at time t_i:

$$\boldsymbol{X}(t_i) = \begin{pmatrix} \rho_i \\ \eta_i \\ \phi_i \\ \dot{\rho}_i \\ \dot{\eta}_i \\ \dot{\phi}_i \end{pmatrix} = \begin{pmatrix} [0.5d + 0.5\sqrt{d^2 + 4c^2(z_i + \delta)^2}]^{\frac{1}{2}} \\ (z + \delta)/\rho_i \\ \tan^{-1}(y_i/x_i) \\ \sqrt{F}/(\rho_i^2 + c^2\eta_i^2) \\ \sqrt{G}/(\rho_i^2 + c^2\eta_i^2) \\ (x_i\dot{y}_i + \dot{x}_i y_i)/D^2 \end{pmatrix}$$

where x_i, y_i, z_i, $\dot{x}_i$, $\dot{y}_i$, and $\dot{z}_i$ are the components of $\boldsymbol{x}(t_i)$, and

$$d = \left(r_i^2 - c^2\right) + \delta(2z_i + \delta)$$

$$r_i^2 = x_i^2 + y_i^2 + z_i^2$$

$$r_i\dot{r}_i = x_i\dot{x}_i + y_i\dot{y}_i + z_i\dot{z}_i$$

$$F = c^2\alpha_3^2 + \left(\rho_i^2 + c^2\right)\left(-\alpha_2^2 + 2\mu\rho_i + 2\alpha_1\rho_i^2\right)$$

$$G = -\alpha_3^2 + \left(1 - \eta_1^2\right)\left(\alpha_2^2 + 2\mu\delta\eta_i + 2\alpha_1 c^2\eta_i^2\right)$$

$$D^2 = \left(\rho_i^2 + c^2\right)\left(1 - \eta_i^2\right)$$

2) Compute the first half of the Jacobi constants $(\alpha_1, \alpha_3, \alpha_2)$:

$$\alpha_1 = \frac{1}{2}\left(\dot{x}_i^2 + \dot{y}_i^2 + \dot{z}_i^2\right) - \frac{\mu(\rho_i + \delta\eta_i)}{\rho_i^2 + c^2\eta_i^2}$$

$$\alpha_3 = x_i\dot{y}_i - y_i\dot{x}_i$$

$$\alpha_2^2 = -2\alpha_1 c^2\eta_i^2 - 2\mu\delta\eta_i + \left(1 - \eta_i^2\right)^{-1}\left[\left(\rho_i^2 + c^2\eta_i^2\right)^2\dot{\eta}_i^2 + \alpha_3^2\right]$$

3) Factor the quartics, $F(\rho)$ and $G(\eta)$:

$$F(\rho) = c^2\alpha_3^2 + (\rho^2 + c^2)\left(-\alpha_2^2 + 2\mu\rho + 2\alpha_1\rho^2\right)$$

$$F(\rho) = -2\alpha_1(\rho - \rho_1)(\rho_2 - \rho)(\rho^2 + A\rho + B)$$

$$G(\eta) = -\alpha_3^2 + (1 - \eta^2)\left(\alpha_2^2 + 2\mu\delta\eta + 2\alpha_1 c^2\eta^2\right)$$

$$G(\eta) = \left(\alpha_2^2 - \alpha_3^2\right)\eta^4\left(\eta^{-2} - \eta_0^{-2}\right)\left(\eta^{-2} - \eta_2^{-2}\right)$$

Comparing the two equations for $F(\rho)$, we find $\rho_1 + \rho_2$, $\rho_1\rho_2$, A, and B, while comparing the two equations for $G(\eta)$, we find $\pm\eta_0$ and $\pm\eta_2$. After factorization, the prime constants ($a_0, p_0, e_0, i_0, \ldots$) and the mutual constants ($a, p, e, I, \ldots$) are completely determined. The conversion of the eight constants from factorization between Bonavito[2] and Getchell[4] (**vinti3** and **vinti5/vinti6**) are listed as follows:

$$-(\rho_1 + \rho_2) = \frac{2}{\gamma} \qquad \rho_1\rho_2 = \frac{\rho}{\gamma}$$

$$A = -2A_1 \qquad B = B_1$$

$$S = S_0/S_1 \qquad P = P$$

$$C_1 = P_1 \qquad C_2 = Q_1$$

The constants on the left side are from Bonavito[2] and on the right side are from Getchell.[4] Note that $2\alpha_1 = \mu\gamma\gamma_1 = \mu\gamma_0$ and $\alpha_2^2 - \alpha_3^2 = \mu p_0 S_0$.

4) Initialize the coefficients of the six integrals ($R_1, R_2, R_3, N_1, N_2, N_3$)

$$R_1 = (-2\alpha_1)^{-\frac{1}{2}}\left[b_1 E + a(E - e\sin E) + vA_1 + \sum_{j=1}^{2} A_{1j}\sin jv\right]$$

$$R_2 = (-2\alpha_1)^{-\frac{1}{2}}\left[vA_2 + \sum_{j=1}^{4} A_{2j}\sin jv\right]$$

$$R_3 = (-2\alpha_1)^{-\frac{1}{2}}\left[vA_3 + \sum_{j=1}^{4} A_{3j}\sin jv\right]$$

For R's, we need the secular coefficients A_1, A_2, and A_3 and the periodic coefficients A_{1j}, A_{2j}, and A_{3j}. Because $b_1 = -(A/2)$, the R integrals can be determined if the eccentric anomaly E is known.

$$N_1 = \left(\alpha_3^2 - \alpha_3^2\right)^{-\frac{1}{2}}\eta_0^3\left[B_1\psi - \left(\frac{2+q^2}{8}\right)\sin 2\psi + \frac{q^2}{64}\sin 4\psi + \cdots\right]$$

$$N_2 = \left(\alpha_2^2 - \alpha_3^2\right)^{-\frac{1}{2}}\eta_0^3\left[B_2\psi - \frac{q^2}{32}(4+3q^2)\sin 2\psi + \frac{3q^4}{256}\sin 4\psi + \cdots\right]$$

$$N_3 = \left(\alpha_2^2 - \alpha_3^2\right)^{-\frac{1}{2}}\left[\chi\eta_0\left(1-\eta_0^2\right)^{-\frac{1}{2}}\left(1-\eta_2^{-2}\right)^{-\frac{1}{2}}\right.$$
$$\left. + B_3\psi + \frac{3}{32}\eta_0^3\eta_2^{-4}\sin 2\psi + \cdots\right]$$

For N's, we need the variables $\psi, q, B_1, B_2, B_3, \chi, \gamma_m, \nu_1, \nu_2, \ldots$.

5) Compute the second half of the Jacobi constants ($\beta_1, \beta_2, \beta_3$):

$$t_i + \beta_1 = R_1(\rho_i) + c^2 N_1(\eta_i)$$

$$\beta_2 = -\alpha_2 R_2(\rho_i) + \alpha_2 N_2(\eta_i)$$

$$\beta_3 = \phi_i + c^2\alpha_3 R_3(\rho_i) - \alpha_3 N_3(\eta_i)$$

using initial conditions and initialized coefficients. In the computer routine,

$$\beta_1 = -T \qquad \text{(capt), time of perigee passage}$$

$$\beta_2 = \omega \qquad \text{(somega), argument of perigee}$$

$$\beta_3 = \Omega \qquad \text{(comega), longitude of the ascending node}$$

It is critical that the eccentric anomaly E_i be determined exactly at t_i. Here, E_i is not the Kepler or two-body solution at t_i.

6) Substitute the Jacobi constants ($\alpha_1, \alpha_2, \alpha_3, \beta_1, \beta_2, \beta_3$) back into the kinematical equations and solve for ρ_f, η_f, and ϕ_f at the given final time t_f.

$$t_f + \beta_1 = R_1(\rho_f) + c^2 N_1(\eta_f)$$

$$\beta_2 = -\alpha_2 R_2(\rho_f) + \alpha_2 N_2(\eta_f)$$

$$\beta_3 = \phi_f + c^2 \alpha_3 R_3(\rho_f) - \alpha_3 N_3(\eta_f)$$

We have three equations for three unknowns. The first kinematical equation is the generalized Keplerian equation. The initial guess of the anomaly $E = E_f$ is critical to guarantee an accurately converged Keplerian solution.

7) Transform the oblate spheroidal state vector $\boldsymbol{X}(t_f)$ back to the ECI state vector $\boldsymbol{x}(t_f)$ at time t_f:

$$\boldsymbol{x}(t_f) = \begin{pmatrix} x_f \\ y_f \\ z_f \\ \dot{x}_f \\ \dot{y}_f \\ \dot{z}_f \end{pmatrix} = \begin{pmatrix} D\cos\phi_f \\ D\sin\phi_f \\ \rho_f \eta_f - \delta \\ \dot{D}\cos\phi_f - \dot{\phi}_f D \sin\phi_f \\ \dot{D}\sin\phi_f + \dot{\phi}_f D\cos\phi_f \\ \dot{\rho}_f \eta_f + \rho_f \dot{\eta}_f \end{pmatrix}$$

where

$$D = \sqrt{(\rho_f^2 + c^2)(1 - \eta_f^2)}$$

$$\dot{D} = \left[\rho_f \dot{\rho}_f (1 - \eta_f^2) - \eta_f \dot{\eta}_f (\rho_f^2 + c^2)\right] / D$$

IV. The vinti6 Computation Procedure

1) Transform the given ECI state vector $\boldsymbol{x}(t_i)$ to the oblate spheroidal form at time t_i:

$$\boldsymbol{X}(t_i) = \begin{pmatrix} \rho_i \\ \eta_i \\ \phi_i \\ \dot{\rho}_i \\ \dot{\eta}_i \\ \dot{\phi}_{i_i} \end{pmatrix} = \begin{pmatrix} [0.5 + 0.5\sqrt{d^2 + 4c^2(z_i + \delta)^2}]^{\frac{1}{2}} \\ (z + \delta)/\rho_i \\ \tan^{-1}(y_i/x_i) \\ \sqrt{F}/(\rho_i^2 + c^2\eta_i^2) \\ \sqrt{G}/(\rho_i^2 + c^2\eta_i^2) \\ (x_i\dot{y}_i + \dot{x}_i y_i)/D^2 \end{pmatrix}$$

where x_i, y_i, z_i, $\dot{x}_i$, $\dot{y}_i$, and $\dot{z}_i$ are the components of $\boldsymbol{x}(t_i)$, and

$$d = \left(r_i^2 - c^2\right) + \delta(2z_i + \delta)$$

$$r_i^2 = x_i^2 + y_i^2 + z_i^2$$

$$\sqrt{F} = \rho_i r_i \dot{r}_i + (c^2\eta_i + \delta\rho_i)\,\dot{z}_i$$

$$\sqrt{G} = -\eta_i r_i \dot{r}_i - (\delta\eta_i - \rho_i)\,\dot{z}_i$$

$$r_i\dot{r}_i = x_i\dot{x}_i + y_i\dot{y}_i + z_i\dot{z}_i$$

$$D^2 = \left(\rho_i^2 + c^2\right)\left(1 - \eta_i^2\right)$$

2) Compute the first half of the Jacobi constants ($\alpha_1, \alpha_3, \alpha_2$):

$$\alpha_1 = \frac{1}{2}\left(\dot{x}_i^2 + \dot{y}_i^2 + \dot{z}_i^2\right) - \frac{\mu(\rho_i + \delta\eta_i)}{\rho_i^2 + c^2\eta_i^2}$$

$$\alpha_3 = x_i\dot{y}_i - y_i\dot{x}_i$$

$$\alpha_2^2 = -2c^2\eta_i^2\alpha_1 - 2\mu\delta\eta_i + \left(1 - \eta_i^2\right)^{-1}\left[\left(\rho_i^2 + c^2\eta_i^2\right)^2\dot{\eta}_i^2 + \alpha_3^2\right]$$

3) Factor the quartics $F(\rho)$ and $G(\eta)$:

$$F(\rho) = \mu\left[c^2 p_0(1 - S_0) + (\rho^2 + c^2)\left(\gamma_0\rho^2 + 2\rho - p_0\right)\right]$$

$$F(\rho) = \mu\gamma_1(\gamma\rho^2 + 2\rho - p)\left(\rho^2 + 2A_1\rho + B_1\right)$$

$$G(\eta) = \mu\left[-p_0(1 - S_0) + (1 - \eta^2)\left(p_0 + 2\delta\eta + c^2\gamma_0\eta^2\right)\right]$$

$$G(\eta) = \mu S_1 p_0(S + 2P\eta - \eta^2)\left(1 + P_1\eta - Q_1\eta^2\right)$$

Comparing the two equations for $F(\rho)$, we find γ_1, $p\gamma_1$, A_1, and B_1, while comparing the two equations for $G(\eta)$, we find Q_1, P_1, P, and S_1.

4) Initialize the coefficients of the six integrals ($R_1, R_2, R_3, N_1, N_2, N_3$)

$$R_1 = (\mu\gamma_1)^{-\frac{1}{2}}\left[(\rho_1 + A_1)\hat{x} + e\hat{U} + p^{-\frac{1}{2}}\sum_{k=0}^{4} A_{k+2}W_k\right]$$

$$R_2 = (p_0/\gamma_1)^{-\frac{1}{2}}\left[p^{-\frac{1}{2}}\sum_{k=0}^{6} A_k W_k\right]$$

$$\begin{aligned} R_3 = \alpha_3(\mu p\gamma_1)^{-\frac{1}{2}}[&W_2 + A_1W_3 + (A_2 - c^2)W_4 \\ &+ (A_3 - A_1c^2)W_5 + (A_4 - A_2c^2 + c^4)W_6] \end{aligned}$$

For R's, we need A_k and W_k, where $A_0 = 1$, A_1 found by factorization, etc., and

$W_0 = V_0 = W$ is the true anomaly, $W_1 = (W + eV_1)/p$, $V_1 = \sin W$, etc.

$$N_1 = (D_1/\alpha_2)\left[\sum_{k=0}^{6} \bar{C}_k Q^k T_k\right]$$

$$N_2 = D_1\left[u + k_1 T_2/2 + k_1 T_2/2 + 3k_1^2 T_4/8 + 5k_1^3 T_6/16\right]$$

$$N_3 = d_{10}\psi_1 + d_{20}\psi_2 - \frac{D_4}{1-a}\sum_{k=0}^{5} C_{1k}(\beta_1 Q)^k T_k - \frac{D_4}{1+a}\sum_{k=0}^{5} C_{2k}(\beta_2 Q)^k T_k$$

For N's, we need $\bar{C}_k$, C_{1k}, C_{2k}, and T_k, where

$$\bar{C}_0 = a^2 \qquad \bar{C}_1 = 2ab, \ldots$$

$$C_{1k} = \sum_{\alpha=k+1}^{6} \left(d_\alpha/\beta_1^\alpha\right)$$

$$C_{2k} = \sum_{\alpha=k+1}^{6} \left(d_\alpha/\beta_2^\alpha\right)$$

$$T_0 = u \qquad T_1 = 1 - \cos u$$

$$T_k = \left[(k-1)T_{k-2} - \cos u \sin^{k-1} u\right]/k \qquad k = 2, \ldots, 6$$

It is critical that the amplitude u of the elliptic integral and the universal variable $\hat{x}$ are determined exactly at t_i. Note that this $\hat{x}$ is the Vinti solution at t_i.

5) Compute the second half of the Jacobi constants ($\beta_1, \beta_2, \beta_3$):

$$t_i + \beta_1 = R_1(\rho_i) + c^2 N_1(\eta_i)$$
$$\beta_2 = -\alpha_2 R_2(\rho_i) + \alpha_2 N_2(\eta_i)$$
$$\beta_3 = \phi_i + c^2\alpha_3 R_3(\rho_i) - \alpha_3 N_3(\eta_i)$$

using initial conditions and initialized coefficients. In the computer routine,

$\beta_1 = -T$ (capt), time of perigee passage

$\beta_2 = \omega$ (somega), argument of perigee

$\beta_3 = \Omega$ (comega), longitude of the ascending node

6) Substitute the Jacobi constants ($\alpha_1, \alpha_2, \alpha_3, \beta_1, \beta_2, \beta_3$) back into the kinematical equations and solve for ρ_f, η_f, and ϕ_f at the given final time t_f.

$$t_f + \beta_1 = R_1(\rho_f) + c^2 N_1(\eta_f)$$
$$\beta_2 = -\alpha_2 R_2(\rho_f) + \alpha_2 N_2(\eta_f)$$
$$\beta_3 = \phi_f + c^2\alpha_3 R_3(\rho_f) - \alpha_3 N_3(\eta_f)$$

The first kinematical equation is the generalized Keplerian equation. The initial guess of the universal variable (xhat0) is critical and is computed by the routine **Kepler1**, which guarantees an accurately converged Keplerian solution. Once the oblate spheroidal coordinates are computed, the derivatives are calculated from

$$\begin{pmatrix} \dot{\rho}_f \\ \dot{\eta}_f \\ \dot{\phi}_f \end{pmatrix} = \begin{pmatrix} \sqrt{F}/(\rho_f^2 + c^2\eta_f^2) \\ \sqrt{G}/(\rho_f^2 + c^2\eta_f^2) \\ \alpha_3/[(\rho_f^2 + c^2)(1 - \eta_f^2)] \end{pmatrix}$$

giving the oblate spheroidal state vector $\boldsymbol{X}(t_f)$ at the given final time t_f. Here, the quartics are computed from

$$F = c^2\alpha_3^2 + (\rho_f^2 + c^2)(-\alpha_2^2 + 2\mu\rho_f + 2\alpha_1\rho_f^2)$$

$$G = -\alpha_3^2 + (1 - \eta_f^2)(\alpha_2^2 + 2\mu\delta\eta_f + 2\alpha_1 c^2\eta_f^2)$$

7) Transform the oblate spheroidal state vector $\boldsymbol{X}(t_f)$ back to the ECI state vector $\boldsymbol{x}(t_f)$ at time t_f:

$$\boldsymbol{x}(t_f) = \begin{pmatrix} x_f \\ y_f \\ z_f \\ \dot{x}_f \\ \dot{y}_f \\ \dot{z}_f \end{pmatrix} = \begin{pmatrix} D\cos\phi_f \\ D\sin\phi_f \\ \rho_f\eta_f - \delta \\ \dot{D}\cos\phi_f - \dot{\phi}_f D\sin\phi_f \\ \dot{D}\sin\phi_f + \dot{\phi}_f D\cos\phi_f \\ \dot{\rho}_f\eta_f + \rho_f\dot{\eta}_f \end{pmatrix}$$

where

$$D = \sqrt{(\rho_f^2 + c^2)(1 - \eta_f^2)}$$

$$\dot{D} = [\rho_f\dot{\rho}_f(1 - \eta_f^2) - \eta_f\dot{\eta}_f(\rho_f^2 + c^2)]/D$$

V. Summary of the Vinti Trajectory Propagators

Six computer routines, which have been developed by different organizations and individuals since the early 1960s, are listed as follows: 1) **vinti1**: Wadsworth (Bell Laboratory, 1963), 2) **vinti2**: Izsak–Borchers (unknown location, no date), 3) **vinti3**: Bonavito (Goddard Space Flight Center and TRW, 1966), 4) **vinti4**: Lang (MIT, 1968), 5) **vinti5**: Getchell (National Security Agency and TRW, 1970), and 6) **vinti6**: Der–Monuki (TRW, 1996).

Table B.1 shows the regions of applicability and singularities (or limitations) of our six computer routines. The incomplete source code of the **vinti1** and **vinti4** computer routines are included as an exercise for the interested readers. Because there are published papers on Vinti's method by others in China, France, Japan, and Russia, we may assume that Vinti computer routines in these countries exist, but their capabilities are unknown.

The first three routines are formulated in terms of classical orbital elements for circular and elliptic trajectories. The last three routines are formulated in terms of universal variables, and theoretically they are applicable to all conic trajectories.

Table B.1 Regions of applicability and singularities of vinti1 to vinti6

Formulation	Computer routine	Circle	Ellipse	Parabola	Hyperbola	Singularity
Classical elements	**vinti1**	—	—	No	No	—
	vinti2	Yes	Yes	No	No	$i \approx 0, e \approx 1$
	vinti3	Yes	Yes	No	No	$e \approx 1$
Universal variables	**vinti4**	—	—	—	Yes	—
	vinti5	Yes	Yes	No	Yes	$e \approx 1$
	vinti6	Yes	Yes	Yes	Yes	None

However, **vinti4** and **vinti5** have never been applied to parabolic trajectories; therefore, **vinti6** was initiated to solve this problem. Note that a parabolic trajectory in the oblate spheroidal coordinate system is rare. Except for **vinti2**, all the source code was intended to be readable. Even though these routines are not optimized for computational efficiency, they are more computationally efficient than other types of analytic trajectory predictors.

The first five routines were coded before 1970, while the sixth routine was completed in 1997. The last routine not only applies to parabolic orbits but guarantees the convergence of the generalized Kepler equation and avoids all possible singularities including the Vinti forbidden zone. Simulation results of these routines, which are given in Appendix C, demonstrate that a Vinti trajectory is always more accurate than the corresponding Kepler trajectory.

vinti1

Wadsworth[7] (1963) developed this routine to predict accurately the free-flight motion of a rocket near the surface of the Earth. Part of the original source code from Wadsworth is included. Developing a Vinti computer routine is not a simple task. However, we present this as a challenging exercise for the interested reader to complete this computer routine.

vinti2

This Izsak–Borchers[8] routine was developed for the onboard targeting software of a long-range rocket. Onboard computer memory was very limited 30 years ago, and the source code was written to save memory. The reader can immediately find out that the source code is practically useless without a programmer's cookbook. This computer routine is included to illustrate how a simple, elegant theory can be developed into a sophisticated algorithm and then programmed to be unreadable.

vinti3

Bonavito[2] followed almost exactly the algorithm Vinti might have written for his 1966 model. This routine has been used as a workhorse at TRW for drag-free satellite trajectory propagation over long time spans.

vinti4

Lang[3] (1968) developed this routine under Vinti at the Massachusetts Institute of Technology to begin the universal variable approach to the Vinti spheroidal method. Lang's excellent thesis, which complements the short paper of Getchell, presents the fundamental concepts, useful formulas, and the analytic method of factorization. Note that a simple numerical method of factorization as used in both **vinti3** and **vinti6** gives more accurate results. Part of the original source code from Lang is included as an exercise.

vinti5

Getchell[4] developed this routine to predict accurately the motion of satellites using universal variables and the Vinti 1966 potential model. Vinti and many others computed the Jacobi constants α_2 before the factorization process or the computation of the F and G quartics. However, Getchell reversed the order to simplify the coordinate transformation at the initial time. Getchell also used elliptic integrals to simplify the iterative solution of the generalized Kepler equation.

vinti6

Der[6] and Monuki[5] developed this routine to remove the remaining limitations in the previous five computer routines. No new algorithms were invented, but old tricks were applied. This routine guarantees a Vinti solution for circular, elliptic, parabolic, and hyperbolic trajectories.

References

[1]Vinti, J. P., *Astronomical Journal,* Vol. 74, 1969, pp. 25–34.

[2]Bonavito, N. L., NASA Technical Note, D-3562, 1966.

[3]Lang, T., "Vinti Unbounded Trajectories," master thesis, Dept. of Aeronautics and Astronautics, Massachusetts Institute of Technology, Cambridge, MA, 1966.

[4]Getchell, B. C., *Journal of Spacecraft and Rockets,* Vol. 7, No. 4, 1970.

[5]Monuki, A. T., *Vinti Potential,* Unpublished TRW Rept., 1979.

[6]Der, G. J., *Vinti Trajectory Propagators,* Unpublished Paper, 1996.

[7]Wadsworth, D.V., "Vinti Solution for Free-Flight Rocket Trajectories," *AIAA Journal,* Vol. 1, June 1963, pp. 1351–1354.

[8]Borchers, R. V., "A Satellite Orbit Computation Program for Izsak's Second-Order Solution of Vinti's Dynamical Problem," Goddard Space Flight Center Technical Note, NASA TN D-1539, Feb. 1965.

Examples

FOR satellite orbits and ballistic trajectories disturbed by the oblateness of the Earth, it is well known that the Keplerian solution is the first-order approximation to the Vinti solution. It is less obvious that the Keplerian solution is also the key to the Vinti solution. The **kepler1** routine provided by this text guarantees an accurately converged Keplerian solution. The **vinti6** routine, which takes advantage of the Keplerian solution, always computes an accurate Vinti solution for any conic trajectory.

The primary purpose of this appendix is to provide numerical solutions of the analytic **vinti6** routine with which other analytic or numerical solutions can be compared. Though only a small sample of our computer test cases is provided in this appendix, they include circular, elliptic, parabolic, and hyperbolic trajectories at various inclinations (0, 63.4, 90°). These example trajectories, except the first one, exhibit singularities or difficulties for all other analytic method. The reader can, of course, use the computer routines to solve easier problems by changing the input conditions in the given input data files.

In the following examples, the initial time t_i is assumed to be zero without loss of generality. In a Vinti routine, the initial and final time, t_i and t_f, are both arbitrary. The given initial state is $\boldsymbol{x}(t_i)$. The computed final states for the Keplerian, Vinti, and numerical trajectories are, respectively, $\boldsymbol{x}_K(t_f)$, $\mathbf{x}_V(t_f)$, and $\boldsymbol{x}_N(t_f)$. The numerical solution is computed by a seventh-order, 11-iterations-per-step, classical Runge–Kutta integrator (RK711) using a WGS84 Earth gravity model with only the zonal harmonics J_2, J_3, and J_4. In other words, we are comparing the analytic solutions against the numerical solutions with the Earth potential model. We shall also introduce a Vinti numerical exact solution, so that the analytic Vinti solutions can be evaluated against the Vinti potential model.

Again, the Vinti potential, $V = -\mu(\rho+\delta\eta)(\rho^2+c^2\eta^2)^{-1}$, in the oblate spheroidal coordinate system is well known, but the Vinti potential in the ECI coordinate system is not. The ECI Vinti potential has little value if the analytic Vinti solutions cannot be computed. If the equations of motion are numerically integrated in the ECI coordinate system using a particular acceleration model, then the solution is "almost" exact for that model. We may use the Vinti oblate spheroidal potential, but the gravitational acceleration vector $-\nabla V$ must be expressed in the ECI form for numerical integration. The complete expression of the ECI gravitational acceleration due to the Vinti potential will be given in a future paper. All we need is the 3×3 Jacobian of the position components of the ECI and oblate spheroidal coordinate systems. A numerical exact Vinti solution represents the best Vinti solution that any analytic Vinti solution can achieve. The neglected or truncated terms in the formulation of an analytic Vinti solution are represented by the difference between the analytic and numerical exact Vinti solutions. Simulation results show that each component of the state vector of a **vinti6** solution and the corresponding numerical exact Vinti solution match to at least 12 significant digits in all of our test cases. Therefore, we conclude that the neglected terms in Getchell's formulation are insignificant.

In addition to presenting the 10 examples, we use four tables to compare the accuracy of analytic trajectories against numerical reference trajectories. A reference trajectory is computed by the classical Runge–Kutta integrator (RK711) using a WGS84 Earth gravity model with only the zonal harmonics, J_2, J_3, and J_4. The numerical exact Vinti solution is also given for completeness. A number in the matrix represents the averaged number of significant digits matched with the reference position vector components.

If the computer CPU time for a Kepler solution is defined as one time-unit, then the Vinti and SGP solutions take, on the average, 5 and 10 time-units, respectively. In a 143-Mhz Sun workstation with an ULTRA SPARC processor, one time-unit is approximately 20 μs, while on a 200-Mhz Pentium-Pro personal computer, it is 10 μs. Although every computer routine is likely to have bugs just as every book probably has typographical errors, our simulation results show that the **kepler1** and **vinti6** are extremely accurate and reliable. A Vinti solution, which is at least a few orders of magnitude more accurate than a Kepler solution, is just a few microseconds slower in real time.

I. Low-Earth Orbit

This simple example is provided so that the SGP4 routine can compute a solution without any problem. We shall use this low-Earth orbit to compare numerical accuracy in Table C.1. The given initial and final times are $t_i = 0$, $t_f = 10{,}000$ s, where

osculating classical orbital elements at $t_0 = 0$

semi-major axis	= 6640.262815499317000 km
eccentricity	= 9.496210216913872E−003
inclination	= 72.853838974525400 deg
ascending node	= 115.962302753882600 deg
argument of perigee	= 57.735018723715720 deg
mean anomaly	= 105.534231958634600 deg

$$x(t_i) = \begin{bmatrix} 2328.96594 \\ -5995.21600 \\ 1719.97894 \\ 2.91110113 \\ -0.98164053 \\ -7.09049922 \end{bmatrix} \qquad x_K(t_f) = \begin{bmatrix} -500.5832559961 \\ -3075.2376202228 \\ 5822.4061243021 \\ 3.9383267135 \\ -6.1032449766 \\ -2.8166618485 \end{bmatrix}$$

$$x_V(t_f) = \begin{bmatrix} -485.5222682585 \\ -3123.5190458862 \\ 5796.3841118105 \\ 3.9097618929 \\ -6.0846992371 \\ -2.8777002798 \end{bmatrix} \qquad x_N(t_f) = \begin{bmatrix} -479.1990953029 \\ -3132.5319528031 \\ 5790.4839771675 \\ 3.9111905123 \\ -6.0775687486 \\ -2.8918513134 \end{bmatrix}$$

Table C.1 Elliptic low-Earth orbit, 72° inclination[a,b]

Propagation time, s	Analytic predictors						Numerical extra Vinti
	Kepler	**SGP4**	**Vinti2**	**Vinti3**	**Vinti5**	**Vinti6**	RK711
1	9	8	8	9	11	11	13
10	6	7	8	9	9	9	11
100	5	6	6	7	7	7	8
1,000	2	5	6	6	6	6	6
10,000	1	2	2	2	2	2	2

[a] The inaccurate solutions at 10,000 s are due to atmospheric drag.
[b] A number in the matrix represents the averaged number of significant digits matched with the reference position vector components.

II. High-Earth Orbit

When the eccentricity is zero, the SGP4 routine must be replaced by the less accurate SGP routine to avoid the singularity. We shall also use this high-Earth orbit to compare numerical accuracy in Table C.2. The given initial and final times are $t_i = 0$, $t_f = 10{,}000$ s, where

osculating classical orbital elements at $t_0 = 0$
semi-major axis = 7878.135704119925000 km
eccentricity = 0.000000000000000E+000
inclination = 29.999999981223680 deg
ascending node = 137.217976698769400 deg
argument of perigee = 0.000000000000000E+000 deg
mean anomaly = 35.999999974203660 deg

$$x(t_i) = \begin{bmatrix} 2328.96594 \\ -5995.21600 \\ 1719.97894 \\ 2.91110113 \\ -0.98164053 \\ -7.09049922 \end{bmatrix} \qquad x_K(t_f) = \begin{bmatrix} 6693.9937332156 \\ -4053.6749275797 \\ -907.2876049643 \\ 2.8690496198 \\ 5.5123917721 \\ -3.4609097997 \end{bmatrix}$$

$$x_V(t_f) = \begin{bmatrix} 6712.0609670035 \\ -3985.3574556181 \\ -981.32635365161 \\ 2.7986992751 \\ 5.5685271109 \\ -3.4494924890 \end{bmatrix} \qquad \mathrm{x}_N(t_f) = \begin{bmatrix} 6712.0572667907 \\ -3985.361473247 \\ -981.3375535940 \\ 2.7986983307 \\ 5.5685290662 \\ -3.4494902230 \end{bmatrix}$$

III. Molniya Orbit

This example tests the critical inclination on a 12-h satellite orbit. The given initial and final times are $t_i = 0$, $t_f = 86{,}400$ s, where

Table C.2 Circular high-Earth orbit, 0° inclination[a]

Propagation time, s	Analytic predictors: Kepler	SGP	Vinti2	Vinti3	Vinti5	Vinti6	Numerical extra Vinti RK711
1	9	3	7	5	11	11	11
10	7	3	6	5	10	10	10
100	4	3	5	5	8	8	8
1,000	2	3	5	5	6	6	6
10,000	1	1	2	5	5	5	5

[a]A number in the matrix represents the averaged number of significant digits matched with the reference position vector components.

osculating classical orbital elements at $t_0 = 0$
semi-major axis = 26,628.136194743230000 km
eccentricity = 7.416966410816510E−001
inclination = 63.400000000279700 deg
ascending node = 119.999999995627700 deg
argument of perigee = 359.999998521220600 deg
mean anomaly = 144.008864736199700 deg

$$\boldsymbol{x}(t_i) = \begin{bmatrix} 19{,}850.34032 \\ -40{,}076.98531 \\ 5{,}686.51314 \\ 0.9622473922 \\ -0.3840200243 \\ -1.2806877932 \end{bmatrix} \qquad \boldsymbol{x}_K(t_f) = \begin{bmatrix} 19{,}766.0536122 \\ -40{,}042.8145765 \\ 5{,}798.16095975 \\ 0.96977866348 \\ -0.39925120750 \\ -1.27850448490 \end{bmatrix}$$

$$\boldsymbol{x}_V(t_f) = \begin{bmatrix} 19{,}663.9353084 \\ -40{,}094.4781151 \\ 5{,}795.9262619 \\ 0.9686039103 \\ -0.4014772083 \\ -1.2785482612 \end{bmatrix} \qquad \boldsymbol{x}_N(t_f) = \begin{bmatrix} 19{,}663.9664163 \\ -40{,}094.4541867 \\ 5{,}795.8734577 \\ 0.96860045712 \\ -0.40146845120 \\ -1.2785505095 \end{bmatrix}$$

IV. Geosynchronous Orbit

This example tests both the zero eccentricity and inclination. Note that there is no "equatorial" orbit in the Vinti solution, which implies that equatorial orbits do not exist in real motion. The given initial and final times are $t_i = 0$, $t_f = 86{,}400$ s, where

osculating classical orbital elements at $t_0 = 0$
semi-major axis = 42,164.171587425180000 km
eccentricity = 0.000000000000000E+000
inclination = 0.000000000000000E+000 deg
ascending node = 0.000000000000000E+000 deg
argument of perigee = 0.000000000000000E+000 deg
mean anomaly = 250.000000003011600 deg

$$x(t_i) = \begin{bmatrix} -14{,}420.99601 \\ -39{,}621.36091 \\ 0. \\ 2.8892355501 \\ -1.0515957400 \\ 0. \end{bmatrix} \qquad x_K(t_f) = \begin{bmatrix} -13{,}737.29692824 \\ -39{,}863.56782061 \\ 0. \\ 2.9068975587 \\ -1.0017396107 \\ 0. \end{bmatrix}$$

$$x_V(t_f) = \begin{bmatrix} -13{,}727.98920329 \\ -39{,}866.77387685 \\ -0.0002772068 \\ 2.9071320167 \\ -1.0010590414 \\ -0.0000000003 \end{bmatrix} \qquad x_N(t_f) = \begin{bmatrix} -13{,}718.67926054 \\ -39{,}869.97849942 \\ -0.000000086551 \\ 2.90736571383 \\ -1.00038011634 \\ -0.0000000007 \end{bmatrix}$$

V. Parabolic Orbit of 0° Inclination

If ECI input is parabolic, then there is no "parabolic" orbit in the spheroidal coordinate system. The final orbit is highly eccentric. The given initial and final times are $t_i = 0$, $t_f = 21{,}600$ s, where

osculating classical orbital elements at $t_0 = 0$
semi-major axis = 1.000000000000000E+030 km
eccentricity = 1.000000000000000
inclination = 0.000000000000000E+000 deg
ascending node = 0.000000000000000E+000 deg
argument of perigee = 0.000000000000000E+000 deg
mean anomaly = 0.000000000000000E+000 deg

$$x(t_i) = \begin{bmatrix} 10{,}000. \\ 0. \\ 0. \\ 0. \\ 8.9286113142 \\ 0. \end{bmatrix} \qquad x_K(t_f) = \begin{bmatrix} -65{,}371.81216572 \\ 54{,}907.85450761 \\ 0. \\ -2.8712690908 \\ 1.0458500397 \\ 0. \end{bmatrix}$$

$$x_V(t_f) = \begin{bmatrix} -65{,}386.51048664 \\ 54{,}824.07404366 \\ -0.0427413796 \\ -2.8706415782 \\ 1.0414098075 \\ -0.0000013464 \end{bmatrix} \qquad x_N(t_f) = \begin{bmatrix} -65{,}386.51377768 \\ 54{,}824.06154128 \\ -0.04270679538 \\ -2.87064153247 \\ 1.04140916778 \\ -0.00000134538 \end{bmatrix}$$

VI. "Parabolic Orbit" of 0° Inclination in the Oblate Spheroidal System

The ECI input is slightly hyperbolic, so that it is "parabolic" in the Vinti oblate spheroidal coordinate system (see Table C.3). That is, the total energy is zero or

Table C.3 Parabolic orbit, 0° inclination[a]

Propagation time, s	Analytic predictors: Kepler	SGP	Vinti2	Vinti3	Vinti5	Vinti6	Numerical extra Vinti RK711
1	9	—	—	—	—	13	13
100	5	—	—	—	—	9	9
10,000	2	—	—	—	—	8	8
86,400 (1 day)	2	—	—	—	—	7	7
864,000 (10 day)	2	—	—	—	—	6	6

[a]A number in the matrix represents the averaged number of significant digits matched with the reference position vector components.

$\alpha_1 = 0$. The given initial and final times are $t_i = 0, t_f = 21,600$ s, where

osculating classical orbital elements at $t_0 = 0$
semi-major axis = −2.269809983628260E+007 km
eccentricity = 1.000440565513066
inclination = 0.000000000000000E+000 deg
ascending node = 0.000000000000000E+000 deg
argument of perigee = 0.000000000000000E+000 deg
mean anomaly = 0.000000000000000E+000 deg

$$x(t_i) = \begin{bmatrix} 10{,}000. \\ 0. \\ 0. \\ 0. \\ 8.9295946696017 \\ 0. \end{bmatrix} \qquad x_K(t_f) = \begin{bmatrix} -65{,}379.23990243 \\ 54{,}962.18246752 \\ 0. \\ -2.87242624638 \\ 1.04893952398 \\ 0. \end{bmatrix}$$

$$x_V(t_f) = \begin{bmatrix} -65{,}393.97186689 \\ 54{,}878.43471233 \\ -0.042750659016 \\ -2.87180213163 \\ 1.044500848346 \\ -0.00000134746 \end{bmatrix} \qquad x_N(t_f) = \begin{bmatrix} -65{,}393.97516284 \\ 54{,}878.42221500 \\ -0.042716099436 \\ -2.87180208635 \\ 1.04450020887 \\ -0.0000034645 \end{bmatrix}$$

VII. Hyperbolic Orbit of 0° Inclination

This hyperbolic trajectory is artificial, because the satellite has reached a distance far from the sphere of influence of the Earth. This example demonstrates the robustness of the **vinti6** routine. The given initial and final times are $t_i = 0, t_f =$ 864,000 s (10 days), where

osculating classical orbital elements at $t_0 = 0$
semi-major axis = −81,018.008496107870000 km
eccentricity = 1.123429348432829
inclination = 0.000000000000000E+000 deg
ascending node = 0.000000000000000E+000 deg
argument of perigee = 0.000000000000000E+000 deg
mean anomaly = 0.000000000000000E+000 deg

$$x(t_i) = \begin{bmatrix} 10{,}000. \\ 0. \\ 0. \\ 0. \\ 9.2 \\ 0. \end{bmatrix} \qquad x_K(t_f) = \begin{bmatrix} -1{,}897{,}260.450641 \\ 1{,}017{,}055.109125 \\ 0. \\ -2.0469939635 \\ 1.0488310491 \\ 0. \end{bmatrix}$$

$$x_V(t_i) = \begin{bmatrix} -1{,}895{,}825.589375 \\ 1{,}013{,}534.429643 \\ -0.9236691031 \\ -2.0449291200 \\ 1.0447195567 \\ -0.0000009786 \end{bmatrix} \qquad x_N(t_f) = \begin{bmatrix} -1{,}895{,}825.434780 \\ 1{,}013{,}533.940893 \\ -0.92295381665 \\ -2.04492888725 \\ 1.04471899026 \\ -0.000000977894 \end{bmatrix}$$

VIII. Hyperbolic Orbit of 90° Inclination

The given initial and final times are $t_i = 0$, $t_f = 864{,}000$ s (10 days), where

osculating classical orbital elements at $t_0 = 0$
semi-major axis = −81,018.008496107870000 km
eccentricity = 1.123429348432829
inclination = 90.00000000000000E+000 deg
ascending node = 0.000000000000000E+000 deg
argument of perigee = 0.000000000000000E+000 deg
mean anomaly = 0.000000000000000E+000 deg

$$x(t_i) = \begin{bmatrix} 10{,}000. \\ 0. \\ 0. \\ 0. \\ 0. \\ 9.2 \end{bmatrix} \qquad x_K(t_f) = \begin{bmatrix} -1{,}897{,}260.45064 \\ 0. \\ 1{,}017{,}055.10912 \\ -2.0469939634 \\ 0. \\ 1.0488310491 \end{bmatrix}$$

$$x_V(t_f) = \begin{bmatrix} -1{,}895{,}222.00657 \\ 0. \\ 1{,}014{,}670.41072 \\ -2.0442992160 \\ 0. \\ 1.0459513077 \end{bmatrix} \qquad x_N(t_f) = \begin{bmatrix} -1{,}895{,}221.78154 \\ 0. \\ 1{,}014{,}670.05463 \\ -2.0442989103 \\ 0 \\ 1.0459508846 \end{bmatrix}$$

Table C.4 Hyperbolic orbit, 90° inclination[a]

Propagation time, s	Analytic predictors						Numerical extra Vinti
	Kepler	**SGP**	**Vinti2**	**Vinti3**	**Vinti5**	**Vinti6**	RK711
1	9	—	—	—	—	13	13
100	5	—	—	—	—	9	9
10,000	3	—	—	—	—	6	6
86,400 (1 day)	2	—	—	—	—	6	6
864,000 (1 day)	0	—	—	—	—	6	6

[a]A number in the matrix represents the averaged number of significant digits matched with the reference position vector components.

IX. Long-Range Ballistic Missile Trajectory

This example illustrates a long-range ballistic missile in a retrograde trajectory (see Table C.4). The given initial and final times are $t_i = 0$, $t_f = 1000$ s, where

osculating classical orbital elements at $t_0 = 0$
semi-major axis = 4687.953562723175000 km
eccentricity = 6.156073264729958E−001
inclination = 133.914685183962600 deg
ascending node = 18.107803794189210 deg
argument of perigee = 335.867839344461500 deg
mean anomaly = 107.185803129158600 deg

$$\boldsymbol{x}(t_i) = \begin{bmatrix} -3158.00000 \\ -4647.00000 \\ 3568.00000 \\ -5.74500000 \\ -0.97200000 \\ -0.89500000 \end{bmatrix} \qquad \boldsymbol{x}_K(t_f) = \begin{bmatrix} -6473.6112958366 \\ -3206.4212088435 \\ 1075.5765925537 \\ -0.526409920884 \\ 3.389073897476 \\ -3.515561063365 \end{bmatrix}$$

$$\boldsymbol{x}_V(t_f) = \begin{bmatrix} -6473.0551629885 \\ -3206.1626988526 \\ 1071.7467222969 \\ -0.523319895600 \\ 3.390916610237 \\ -3.521575157896 \end{bmatrix} \qquad \boldsymbol{x}_N(t_f) = \begin{bmatrix} -6473.0557229332 \\ -3206.1637491443 \\ 1071.7459270133 \\ -0.523320813163 \\ 3.390915437503 \\ -3.521576882969 \end{bmatrix}$$

X. Exo-Atmospheric Interceptor Trajectory

This example illustrates an exo-atmospheric interceptor that has a perigee radius of 19 km, and, therefore, the extension of the trajectory passes the focal circle or the Vinti forbidden zone. The eccentricity is approximately 0.994. By default, **vinti6** gives the Keplerian solution, but the numerical exact Vinti solution is given below in $\boldsymbol{x}_V(t_f)$. The given initial and final times are $t_i = 0$, $t_f = 100$ s, where

osculating classical orbital elements at $t_0 = 0$
semi-major axis = 3251.548870391171000 km
eccentricity = 9.940795562606448E−001
inclination = 96.057156000898360 deg
ascending node = 106.928715972110900 deg
argument of perigee = 213.426009474111200 deg
mean anomaly = 166.006964173314400 deg

$$\boldsymbol{x}(t_i) = \begin{bmatrix} -1221.14362 \\ 5288.41648 \\ 3502.50807 \\ 0.0192755409 \\ 0.2545356003 \\ 0.8722443619 \end{bmatrix} \qquad \boldsymbol{x}_K(t_f) = \begin{bmatrix} -1210.2635448748 \\ 5275.0167907335 \\ 3563.8283386621 \\ 0.1977767393 \\ -0.5209724863 \\ 0.3534817097 \end{bmatrix}$$

$$\boldsymbol{x}_V(t_f) = \begin{bmatrix} -1210.2762310557 \\ 5275.0431119839 \\ 3563.7555575245 \\ 0.197565823142 \\ -0.520405921315 \\ 0.352124771187 \end{bmatrix} \qquad \boldsymbol{x}_N(t_f) = \begin{bmatrix} -1210.2754882091 \\ 5275.0427800469 \\ 3563.7548164270 \\ 0.19758054893989 \\ -0.52041272393527 \\ 0.35210988750489 \end{bmatrix}$$

Appendix D

How to Use the Vinti Routines

ON THE floppy, there are two folders: **source** and **examples**. The **source** folder has all the source code, the input data files, and a subfolder named **others**. The **examples** folder has the output data files that were generated for the 10 examples in Appendix C. These routines were originally developed on a UNIX workstation and then ported to a personal computer (PC) with the WINDOWS 95 operating system. On the PC, we use the Microsoft Fortran PowerStation 4.0 to compile, link, and run the program, and this is the preferred configuration.

In this book, we provide two extremely accurate and robust Kepler and Vinti routines (**kepler1.f** and **vinti6.f**). These routines, which use universal variables to simplify the conic trajectories, are also free of singularities.

I. The Source Folder

The computer routines that are illustrated in Fig. D.1 are listed as follows.

1) Input data file: **input.txt**
2) Output data file: **v_prop.log**
3) Main program: **propagate.for** (which calls the following subroutines):

kepler.f
sgp_driver.f (which calls **sgp.f**, **sgp4.f**, **sgp8.f**, **sdp4.f**, **sdp8**)
vinti2.f
vinti3.f
vinti5.f
vinti6.f (which calls **kepler1.f**)

Note that the **input.txt** file is created using the Notepad utility. The 10 input data files are named accordingly. The routines **kepler.f** and **kepler1.f** are identical except that **kepler1.f** returns with the universal variable needed by **vinti6.f**. The remaining routines are utility codes for output purposes.

The subfolder **others** contains all the C routines and the two partially completed Vinti Fortran routines (**vinti1.f** and **vinti4.f**).

II. The Examples Folder

This folder contains all the output data files of the 10 examples. The v_prop.log output files are renamed to:

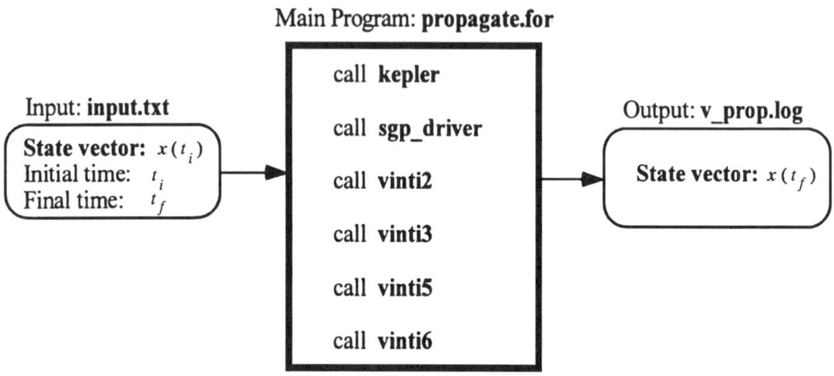

Fig. D.1 An overview of the source code.

Ex1_leo.log	= output file of Example 1 for the low-Earth orbit satellite
Ex2_heo.log	= output file of Example 2 for the high-Earth orbit satellite of zero eccentricity
Ex3_mol.log	= output file of Example 3 for the Molniya orbit satellite of critical inclination
Ex4_geo.log	= output file of Example 4 for the geosynchronous Earth orbit satellite
Ex5_par0.log	= output file of Example 5 for the parabolic orbit satellite of zero inclination
Ex6_par0x.log	= output file of Example 6 for the "Vinti" parabolic orbit satellite of zero inclination
Ex7_hyp0.log	= output file of Example 7 for the hyperbolic orbit satellite of zero inclination
Ex8_hyp90.log	= output file of Example 8 for the hyperbolic orbit satellite of 90° inclination
Ex9_kwaj.log	= output file of Example 9 for a ballistic missile trajectory
Ex10_intr.log	= output file of Example 10 for an exo-atmospheric interceptor trajectory at approximately 80 km altitude

III. The Users

We envision three types of users. A user may follow the relevant procedures to retrieve the source code and example results from the floppy.

User Who Wants to Use All the Fortran Source Code on the Floppy

1) Create a new directory (UNIX) or folder (PC) and name it **Vinti**.

2) Copy or drag the **examples** and **source** folders into the **Vinti** folder and go to the source folder.

3) Create a makefile (UNIX) or a workspace (PC, Microsoft Fortran) to compile the code.

4) Compile and link all the source code.

5) Copy the appropriate input file (i.e., inputgeo.txt) to replace input.txt (inputleo.txt is the input.txt by default).

6) Run the main program.

7) Compare the newly generated output data file v_prop.log with that in the **examples** folder (i.e., ex4_geo.log).

User Who Wants to Replace His/Her kepler Fortran Subroutine with kepler1 or vinti6

1) Copy or drag the **vinti6.f** and **kepler1.f** of the **source** folder into the directory or workspace that has the **kepler** subroutine. (Note that **vinti6.f** calls **kepler1.f** for the Keplerian final state and universal variable at the given final time; **kepler1.f** is the only external routine called by **vinti6.f**.)

2) Make sure the calling parameters match with those of **kepler1.f** or **vinti6.f**.

3) Replace the **kepler** subroutine by **kepler1.f** or **vinti6.f**; recompile and run the program.

User Who Wants to Replace His/Her kepler C Subroutine with kepler1 or vinti6

The C routines were originally developed on a PC with the WINDOWS 95 operating system. On the PC, we use the Microsoft Visual-C++ PowerStation 4.0 to compile, link, and run.

1) Open the subdirectory or subfolder (PC) **others source** in the **source** folder.

2) Copy or drag **kepler1.cpp** or **Vinti6.cpp** into your directory or folder.

3) Make sure the calling parameters match with those of **kepler1.cpp** or **Vinti6.cpp**.

4) Replace the **kepler** subroutine with **kepler1.cpp** or **Vinti6.cpp**; recompile and run the program.

IV. Some Editing Problems

Some difficulties were encountered in using the Layhey Fortran compiler on a PC. To circumvent such problems, the user may open the problem routine by Microsoft Word and then immediately save the routine as text with line breaks. The newly saved routine should work.

When a computer routine is copied into a directory of a UNIX workstation, using a vi editor, the user may see a strange symbol "^M." This is also due to a line break in a DOS routine, which is not recognized by the UNIX vi editor. The user may delete this by using a vi editor command ":1, $s/^M//g." To type the symbol "^M," the keys **control** and **v** are pressed simultaneously for ^, and then the keys **control** and **m** are pressed simultaneously for M.

Appendix E

Bibliography

I. Papers Published by the Author

[1] Vinti, J. P., "Energies and Waves-Functions of the State (1s)(2s)'S in Helium-Like Atoms," *Physics Review,* Vol. 37, No. 4, Feb. 15, 1931, pp. 448–449.

[2] Vinti, J. P., "Sum Rules for Atomic Transition Probabilities," *Physics Review,* Vol. 41, No. 4, Aug. 15, 1932, pp. 432–442.

[3] Vinti, J. P., "Relation Between the Electric and Diamagnetic Susceptibilities of Monatomic Gases," *Physics Review,* Vol. 41, No. 6, Sept. 15, 1932, pp. 813–817.

[4] Vinti, J. P., "Dispersion and Absorption of Helium," *Physics Review,* Vol. 42, No. 5, Dec. 1, 1932, pp. 632–640.

[5] Vinti, J. P., and Morse, P. M., "Variable Scale Atomic Waves Functions," *Physics Review,* Vol. 43, March 1, 1933, pp. 337–340.

[6] Witmer, E. E., and Vinti, J. P., "Symmetry Properties and the Indistinguishability of Similar Particles," *Physics Review,* Vol. 43, March 1, 1933, p. 780.

[7] Vinti, J. P., "The Continuous Absorption of Helium," *Physics Review,* Vol. 44, Oct. 1, 1933, pp. 524–528.

[8] Vinti, J. P., and Fender, F. G., "Doubly-Excited States in Helium," *Physics Review,* Vol. 46, July 1, 1934, pp. 77–78.

[9] Witmer, E. E., and Vinti, J. P., "Symmetry Properties and the Identity of Similar Particles," *Physics Review,* Vol. 47, April 1, 1935, pp. 538–541.

[10] Vinti, J. P., "Isotope Shift in Magnesium," *Physics Review,* Vol. 56, Dec. 1, 1939, pp. 1120–1132.

[11] Vinti, J. P., "Isotope Shift in Boron," *Physics Review,* Vol. 58, No. 10, Nov. 15, 1940, pp. 879–881.

[12] Vinti, J. P., "A Theorem on Nuclear Motion in Atomic Spectra," *Physics Review,* Vol. 58, No. 10, Nov. 15, 1940, pp. 882–883.

[13] Vinti, J. P., "Some Developments in the Theory of the Solenoid Chronograph," Ballistic Research Laboratories, Rept. No. 271, Aberdeen Proving Ground, MD, March 1942.

[14] Vinti, J. P., and Hodge, A. H., "Addendum to Ballistic Research Laboratory Report No. 271 Experimental Determination of Solenoid E.m.f. Curves," Ballistic Research Laboratories, Rept. No. 271, Aberdeen Proving Ground, MD.

[15] Vinti, J. P., and Kent, R. H., "Comments on Report by Lt. P. W. Klipsch on Amplifier Design for the Solenoid Chronograph," Ballistic Research Laboratories, Rept. No. 39, Aberdeen Proving Ground, MD, April 1942.

[16] Vinti, J. P., " The Equation of State of the Powder Gas," Ballistic Research Laboratories, Rept. No. 288, Aberdeen Proving Ground, MD, July 1942.

[17] Vinti, J. P., "Interior Ballistic Trajectories for the Case of Instantaneous Combustion," Ballistic Research Laboratories, Rept. No. 299, Aberdeen Proving Ground, MD, Aug. 1942.

[18]Vinti, J. P., and Kent, R. H., "Cooling Corrections for Closed Chamber Firing," Ballistic Research Laboratories, Rept. No. 281, Aberdeen Proving Ground, MD, Sept. 1942.

[19]Vinti, J. P., "The Practical Estimation of Lag in the Solenoid Chronograph," Ballistic Research Laboratories, Rept. No. 305, Aberdeen Proving Ground, MD, Sept. 1942.

[20]Vinti, J. P., and Hodge, A. H., "The Equations of Interior Ballistics," Ballistic Research Laboratories, Rept. No. 307, Aberdeen Proving Ground, MD, Oct. 1942.

[21]Vinti, J. P., and Hodge, A. H., "Can Flareback in a Gun Be Prevented by Opening the Breech at a Proper Moment?" Ballistic Research Laboratories, Rept. No. 83, Aberdeen Proving Ground, MD, Nov. 1942.

[22]Vinti, J. P., "Available Energy of a Typical Caliber .50 Powder," Ballistic Research Laboratories, Rept. No. 164, Aberdeen Proving Ground, MD, May 1943.

[23]Vinti, J. P., "The Dependence of Mortar Velocities on Projectile Weight," Ballistic Research Laboratories, Rept. No. 358, Aberdeen Proving Ground, MD, May 1943.

[24]Vinti, J. P., "Investigation of a Vented 81 MM Mortar for a Tank," Ballistic Research Laboratories, Rept. No. 372, Aberdeen Proving Ground, MD, June 1943.

[25]Vinti, J. P., "Project for a New Table for Interior Ballistics for Multi-Perforated Powder," Ballistic Research Laboratories, Rept. No. 402, Aberdeen Proving Ground, MD, Sept. 1943.

[26]Vinti, J. P., "A Simple Theory of Vented Mortars," Ballistic Research Laboratories, Rept. No. 414, Aberdeen Proving Ground, MD, Oct. 1943.

[27]Vinti, J. P., "The Flow of Compressible Gases Through Nozzles," Ballistic Research Laboratories, Rept. No. 476, Aberdeen Proving Ground, MD, July 1944.

[28]Vinti, J. P., and Hodge, A. H., "The Effective Velocity of Escape of the Powder Gas from a Gun," Ballistic Research Laboratories, Rept. No. 538, Aberdeen Proving Ground, MD, April 1945.

[29]Vinti, J. P., "Theory and Design of a Grooved Projectile for a 105 MM Smooth Bore Howitzer," Ballistic Research Laboratories, Rept. No. 580, Aberdeen Proving Ground, MD, Oct. 1945.

[30]Vinti, J. P., "The Dependence of Muzzle Velocity on the Potential and the Force of the Powder," Ballistic Research Laboratories, Rept. No. 583, Aberdeen Proving Ground, MD, Oct. 1945.

[31]Vinti, J. P., and Hodge, A. H., "Interior Ballistics of the Post-Combustion Period," Ballistic Research Laboratories, Rept. No. 600, Aberdeen Proving Ground, MD, Jan. 1946.

[32]Vinti, J. P., "The Possible Effects of Fall-Back in a 16-inch Gun," Ballistic Research Laboratories, Rept. No. 431, Aberdeen Proving Ground, MD, May 1946.

[33]Vinti, J. P., and Chernick, J., "Interior Ballistics for Powder of Constant Burning Surface," Ballistic Research Laboratories, Rept. No. 625, Aberdeen Proving Ground, MD, Feb. 1947.

[34]Vinti, J. P., and Kravitz, S., "Tables for the Pidduck-Kent Special Solutions for the Motion of the Powder Gas in a Gun," Ballistic Research Laboratories, Rept. No. 693, Aberdeen Proving Ground, MD, Jan. 1949.

[35]Montogomery, D. J., and Vinti, J. P., "Note on the Presentation of Maxwell's Equations," *American Journal of Physics,* Vol. 17, 1949, p. 298.

[36]Vinti, J. P., and Kravitz, S., "Theory of the Transient State of a Rocket Motor or of the Reservoir of a High-Low Pressure Gun," Ballistic Research Laboratories, Rept. No. 772, Aberdeen Proving Ground, MD, Dec. 1951.

[37]Vinti, J. P., "Note on a Series of Products of Three Legendre Polynomials," Proceedings of the American Mathematical Society, Vol. 2, No. 1, Feb. 1951, pp. 19–23.

[38]Vinti, J. P., "Theory of the Rapid Burning of Propellants," Ballistic Research Laboratories, Rept. No. 841, Aberdeen Proving Ground, MD, Dec. 1952.

[39]Vinti, J. P., "The Multiple Scattering of Hard X-Rays," Ballistic Research Laboratories, Rept. No. 842, Aberdeen Proving Ground, MD, Dec. 1952.

[40]Vinti, J. P., "Theory of the Penetration of γ-Rays Through Thin Barriers," *Physics Review,* Vol. 91, July 15, 1953, pp. 345–348.

[41]Vinti, J. P., "The Reflection of Plane Electromagnetic Waves at a Plane Boundary of a Conducting Medium," Ballistic Research Laboratories, Rept. No. 827, Aberdeen Proving Ground, MD, Oct. 1953.

[42]Vinti, J. P., "Theory of the Electromagnetic Field About an Antenna, According to the Gap Model," Ballistic Research Laboratories, Rept. No. 914, Aberdeen Proving Ground, MD, Sept. 1954.

[43]Vinti, J. P., "Theory of the Electromagnetic Field About an Antenna, According to the Gap Model," Monograph, Applied Science Research Section B, Vol. 6.

[44]Vinti, J. P., "Phase and Doran Centers of a Thin Half-Wave Antenna, Fed Off-Center," Ballistic Research Laboratories, Rept. No. 919, Aberdeen Proving Ground, MD, Sept. 1954.

[45]Vinti, J. P., "The Determination of Range and Direction with a Five-Station System for Electronic Missile Acquisition," Ballistic Research Laboratories, Rept. No. 967, Aberdeen Proving Ground, MD, Nov. 1955.

[46]Vinti, J. P., "Theory of the Spin of a Conducting Satellite in the Magnetic Field of the Earth," Ballistic Research Laboratories, Rept. No. 1020, Aberdeen Proving Ground, MD, July 1957.

[47]Vinti, J. P., "Theory of the Multipath Propagation of Frequency Modulated Waves," Ballistic Research Laboratories, Rept. No. 1025, Aberdeen Proving Ground, MD, Sept. 1957.

[48]Vinti, J. P., "Theory of the Spin of a Conducting Satellite in Non-Equatorial Orbits," Ballistic Research Laboratories, Rept. No. 1031, Aberdeen Proving Ground, MD, Oct. 1957.

[49]Vinti, J. P., and Leser, T., "The Sums of Certain Series Involving Bessel Functions," *Journal of Social Industry and Applied Mathematics,* Vol. 5, No. 1, March 1957, pp. 15–31.

[50]Vinti, J. P., "Theory of the Effect of Drag on the Orbital Inclination of an Earth Satellite," *Journal of Research of the National Bureau of Standards,* Vol. 62, No. 2, Feb. 1959, pp. 79–88.

[51]Vinti, J. P., "New Method of Solution for Unretarded Satellite Orbits," *Journal of Research of the National Bureau of Standards,* Vol. 63B, No. 2, Oct.–Dec. 1959, pp. 105–116.

[52]Vinti, J. P., "Theory of the Multipath Propagation of Frequency Modulated Waves," U.S. Dept. of Commerce, Monograph, PB 151117, 1959.

[53]Vinti, J. P., "Satellite Frequencies with a New Gravitational Potential," *Bulletin of the American Journal of the Physics Society,* Vol. 15, No. 8, 1960.

[54]Vinti, J. P., "Theory of Orbits of an Artificial Satellite, with Use of Spheroidal Coordinates," *Astronautical Journal,* Vol. 65, No. 353, 1960.

[55]Vinti, J. P., and Dressler, R. F., "The Functional Synthesis of Linear Plots," *Journal of Research of the National Bureau of Standards,* Vol. 64C, No. 2, April–June 1960, pp. 115–119.

[56]Vinti, J. P., "Mean Motions in Conditionally Periodic Separable System," *Journal of Research of the National Bureau of Standards,* Vol. 65B, No. 2, April–June 1961, pp. 131–135.

[57]Vinti, J. P., "Theory of an Accurate Intermediary Orbit for Satellite Astronomy," *Journal of Research of the National Bureau of Standards,* Vol. 65B, No. 3, July–Sept. 1961, pp. 169–201.

[58]Vinti, J. P., "Formulae for an Accurate Intermediary Orbit of an Artificial Satellite," *Astronautical Journal,* Vol. 66, 1961, pp. 514–516.

[59]Vinti, J. P., "Intermediary Equatorial Orbits of an Artificial Satellite," *Journal of Research of the National Bureau of Standards,* Vol. 66B, No. 1, Jan.–March 1962, pp. 5–13.

[60]Vinti, J. P., "Zonal Harmonic Perturbations of an Accurate Reference Orbit of an Artificial Satellite," *Journal of Research of the National Bureau of Standards,* Vol. 67B, No. 4, Oct.–Dec. 1963, pp. 191–222.

[61]Vinti, J. P., and Langer, R. M., "Drag Compensation and Measurement in Manned Satellites," *Journal of Research of the National Bureau of Standards,* Vol. 67C, 1963, pp. 247–249.

[62]Vinti, J. P., "The Spheroidal Method in the Theory of the Orbit of an Artificial Satellite," *Satellitenmechanik Im Vakuum Und In Der Atmosphäre,* 7 Himmelsmechanik, July 1964.

[63]Vinti, J. P., "Invariant Properties of the Spheroidal Potential of an Oblate Planet," *Journal of Research of the National Bureau of Standards,* Vol. 70B, No. 1, Jan.–March 1966, pp. 1–16.

[64]Vinti, J. P., "Inclusion of the Third Zonal Harmonic in an Accurate Reference Orbit of an Artificial Satellite," *Journal of Research of the National Bureau of Standards,* Vol. 70B, No. 1, Jan.–March 1966, pp. 17–46.

[65]Vinti, J. P., "Effects of a Constant Force on a Keplerian Orbit," International Astronomical Union, Symposium No. 25, 1966, pp. 355–362.

[66]Vinti, J. P., "Proposed Representation of the Earth's Gravitational Potential," *Transactions of the American Geophysical Union,* Vol. 49, No. 506, 1968.

[67]Vinti, J. P., "Improvement of the Spheroidal Method for Artificial Satellite," *The Astronomical Journal,* Vol. 74, No. 1, Feb. 1969, pp. 25–34.

[68]Vinti, J. P., "Conservation Laws and Liapounov Stability of the Free Rotation of a Rigid Body," *Celestial Mechanics,* Vol. 1, 1969, pp. 59–71.

[69]Vinti, J. P., "Stability of Free Rotation of a Rigid Body," M.I.T. Measurement Systems Laboratory, Rept., Cambridge, MA, 1970.

[70]Vinti, J. P., "Gauss Variational Equations for Osculating Elements of an Arbitrary Separable Reference Orbit," *Celestial Mechanics,* Vol. 7, 1973, pp. 367–375.

II. Papers Acknowledging Vinti's Work

[1]Brouwer, D., "Solution of Problem of Artificial Satellite Theory Without Drag," *Astronautical Journal,* Vol. 64, No. 9, Nov. 1959, pp. 378–397.

[2]Izsak, I. G., "A Theory of Satellite Motion About an Oblate Planet — I. A Second-Order Solution of Vinti's Dynamical Problem," Smithsonian Astrophysics Observatory, Special Rept. No. 52, Nov. 1960.

[3]Koskela, P. E., and Arsenault, J. L., "Methods for Orbit Computation," Aeronutronics, Rept. S-981, 1960, pp. 126–185.

[4]Callender, E. D., "On the Almost Periodicity of Satellite Motion," *Astronautical Journal,* Vol. 66, 1961, pp. 134–137.

[5]Aksenov, E. P., Grebenikov, E. A., and Demin, V. G., *Planetary Sciences,* Vol. 9, 1962, p. 491.

[6]Bonavito, N. L., "Computational Procedure for Vinti's Theory of an Accurate Intermediary Orbit," Technical Note NASA D-1177, March 1962.

[7]Kyner, W. T., "A Mathematical Theory of the Orbits About an Oblate Planet," Dept. of Mathematics, University of Southern California, Technical Rept. for the Office of Naval Research, NR041-152, Feb. 1963.

[8]Wadsworth, D. V., "Vinti Solution for Free-Flight Rocket Trajectories," *AIAA Journal,* Vol. 1, June 1963, pp. 1351–1354.

[9]Spies, O. R., "Two Notes in Orbit Theory," *12th International Astronautical Congress,* Academic Press, New York, 1963, pp. 121–130.

[10]Bonavito, N. L., "Determination of Mean Elements for Vinti's Satellite Theory," Technical Note NASA D-2312, June 1964.

[11]Wadsworth, D. V., and Wagoner, A. L., "Computer Program for the Vinti Equations for Free-Flight Motion of a Rocket," IBM, Monograph, May 29, 1964.

[12]Arsenault, J. L., Enright, J. D., and Purcell, C., "General Perturbation Techniques for Satellite Orbit Prediction Studies," Defense Documentation Center for Scientific and Technical Information, Technical Rept. No. AL-TDR-64-70, Vols. 1 and 2, AD 437-476, AD 437-475, April 1964.

[13]Marchal, C., "Mouvement des Satellites Artificiels Calcul a Partir des Solutions Exactes du Probleme des Deux Centres Fixes," ONERA, Publication No. 111, 1964.

[14]Allen, W. A., and Knolle, W. E., *Astronautical Journal,* Vol. 69, 1964, pp. 393–401.

[15]Allen, W. A., and Knolle, W. E., *Astronautical Journal,* Vol. 70, 1965, pp. 152–154.

[16]Gormally, J. M., "On Bohlin Series Solutions to the Critical Inclination Problem," Lecture notes for the 1965 Summer Institute in Dynamical Astronomy at Stanford University, Bell Telephone Laboratories, Whippany, NJ, 1965.

[17]Borchers, R. V., "A Satellite Orbit Computation Program for Izsak's Second-Order Solution of Vinti's Dynamical Problem," Goddard Space Flight Center, Technical Note NASA D-1539, Feb. 1965.

[18]Lubowe, A. G., "TELSTAR Orbit Prediction Method—Accuracy Comparison with Other Orbit Prediction Methods," *AIAA Journal,* Vol. 4, No. 2, Feb. 1966, pp. 361–363.

[19]Cook, A. H., "Exact Solutions in the Theory of Orbits," *Monthly Notices of the Royal Astronomical Society,* Vol. 134, 1966, pp. 253–268.

[20]Lang, T., "Vinti Unbounded Trajectories," master thesis, Dept. of Aeronautics and Astronautics, Massachussetts Institute of Technology, Cambridge, MA, 1966.

[21]Bonavito, N. L., "Computational Procedure for Vinti's Accurate Reference Orbit with Inclusion of the Third Zonal Harmonic," Technical Note NASA D-3562, 1966.

[22]Sherrill, T. J., "Development of a Satellite Drag Theory Based on the Vinti Formulation," Ph.D. thesis in Astronomy, University of California, Berkeley, 1966.

[23]Lavnik, V. N., "Vinti Potential for the Lambert Problem," translation from *Kosmicheskie Issledovaniya,* Vol. 5, 1967, pp. 194–199.

[24]Madden, S. J., "Theory of a Potential for Satellite Motion About a Triaxially Ellipsoidal Planet," Experimental Astronomy Laboratory, RE-34, Massachusetts Institute of Technology, Cambridge, MA, May 1968.

[25]Walden, H., "Improvement of Mean Orbital Elements for Vinti's the Spheroidal Satellite Theory," *AIAA Journal,* Vol. 6, No. 7, July 1968, pp. 1305–1308.

[26]Kyner, W. T., "Memoirs of the American Mathematical Society," No. 81, Sec. 5, 1968.

[27]Bonavito, N. L., Watson, J. S., and Walden, H., "An Accuracy and Speed Comparison of the Vinti and Brouwer Orbit Prediction Methods," Technical Note NASA D-5203, May 1969.

[28]Getchell, B. C., "Orbit Computation with the Vinti Potential and Universal Variables," *Journal of Spacecraft,* Vol. 7, No. 4, April 1970, pp. 405–408.

[29]Walden, H., "Determination of Mean Orbital Elements from Initial Conditions for a Vinti Ballistic Trajectory," Technical Note NASA D-6069, Nov. 1970.

[30]Watson, J. S., Mistretta, G. D., and Bonavito, N. L., "An Analytic Method to Account for Drag in the Vinti Satellite Theory," *Celestial Mechanics,* Vol. 11, 1975, pp. 145–177.

[31]Monuki, A. T., "Vinti Potential," orbital mechanics lecture notes, TRW, Redondo Beach, CA, 1979.

[32]Wu, Lian-Da, and Tong, Fu, "The Third Order Solution of Vinti's Problem and the Poisson's Brackets of Elements," *Acta Astronomica Sinica,* Vol. 21, No. 4, Dec. 1980.

[33]Tong, Fu, and Wu, Lian-Da, "The Perturbation Equations of the Elements of Vinti's Intermediate," *Acta Astronomica Sinica,* Vol. 22, No. 1, March 1981.

[34]Tong, Fu, and Chen, Zhen, "On the Third Order Artificial Satellite Theory," *Acta Astronomica Sinica,* Vol. 22, No. 2, June 1981.

[35]Tong Fu, and Chen, Zhen, "Second Order Solution of Lindstedt's Equation with Constant Coefficients," *Acta Astronomica Sinica,* Vol. 22, No. 3, Sept. 1981.

[36]Moser, J., "The Stability Behavior of the Solutions of Hamiltonian Systems," New York University, New York, (date unknown).

[37]Der, G. J., and Newnam, G., "Brilliant Pebbles Orbit Mechanics Issues for Space vs Ground," System Integration Group, TRW, IOC BP.92.500 020, March 1992.

III. Books Acknowledging Vinti's Work

[1]Goldstein, H., *Classical Mechanics,* 2nd ed., Addison–Wesley, Reading, MA, 1980.

[2]Escobal, P. R., *Methods of Orbit Determination,* Krieger, Malabar, FL, 1965.

[3]Baker, R. M. L, Jr., *Astrodynamics, Applications and Advanced Topics,* Academic, New York, 1967.

[4]Caputo, M., *The Gravity Field of the Earth,* Academic, New York, 1967.

[5]Cook, A. H., "The Motion of Artificial Satellites in the Gravitational Field of the Earth," lectures at Trieste, 1967.

[6]Herrick, S., *Astrodynamics, Orbit Correction, Perturbation Theory, Integration,* Vol. 2, Reinhold, London, 1972.

[7]Kaplan, M. H., *Modern Spacecraft Dynamics & Control,* Wiley, New York, 1976.

Index

PROGRESS IN ASTRONAUTICS AND AERONAUTICS SERIES VOLUMES

***1. Solid Propellant Rocket Research (1960)**
Martin Summerfield
Princeton University

***2. Liquid Rockets and Propellants (1960)**
Loren E. Bollinger
Ohio State University
Martin Goldsmith
The Rand Corp.
Alexis W. Lemmon Jr.
Battelle Memorial Institute

***3. Energy Conversion for Space Power (1961)**
Nathan W. Snyder
Institute for Defense Analyses

***4. Space Power Systems (1961)**
Nathan W. Snyder
Institute for Defense Analyses

***5. Electrostatic Propulsion (1961)**
David B. Langmuir
Space Technology Laboratories, Inc.
Ernst Stuhlinger
NASA George C. Marshall Space Flight Center
J. M. Sellen Jr.
Space Technology Laboratories, Inc.

***6. Detonation and Two-Phase Flow (1962)**
S. S. Penner
California Institute of Technology
F. A. Williams
Harvard University

***7. Hypersonic Flow Research (1962)**
Frederick R. Riddell
AVCO Corp.

***8. Guidance and Control (1962)**
Robert E. Roberson
Consultant
James S. Farrior
Lockheed Missiles and Space Co.

***9. Electric Propulsion Development (1963)**
Ernst Stuhlinger
NASA George C. Marshall Space Flight Center

***10. Technology of Lunar Exploration (1963)**
Clifford I. Cumming
Harold R. Lawrence
Jet Propulsion Laboratory

***11. Power Systems for Space Flight (1963)**
Morris A. Zipkin
Russell N. Edwards
General Electric Co.

***12. Ionization in High-Temperature Gases (1963)**
Kurt E. Shuler, Editor
National Bureau of Standards
John B. Fenn, Associate Editor
Princeton University

***13. Guidance and Control–II (1964)**
Robert C. Langford
General Precision Inc.
Charles J. Mundo
Institute of Naval Studies

***14. Celestial Mechanics and Astrodynamics (1964)**
Victor G. Szebehely
Yale University Observatory

***15. Heterogeneous Combustion (1964)**
Hans G. Wolfhard
Institute for Defense Analyses
Irvin Glassman
Princeton University
Leon Green Jr.
Air Force Systems Command

***16. Space Power Systems Engineering (1966)**
George C. Szego
Institute for Defense Analyses
J. Edward Taylor
TRW Inc.

***17. Methods in Astrodynamics and Celestial Mechanics (1966)**
Raynor L. Duncombe
U.S. Naval Observatory
Victor G. Szebehely
Yale University Observatory

***18. Thermophysics and Temperature Control of Spacecraft and Entry Vehicles (1966)**
Gerhard B. Heller
NASA George C. Marshall Space Flight Center

***19. Communication Satellite Systems Technology (1966)**
Richard B. Marsten
Radio Corporation of America

***20. Thermophysics of Spacecraft and Planetary Bodies: Radiation Properties of Solids and the Electromagnetic Radiation Environment in Space (1967)**
Gerhard B. Heller
NASA George C. Marshall Space Flight Center

*Out of print.

***21. Thermal Design Principles of Spacecraft and Entry Bodies (1969)**
Jerry T. Bevans
TRW Systems

***22. Stratospheric Circulation (1969)**
Willis L. Webb
Atmospheric Sciences Laboratory, White Sands, and University of Texas at El Paso

***23. Thermophysics: Applications to Thermal Design of Spacecraft (1970)**
Jerry T. Bevans
TRW Systems

***24. Heat Transfer and Spacecraft Thermal Control (1971)**
John W. Lucas
Jet Propulsion Laboratory

25. Communication Satellites for the 70's: Technology (1971)
Nathaniel E. Feldman
The Rand Corp.
Charles M. Kelly
The Aerospace Corp.

26. Communication Satellites for the 70's: Systems (1971)
Nathaniel E. Feldman
The Rand Corp.
Charles M. Kelly
The Aerospace Corp.

27. Thermospheric Circulation (1972)
Willis L. Webb
Atmospheric Sciences Laboratory, White Sands, and University of Texas at El Paso

28. Thermal Characteristics of the Moon (1972)
John W. Lucas
Jet Propulsion Laboratory

***29. Fundamentals of Spacecraft Thermal Design (1972)**
John W. Lucas
Jet Propulsion Laboratory

***30. Solar Activity Observations and Predictions (1972)**
Patrick S. McIntosh
Murray Dryer
Environmental Research Laboratories, National Oceanic and Atmospheric Administration

31. Thermal Control and Radiation (1973)
Chang-Lin Tien
University of California at Berkeley

***32. Communications Satellite Systems (1974)**
P. L. Bargellini
COMSAT Laboratories

33. Communications Satellite Technology (1974)
P. L. Bargellini
COMSAT Laboratories

***34. Instrumentation for Airbreathing Propulsion (1974)**
Allen E. Fuhs
Naval Postgraduate School
Marshall Kingery
Arnold Engineering Development Center

35. Thermophysics and Spacecraft Thermal Control (1974)
Robert G. Hering
University of Iowa

36. Thermal Pollution Analysis (1975)
Joseph A. Schetz
Virginia Polytechnic Institute
ISBN 0-915928-00-0

***37. Aeroacoustics: Jet and Combustion Noise; Duct Acoustics (1975)**
Henry T. Nagamatsu, Editor
General Electric Research and Development Center
Jack V. O'Keefe, Associate Editor
The Boeing Co.
Ira R. Schwartz, Associate Editor
NASA Ames Research Center
ISBN 0-915928-01-9

***38. Aeroacoustics: Fan, STOL, and Boundary Layer Noise; Sonic Boom; Aeroacoustics Instrumentation (1975)**
Henry T. Nagamatsu, Editor
General Electric Research and Development Center
Jack V. O'Keefe, Associate Editor
The Boeing Co.
Ira R. Schwartz, Associate Editor
NASA Ames Research Center
ISBN 0-915928-02-7

39. Heat Transfer with Thermal Control Applications (1975)
M. Michael Yovanovich
University of Waterloo
ISBN 0-915928-03-5

***40. Aerodynamics of Base Combustion (1976)**
S. N. B. Murthy, Editor
J. R. Osborn, Associate Editor
Purdue University
A. W. Barrows
J. R. Ward, Associate Editors
Ballistics Research Laboratories
ISBN 0-915928-04-3

---
*Out of print.

41. Communications Satellite Developments: Systems (1976)
Gilbert E. LaVean
Defense Communications Agency
William G. Schmidt
CML Satellite Corp.
ISBN 0-915928-05-1

***42. Communications Satellite Developments: Technology (1976)**
William G. Schmidt
CML Satellite Corp.
Gilbert E. LaVean
Defense Communications Agency
ISBN 0-915928-06-X

***43. Aeroacoustics: Jet Noise, Combustion and Core Engine Noise (1976)**
Ira R. Schwartz, Editor
NASA Ames Research Center
Henry T. Nagamatsu, Associate Editor
General Electric Research and Development Center
Warren C. Strahle, Associate Editor
Georgia Institute of Technology
ISBN 0-915928-07-8

***44. Aeroacoustics: Fan Noise and Control; Duct Acoustics; Rotor Noise (1976)**
Ira R. Schwartz, Editor
NASA Ames Research Center
Henry T. Nagamatsu, Associate Editor
General Electric Research and Development Center
Warren C. Strahle, Associate Editor
Georgia Institute of Technology
ISBN 0-915928-08-6

***45. Aeroacoustics: STOL Noise; Airframe and Airfoil Noise (1976)**
Ira R. Schwartz, Editor
NASA Ames Research Center
Henry T. Nagamatsu, Associate Editor
General Electric Research and Development Center
Warren C. Strahle, Associate Editor
Georgia Institute of Technology
ISBN 0-915928-09-4

***46. Aeroacoustics: Acoustic Wave Propagation; Aircraft Noise Prediction; Aeroacoustic Instrumentation (1976)**
Ira R. Schwartz, Editor
NASA Ames Research Center
Henry T. Nagamatsu, Associate Editor
General Electric Research and Development Center
Warren C. Strahle, Associate Editor
Georgia Institute of Technology
ISBN 0-915928-10-8

***47. Spacecraft Charging by Magnetospheric Plasmas (1976)**
Alan Rosen
TRW Inc.
ISBN 0-915928-11-6

48. Scientific Investigations on the Skylab Satellite (1976)
Marion I. Kent
Ernst Stuhlinger
NASA George C. Marshall Space Flight Center
Shi-Tsan Wu
University of Alabama
ISBN 0-915928-12-4

49. Radiative Transfer and Thermal Control (1976)
Allie M. Smith
ARO Inc.
ISBN 0-915928-13-2

***50. Exploration of the Outer Solar System (1976)**
Eugene W. Greenstadt
TRW Inc.
Murray Dryer
National Oceanic and Atmospheric Administration
Devrie S. Intriligator
University of Southern California
ISBN 0-915928-14-0

***51. Rarefied Gas Dynamics, Parts I and II (two volumes) (1977)**
J. Leith Potter
ARO Inc.
ISBN 0-915928-15-9

***52. Materials Sciences in Space with Application to Space Processing (1977)**
Leo Steg
General Electric Co.
ISBN 0-915928-16-7

***53. Experimental Diagnostics in Gas Phase Combustion Systems (1977)**
Ben T. Zinn, Editor
Georgia Institute of Technology
Craig T. Bowman, Associate Editor
Stanford University
Daniel L. Hartley, Associate Editor
Sandia Laboratories
Edward W. Price, Associate Editor
Georgia Institute of Technology
James G. Skifstad, Associate Editor
Purdue University
ISBN 0-915928-18-3

---
*Out of print.

54. Satellite Communication: Future Systems (1977)
David Jarett
TRW Inc.
ISBN 0-915928-18-3

***55. Satellite Communications: Advanced Technologies (1977)**
David Jarett
TRW Inc.
ISBN 0-915928-19-1

***56. Thermophysics of Spacecraft and Outer Planet Entry Probes (1977)**
Allie M. Smith
ARO Inc.
ISBN 0-915928-20-5

***57. Space-Based Manufacturing from Nonterrestrial Materials (1977)**
Gerald K. O'Neill, Editor
Brian O'Leary, Assistant Editor
Princeton University
ISBN 0-915928-21-3

***58. Turbulent Combustion (1978)**
Lawrence A. Kennedy
State University of New York at Buffalo
ISBN 0-915928-22-1

***59. Aerodynamic Heating and Thermal Protection Systems (1978)**
Leroy S. Fletcher
University of Virginia
ISBN 0-915928-23-X

***60. Heat Transfer and Thermal Control Systems (1978)**
Leroy S. Fletcher
University of Virginia
ISBN 0-915928-24-8

***61. Radiation Energy Conversion in Space (1978)**
Kenneth W. Billman
NASA Ames Research Center
ISBN 0-915928-26-4

***62. Alternative Hydrocarbon Fuels: Combustion and Chemical Kinetics (1978)**
Craig T. Bowman
Stanford University
Jorgen Birkeland
Department of Energy
ISBN 0-915928-25-6

***63. Experimental Diagnostics in Combustion of Solids (1978)**
Thomas L. Boggs
Naval Weapons Center
Ben T. Zinn
Georgia Institute of Technology
ISBN 0-915928-28-0

***64. Outer Planet Entry Heating and Thermal Protection (1979)**
Raymond Viskanta
Purdue University
ISBN 0-915928-29-9

***65. Thermophysics and Thermal Control (1979)**
Raymond Viskanta
Purdue University
ISBN 0-915928-30-2

***66. Interior Ballistics of Guns (1979)**
Herman Krier
University of Illinois at Urbana–Champaign
Martin Summerfield
New York University
ISBN 0-915928-32-9

***67. Remote Sensing of Earth from Space: Role of "Smart Sensors" (1979)**
Roger A. Breckenridge
NASA Langley Research Center
ISBN 0-915928-33-7

68. Injection and Mixing in Turbulent Flow (1980)
Joseph A. Schetz
Virginia Polytechnic Institute and State University
ISBN 0-915928-35-3

***69. Entry Heating and Thermal Protection (1980)**
Walter B. Olstad
NASA Headquarters
ISBN 0-915928-38-8

***70. Heat Transfer, Thermal Control, and Heat Pipes (1980)**
Walter B. Olstad
NASA Headquarters
ISBN 0-915928-39-6

***71. Space Systems and Their Interactions with Earth's Space Environment (1980)**
Henry B. Garrett
Charles P. Pike
Hanscom Air Force Base
ISBN 0-915928-41-8

***72. Viscous Flow Drag Reduction (1980)**
Gary R. Hough
Vought Advanced Technology Center
ISBN 0-915928-44-2

***73. Combustion Experiments in a Zero-Gravity Laboratory (1981)**
Thomas H. Cochran
NASA Lewis Research Center
ISBN 0-915928-48-5

***74. Rarefied Gas Dynamics, Parts I and II (two volumes) (1981)**
Sam S. Fisher
University of Virginia
ISBN 0-915928-51-5

*Out of print.

***75. Gasdynamics of Detonations and Explosions (1981)**
J. R. Bowen
University of Wisconsin at Madison
N. Manson
Universite de Poitiers
A. K. Oppenheim
University of California at Berkeley
R. I. Soloukhin
Institute of Heat and Mass Transfer, BSSR Academy of Sciences
ISBN 0-915928-46-9

76. Combustion in Reactive Systems (1981)
J. R. Bowen
University of Wisconsin at Madison
N. Manson
Universite de Poitiers
A. K. Oppenheim
University of California at Berkeley
R. I. Soloukhin
Institute of Heat and Mass Transfer, BSSR Academy of Sciences
ISBN 0-915928-47-7

***77. Aerothermodynamics and Planetary Entry (1981)**
A. L. Crosbie
University of Missouri-Rolla
ISBN 0-915928-52-3

***78. Heat Transfer and Thermal Control (1981)**
A. L. Crosbie
University of Missouri-Rolla
ISBN 0-915928-53-1

***79. Electric Propulsion and Its Applications to Space Missions (1981)**
Robert C. Finke
NASA Lewis Research Center
ISBN 0-915928-55-8

***80. Aero-Optical Phenomena (1982)**
Keith G. Gilbert
Leonard J. Otten
Air Force Weapons Laboratory
ISBN 0-915928-60-4

***81. Transonic Aerodynamics (1982)**
David Nixon
Nielsen Engineering & Research, Inc.
ISBN 0-915928-65-5

***82. Thermophysics of Atmospheric Entry (1982)**
T. E. Horton
University of Mississippi
ISBN 0-915928-66-3

***83. Spacecraft Radiative Transfer and Temperature Control (1982)**
T. E. Horton
University of Mississippi
ISBN 0-915928-67-1

84. Liquid-Metal Flows and Magneto-hydrodynamics (1983)
H. Branover
Ben-Gurion University of the Negev
P. S. Lykoudis
Purdue University
A. Yakhot
Ben-Gurion University of the Negev
ISBN 0-915928-70-1

***85. Entry Vehicle Heating and Thermal Protection Systems: Space Shuttle, Solar Starprobe, Jupiter Galileo Probe (1983)**
Paul E. Bauer
McDonnell Douglas Astronautics Co.
Howard E. Collicott
The Boeing Co.
ISBN 0-915928-74-4

***86. Spacecraft Thermal Control, Design, and Operation (1983)**
Howard E. Collicott
The Boeing Co.
Paul E. Bauer
McDonnell Douglas Astronautics Co.
ISBN 0-915928-75-2

***87. Shock Waves, Explosions, and Detonations (1983)**
J. R. Bowen
University of Washington
N. Manson
Universite de Poitiers
A. K. Oppenheim
University of California at Berkeley
R. I. Soloukhin
Institute of Heat and Mass Transfer, BSSR Academy of Sciences
ISBN 0-915928-76-0

88. Flames, Lasers, and Reactive Systems (1983)
J. R. Bowen
University of Washington
N. Manson
Universite de Poitiers
A. K. Oppenheim
University of California at Berkeley
R. I. Soloukhin
Institute of Heat and Mass Transfer, BSSR Academy of Sciences
ISBN 0-915928-77-9

***89. Orbit-Raising and Maneuvering Propulsion: Research Status and Needs (1984)**
Leonard H. Caveny
Air Force Office of Scientific Research
ISBN 0-915928-82-5

*Out of print.

90. Fundamentals of Solid-Propellant Combustion (1984)
Kenneth K. Kuo
Pennsylvania State University
Martin Summerfield
Princeton Combustion Research Laboratories, Inc.
ISBN 0-915928-84-1

91. Spacecraft Contamination: Sources and Prevention (1984)
J. A. Roux
University of Mississippi
T. D. McCay
NASA Marshall Space Flight Center
ISBN 0-915928-85-X

92. Combustion Diagnostics by Nonintrusive Methods (1984)
T. D. McCay
NASA Marshall Space Flight Center
J. A. Roux
University of Mississippi
ISBN 0-915928-86-8

93. The INTELSAT Global Satellite System (1984)
Joel Alper
COMSAT Corp.
Joseph Pelton
INTELSAT
ISBN 0-915928-90-6

94. Dynamics of Shock Waves, Explosions, and Detonations (1984)
J. R. Bowen
University of Washington
N. Manson
Universite de Poitiers
A. K. Oppenheim
University of California at Berkeley
R. I. Soloukhin
Institute of Heat and Mass Transfer, BSSR Academy of Sciences
ISBN 0-915928-91-4

95. Dynamics of Flames and Reactive Systems (1984)
J. R. Bowen
University of Washington
N. Manson
Universite de Poitiers
A. K. Oppenheim
University of California at Berkeley
R. I. Soloukhin
Institute of Heat and Mass Transfer, BSSR Academy of Sciences
ISBN 0-915928-92-2

96. Thermal Design of Aeroassisted Orbital Transfer Vehicles (1985)
H. F. Nelson
University of Missouri-Rolla
ISBN 0-915928-94-9

97. Monitoring Earth's Ocean, Land, and Atmosphere from Space—Sensors, Systems, and Applications (1985)
Abraham Schnapf
Aerospace Systems Engineering
ISBN 0-915928-98-1

98. Thrust and Drag: Its Prediction and Verification (1985)
Eugene E. Covert
Massachusetts Institute of Technology
C. R. James
Vought Corp.
William F. Kimzey
Sverdrup Technology AEDC Group
George K. Richey
U.S. Air Force
Eugene C. Rooney
U.S. Navy Department of Defense
ISBN 0-930403-00-2

99. Space Stations and Space Platforms—Concepts, Design, Infrastructure, and Uses (1985)
Ivan Bekey
Daniel Herman
NASA Headquarters
ISBN 0-930403-01-0

100. Single- and Multi-Phase Flows in an Electromagnetic Field: Energy, Metallurgical, and Solar Applications (1985)
Herman Branover
Ben-Gurion University of the Negev
Paul S. Lykoudis
Purdue University
Michael Mond
Ben-Gurion University of the Negev
ISBN 0-930403-04-5

101. MHD Energy Conversion: Physiotechnical Problems (1986)
V. A. Kirillin
A. E. Sheyndlin
Soviet Academy of Sciences
ISBN 0-930403-05-3

102. Numerical Methods for Engine-Airframe Integration (1986)
S. N. B. Murthy
Purdue University
Gerald C. Paynter
Boeing Airplane Co.
ISBN 0-930403-09-6

103. Thermophysical Aspects of Re-Entry Flows (1986)
James N. Moss
NASA Langley Research Center
Carl D. Scott
NASA Johnson Space Center
ISBN 0-930430-10-X

*Out of print.

***104. Tactical Missile Aerodynamics (1986)**
M. J. Hemsch
PRC Kentron, Inc.
J. N. Nielson
NASA Ames Research Center
ISBN 0-930403-13-4

105. Dynamics of Reactive Systems Part I: Flames and Configurations; Part II: Modeling and Heterogeneous Combustion (1986)
J. R. Bowen
University of Washington
J.-C. Leyer
Universite de Poitiers
R. I. Soloukhin
Institute of Heat and Mass Transfer, BSSR Academy of Sciences
ISBN 0-930403-14-2

106. Dynamics of Explosions (1986)
J. R. Bowen
University of Washington
J.-C. Leyer
Universite de Poitiers
R. I. Soloukhin
Institute of Heat and Mass Transfer, BSSR Academy of Sciences
ISBN 0-930403-15-0

***107. Spacecraft Dielectric Material Properties and Spacecraft Charging (1986)**
A. R. Frederickson
U.S. Air Force Rome Air Development Center
D. B. Cotts
SRI International
J. A. Wall
U.S. Air Force Rome Air Development Center
F. L. Bouquet
Jet Propulsion Laboratory, California Institute of Technology
ISBN 0-930403-17-7

***108. Opportunities for Academic Research in a Low-Gravity Environment (1986)**
George A. Hazelrigg
National Science Foundation
Joseph M. Reynolds
Louisiana State University
ISBN 0-930403-18-5

109. Gun Propulsion Technology (1988)
Ludwig Stiefel
U.S. Army Armament Research, Development and Engineering Center
ISBN 0-930403-20-7

110. Commercial Opportunities in Space (1988)
F. Shahrokhi
K. E. Harwell
University of Tennessee Space Institute
C. C. Chao
National Cheng Kung University
ISBN 0-930403-39-8

111. Liquid-Metal Flows: Magnetohydrodynamics and Application (1988)
Herman Branover
Michael Mond
Yeshajahu Unger
Ben-Gurion University of the Negev
ISBN 0-930403-43-6

112. Current Trends in Turbulence Research (1988)
Herman Branover
Micheal Mond
Yeshajahu Unger
Ben-Gurion University of the Negev
ISBN 0-930403-44-4

113. Dynamics of Reactive Systems Part I: Flames; Part II: Heterogeneous Combustion and Applications (1988)
A. L. Kuhl
R&D Associates
J. R. Bowen
University of Washington
J.-C. Leyer
Universite de Poitiers
A. Borisov
USSR Academy of Sciences
ISBN 0-930403-46-0

114. Dynamics of Explosions (1988)
A. L. Kuhl
R & D Associates
J. R. Bowen
University of Washington
J.-C. Leyer
Universite de Poitiers
A. Borisov
USSR Academy of Sciences
ISBN 0-930403-47-9

115. Machine Intelligence and Autonomy for Aerospace (1988)
E. Heer
Heer Associates, Inc.
H. Lum
NASA Ames Research Center
ISBN 0-930403-48-7

116. Rarefied Gas Dynamics: Space Related Studies (1989)
E. P. Muntz
University of Southern California
D. P. Weaver
U.S. Air Force Astronautics Laboratory (AFSC)
D. H. Campbell
University of Dayton Research Institute
ISBN 0-930403-53-3

---
*Out of print.

117. Rarefied Gas Dynamics: Physical Phenomena (1989)
E. P. Muntz
University of Southern California
D. P. Weaver
U.S. Air Force Astronautics Laboratory (AFSC)
D. H. Campbell
University of Dayton Research Institute
ISBN 0-930403-54-1

118. Rarefied Gas Dynamics: Theoretical and Computational Techniques (1989)
E. P. Muntz
University of Southern California
D. P. Weaver
U.S. Air Force Astronautics Laboratory (AFSC)
D. H. Campbell
University of Dayton Research Institute
ISBN 0-930403-55-X

119. Test and Evaluation of the Tactical Missile (1989)
Emil J. Eichblatt Jr.
Pacific Missile Test Center
ISBN 0-930403-56-8

120. Unsteady Transonic Aerodynamics (1989)
David Nixon
Nielsen Engineering & Research, Inc.
ISBN 0-930403-52-5

121. Orbital Debris from Upper-Stage Breakup (1989)
Joseph P. Loftus Jr.
NASA Johnson Space Center
ISBN 0-930403-58-4

122. Thermal-Hydraulics for Space Power, Propulsion and Thermal Management System Design (1990)
William J. Krotiuk
General Electric Co.
ISBN 0-930403-64-9

123. Viscous Drag Reduction in Boundary Layers (1990)
Dennis M. Bushnell
Jerry N. Hefner
NASA Langley Research Center
ISBN 0-930403-66-5

***124. Tactical and Strategic Missile Guidance (1990)**
Paul Zarchan
Charles Stark Draper Laboratory, Inc.
ISBN 0-930403-68-1

125. Applied Computational Aerodynamics (1990)
P. A. Henne
Douglas Aircraft Company
ISBN 0-930403-69-X

126. Space Commercialization: Launch Vehicles and Programs (1990)
F. Shahrokhi
University of Tennessee Space Institute
J. S. Greenberg
Princeton Synergetics Inc.
T. Al-Saud
Ministry of Defense and Aviation Kingdom of Saudi Arabia
ISBN 0-930403-75-4

127. Space Commercialization: Platforms and Processing (1990)
F. Shahrokhi
University of Tennessee Space Institute
G. Hazelrigg
National Science Foundation
R. Bayuzick
Vanderbilt University
ISBN 0-930403-76-2

128. Space Commercialization: Satellite Technology (1990)
F. Shahrokhi
University of Tennessee Space Institute
N. Jasentuliyana
United Nations
N. Tarabzouni
King Abulaziz City for Science and Technology
ISBN 0-930403-77-0

129. Mechanics and Control of Large Flexible Structures (1990)
John L. Junkins
Texas A&M University
ISBN 0-930403-73-8

130. Low-Gravity Fluid Dynamics and Transport Phenomena (1990)
Jean N. Koster
Robert L. Sani
University of Colorado at Boulder
ISBN 0-930403-74-6

131. Dynamics of Deflagrations and Reactive Systems: Flames (1991)
A. L. Kuhl
Lawrence Livermore National Laboratory
J.-C. Leyer
Universite de Poitiers
A. A. Borisov
USSR Academy of Sciences
W. A. Sirignano
University of California
ISBN 0-930403-95-9

---
*Out of print.

132. Dynamics of Deflagrations and Reactive Systems: Heterogeneous Combustion (1991)
A. L. Kuhl
Lawrence Livermore National Laboratory
J.-C. Leyer
Universite de Poitiers
A. A. Borisov
USSR Academy of Sciences
W. A. Sirignano
University of California
ISBN 0-930403-96-7

133. Dynamics of Detonations and Explosions: Detonations (1991)
A. L. Kuhl
Lawrence Livermore National Laboratory
J.-C. Leyer
Universite de Poitiers
A. A. Borisov
USSR Academy of Sciences
W. A. Sirignano
University of California
ISBN 0-930403-97-5

134. Dynamics of Detonations and Explosions: Explosion Phenomena (1991)
A. L. Kuhl
Lawrence Livermore National Laboratory
J.-C. Leyer
Universite de Poitiers
A. A. Borisov
USSR Academy of Sciences
W. A. Sirignano
University of California
ISBN 0-930403-98-3

135. Numerical Approaches to Combustion Modeling (1991)
Elaine S. Oran
Jay P. Boris
Naval Research Laboratory
ISBN 1-56347-004-7

136. Aerospace Software Engineering (1991)
Christine Anderson
U.S. Air Force Wright Laboratory
Merlin Dorfman
Lockheed Missiles & Space Company, Inc.
ISBN 1-56347-005-0

137. High-Speed Flight Propulsion Systems (1991)
S. N. B. Murthy
Purdue University
E. T. Curran
Wright Laboratory
ISBN 1-56347-011-X

138. Propagation of Intensive Laser Radiation in Clouds (1992)
O. A. Volkovitsky
Yu. S. Sedenov
L. P. Semenov
Institute of Experimental Meteorology
ISBN 1-56347-020-9

139. Gun Muzzle Blast and Flash (1992)
Günter Klingenburg
Fraunhofer-Institut für Kurzzeitdynamik, Ernst-Mach-Institut
Joseph M. Heimerl
U.S. Army Ballistic Research Laboratory
ISBN 1-56347-012-8

140. Thermal Structures and Materials for High-Speed Flight (1992)
Earl. A. Thornton
University of Virginia
ISBN 1-56347-017-9

141. Tactical Missile Aerodynamics: General Topics (1992)
Michael J. Hemsch
Lockheed Engineering & Sciences Company
ISBN 1-56347-015-2

142. Tactical Missile Aerodynamics: Prediction Methodology (1992)
Michael R. Mendenhall
Nielsen Engineering & Research, Inc.
ISBN 1-56347-016-0

143. Nonsteady Burning and Combustion Stability of Solid Propellants (1992)
Luigi De Luca
Politecnico di Milano
Edward W. Price
Georgia Institute of Technology
Martin Summerfield
Princeton Combustion Research Laboratories, Inc.
ISBN 1-56347-014-4

144. Space Economics (1992)
Joel S. Greenberg
Princeton Synergetics, Inc.
Henry R. Hertzfeld
HRH Associates
ISBN 1-56347-042-X

145. Mars: Past, Present, and Future (1992)
E. Brian Pritchard
NASA Langley Research Center
ISBN 1-56347-043-8

146. Computational Nonlinear Mechanics in Aerospace Engineering (1992)
Satya N. Atluri
Georgia Institute of Technology
ISBN 1-56347-044-6

147. Modern Engineering for Design of Liquid-Propellant Rocket Engines (1992)
Dieter K. Huzel
David H. Huang
Rocketdyne Division of Rockwell International
ISBN 1-56347-013-6

*Out of print.

148. Metallurgical Technologies, Energy Conversion, and Magnetohydrodynamic Flows (1993)
Herman Branover
Yeshajahu Unger
Ben-Gurion University of the Negev
ISBN 1-56347-019-5

149. Advances in Turbulence Studies (1993)
Herman Branover
Yeshajahu Unger
Ben-Gurion University of the Negev
ISBN 1-56347-018-7

150. Structural Optimization: Status and Promise (1993)
Manohar P. Kamat
Georgia Institute of Technology
ISBN 1-56347-056-X

151. Dynamics of Gaseous Combustion (1993)
A. L. Kuhl
Lawrence Livermore National Laboratory
J.-C. Leyer
Universite de Poitiers
A. A. Borisov
USSR Academy of Sciences
W. A. Sirignano
University of California
ISBN 1-56347-060-8

152. Dynamics of Heterogeneous Gaseous Combustion and Reacting Systems (1993)
A. L. Kuhl
Lawrence Livermore National Laboratory
J.-C. Leyer
Universite de Poitiers
A. A. Borisov
USSR Academy of Sciences
W. A. Sirignano
University of California
ISBN 1-56347-058-6

153. Dynamic Aspects of Detonations (1993)
A. L. Kuhl
Lawrence Livermore National Laboratory
J.-C. Leyer
Universite de Poitiers
A. A. Borisov
USSR Academy of Sciences
W. A. Sirignano
University of California
ISBN 1-56347-057-8

154. Dynamic Aspects of Explosion Phenomena (1993)
A. L. Kuhl
Lawrence Livermore National Laboratory
J.-C. Leyer
Universite de Poitiers
A. A. Borisov
USSR Academy of Sciences
W. A. Sirignano
University of California
ISBN 1-56347-059-4

155. Tactical Missile Warheads (1993)
Joseph Carleone
Aerojet General Corporation
ISBN 1-56347-067-5

156. Toward a Science of Command, Control, and Communications (1993)
Carl R. Jones
Naval Postgraduate School
ISBN 1-56347-068-3

157. Tactical and Strategic Missile Guidance Second Edition (1994)
Paul Zarchan
Charles Stark Draper Laboratory, Inc.
ISBN 1-56347-077-2

158. Rarefied Gas Dynamics: Experimental Techniques and Physical Systems (1994)
Bernie D. Shizgal
University of British Columbia
David P. Weaver
Phillips Laboratory
ISBN 1-56347-079-9

159. Rarefied Gas Dynamics: Theory and Simulations (1994)
Bernie D. Shizgal
University of British Columbia
David P. Weaver
Phillips Laboratory
ISBN 1-56347-080-2

160. Rarefied Gas Dynamics: Space Sciences and Engineering (1994)
Bernie D. Shizgal
University of British Columbia
David P. Weaver
Phillips Laboratory
ISBN 1-56347-081-0

161. Teleoperation and Robotics in Space (1994)
Steven B. Skaar
University of Notre Dame
Carl F. Ruoff
Jet Propulsion Laboratory, California Institute of Technology
ISBN 1-56347-095-0

162. Progress in Turbulence Research (1994)
Herman Branover
Yeshajahu Unger
Ben-Gurion University of the Negev
ISBN 1-56347-099-3

---
*Out of print.

163. Global Positioning System: Theory and Applications, Volume I (1996)
Bradford W. Parkinson
Stanford University
James J. Spilker Jr.
Stanford Telecom
Penina Axelrad,
Associate Editor
University of Colorado
Per Enge,
Associate Editor
Stanford University
ISBN 1-56347-107-8

164. Global Positioning System: Theory and Applications, Volume II (1996)
Bradford W. Parkinson
Stanford University
James J. Spilker Jr.
Stanford Telecom
Penina Axelrad,
Associate Editor
University of Colorado
Per Enge,
Associate Editor
Stanford University
ISBN 1-56347-106-X

165. Developments in High-Speed Vehicle Propulsion Systems (1996)
S. N. B. Murthy
Purdue University
E. T. Curran
Wright Laboratory
ISBN 1-56347-176-0

166. Recent Advances in Spray Combustion: Spray Atomization and Drop Burning Phenomena, Volume I (1996)
Kenneth K. Kuo
Pennsylvania State University
ISBN 1-56347-175-2

167. Fusion Energy in Space Propulsion (1995)
Terry Kammash
University of Michigan
ISBN 1-56347-184-1

168. Aerospace Thermal Structures and Materials for a New Era (1995)
Earl A. Thornton
University of Virginia
ISBN 1-56347-182-5

169. Liquid Rocket Engine Combustion Instability (1995)
Vigor Yang
William E. Anderson
Pennsylvania State University
ISBN 1-56347-183-3

170. Tactical Missile Propulsion (1996)
G. E. Jensen
United Technologies Corporation
David W. Netzer
Naval Postgraduate School
ISBN 1-56347-118-3

171. Recent Advances in Spray Combustion: Spray Combustion Measurements and Model Simulation, Volume II (1996)
Kenneth K. Kuo
Pennsylvania State University
ISBN 1-56347-181-7

172. Future Aeronautical and Space Systems (1997)
Ahmed K. Noor
NASA Langley Research Center
Samuel L. Venneri
NASA Headquarters
ISBN 1-56347-188-4

173. Advances in Combustion Science: In Honor of Ya. B. Zel'dovich (1997)
William A. Sirignano
University of California
Alexander G. Merzhanov
Russian Academy of Sciences
Luigi De Luca
Politecnico di Milano
ISBN 1-56347-178-7

174. Fundamentals of High Accuracy Inertial Navigation (1997)
Averil B. Chatfield
ISBN 1-56347-243-0

175. Liquid Propellant Gun Technology (1997)
Günter Klingenberg
Fraunhofer-Institut für Kurzzeitdynamik, Ernst-Mach-Institut
John D. Knapton
Walter F. Morrison
Gloria P. Wren
U.S. Army Research Laboratory
ISBN 1-56347-196-5

176. Tactical and Strategic Missile Guidance Third Edition (1997)
Paul Zarchan
Charles Stark Draper Laboratory, Inc.
ISBN 1-56347-279-1

177. Orbital and Celestial Mechanics (1998)
John P. Vinti
Gim J. Der, Editor
TRW
Nino L. Bonavito, Editor
NASA Goddard Space Flight Center
ISBN 1-56347-256-2

*Out of print.

License Agreement for

Orbital and Celestial Mechanics Vinti Software

You should carefully read all of the terms and conditions of this agreement before opening this package.

If you have ordered this program through the mail or by telephone, read the following warranty information before you open the package. If you do not agree to the terms, return the program unopened for a prompt refund. If you open the package, the publisher will assume you have read, understood, and agreed to our warranty terms.

DISCLAIMER AND PURCHASE AGREEMENT:

AIAA and the editor provide this documentation and the accompanying diskette and software for use "as is" without warranty of any kind. AIAA and the editor assume no responsibility for errors, omissions, or damage, including, without limitation, damages caused by these computer programs or from the use of information contained therein. The entire risk of performance of the accompanying software is assumed by the user.

For safety, the purchaser may make one, and only one, archival copy of the original diskette, and that copy must not be sold, rented, lent, or in any other way transferred to another person. Beyond that, any duplication infringes the copyright owner's right to this software, and it violates copyright law.

Government: Please note that if this software is purchased by, or on behalf of, a unit or agency of the United States Government, this license agreement still applies. The Orbital and Celestial Mechanics Vinti Software was developed without government funds at private expense and in all respects is proprietary data copyrighted solely to the editor.

If you are unable to accept this usage agreement please contact AIAA for clarification or a refund.